Contents

ACADEMIC PRESS, INC.
111 Fifth Avenue, New York, New York 10003

United Kingdom Edition published by
ACADEMIC PRESS, INC. (LONDON) LTD.
24/28 Oval Road, London NW1

Library of Congress Cataloging in Publication Data

Main entry under title:

Solar energy engineering.

Bibliography: p.
1. Solar energy. I. Sayigh, A. A. M.
TJ810.S623 621.47 76-27452
ISBN 0-12-620850-6

PRINTED IN THE UNITED STATES OF AMERICA

Solar Energy Engineering

Edited by *A. A. M. SAYIGH*
MECHANICAL ENGINEERING DEPARTMENT
COLLEGE OF ENGINEERING
RIYADH UNIVERSITY
RIYADH, SAUDI ARABIA

ACADEMIC PRESS New York San Francisco London 1977
A Subsidiary of Harcourt Brace Jovanovich, Publishers

Solar Energy Engineering

CONTRIBUTORS

CHARLES E. BACKUS
B. J. BRINKWORTH
W. W. S. CHARTERS
ENRICO COFFARI
E. D. HOWE
S. KAKAC
M. ALI KETTANI
A. B. MEINEL
J. A. SABBAGH
TAKEMARO SAKURAI
S. B. SAVAGE
A. A. M. SAYIGH
M. KUDRET SELÇUK
MATHEW P. THEKAEKARA
B. W. TLEIMAT
T. N. VEZIROGLU
BYARD D. WOOD
JOHN I. YELLOTT

List of Contributors

Numbers in parentheses indicate the pages on which the authors' contributions begin.

CHARLES E. BACKUS (263, 287), Mechanical Engineering Faculty, College of Engineering and Applied Sciences, Arizona State University, Tempe, Arizona

B. J. BRINKWORTH (341), Solar Energy Unit, Department of Mechanical Engineering and Energy Studies, University College, Cardiff, United Kingdom

W. W. S. CHARTERS (105, 137), Mechanical Engineering Department, University of Melbourne, Parkville, Victoria, Australia

ENRICO COFFARI* (5), Istituto di Chimica Fisica, Università degli Studi di Roma, Rome, Italy

E. D. HOWE (431), Sea Water Conversion Laboratory, University of California, Richmond, California

S. KAKAC (385), Middle East Technical University, Ankara, Turkey

M. ALI KETTANI (305, 325), Electrical Engineering Department, University of Petroleum and Minerals, Dhahran, Saudi Arabia

A. B. MEINEL (183), Optical Sciences Center, University of Arizona, Tucson, Arizona

J. A. SABBAGH (83), Mechanical Engineering Department, College of Engineering, Riyadh University, Riyadh, Saudi Arabia

TAKEMARO SAKURAI (233), Solar Energy Laboratory, Research Institute for Scientific Measurements, Tohoku University, Sendai, Japan

S. B. SAVAGE (217), Department of Civil Engineering and Applied Mechanics, McGill University, Montreal, Quebec, Canada

A. A. M. SAYIGH (1, 61, 465), Mechanical Engineering Department, College of Engineering, Riyadh University, Riyadh, Saudi Arabia

* Present address: Solar Energy Division, Kuwait Institute for Scientific Research, Kuwait, Kuwait.

M. KUDRET SELÇUK* (155), Jet Propulsion Laboratory, California Institute of Technology, Pasadena, California

MATHEW P. THEKAEKARA† (37), NASA/Goddard Space Flight Center, Greenbelt, Maryland

B. W. TLEIMAT (431), Sea Water Conversion Laboratory, University of California, Richmond, California

T. N. VEZIROGLU (385), Clean Energy Research Institute, University of Miami, Coral Gables, Florida

BYARD D. WOOD (397), Mechanical Engineering Faculty, College of Engineering and Applied Sciences, Arizona State University, Tempe, Arizona

JOHN I. YELLOTT (365), College of Architecture, Arizona State University, Tempe, Arizona

* Formerly: Mechanical Engineering Department, Middle East Technical University, Ankara, Turkey.

† Deceased.

Preface

In a world of dwindling conventional energy resources, a rising population, and rapid technological advancements, it is inevitable in the near future that the present energy conversion systems will change drastically. No doubt some novel source of energy will be utilized. Within this framework, as I see it, solar energy will play a prominent role. Over the past ten years, research and development activities have accelerated to such an extent that it may be possible that by the turn of this century solar energy will be a major source of power. This may well be advantageous since, on the whole, solar energy does not damage the environment and is abundant in an area of great need, the developing countries.

We hope that this book will help to further promote research and advance the state of knowledge in this field.

I would like to extend my thanks and gratitude to all the contributors, to my friends and colleagues who have provided advice and encouragement, and a special thanks to the staff of the College of Engineering, University of Riyadh, for help with typing and drawings.

General Information

NOMENCLATURE

A	area, azimuth angle
B	width
c	velocity of light, specific heat (wetted perimeter)
C	thermal capacitance, brightness concentration factor, attenuation factor or extinction optical thickness
C_p	specific heat at constant pressure
COP	coefficient of performance of compression refrigerator
d	diameter, number of days
d_h	hydraulic diameter ($4A/C$ = area/wetted perimeter)
D	ratio of actual daily sunshine hours to 12 hours, honeycomb diameter or cell side dimension, diameter, collector effectiveness
D'	ratio of actual daily sunshine hours to the length of the day
E	energy emitted per unit time
E_a	acceptor energy level
E_d	donor energy level
E_f	Fermi energy
E_g	energy gap
f	surface conductance, friction factor
$f_{c,e}$	effective absorptivity transmissivity factor
F	fin efficiency factor, focal length
F'	collector efficiency factor
F_R	collector heat removal factor
g	acceleration due to gravity, glass
G	mass velocity of air
G_c	capital cost
Gr	Grashof number
Gz	Graetz number
h	aperture zone, hour angle, Planck's constant
h, h_c, h_c'	convective heat transfer coefficients
h_{Fg}	latent heat of evaporation
H	total insolation, length
H_c	critical radiation intensity
H_d	direct intensity of solar radiation
H_e	hydraulic head
H_f	diffuse radiation
H_0	solar radiation above the atmosphere
i	inside, angle of diversion
$I(x)$	photon intensity at depth x
J	net current
J_L	light current
J_0	saturation current
k	thermal conductivity, lattice parameter, Boltzmann's constant
K	a factor
K_m	momentum diffusivity
K_n	heat diffusivity
L, l	length, honeycomb cell height, load factor
m	air mass, mass

$\dot{m}$	mass flow rate
n, N	number, nadir, refraction index
Nu	Nusselt number
P	Plate, pressure
Pa	Pascal $= N/m^2$
P_C	plant capital investment cost \$/Kw $(=1/P)$
P_e	power
P_0	partial pressure of atmospheric water vapor
Pr	Prandtl number
q	rate of heat transfer per unit area
q_u	useful heat collected
Q_h	heat supplied to engine or generator
Q_1	cooling effect of refrigerator
Q_2	heat rejected by refrigerator
r	radius, angle of incidence
R, RH	relative humidity (%), total fixed charge rate per ampere
Ra	Rayleigh number
Re	Reynolds number
R_s	internal resistance
S	lake area, sunshine hours
St	Stanton number
STR	system thermal ratio (=cooling effect/incident solar energy)
t	time, top
T	absolute temperature
T_a	ambient temperature
T_b	final equilibrium temperature attained by the collector surface
T_c	cover temperature or absorber temperature
T_g	generator temperature
T_h	engine source temperature
T_1	cold space temperature or evaporator temperature
T_2	hot space temperature or heat rejected temperature
TR	thermal ratio of refrigerator (=cooling effect/heat input)
u, v, w	velocity components in the x, y, z directions
U	overall loss coefficient of collector
U_m	mean velocity
U_L, U_0, U_r	overall heat transfer coefficients: in air stream, between air stream, and outside air stream through the transparent cover and rear heat loss coefficient
v_m	mean air velocity
V	velocity vector, voltage
W	mass flow rate, work input to the refrigerator
WBT	wet bulb temperature
X	radiation balance concentration
z	zenith
Z	average hours of the length of the day
Z_0	surface roughness
α	thermal diffusivity, angle of convergence to the focus, absortivity of collector, angle
β	altitude, expansion coefficient, angle
δ	declination angle, angle
ε	emissivity
η	efficiency
λ	specific wavelength, latitude factor
λ_s'	net evaporation rate
μ	dynamic viscosity
μ_B	bulk viscosity
μ_w	wall viscosity
ρ	density
σ	Stefan–Boltzmann constant
τ	transmissivity of collector cover(s), shear stress
ν	kinematic viscosity
ϕ	latitude, tilt angle, viscous dissipation
ω	sunset hour angle
Ω	Seebeck coefficient

Subscripts

L, l	total value at the total distance
m	mean value
n	in the direction n, at step n
0	value at the surface
r	reference
ss	steady state
x	local value at distance x

UNITS

SI units

Quantity	*Unit*	*Unit symbol*
length	meter	m
mass	kilogram	kg
time	second	s
electric current	ampere	A
thermodynamic temperature	degree Kelvin	°K
luminous intensity	candela	cd
force	newton	$N = kg/s^2$
work, energy, or heat	joule	J = Nm
power	watt	W = J/s
potential	volt	V = W/A
resistance	ohm	Ω = V/A
charge	coulomb	C = AS
capacitance	farad	F = As/V
electric field strength	—	V/m
electric flux density		C/m^2
magnetic flux	weber	Wb = Vs
inductance	henry	H = Vs/A
magnetic field strength	—	A/m
plane angle	radian	rad
solid angle	steradian	sr

Decimal prefixes

Prefix	*Factor by which unit is multiplied*	*Symbol*
tera	10^{12}	T
giga	10^9	G
mega	10^6	M
kilo	10^3	k
hecto	10^2	h
deca	10	da
deci	10^{-1}	d
centi	10^{-2}	c
milli	10^{-3}	m
micro	10^{-6}	μ
nano	10^{-9}	n
pico	10^{-12}	p

English units and other units

Quantity	*Unit*	*Unit symbol*
length	inch	in.
	foot	ft
	angstrom	Å
	micron	μm
capacity	gallon	gal
	US gallon	US gal
mass	pound	lb
	gram	g
	tonne	t
force	poundal	pdl
	pound force	lbf
	ton force	tonf
	dyne	dyn
temperature	degree Celsius	°C
	degree Fahrenheit	°F
	degree Rankine	°R
work, energy, heat	British thermal	Btu
	calorie	cal
	electron volt	eV
power	horsepower	hp
time	hour	h

CONVERSION OF UNITS

Length, area, and volume

1 in.	= 25.44 mm	1 US gal	= 0.1337 ft^3 = 3785 cm^3
1 ft	= 0.3048 m	1 in.^{3}	= 16.39 cm^3
1 mile	= 5280 ft = 1.609344 km	1 ft^3	= 0.02832 m^3
1 μm	= 10^{-6} m = 39.37 μin.	1 pint	= 568.261 cm^3
1 β	= 10^{-10} m	1 ft/min	= 0.00508 m/s
1 ft^2	= 0.09290304 m^2	1 km/h	= 0.277778 m/s
1 in.^{2}	= 0.00064516 m^2	1 ft^3/lb	= 0.062428 m^3/kg
1 acre	= 4046.86 m^2	1 cfm	= 0.471947 liter/s
1 mile^2	= 2.58999 km^2	1 cfm/ft^2	= 5.08000 liter/sm^2
1 hectare	= 10^4 m^2	1 mile/h	= 1.467 ft/s = 1.609 km/h
1 gal	= 0.1605 ft^3 = 4546 cm^3	1 knot	= 1.689 ft/s = 1.853 km/h

Mass, density, pressure, and force

1 lb	= 0.4536 kg	1 dyne/cm	= 1 mN/m (milli N per m)
1 slug	= 32.17 lb = 14.59 kg	1 bar	= 10^5 Pa = 33.9 ft of water (1 Pa = 1 N/m^2) = 14.5 lbf/in.^{2}
1 ton	= 2240 lb = 1016 kg = 1.016 t	1 psi	= 6.89476 kPa
1 oz	= 28.3495 g	1 mm H_2O	= 9.80665 Pa
1 lb/in.^{3}	= 27.68 g/cm^3	1 in. H_2O	= 249.089 Pa
1 lb/ft^3	= 16.02 kg/m^3	1 mm Hg	= 133.322 Pa = 1 torr

1 slug/ft^3 = 515.4 kg/m^3
1 lbf = 4.44822 N
1 kgf = 9.80665 N
1 lbf/ft = 14.5939 N/m

1 at = 1 kgf/cm^2 = 98.0665 kPa
1 atm = 101.325 kPa
= 14.70 lbf/in.^{2}
1 dyne = 10^{-5} N

Energy and power

1 kWh = 3.6 MJ
1 Btu = 1.05506 kJ = 252 cal
= 778.2 ft lbf
1 Therm = 105.506 MJ
1 Btu/lb = 2.326 kJ/kg
1 Btu/lb °F = 4.1868 kJ/kg °C
1 kcal = 4.1868 kJ
1 ft lb = 1.356 J
1 hp h = 2.685 MJ
1 cal/s = 4.187 W

1 Btu/ft^2 = 0.271 cal/cm^2
1 kcal/h = 1.163 W
1 hp = 0.745700 kW
1 ton refrigrant = 3.51685 kW
1 Btu/h ft^2 °F = 5.67826 W/cm^2 °C
1 g cal/cm^2/h = 11.63 W/m^2
1 cal/cm^2 s °C = 7380 Btu/ft^2 h °F
1 langley (ly) = 1 cal/cm^2 = 3.69 Btu/ft^2
1 ly/min = 0.06978 W/cm^2

Other useful conversions

1 poise (g/cm s) = 0.1 kg/ms = 0.1 N s/m^2
1 lb/ft h = 0.4134 g/ms
1 slug/ft s = 1 lbf s/ft^2 = 478.8 poise
1 Btu/ft h °R = 1.731 J/m s °K
1 cal/cm s °K = 418.7 J/m s °K
1 foot-candle = 1 lm/ft^2 (lm = cd sr)
1 lux (lx) = 1 lm/m^2
1 foot-candle = 10.7639 lx
1 metric ton crude oil = 0.984 long ton
= 1.102 short ton
= 7.33 barrels
= 1.16 m^3
= 256 Imp gal
= 308 US gal

1 ton of oil[a] = 41 × 10^6 Btu
= 1.5 tons of coal
= 4.9 tons of lignite
= 3.3 tons of peat
= 1167 m^3 of natural gas (1 ft^3 = 1000 Btu)
= 2500 m^3 of town gas (1 ft^3 = 470 Btu)
= 12000 kWh of electricity

[a] In a modern power station, one ton of oil produces about 4000 kWh of electricity

COMMON PROPERTIES OF SOME MATERIALS

	Density (kg/m^3)	*Thermal conductivity* (W/m °C)
air at normal pressure 20°C	1.204	0.0260
expanded polystyrene at 10°C	16.0	0.0346
foam polyurethane, rigid.	24.0	0.0245
mineral, batts, at −2°C	32.0	0.0346
foam PVC, rigid at 21°C	60.0	0.0360
expanded ebonite at 10°C	64.0	0.0303
cork board, dry at 18°C	144.0	0.0418
sheets of asbestos cement at 30°C	150.0	0.319

	Density (kg/m³)	*Thermal conductivity* (W/m °C)
pine fiberboard at 24°C	256.0	0.0519
pine 15% wet	570.0	0.138
oak, 14% wet	770.0	0.160
dry plaster, gypsum at 23°C	881.0	0.170
ice, at −1°C	918.0	2.26
water, at 4°C	1000.0	0.596[a]
typical building concrete	2400.0	1.73
standard glass	2515.0	1.05
aluminum	2675.0	47.6
steel	7850.0	211
copper	8795.0	385

[a] Water at 20°C.

Specific heat (kJ/kg °C)

water, at 20°C and normal pressure	4.19
ice, at −21°C to −1°C	2.10
steam (C_p), at 100°C and normal pressure	1.95
air (C_p), at 20°C and normal pressure	1.012
concrete, at 18°C	0.837

Heat of vaporization (kJ/kg)

water, at 20°C	2454.0
water, at 100°C	2257.0
R12, at 0°C, sat	151.5
R22, at 0°C, sat	205.4
R11, at 0°C, sat	188.9
R500, at 0°C, sat	183.0
R717, at 0°C, sat	1263.3

Common constants

NTP is normal temperature T_0 of 273.15 K = 0°C and normal pressure P_0 of 101.325 kPa

90 is standard gravity	9.80665 m/S²
velocity of sound in air	344 m/S at P_0, 20°C and 50% RH
gas constants	R_u = 8314.4 J/K mol K Universal
	R_a = 287.045 J/kg K air
	R_v = 461.52 J/kg K water vapor
Boltzmann's constant	1.380×10^{-23} J/°K
Faraday constant	9.65×10^{7} C/kg mole
Planck's constant	6.626×10^{-34} Js
Stefan–Boltzmann constant	5.67×10^{-8} J/m² s °K⁴
volume of 1 kg mole of ideal gas at NTP	22.42 m³

1

Introduction

A. A. M. SAYIGH

MECHANICAL ENGINEERING DEPARTMENT
COLLEGE OF ENGINEERING
RIYADH UNIVERSITY
RIYADH, SAUDI ARABIA

Worship of the sun can be traced back to the very limits of history. That it was a fundamental and ubiquitous stage in the development of man can be seen by the fact that in the Americas the Aztecs, and even earlier peoples, worshipped the sun; the people of Oceania, the Druids of Europe, the peoples of China and ancient Egypt worshipped the sun—in fact all the great early agrarian civilizations have practiced sun worship in one form or another. Whenever man has started to use the soil and realized (consciously or unconsciously) the effect of the sun on his life, he has come to worship it for its power. So today, we do not worship the sun, but we know and appreciate the extent of its influence on our lives. Perhaps, what is not as yet fully appreciated is the sun's power and its place in technology as a source of energy. We have only recently, through the political maneuverings of man, come to realize that there has to be an alternative to fossil fuels. Those who scoffed at the pioneers in the field can no longer do so. As a science and as an intrinsic part of technology, the field of solar energy has arrived. Like the worshippers of Ra and the men gathered at Stonehenge, we are trying to harness the power of the sun. We need energy to sustain life on earth, to fulfill our basic needs, as well as to travel and communicate. Everything, whether directly or indirectly, is related to energy. The standard by which we judge the economic and social advancements of nations is the energy per capita used in individual nations. Food production per person alone consumes 0.15 kW per person.

In order to maintain the standard of living and comfort enjoyed by the so-called advanced nations, 10 kW per person is required (USA, 1970), with 5 kW per person in Europe and 0.5 kW per person in the developing nations. The world population is increasing rapidly and by the year 2000 it will have reached approximately 6000 million (an increase of 2%), while the corresponding increase in energy demand is about 6%, reaching a figure of 2000×10^{12} kWh. Obviously, the exhaustible forms of energy, such as wood, coal, natural gas, and fossil fuels, will not meet this growing demand since these are mostly utilized for the advancement of technology in the industrialized countries and the rapid catching up of the developing countries. This leaves no alternative but to use the inexhaustible solar form of energy directly and indirectly. Indirect solar energy consists of biomass, tidal, ocean thermal, and wind energies, which are all related to the sun.

Modern scientific research in the utilization of solar energy commenced in 1845 when C. Gunter, an Austrian, invented a solar boiler using mirrors. In 1876 an American inventor, John Ericsson, who invented several types of hot air engines prior to this date, envisaged that at some time in the near future, a chain of solar power stations across North Africa, the Middle East, India, Australia, and Central America would be set up. In the latter half of the last century and during the first of this century, progress in the field of solar energy research was fairly slow. This was mainly due to the availability of cheap fossil fuels. The waning solar energy research was revived in 1940 when Godfray Cabot left a large sum of money for research projects at the Massachusetts Institute of Technology. During the recent energy panic, research in the utilization of solar energy has gathered considerable momentum, especially in industrialized countries such as USA, USSR, France, Australia, and Canada. The near future will certainly show some major breakthrough in solar energy technology.

All countries in the world receive some solar energy. This amount varies from a few hundred hours per year as in the northern countries and the lower part of South America, to four thousand hours per year as is the case in most of the Arabian Peninsula and the Sahara Desert. In estimating the amount of solar energy falling on the earth, let us consider first all the natural deserts of the world. Their area is about 20×10^6 km^2 with average solar insolation of 583.30 W/m^2/day (500 gm cal/cm^2/day). Another 30×10^6 km^2 receive about 291.65 W/m^2/day (250 gm cal/cm^2/day). Let us ignore all the areas of sea and the rest of the land. Therefore the amount of solar energy received by this 50×10^6 km^2 is 163.2×10^{12} kWh/day, assuming eight hours of sunshine, or approximately 60×10^{15} kWh/year. Using 5% efficiency, this energy will result in 300×10^{13} kWh and comparing this with the estimated world energy demand in the year 2000 (50×10^{12} kWh/year), it can be seen that it is about 60 times what the world will require then. Furthermore, comparing this figure with USA demand for energy at the present time which is about 20×10^{12} kWh, it is about 150 times what is needed by the USA now. Solar energy, which is the ultimate source of most forms of energy used now, is clean, safe, free, and exists in viable quantities in many countries. The drawbacks in using solar radiation as energy, as have been pointed out, are that it cannot be stored and it is a dilute form of energy. This is, however, an outdated argument since the energy can be stored by producing hydrogen, or by storing in other mechanical or electrical storage devices; the energy can be concentrated in solar furnaces, for example, which can achieve temperatures in the region of 5000°C.

The idea of putting together this book occurred to me when I was in Italy participating in the first Solar Energy Conversion Course, which is run by the International College of Applied Physics, in the summer of 1974. I realized that there is a scarcity of texts for scientists and engineers on solar energy. Several of the contributors to this book were also in Italy, Dr. Kettani, Dr. Backus, and Dr. Coffari, for example. The rest of the authors were contacted, bearing in mind the aim of making this a truly international book so that it will not only serve as a textbook but also as a compilation of information from the major countries concerned with research into utilizing solar energy.

The aim of the book is to provide ample information concerning all forms and topics of solar energy, being, we hope, both a textbook as well as a work of reference. All the information is as up to date as we can make it and ideas for further research are also included. However, no book is perfect and as the reader will appreciate, a book with such a wide scope is bound to contain some errors and repetition.

The book consists of 21 chapters, written by 16 different authors from eight different countries. Chapter 1 deals with the scope and advantages of solar energy and serves as an introduction to the rest of the chapters. Chapters 2–5 explain the nature of the sun, the solar radiation spectrum, the estimation of total, direct, and diffuse radiation, and the heat transfer fundamentals for solar energy application. Chapters 6–8 outline the fundamentals, fabrication, and uses of various water and air heaters. Optics and concentrating collectors, as well as solar furnaces are dealt with in Chapters 9 and 11, while Chapters 10 and 12–20 discuss the various applications of solar energy, namely, solar pond, solar distillation, photovoltaic conversion of solar energy, solar refrigeration, solar hydrogen production, space applications, and solar measuring equipment. Finally, Chapter 21 discusses the cost of solar appliances.

2

The Sun and the Celestial Vault

*ENRICO COFFARI**

ISTITUTO DI CHIMICA FISICA
UNIVERSITÀ DEGLI STUDI DI ROMA
ROME, ITALY

2.1 INTRODUCTION

If we observed the universe from "outside," we would see a vast number of bright areas that seem to dot the heavens. Years ago, when the resolving power of telescopes was less than that of the present (a few hundredths of a second of an arc), the images of these bright areas in the universe appeared to be great luminous clouds with blurred outlines. These formations were called nebulas, a name that today is out of date even if still in common usage. But at present, upon closer inspection, these luminous areas appear as galaxies, i.e., systems of millions of stars. Many of these formations show a spiral structure (spiral galaxies), others a spherical structure (galactic clusters), others an ellipsoidal structure (ellipsoidal galaxies), and others a less regular structure (irregular galaxies).

One of the galaxies, our galaxy, is the Milky Way. One can see it at night as a band of light that goes right around the celestial sphere. This gives the first clue as to the structure of our galaxy: a spiral structure. Its diameter is approximately 80×10^3 light years and it has a central bulge of ellipsoidal structure with the two major semiaxes on the galactic plane of about 10×10^3 light years each and the minor semiaxis of approximately 3.5×10^3 light years. Our sun is located at approximately 27×10^3 light years from the galactic center and quite near to the galactic plane, i.e., within 100 light years as shown in Fig. 2.1.

* *Present address*: Solar Energy Division, Kuwait Institute for Scientific Research, Kuwait, Kuwait.

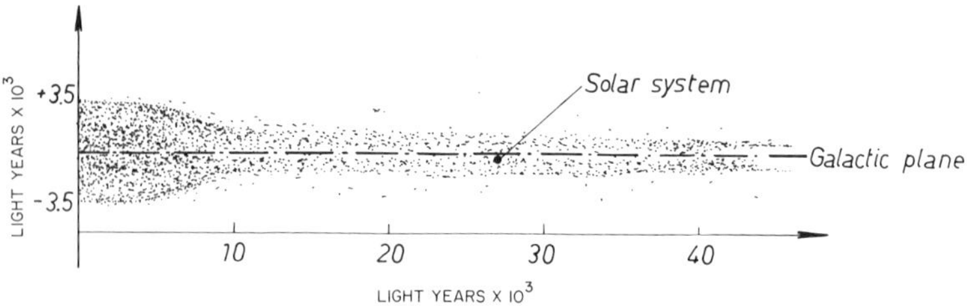

FIG. 2.1 The Milky Way, a cross section normal to the galactic plane.

2.2 THE SUN: A GRAIN OF SAND IN A WHIRLING DESERT STORM

The sun is a part of the Milky Way which is a beautiful spiral composed of over 10^{10} stars. The galaxy is rotating with velocities that decrease as the distance from the galactic center increases (see Table 2.1). The part of the galaxy nearest to the sun has a circular velocity of at least 2.2×10^5 m/s relative to the center of our galaxy. The translational motion of our galaxy cannot be measured in absolute terms because of the absence of any frame of reference. Spectroscopic studies of the Doppler effect* reveal that we are moving away from the other galaxies, or they are moving away from us, at a velocity in the order of 10^6 m/s (only 100 times less than light). This motion is the base of the theory of the expanding universe. Within the galaxy there are groups of stars that share a common motion converging on a common area. These loosely connected groupings are known as stellar associations and aggregates. Some constellations, the groups of stars in our galaxy observed from our point of view, such as the Pleiades (Seven Sisters), have a motion of their own. This suggests that the stars of these constellations are close together and form a group in space, not simply an apparent group due to their being in the same direction for the observer. These groups are "moving

* The Doppler effect causes the change of wavelength received from a moving source with respect to its speed of approach or recession in relation to the observer-source system. This is similar to the change in pitch of the sound of a siren which suddenly drops as it passes by. Therefore, in electromagnetic waves, if a star moves rapidly toward us, its spectrum of emission is shifted toward shorter wavelengths, and if it recedes, its spectrum is shifted toward longer wavelengths. This effect does not change whether it is the star or the solar system or both that move. The wavelength shift $(\lambda - \lambda_0)$ is expressed by

$$\lambda - \lambda_0 = \lambda(v/c) \tag{2.1}$$

where λ_0 is the standard wavelength of a line in an emission spectrum, λ the observed wavelength for the same line in the spectrum of a star, v the velocity of the star, relative to the observer, and c the velocity of light.

TABLE 2.1

The Rotational Velocities of Various Galaxies from the Galactic Center of the Milky Way

Distance (light years)	Velocity (10^3 m/s)	Distance (light years)	Velocity (10^3 m/s)
0	0	25,000	224
5000	150[a]	30,000	213
10,000	190[a]	35,000	197
15,000	220[a]	40,000	180
20,000	230		

[a] These values are very uncertain.

clusters." Finally the stars, once called "fixed stars" as distinct from the planets, have individual motions within the general movement. Although the star-mobility distribution is essentially a Maxwellian distribution, there is an overall motion toward the galactic center: the star streaming. In other words, the distribution of velocities has an elliptical symmetry suggesting that the individual stars move on almost elliptical orbits. Therefore, the sun itself is moving, and when its motion is referred to the average position of the surrounding stars, its velocity is approximately 2×10^4 m/s in the direction of the constellation of Hercules. In turn our planet moves around the sun in an elliptical orbit. The sun is at one focus and the earth–sun vector sweeps equal areas in equal times (Keplerian motion). Hence both the distance between the sun and the earth and the circular velocity of our planet around the sun vary periodically. The period of revolution is 365.2564 . . . days and is called the sidereal year. Also, there is a terrestrial spin that varies with time.

Considering the overall motion of the earth around the sun—the grain of sand whirling in the desert storm—we select, for simplicity of calculation, the point we stand on (latitude, longitude, and altitude) as the origin of our frame of reference. In this system polar coordinates indicate the azimuth (the bearing), the altitude (above the horizon), and our distance from the celestial bodies in their "apparent"* motion around the earth.

* Astronomical distances are extremely large, even in relation to the tremendous velocity of light. We receive the solar radiation eight minutes after it has been emitted; the light from the closest star, Proxima Centauri, 4 years after; the light from the Nebula of Andromeda, the most distant object the naked eye can see and our nearest neighboring galaxy, 1,800,000 years after. This means that we see celestial bodies where they were and not where they are! This explains our interest in the apparent positions of the celestial bodies and our calculations of apparent positions when dealing with astronomical coordinates for terrestrial purposes, such as the energy income to the earth, . . . but not for computing the trajectories of space ships!

A. Determination of the Sun's Position

The position of the sun in the sky varies throughout the day and season due to the spin of the earth around its axis and to its orbiting around the sun. Knowledge of spherical and plane trigonometry as well as some elementary notions of astronomy are required in order to understand, and not just apply, the formulas for the computation of the times and directions of the direct incoming radiation on any locality of a planet or satellite, and the accuracy to which the calculations should be limited.

B. The Celestial Vault

In order to determine the position of the celestial bodies in the sky, they are assumed to lie on a single sphere, the Celestial sphere. The radius of the sphere must be large enough to identify the celestial bodies as points on the sphere, and the center, depending on the different conventions, coincides with the position of the observer (horizontal system), the center of the earth (equatorial system), the center of the sun (ecliptic system), or the center of the galaxy (galactic system).

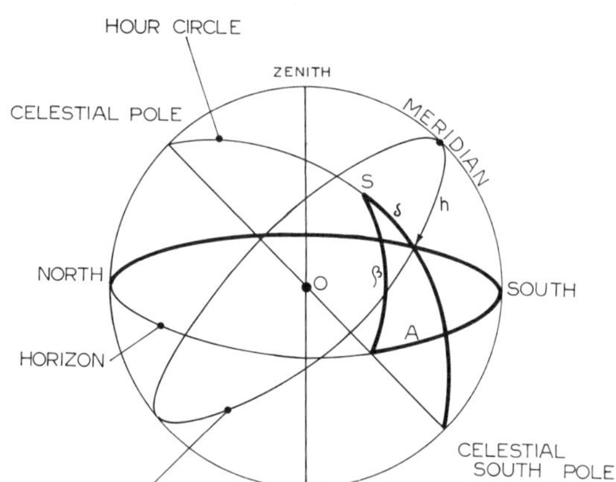

FIG. 2.2 The celestial vault represented in the horizontal system.

The horizontal system (Figs. 2.2 and 2.3) is preferred when calculating the position of the sun with respect to the geographical coordinates on the earth. In this system the reference plane is the horizon of the observer, i.e., the plane passing through the observer and normal to the vertical. The

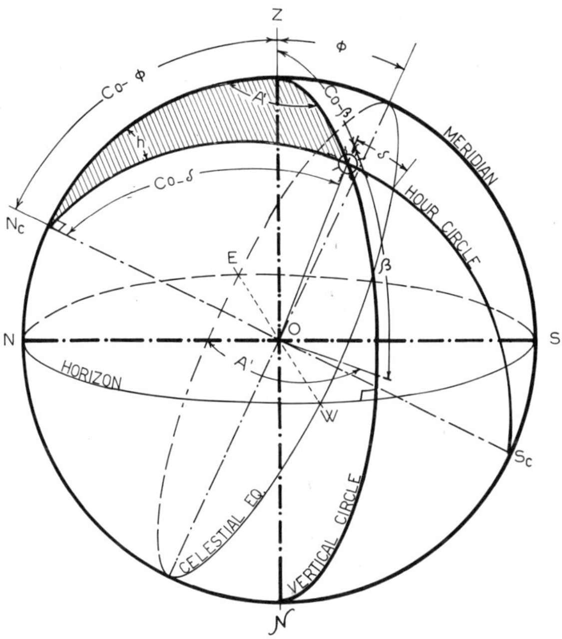

FIG. 2.3 The astronomical spherical triangle N_cZS, relative to the earth's coordinates. Point S, the interception of the vertical circle with the hour circle, is not labeled for graphical reasons.

elements in this system are:

Zenith, $\bar{Z}$ The point at which the sphere intercepts the upward vertical axis passing through the observer.

Nadir, N The point of the celestial sphere diametrically opposite to the zenith.

Celestial poles, N_c, and S_c The zeniths of the terrestrial poles.

Vertical circles Any of the great (maximum) circles of the sphere passing through the zenith of the observer. A particular vertical circle is the meridian: the vertical circle passing also through the celestial poles.

Celestial equator The great circle normal to the earth's axis.

Hour circle The great circle normal to the celestial equator and passing through the sun; it is also called the declination circle as the angular distance from the celestial equator to the sun measured on this circle corresponds to the declination. This distance is constant during the day because the apparent path of the sun is always a circle parallel to the celestial equator.

Almucantar The altitude parallels, i.e., every circle parallel to the horizon.

The spherical coordinates of the system are:

Altitude, β The almucantar, i.e., the angular elevation above the horizon, positive toward the zenith, negative toward the nadir.

Azimuth, A The bearing, i.e., the angular distance from the meridian to the great circle passing through the zenith of the observer and the celestial body. It is measured on the horizontal from north towards east, from 0° to 360°, positively, or negatively in the opposite direction.

The earth's equator is the projection on earth (or vice versa) of the celestial equator, the maximum circle equally distant from the north and south celestial poles. The corresponding plane is normal to the terrestrial rotation axis. This axis is constantly* inclined 66°43′ with respect to the elliptic plane, i.e., the plane defined by the earth's orbit around the sun. Consequently the equatorial plane is inclined 23°27′ with respect to this plane and the apparent path of the sun changes progressively throughout the year.

The solar altitude angle measured at noon will differ from the corresponding equinoctial angle by an angle of up to ±23°27′. This angle, called the solar declination, is generally defined as the angular distance from the zenith of the observer at the equator and the sun at solar noon, or the sun's angular distance north (+) of south (−) from the celestial equator.

The declination reaches its maximum value, +23°17′ on June 21 (the summer solstice in the Northern Hemisphere, the winter solstice in the Southern Hemisphere) when the earth is near aphelion, i.e., at 152,184,000 km from the sun. The minimum value, −23°27′ is reached on December 20 (the winter solstice in the Northern Hemisphere, the summer solstice in the other) when the earth is near perihelion, i.e., at 147,176,000 km from the sun. On March 21 and September 22, the spring and the autumn equinoxes, the declination is 0°.

* The earth's axis is slowly shifting. This motion is known as the precession of the earth's axis and is the sum of the precessional nutation and the Eulerian nutation. The former is caused by the position of the lunar orbit around the earth, which varies continually inducing the earth's axis to describe an elliptical cone whose major semiaxis is 9.2 minutes of a degree, the period being about 19 years.

The latter is caused by the lunar–solar precession and the planetary precession which amount to a yearly movement of 50.27 seconds of a degree. The luni–solar precession is caused by the gravitational attraction of the sun and the moon, which, because of the nonsphericity of the earth and the inclination of the earth's axis, induces a small torque leading to an oscillation of the earth about the plane of the ecliptic. The planetary precession is a movement of the same type, but opposite in direction and quite small if compared to the first. It is caused by the planets of the solar system. The effect of the eulerian nutation, whose period is approximately 26,000 years, is an appreciable variation of the latitude. When it was first observed by Hipparchus (125 B.C.), the polar star was 12° off the North Pole. At present it is 1° off and the distance will slowly decrease to 30′, then it will increase again up to approximately 46° in 13,000 years!

The declination, in degrees, for any given day may be calculated in first approximation with the equation

$$\delta = \left(23 + \frac{27}{60}\right)^{\circ} \cdot \sin\left(\frac{360d}{365.25}\right) \tag{2.2}$$

where d represents the number of days passed after the spring equinox (spring referred to the Northern Hemisphere). The value of δ will be correct within $+0°22'$ (the maximum deviation, on May 1) and $-1°42'$ (the minimum deviation, on October 9) as shown in Fig. 2.4.

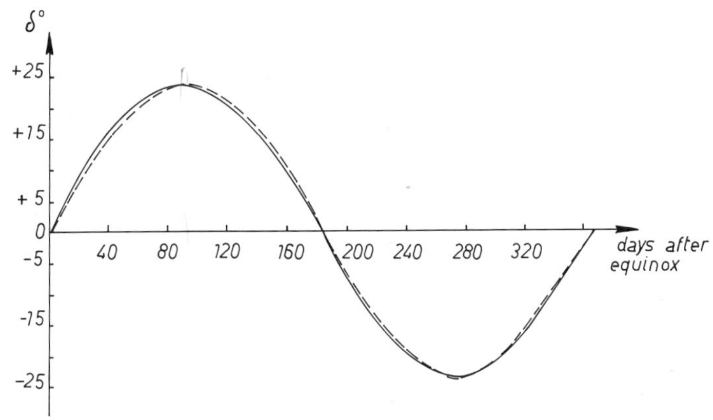

FIG. 2.4 The declination, calculated (—) and tabulated. (----).

Linear interpolation from Table 2.2, which gives the declination at intervals of four days within every month of the year, will give much more accurate values and certainly within the half-degree precision required in the most accurate cases of solar energy engineering computations. In fact the angle under which we observe the solar disk is approximately 32′ (0.0093 rad) corresponding to a solid angle of 6.783×10^{-5} sr. The variation of the angle subtended by the solar disk throughout the year is negligible. In all the following calculations we shall determine the apparent position of the center of the sun. Therefore all the values of the angles should be looked at with a "tolerance" of $\pm 0.25°$ for the dimensions of the solar disk, considering also the corona, and the computations should be limited to the first decimal. Special attention is needed when calculating the shape of the targets and the attainable temperature of solar concentrating systems, not only because of the solid angle subtended by the solar disk and corona, but also because of the nonuniform brightness distribution of the solar disk.

TABLE 2.2

The Solar Declination[a]

January	1	−23°	4′	February	1	−17°	19′	March	1	−7°	53′
	5	22	42		5	16	10		5	6	21
	9	22	13		9	14	55		9	4	48
	13	21	37		13	13	37		13	3	14
	17	20	54		17	12	15		17	1	30
	21	20	5		21	10	50		21	0	5
	25	19	9		25	9	23		25	+1	30
	29	18	8						29	3	4
April	1	+4°	14′	May	1	+14°	50′	June	1	+21°	57′
	5	5	46		5	16	2		5	22	28
	9	7	17		9	17	9		9	22	52
	13	8	46		13	18	11		13	23	10
	17	10	12		17	19	9		17	23	22
	21	11	35		21	20	2		21	23	27
	25	12	26		25	20	49		23	23	25
	29	14	13		29	21	30		29	23	17
July	1	+23°	10′	August	1	+18°	14′	September	1	+8°	35′
	5	22	52		5	17	12		5	7	7
	9	22	28		9	17	6		9	5	37
	13	21	57		13	14	55		13	4	6
	17	21	21		17	13	41		17	2	34
	21	20	38		21	12	23		21	1	1
	25	19	50		25	11	2		25	−0	32
	29	18	57		29	9	39		29	2	6
October	1	−2°	53′	November	1	−14°	11′	December	1	−21°	41′
	5	4	26		5	15	27		5	22	16
	9	5	58		9	16	38		9	22	45
	13	7	29		13	17	45		13	22	45
	17	8	58		17	18	48		17	23	20
	21	10	25		21	19	45		21	23	26
	25	11	50		25	20	36		25	23	25
	29	13	12		29	21	21		29	23	17

[a] US Naval Observatory, The American Ephemeris and Nautical Almanac for the Year 1950, Washington, 1948. All the above data are for 0 h Greenwich civil time in the year 1950; the variations of these data from year to year are negligible for solar engineering purposes; the largest variation occurs through the four-year leap-year cycle.

C. Hour Angle, Apparent Solar Time, Clock Time

The apparent daily rotation of the celestial sphere about the earth's axis may be expressed in terms of the hour angle: the angular distance between the hour circle and the observer's meridian. It is measured positively west-

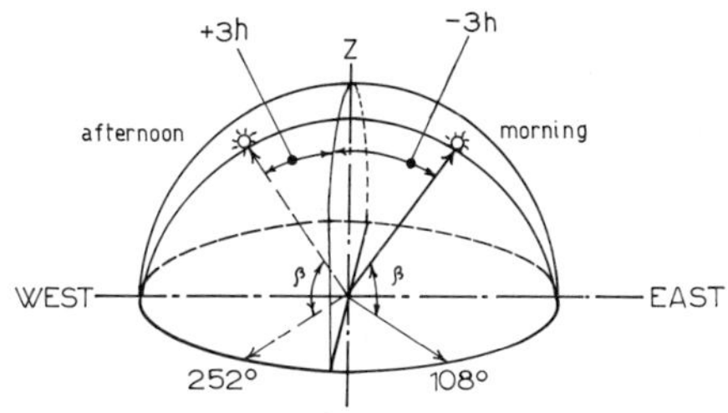

FIG. 2.5 The apparent path of the sun in the northern hemisphere at equinox. The specular symmetry of the path with respect to the plane defined by the meridian may be observed.

ward from the observer and it may be expressed in

hours, minutes, and seconds
degrees, minutes
radians

One hour is equivalent to $2\pi/24 = 0.262$ rad, or $360°/24 = 15°$, and consequently 1 min $= 15'$, 1 s $= 15''$ of the rotation of the celestial vault. It should be noted that 12 noon in everyday language corresponds to 0 h, morning hours are negative, afternoon hours positive, midnight ± 12 h. The spherical coordinates of the sun's ephemeris and the hour angle h, from an observer's point of view, are shown in Fig. 2.5.

All the values of time in solar energy computations are expressed in terms of apparent solar time (this is also known as true solar time). But in certain cases, it may be necessary to express the successive positions of the sun relative to a fixed point on the earth's surface in terms of the local clock time that differs from the apparent solar time because the latter is defined as the angular distance from the meridian to the hour circle. But, due to the elliptical shape of the earth's orbit and to its increase in velocity at the perihelion, the length of the apparent solar day, i.e., the interval between two successive passages of the sun through the meridian, is not constant. In order to provide a basis for a uniform time scale, reference is made to the mean sun, which is an imaginary point traveling not on the earth's elliptical orbit, but on a circular orbit (the celestial equator) with an angular velocity ω, equal to 2π rad/year $= 1.99 \times 10^{-8}$ rad/s.

The mean solar time is the hour angle of the mean sun and, similarly to true solar time, it is measured from the instant it intercepts the meridian of the observer. Mean solar time may deviate from true solar time by as much as 4.5° because, even if the length of any apparent solar day and its corresponding mean solar day differ little, the effect is cumulative. The differences between apparent solar time H_{sol} and the mean solar time are known as the equation of time and are listed in Table 2.3 at intervals of four days for every month of the year, and may be linearly interpolated.

TABLE 2.3

The Equation of Time (Apparent Mean)[a]

Month	Day	min	sec	Month	Day	min	sec	Month	Day	min	sec
January	1	−3′	14″	February	1	−13′	19″	March	1	−12′	38″
	5	5	6		5	14	2		5	11	48
	9	6	50		9	14	17		9	10	51
	13	8	27		13	14	20		13	9	49
	17	9	54		17	14	10		17	8	42
	21	11	10		21	13	50		21	7	32
	25	12	14		25	13	19		25	6	20
	29	13	5						29	5	7
April	1	−4′	12″	May	1	+2′	50″	June	1	+2′	27″
	5	3	1		5	3	17		5	1	49
	9	1	52		9	3	35		9	1	6
	13	0	47		13	3	44		13	0	18
	17	+0	13		17	3	44		17	−0	33
	21	1	6		21	3	34		21	1	25
	25	1	53		25	3	16		25	2	17
	29	2	33		29	2	51		29	3	7
July	1	−3′	31″	August	1	−6′	17″	September	1	−0′	15″
	5	4	16		5	5	59		5	+1	2
	9	4	56		9	5	33		9	2	22
	13	5	30		13	4	57		13	3	45
	17	5	57		17	4	12		17	5	10
	21	6	15		21	3	19		21	6	35
	25	6	24		25	2	18		25	8	0
	29	6	23		29	1	10		29	9	22
October	1	+10′	1″	November	1	+16′	21″	December	1	+11′	16″
	5	11	17		5	16	23		5	9	43
	9	12	27		9	16	12		9	8	1
	13	13	30		13	15	47		13	6	12
	17	14	25		17	15	10		17	4	17
	21	15	10		21	14	18		21	2	19
	25	15	46		25	13	15		25	0	20
	29	16	10		29	11	59		29	−1	39

[a] US Naval Observatory, The American Ephemeris and Almanac for the Year 1950, Washington, 1948. All the above data are for 0 h Greenwich civil time in the year 1950; the variations of these data from year to year are negligible for solar engineering purposes; the largest variation occurs through the four-year leap-year cycle.

D. The Astronomical Spherical Triangle Related to the Earth's Coordinates

The relation of the celestial sphere to a particular location on the earth, O, is shown in Fig. 2.3.

The cardinal points of the observer's horizon are indicated by the points N, E, S, and W on the great circle described by the horizon on the celestial sphere. The plane of the diagram is the plane defined by the meridian of the observer. N_c and S_c indicate, respectively, the North and South celestial poles, Z and N, the zenith and the nadir. The celestial equator, the vertical circle intercepting the sun, S, and the hour circle are also indicated. The hour angle h, the declination δ, the latitude (the angular distance from the equator), and the solar altitude β are indicated as well as their complementary angles, when necessary. The angle A' is the azimuth in the direction from north to west, hence the azimuth angle A will be equal to the opposite of A' ($-A'$ or $360° - A'$).

The astronomical spherical triangle N_cZS, relative to the earth's coordinates, is shown shaded in Fig. 2.3 where each side and angle are labeled. This spherical triangle is an oblique (nonright) spherical triangle and its orthogonal projection on the plane defined by the meridian of the observer is shown in Fig. 2.6. The triangle is resolved into two right spherical triangles by constructing from S a line normal to N_cZ (Fig. 2.7). The interception of this line with the meridian is indicated as Q. From Fig. 2.7,

$$QZ = (\text{Co}-\phi) - N_cQ = 90 - \phi - N_cQ = \text{Co}-(\phi + N_cQ),$$

where ϕ is the latitude and $\text{Co}-\phi$ is the complement of ϕ. The two projection triangles for the application of Napier's rules, with Napier's circular parts indicated, are shown in Fig. 2.8.

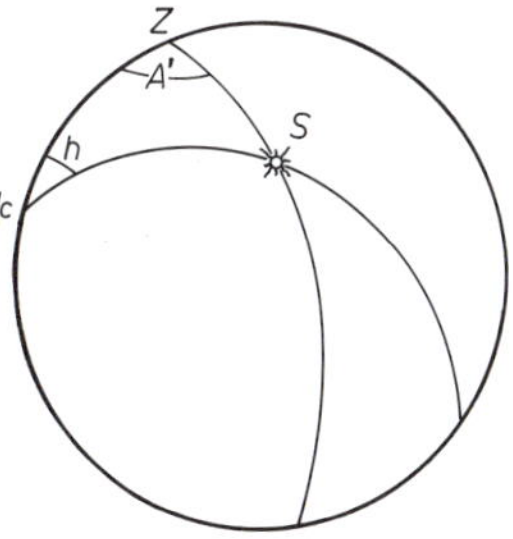

FIG. 2.6

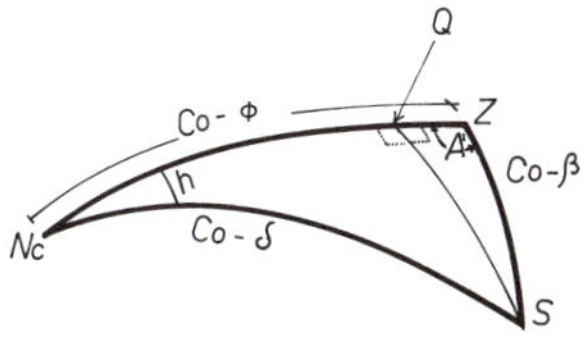

FIG. 2.7

FIG. 2.6 The projection of the oblique spherical triangle SN_cZ of Fig. 2.3 on the plane defined by the meridian.

FIG. 2.7 The resolution of the oblique spherical triangle N_cZS into two right-spherical triangles.

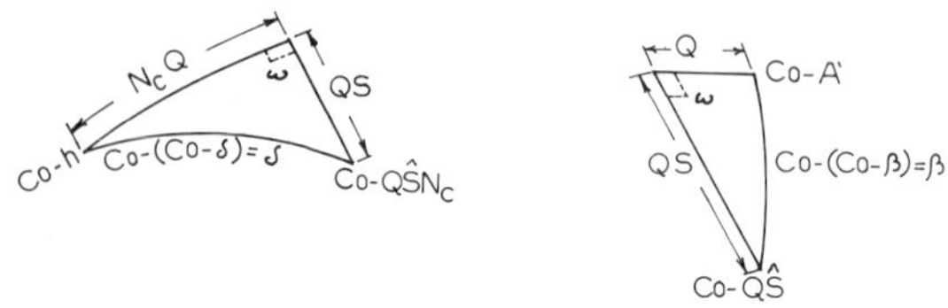

FIG. 2.8 Napier's circular parts of the two right-spherical triangles N_cZS and QZS.

By applying Napier's rules to triangle N_cQS, the following equations may be obtained:

$$\sin QS = \cos(\text{Co}-h)\cos\delta = \sin h\cos\delta \tag{2.4}$$

$$\sin N_cQ = \tan QS\tan(\text{Co}-h)$$

or

$$\tan QS = \sin N_cQ\tan h \tag{2.5}$$

$$\sin(\text{Co}-h) = \tan N_cQ\tan\delta$$

or

$$\tan N_cQ = \cos h\cot\delta \tag{2.6}$$

$$\sin\delta = \cos N_cQ\cos QS \tag{2.7}$$

Similarly, by applying Napier's rules to the right spherical triangle SQZ, the following relations may be obtained:

$$\sin QS = \cos\beta\cos(\text{Co}-A') \quad \text{or} \quad \sin QS = \cos\beta\sin A' \tag{2.8}$$

$$\sin QZ = \tan QS\tan(\text{Co}-A') \quad \text{or} \quad \cot A' = \sin QZ\cot QS \tag{2.9}$$

$$\sin\beta = \cos QS\cos QZ \tag{2.10}$$

All the general expressions for the altitude, the azimuth, the declination, the latitude, and the hour angle may be derived from the above eight equations.

E. Equations for Solar Positions

1. Solar Altitude, β

Solving Eqs. (2.7) and (2.10) simultaneously by substituting for $\cos QS$, we obtain

$$\sin\beta = \sin\delta\cos QZ/\cos N_cQ$$

and substituting for QZ expression (2.3),

$$\sin\beta = \sin\delta\frac{\cos(\text{Co}-(\phi + N_cQ))}{\cos N_cQ} = \frac{\sin\delta}{\cos N_cQ}\sin(\phi + N_cQ)$$
$$= (\sin\delta/\cos N_cQ)[\sin(\phi)\cos N_cQ + \cos(\phi)\sin N_cQ]$$
$$= \sin\delta\sin\phi + \sin\delta\cos\phi\tan N_cQ$$

and by substituting the expression for $\tan N_cQ$ of Eq. (2.6),

$$\sin\beta = \sin\delta\sin\phi + \sin\delta\sin\phi\cos h\cot\delta$$
$$= \sin\delta\sin\phi + \cos\delta\cos h\cos\phi \tag{2.11}$$

At solar noon the hour angle h is zero and Eq. (2.11) is reduced to

$$\sin\beta = \sin\delta\sin\phi + \cos\delta\cos\phi = \cos(\delta - \phi) = \sin(\text{Co}-(\delta - \phi)) \tag{2.12}$$

The "safest way" of solving Eq. (2.12) is to solve for β as follows:

$$\beta = \arcsin(\cos(\delta - \phi)) \tag{2.13}$$

because $\beta = 0° \ldots 90°$, negative values of β simply meaning that the sun is below the horizon (polar night). The "simplest way" of solving Eq. 2.12 is as follows:

$$\sin\beta = \sin(\text{Co}-(\delta - \phi))$$
$$\beta = \text{Co}-(\delta - \phi) = 90° - |\delta - \phi| = (\pi/2) - |\delta - \phi| \tag{2.14}$$

It should be noted that the equation ($\beta = 90 - \delta + \phi$), which is indicated in some textbooks on solar energy, is valid only in the northern temperate zone for all year, as the angle β is not the altitude of the sun, but the angle the direct solar radiation forms with the south direction of the projection of the observer's meridian on the horizon. β obtained in this way may be converted into solar altitude if when $\beta > 90$ we take the specular image of this angle as β ($\beta = 180 - \beta$).

Other equations for the solar altitude valid for special cases such as at the equator ($\phi = 0$), at equinox ($\delta = 0$), etc. may be obtained from Eq. (2.11).

2. The Solar Azimuth Angle, A

The solar azimuth angle A measured from true (geographical) north positively in the eastward direction (as in nautical compasses) may be expressed by comparing Eqs. (2.4) and (2.8):

$$\sin h\cos\delta = \cos\beta\sin A'$$
$$\sin A' = \sin h\cos\delta/\cos\beta$$
$$\sin A = -\sin h\cos\delta/\cos\beta \tag{2.15}$$

or by substituting in Eq. (2.9) the value of $\tan QS$ from Eq. (2.5):

$$\cot A' = \sin QZ \cot QS$$

$$\tan QS = \sin N_c Q \tan h$$

$$\cot A' = \sin QZ \cot h / \sin N_c Q$$

and by substituting in $\sin QZ$ the expression for QZ of Eq. (2.3):

$$\sin QZ = \sin(Co - (\phi + N_c Q)) = \cos(\phi + N_c Q)$$

$$\begin{aligned}\cot A' &= [\cos(\phi + N_c Q)/\sin N_c Q] \cot h \\ &= [(\cos\phi \cos N_c Q - \sin\phi \sin N_c Q)/\sin N_c Q] \cot h \\ &= (\cot\phi \cot N_c Q - \sin\phi) \cot h\end{aligned}$$

but from Eq. (2.6):

$$\tan N_c Q = \cos h \cot \delta$$

hence,

$$\cot A' = [(\cos\phi/\cos h) \tan\delta - \sin\phi] \cot h$$

$$\cot A = (-\cos\phi \tan\delta + \sin\phi \cos h)/\sin h \qquad (2.16)$$

Equation (2.16) is to be preferred to Eq. (2.15) because solving it, we have

$$A = \arctan\left(\frac{\sin h}{-\cos\phi \tan\delta + \sin\phi \cos h}\right)$$

with $-90^\circ \leqslant A \leqslant +90^\circ$, and if we are calculating the azimuth for an afternoon ($h > 0$) and from the solution of Eq. (2.16) $A < 180^\circ$, we simply add a period of the tangent 180°; and similarly, if we are calculating a morning ($h < 0$) azimuth and the computed solution $A > 180^\circ$, we subtract 180°.

Instead, by solving Eq. (2.15), we have

$$A = -\arcsin(\sin h \cos\delta/\cos\beta) \qquad \text{with} \quad -90^\circ \leqslant A \leqslant +90^\circ$$

but recalling from elementary trigonometry that $\sin A = \sin(180^\circ - A)$, we see that we have two possible solutions for every A and that they lie in the same part of the day (morning or afternoon). There may be difficulties in selecting the right solution. Fortunately, the two solutions lying on the same side of the meridian are supplementary and hence lie in different quadrants. In winter there is no problem as the sun will shine from the two southern quadrants in the northern hemisphere, and only from two northern quadrants in the southern hemisphere. In summer we must find the hour angle for which the sun is due west, h_w (the time it is due east $= -h_w$) by setting in Eq. (2.16)

$$A = (2N + 1)\pi/2, \qquad N = 0, 1, 2, \ldots$$

equivalent to $\cot A = 0$, we have

$$\cos h \sin \phi = \cos \phi \tan \delta, \qquad \cos h = \cot \phi \tan \delta$$

and solving for the absolute (afternoon) value of h, we obtain

$$h = \arccos(\cot \phi \tan \delta) \tag{2.17}$$

If the hour h, for which we are computing the solar azimuth, is closer to noon than h_w, i.e., if $|h| < h_w$, the solution will be the one lying in one of the quadrants of the center of the day; if instead, $|h| > h_w$, the solution will lie in one of the others.

Schematic sets of diagrams of the solar azimuth from sunrise to sunset in the different seasons and at different latitudes for arctic, temperate, and equator zones are shown in Fig. 2.9. The azimuth is not given for the poles because at those particular points of the globe the cardinal points lose definition.

3. *Sunrise and Sunset Times and Azimuths*

The hour angles for sunrise and sunset are given by the equation obtained setting the solar altitude β of Eq. (2.11) equal to zero:

$$\cos \delta \cos h \cos \phi = -\sin \delta \sin \phi$$

$$\cos h = -\sin \delta \sin \phi / \cos \delta \cos \phi = -\tan \delta \tan \phi \tag{2.18}$$

Equation (2.18) may be solved for h if $-1 \leqslant -\tan \delta \tan \phi \leqslant +1$; then

$$h = \arccos(-\tan \delta \tan \phi)$$

and, as $\cos h = \cos(-h)$, the two solutions will be

$$h_{ss} = h, \qquad h_{sr} = -h$$

But if

$-\tan \delta \tan \phi > +1$,	the sun will neither rise nor set for the day: polar night.
$= +1$,	the sun will be on the horizon for an instant only, i.e., at solar noon. This occurs the last day before a polar night and the first day after.
$= -1$,	the sun will be on the horizon for an instant only, i.e., at midnight ($h = \pm 180°$). This occurs the day before and the day after a polar day.
< -1,	the sun will neither rise nor set for the day: polar day.

Of the above four statements the two central ones are only of theoretical interest because, as indicated previously, the calculations are referred to the center of the solar disk.

The azimuths of the sunrise and the sunset may be obtained from Eq. (2.15) by setting the solar altitude β to zero:

$$\sin A = -\sin h \cos \delta \tag{2.19}$$

and by substituting the values of h_{sr} and h_{ss} to h in Eq. (2.19), we shall have two solutions for the morning and two for the afternoon. The correct solutions will be selected by recalling that when the declination is positive the sun rises and sets in the northern quadrants; when zero, it rises due east and sets due west; and when negative, it rises and sets in the southern quadrants (see Fig. 2.9).

An example of solar computation is to calculate the altitude and the azimuth of the sun at solar noon and the times and azimuths of sunrise and sunset for a location at 42°35′ N on March 15. The declination, interpolated linearly from Table 2.2, is $\delta = -2.442°$. The solar altitude at solar noon is

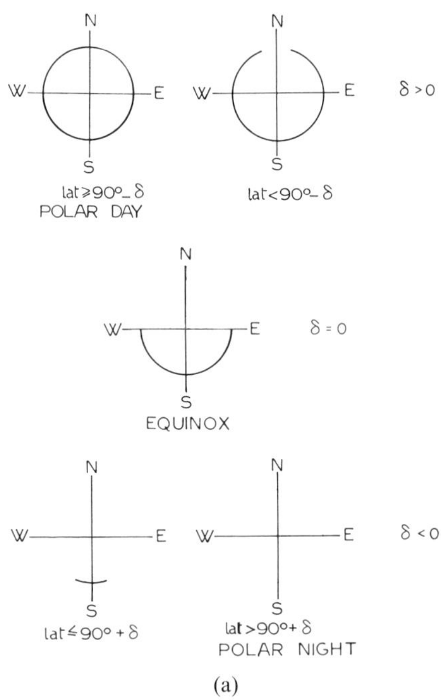

FIG. 2.9 The solar azimuth (a) Arctic zone: 66°43′ < latitude ⩽ 90°. (b) North temperate zone: 23°17′ < latitude < 66°43′. (c) North tropical zone: 0° < latitude < 23°17′. (d) Equator: latitude = 0°.

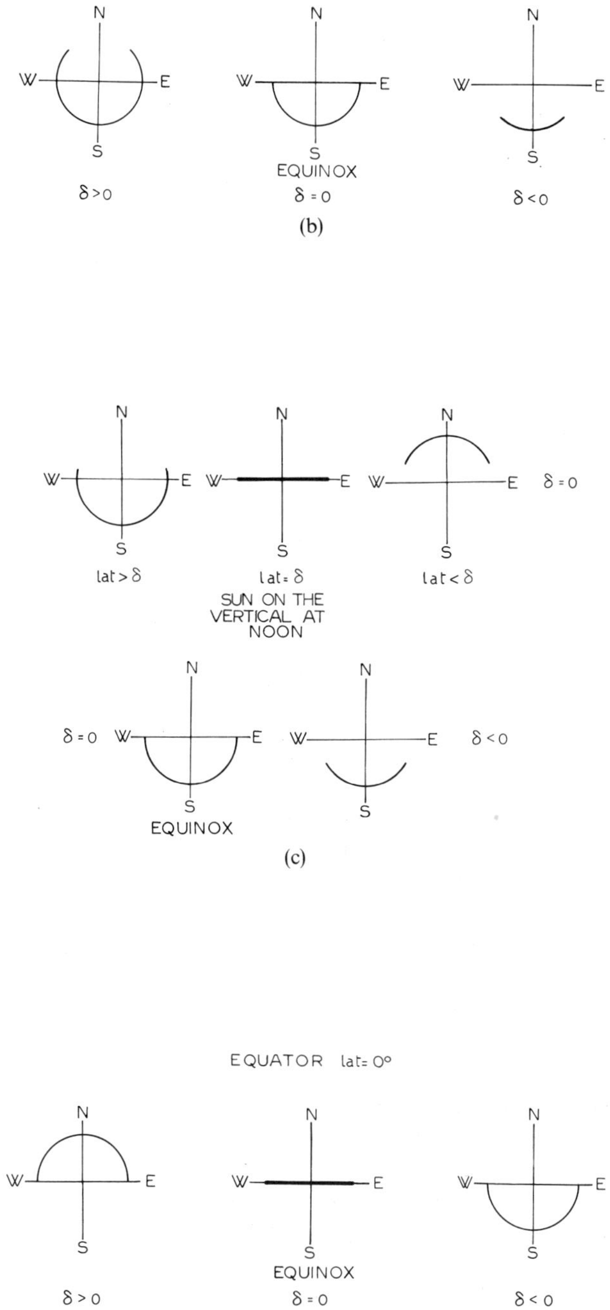

FIG. 2.9 (b), (c), and (d)

given by Eq. (2.14) as

$$\begin{aligned}\beta &= 90^\circ - |\delta - \phi| = 90^\circ - |-2.442^\circ - 42.538| \\ &= (90. - 45.025)^\circ = 44.97^\circ = 44^\circ 58'\end{aligned}$$

The solar azimuth is 180° because at noon ($h = 0$) the sun lies on the meridian and the location is in the northern temperate zone (see Fig. 2.9).

The times of sunrise and sunset are given by Eq. (2.18) as

$$\cos h = -\tan(-2^\circ 26') \tan 42^\circ 35' = 0.0391$$

$$h = 87^\circ 46' = 1.53 \text{ rad} = 5 \text{ h } 51 \text{ min}$$

Hence the sunrise and the sunset occur at an angular distance (− and +, respectively) of 5 h 51 min from solar noon.

The azimuth of the sunrise is given by Eq. (2.19) as

$$\sin A = -\sin 87^\circ 46' \cos(-2^\circ 26') = -0.9992$$

$$A = -87^\circ 42' = 272^\circ 18', \qquad A_1 = 272^\circ 18',$$

$$A_2 = 180^\circ - A_1 = 92^\circ 18' = 267^\circ 42'$$

The declination is negative and therefore the azimuth of the sunset must be in the third quadrant (south–west), the correct answer is

$$A_{ss} = A_2 = 272^\circ 18'$$

The two possible solutions for the sunrise are the two solutions symmetrical (σ_v) to the afternoon solutions:

$$A_1 = 360^\circ - 272^\circ 18' = 87^\circ 42', \qquad A_2 = 360^\circ - 267^\circ 42' = 92^\circ 18'$$

The correct solution will be, again, A_2 because $\delta < 0$:

$$A_{sr} = 92^\circ 18'$$

2.3 THE CALCULATION OF THE ANGLE OF INCIDENCE OF DIRECT RADIATION ON AN INCLINED PLANE

The orientation α of a plane is described as the angular distance from true north to the projection of the profile of the slope on the horizontal plane, it varies from -90° to 0° to $+90^\circ$, negatively westward, positively eastward. The tilt γ is described in terms of the vertical angle below the horizon and facing south, for both northern and southern hemispheres (i.e., the angle of the profile of the slope taken as positive if the slope faces south, and taken as negative if it faces north). The tilt and the orientation of a plane are illustrated in Fig. 2.10. In the figure I_n represents the normal to the plane passing through

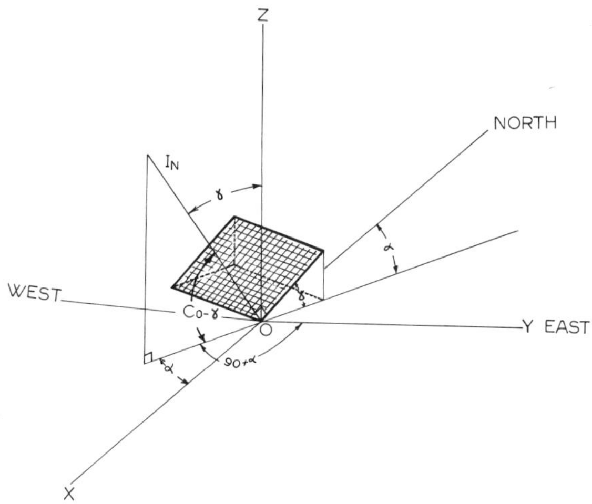

FIG. 2.10 The orientation and the tilt of an inclined plane.

the origin. The altitude of vector I_n is the complement of the tilt of the plane; its orientation α coincides with the orientation of the plane. The three direction-cosines of vector I_n are

$$l' = \cos(\mathrm{Co}-\gamma)\cos\alpha = \sin\gamma\cos\alpha$$

$$m' = \cos(\mathrm{Co}-\gamma)\cos(90^\circ + \alpha) = -\sin\gamma\sin\alpha$$

$$n' = \cos\gamma$$

Figure 2.11 represents the sun with its coordinates referring to the usual geographical orientation and to a coincidental Cartesian reference system. The three direction-cosines of vector OP representing the direction of the incoming direct radiation are

$$l = \cos\beta\cos(180^\circ - A) = -\cos\beta\cos A$$

$$m = \cos\beta\cos(A - 90^\circ) = \cos\beta\sin A$$

$$n = \cos(\mathrm{Co}-\beta) = \sin\beta$$

The angle θ between two straight lines that intersect is given by

$$\cos\theta = l\cdot l' + m\cdot m' + n\cdot n'$$

and in our case, the straight lines are the normal to the plane and the direction of the incoming direct radiation. The intersect at the origin in the graph is obtained by superimposing Fig. 2.10 on Fig. 2.11. The cosine of their angle of

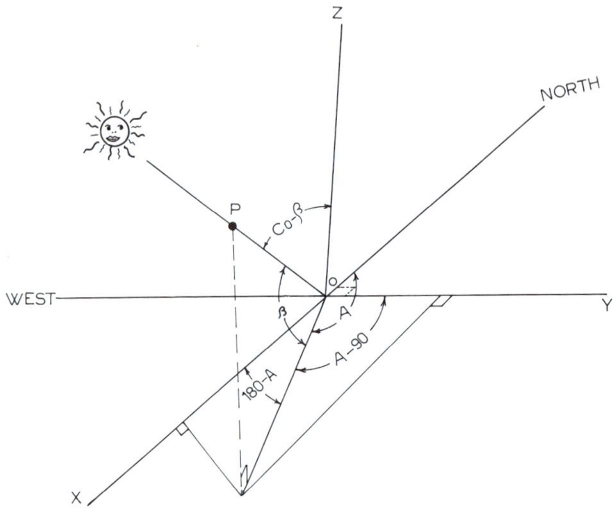

FIG. 2.11 The altitude and the azimuth of the sun.

diversion i is

$$\cos i = -\cos\beta\cos A\sin\gamma\cos\alpha - \cos\beta\sin A\sin\gamma\sin\alpha + \sin\beta\cos\gamma \qquad (2.20)$$

and rearranging and factoring, we get

$$\cos i = \sin\beta\cos\gamma - \sin\gamma\cos\beta\cos(A-\alpha) \qquad (2.21)$$

Although Eq. (2.21) is suitable for the calculation of the angle of incidence, it is convenient to have

$$\cos i = f(\alpha, \gamma, \delta, h, \phi)$$

This can be achieved as follows. From Eq. (2.16),

$$\cot A = \cos A/\sin A = (-\cos\phi\tan\delta + \sin\phi\cos h)/\sin h$$

and substituting Eq. (2.15) for sin A, we obtain

$$\begin{aligned}\cos A &= (\sin h\cos\delta/\cos\beta)\cdot[(\cos\phi\tan\delta - \sin\phi\cos h)/\sin h]\\ &= (1/\cos\beta)\cdot(\cos\phi\sin\delta - \sin\phi\cos\delta\cos h)\end{aligned}$$

Substituting the above expression for cos A, Eq. (2.11) for sin β, and Eq. (2.15) for sin A into Eq. (2.20), we have

$$\begin{aligned}\cos i = {} & (-\cos\beta/\cos\beta)\,[\cos\phi\sin\delta - \sin\phi\cos\delta\cos h]\sin\gamma\cos\alpha\\ & + \cos\beta(\sin h\cos\delta/\cos\beta)\sin\gamma\sin\alpha + \sin\delta\sin\phi\cos\gamma\\ & + \cos\delta\cos h\cos\phi\cos\gamma\end{aligned}$$

Rearranging, we obtain

$$\begin{aligned}\cos i = {} & (\cos\gamma \sin\phi - \sin\gamma \cos\phi \cos\alpha) \sin\delta \\ & + (\cos\gamma \cos\phi + \sin\gamma \sin\phi \cos\alpha) \cos\delta \cos h \\ & + \sin\gamma \sin\alpha \cos\delta \sin h \end{aligned} \tag{2.22}$$

The expressions in parentheses in the first two lines of Eq. (2.22) and the first part of the third line are constants characteristic of the inclination and orientation of the plane, and of the latitude. Hence Eq. (2.22) may be regarded as

$$\cos i = C_1 \sin\delta + C_2 \cos\delta \cos h + C_3 \cos\delta \sin h \tag{2.23}$$

with

$$\begin{aligned} C_1 &= \cos\gamma \sin\phi - \sin\gamma \cos\phi \cos\alpha \\ C_2 &= \cos\gamma \cos\phi + \sin\gamma \sin\phi \cos\alpha \\ C_3 &= \sin\gamma \sin\alpha \end{aligned}$$

Zero-Irradiation Solar Times, Altitudes, and Azimuths of an Inclined Plane

Zero-irradiation times are usually calculated from second-degree equations in sine form, or from trigonometrical equations in sines and cosines that may be solved only by "trial and error." The solutions having been found in sine form, a careful selection of the possible solutions of the inverse functions (arcsine) must be made. In fact, the four possible solutions must be reduced to two and then compared with the sunrise and the sunset times. This is because the condition of parallelism between the solar radiation beam and the plane can be reached also when the sun is below the horizon. Finally, one must have a clear mental image of the sun's ephemeris to determine whether the inclined plane is exposed to the sun *until* or *from* the calculated zero-irradiation time(s).

The author wishes to present a rigorous mathematical model he developed from Eq. (2.22). In this model the solution is obtained by transforming the sines and cosines into tangents, and because of the one-to-one correspondence of tangent functions within their period, the mathematical solutions coincide with the physical positions of the hour angle of the sun at zero-irradiation times.

The two zero-irradiation hour angles of a given day and for a given tilted plane divide the circle representing the 24 h of the day into two sectors: in one the sun is above the plane, in the other it lies under the plane. The lack of real solutions indicate that the sun is above or below the tilted plane for the whole day. To determine which sector represents the time of the day when the sun is above (or under) the tilted plane, or, in the latter case, to determine whether the plane is in the sun or in the shade for the day, it is sufficient to calculate the angle of incidence of the solar radiation for any hour angle except for the hour angles of the zero-irradiation times. This is most conveniently done for solar noon as Eq. (2.22) reduces to

$$\cos i = (\cos\gamma \sin\phi - \sin\gamma \cos\phi \cos\alpha) \sin\delta + (\cos\gamma \cos\phi + \sin\gamma \sin\phi \cos\alpha) \cos\delta$$

and if $\cos i > 0$, the sector in which the selected hour angle lies will be the sector in which the sun is above the tilted plane. Then the sun will be under the tilted plane in the other sector. If $\cos i < 0$, the positions are reversed.

Sunrise and sunset also divide the circle representing the 24 h of a day into two sectors. The sector(s) in which the sun is above both the tilted plane and the horizon is (are) the one(s) in which the surface is exposed to the sun.

Equation (2.22) may be reduced to a form simpler to handle by letting

$$x = \sin\gamma \sin\alpha \cos\delta$$
$$y = (\cos\gamma \cos\phi + \sin\gamma \sin\phi \cos\alpha) \cos\delta$$
$$z = (\cos\gamma \sin\phi - \sin\gamma \cos\phi \cos\alpha) \sin\delta$$
$$\cos i = x \sin h + y \cos h + z \tag{2.24}$$

The condition of parallelism is reached when the solar beam lies on the plane of the slope, i.e., the angle of incidence is equal to 90°, or $\cos i = 0$:

$$x \sin h + y \cos h + z = 0 \tag{2.25}$$

and the hour angles h of the zero-irradiation times may be calculated as follows, letting $h = 2\omega$:

$$\sin h = \sin 2\omega = 2 \sin\omega \cos\omega = 2 \frac{\tan\omega}{\sqrt{1+\tan^2\omega}} \cdot \frac{1}{\sqrt{1+\tan^2\omega}} = \frac{2\tan\omega}{1+\tan^2\omega}$$

$$\cos h = \cos 2\omega = \cos^2\omega - \sin^2\omega = \frac{1}{1+\tan^2\omega} - \frac{\tan^2\omega}{1+\tan^2\omega} = \frac{1-\tan^2\omega}{1+\tan^2\omega}$$

Substituting the above equations in Eq. (2.25), we obtain

$$2x\frac{\tan\omega}{1+\tan^2\omega} + y\frac{1-\tan^2\omega}{1+\tan^2\omega} + z = 0$$

and rearranging, we have

$$(z - y)\tan^2\omega + 2x\tan\omega + (y + z) = 0$$

$$\tan\omega = \frac{-x \pm \sqrt{x^2 - (z^2 - y^2)}}{z - y}$$

$$\tan\omega_1 = \frac{-x + \sqrt{x^2 + y^2 - z^2}}{z - y}, \qquad \tan\omega_2 = \frac{-x - \sqrt{x^2 + y^2 - z^2}}{z - y} \tag{2.26}$$

$$h_1 = 2\arctan(\tan\omega_1), \qquad h_2 = 2\arctan(\tan\omega_2) \tag{2.27}$$

The hour angles, h_1 and h_2, are the angles at which the sun's rays are parallel to the inclined plane; complex and conjugate solutions indicate that the sun's rays are never parallel to the given plane for the day.

The next step involves calculating the cosine of the angle of incidence for an hour angle that is not a zero- irradiation hour angle, by using Eq. (2.24) For simplicity, we shall calculate the angle of incidence of the solar radiation for an hour angle of 0°, corresponding to solar noon.

If $\cos i > 0$, the plane will be exposed to solar radiation at solar noon and hence, the sun will shine on it from h_1 or from sunrise, whichever occurs later, and until h_2 or sunset, whichever occurs earlier (from sunrise to sunset if h_1 and h_2 are complex numbers).

If $\cos i < 0$, the tilted plane is in the shade at solar noon (all day if h_1 and h_2 are complex numbers), and will be exposed to the sun from h_2 to h_1 (clockwise direction) except, of course, for the interval of time between sunset and sunrise.

The solar altitude and azimuth at the hour angle corresponding to the zero-irradiation times may be calculated from Eqs. (2.11) and (2.16), respectively.

2.4 THE SUN: A SPARK IN A VOLCANIC ERUPTION

Brightness is far from being the only feature that distinguishes one star from another. In fact the most important classification of stars is the spectral classification. Generally, the spectrum of stars is continuous with the exception of some absorption lines, the Fraunhofer lines. Almost all the radiation we receive from a star is emitted by the photosphere, the inner atmosphere of the stars. This emission is close to the emission of a blackbody at high temperatures, but part of it is partially and selectively absorbed by the "outer" stellar atmosphere thus making the absorption lines noticeable in the spectra. The spectral classification includes all the stars with a few exceptions and represents a sequence of decreasing atmospheric temperatures. It gives in a simple and symbolic form the characteristics of the different stellar groups: 0, B, A, F, G, K, M, R, N, S.

Table 2.4 illustrates these characteristics. Every spectral class is divided into 10 subclasses because of the large difference between the *cool* stars and the *hot* stars of the same group. The hottest B star is B0, one with a lower temperature is B4, and the coldest B star is B9 which is followed by A0.

TABLE 2.4

The Stellar Spectral Classification

Class	Typical star	Color	Temperature (10^{-3} °K)	Notes
0		blue	60	
B	Achernar	blue	21	
A	Altair	white	8.7	
F	Procione	yellow–white	6.6	
G	Cappella	yellow	6.0	
K	Aldebaran	pale red	4.4	
M	Antares	red	3.3	
R		pale red		Same temperature as the K stars, but containing carbon absorption lines and having high magnetic fields.
N		pale red		
S		red		Same temperature of M stars, but have high magnetic fields.

Another way of classifying stars is done according to their mass. Unfortunately, except for a few special cases, it is difficult, if not impossible, to determine the mass of a star. Most of those with known mass are of the same order as the sun's, and are known as dwarf stars.

The smallest stars are the *white dwarfs*. They emit such a small quantity of light that they easily escape observation. At present only about 180 are known, and they are all quite close to us. They are probably numerous but beyond our present observational reach. White dwarfs, as well as the dwarfs, usually have a large proper motion.

The largest stars are known as *red* and *super* giants. The latter reach diameters that are 300 times the diameter of the sun, and, just to give an idea of the size, we can compare it with the diameter of the earth's orbit which is 215 times the diameter of the sun. The emission of these giant stars reaches values of brightness 50,000 times the emission of the sun!

TABLE 2.5

The Spectral Class and the Average Diameter of the Normal Stars

Spectral class		Diameter (× sun)
	M	42
Giants	K	15
	G	9
	F	4
	B	8
	A	2
	F	1.4
Dwarfs	G	1.0
	K	0.7
	M	0.5
White dwarfs		10^{-4} 10^{-5}

The other stars, the *normal* stars, are dwarfs when compared to the giants although they vary in size and brightness. The spectral class and the average diameter of the normal stars are indicated in Table 2.5.

If the spectral class of a star is plotted against its absolute magnitude, a very interesting diagram is obtained: the Hertzsprung–Russell diagram (Fig. 2.12). Here it can be observed that the distribution of stars is not random and that only one or at the most two luminosities correspond to each spectral class. The diagonal area in the diagram is called the main series. On the

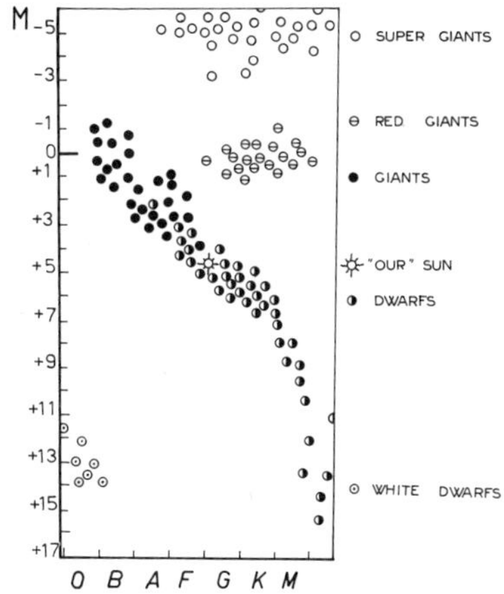

FIG. 2.12 Hertzsprung–Russell diagram.

(top) left part of this series are the giants and on the (bottom) right part the dwarfs, giants and dwarfs compared to each other. The real giants, i.e., the red giants and the super giants are grouped in the upper right part of the diagram and in the lower left we find the white dwarfs.

It is also indicated in the diagram that in the same spectral class, and especially for the classes from M to G, there are stars that differ up to 1000 times in luminosity even though they have the same temperature. This is explained by the difference in surface area.

A. The Characteristics of the Sun

The sun is just an average star. Of the innumerable others some are larger and some are smaller, some are whiter and some show a color more inclined to the red, some are believed to be older and some younger. Its absolute magnitude is 4.48, its spectral type G2, it belongs to the main series, it is a dwarf, and its surface temperature is about 6000°K. But it has something special, unique: the sun is "our" star and the only star whose shape and surface may be observed.

The principal characteristics of the sun are

Mass	$m = (1.991 \pm 0.002) \times 10^{30}$ kg
Radius	$r = (6.960 \pm 0.001) \times 10^{8}$ m
Average density	$\rho = 1.410 \pm 0.002$ g/cm^3
Temperature	
(average on the surface)	$T = 5762 \pm 50°$K
Visual magnitude	$M_v = -27.73 \pm 0.03$
Absolute magnitude	$M = +4.84 \pm 0.03$

The sun may be subdivided into the *inner sun* and the *solar atmosphere.* The inner sun which contains practically all the mass may be investigated only indirectly by computations based on the equilibrium of a "ball of fire" of known mass and luminosity. The calculations tend to determine the values of the temperature, density, and the energy produced, and the structure of the inner layers. The models proposed are far from being final mainly because of our insufficient knowledge of the interactions between matter and radiation at the tremendous temperatures and densities of the sun's core. It is supposed that in the center the temperature exceeds 1.5×10^{7}°K and that the density exceeds 100 g/cm^3. In the central region, i.e., the region of radius $0.23r$ (r = the radius of the sun), that contains 40% of the total mass, a nuclear reaction consumes 4.5×10^3 kg of solar matter per second. The energy created in this *reactor* region is transferred outward by radiation in the form of x and gamma rays up to a distance of about $0.7r$ from the center. Outside this region there is a fluid in which the energy transfer is mainly by convection and is therefore known as the convective zone. Within this zone the density drops to 10^{-8} g/cm^3 and the temperature falls to about 6,000°K.

The surface of the inner sun appears to the observer as a mosaic: a uniform background covered with bright granules. It is called the *photosphere* and is our source of light and heat. The granules are of polygonal shape and have an average diameter of about 1000 km (1.3 seconds of arc if observed from the earth). Their mean life is approximately three minutes. These granular cells are the visible proof of the convection currents in the convective zone. The central and brightest part of each cell is the part in which the solar matter rises from the inner sun to flow back down along the edges, having lost part of its thermal energy by radiation: the difference in luminosity measured in the center of the granules and in the intergranular areas corresponds to a temperature difference of about 200°K. This motion is confirmed by the Doppler effect that shows an overall upward motion of the photosphere, the photosphere being predominantly covered by rising matter.

Above the photosphere there is a transparent layer of rarified gas 10,000 km thick. It is known as the chromosphere because of its red color resulting from the strong Hα line of hydrogen in the spectrum of the layer, a continuous band of low intensity mainly caused by the scattering of solar radiation. Its structure is very complex and not yet fully understood. It is composed of a large number of jets of matter called spicula that intermingle as flames do but at supersonic speeds. They travel through the lower chromosphere at a velocity of about 20–30 km/s, gradually slowing with altitude. The photographs of the limb of the solar disk show that the chromosphere is continuous. Its density gradually decreases from 10^{-8} g/cm^3 at high altitudes. Its luminosity fades at 6000–7000 km above the photosphere. Spicula projected above this altitude have been observed and on occasions to a height of 15,000 km. The temperature of the chromosphere is about 4500°K at the base and 10^6°K at an elevation of 12,000 km where it ends and the isothermal corona starts. The light emitted by the chromosphere is of short wavelength because of the high temperature and is very weak because of the rarification of its gases. It contributes by a factor of 10^{-3} to the total energy emitted by the sun.

Finally there is the corona, a whitish glowing layer that may be observed in all its beauty only during total solar eclipses. It is made of highly ionized gases that are chemically the same as those composing the rest of the sun, but are extremely rarified. Probably it is made up of interplanetary matter flowing by gravity towards the sun and partially balanced by solar evaporation into space. The corona assumes a different form with every eclipse and it extends for millions of kilometers into space. Its temperature, supposed to be raised by shockwaves, is about 10^6°K as confirmed by the recent x-ray coronial spectra taken on the Skylab. The energy radiated by the corona is, again because of its high rarefaction, insignificant in comparison with the rest of the solar radiation: 10^{-6} of the total, and it cannot reach us because of our atmosphere.

The characteristics of the sun that we have examined above are all of a dynamically stable nature and it is quite unlikely that they will change within billions of years. Nevertheless, many relatively small changes take place both in the solar atmosphere and in the central core. The number of these events is quite numerous and their study is fascinating, but it is not within the scope of this publication. Therefore we shall limit this interesting field of solar physics to a partial list with a brief explanation.

Sunspots Areas of the sun that appear to be dark in contrast to the rest of the sun. They present two parts: umbra and penumbra. The temperature of the sunspots is considerably lower than the rest of the solar surface, i.e., it drops to about 4500°K, and the luminosity to about 18% of that of the surrounding area. Sunspots are probably formed by magnetic fields that contract the convection currents that continuously bring up to the surface hot matter. Their lives vary from a few days for the smallest to 2–3 months for the largest, and these may reach diameters of up to 120,000 km. It is by following the sunspots that the rotation of the sun may be observed and studied. The sun does not rotate like a solid body, but its rotational velocity varies with the heliographic latitude, the equator, which is 7° off the ecliptic, rotating faster than the polar caps.

Flacula Self-luminous white glaring calcium clouds with a temperature higher than the photosphere's. Such clouds are always found near the sunspots. They are probably heated by microwave discharges.

Flares, also called Chromospheric Eruptions They are quite a common phenomenon and, although easily observable, a reliable physical interpretation is not yet possible. Flares appear and spread at fantastic velocities (up to 1000 km/s); seen in the Hα light they are luminous areas that suddenly appear and develop to a size of the order of the sunspots while their luminosity rapidly increases to its peak within a few minutes, then they decline more slowly to disappear within 15 min. Flares are also known as *filaments* when they present a particular structure and as *prominences* when observed on the limb of the solar disk. They emit intense UV radiation, x rays, and radiowaves that affect our communications. They also emit corpuscular radiation which causes the boreal and the astral auroras when it reaches our atmosphere. The larger flares also emit cosmic radiation which reaches us from one to four days after the eruption, depending on the velocity with which they have been emitted. (See Figs. 2.13–2.16.)

Center of Activity This is the totality of all the phenomena that occurs in the vicinity of a sunspot area and are first detected by magnetic observations.

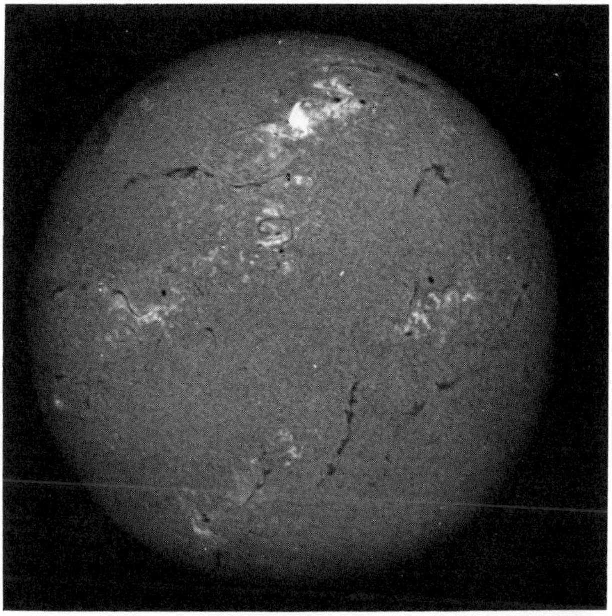

FIG. 2.13 Photograph of the higher photosphere and lower chromosphere in a period of high activity (July 1972) taken in Hα light. The North pole is 40° from the vertical, counter-clockwise. The bright white spot on the top right is an explosion in the chromosphere. Filaments and sun spots are observable in the two belts of activity, parallel to the equator. The network of the photospheric granulation seen through the chromosphere may be recognized under careful observation. (Courtesy of the Swedish Solar Observatory, Capri, Italy.)

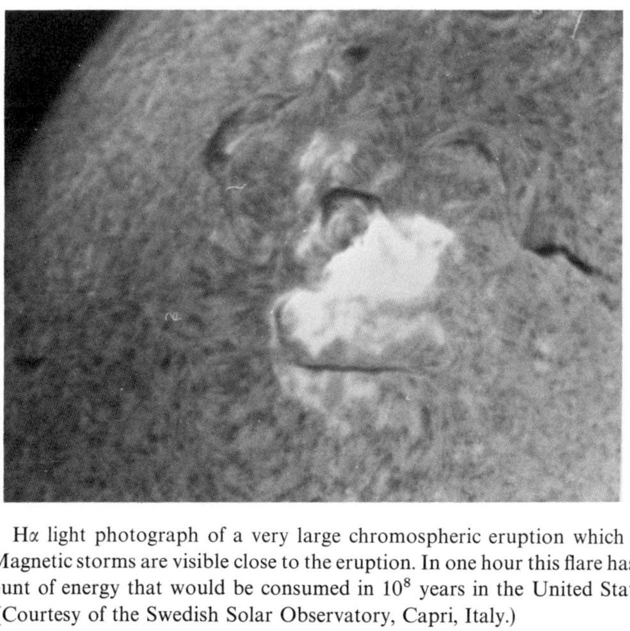

FIG. 2.14 Hα light photograph of a very large chromospheric eruption which occurred in July 1972. Magnetic storms are visible close to the eruption. In one hour this flare has released the same amount of energy that would be consumed in 10^8 years in the United States at the present rate! (Courtesy of the Swedish Solar Observatory, Capri, Italy.)

FIG. 2.15 Coronagraph of the large eruption of Fig. 2.14 when it reached the limb of the sun in a late phase of its life. The photograph was taken in a broad (5A) Hα light band. (Courtesy of the Swedish Solar Observatory, Capri, Italy.)

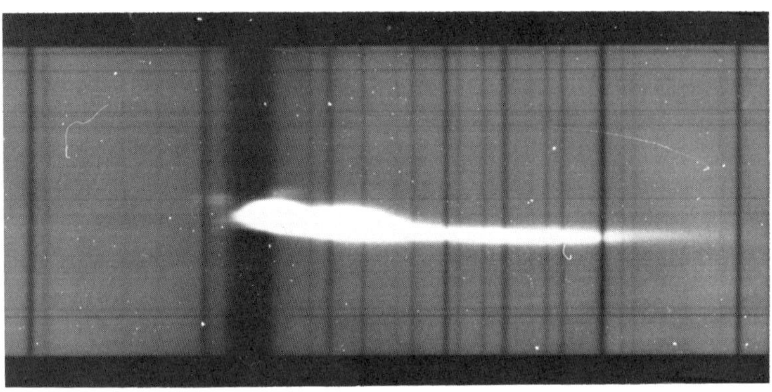

FIG. 2.16 Spectrum of the eruption shown in Fig. 2.14. The dark lines are Fraunhofer absorption lines. The large dark band is the H absorption band. The bright spot is the Hα emission of the eruption; instead of just being localized over the Hα Fraunhofer band, it is shifted to the right, i.e., toward the blue, showing (Doppler effect) velocities of approach ranging from 0 to 1200 km s^{-1} depending on the different parts of the flare. (Courtesy of the Swedish Solar Observatory, Capri, Italy.)

B. The "Temperature" of the Sun

Studying the characteristics of the sun, we have learned that the temperature distribution in the different layers of the inner sun and in the solar atmosphere varies widely and that the temperature of the sun's surface itself is not uniform because of the photospheric perturbations. Investigating the individual regions of the solar disk, i.e., the image of the solar sphere as we observe it, it is found that the brightness of the disk decreases noticeably toward the limb. This effect is called the limb darkening. The reason for this variation depends on the depth of our observation. When observing the central area of the sun, we are looking vertically through the solar atmosphere and reach the greatest depth; consequently, the brightness is greatest. On the other hand, the more we move our observation towards the limb of the disk, the more we are looking into the solar atmosphere with an increasing angle of incidence, thus observing the photosphere through a thicker light-absorbing atmosphere. This increases the optical path and limits our depth of observation, hence the apparent drop in temperature.

The study of the temperature (average) of the sun in the light of the radiation laws of Stefan–Boltzmann, Wien and Planck* leads to values varying from 5500 to 7000°K, depending on the different laws and on the different

* Stefan–Boltzmann's law states that the energy emitted by a blackbody is proportional to the fourth power of its absolute temperature:

$$E = \sigma T^4$$

where E is the energy emitted in the unit of time, σ the constant of Stefan–Boltzmann equal to $5{,}6697^{-8}$ W K^{-4} m, and T the absolute temperature in degrees Kelvin. Wien's law states that the spectral distribution of the radiation emitted varies in such a way that the maximum emission corresponds to a wavelength that is inversely proportional to the temperature:

$$\lambda_1/\lambda_2 = T_2/T_1$$

where λ_1 and λ_2 are two wavelengths and T_1 and T_2 the absolute temperatures. In other words, raising the temperature the maximum value of the emission shifts toward shorter wavelengths. This is visibly observable where an increase in temperature causes an emission spectrum to shift gradually from the IR to the red toward the violet, the overall emission appearing as *white* light at high temperatures.

Planck's quantum statistical law states that the energy radiated at different wavelengths is given by

$$E_\lambda = 2c^2h \frac{\lambda^{-5}}{\exp(ch/K\lambda T) - 1}$$

where E_λ is the energy emitted in unit time at a specific wavelength (per unit volume), h is known as Planck's constant and is 6.625×10^{-34} J s, λ a specific wavelength, c the velocity of light, K Boltzmann's constant equal to 1.380×10^{-23} J K^{-1}, and T the absolute temperature. The total energy emission is therefore

$$E = \int_0^\infty E_\lambda \, d\lambda$$

spectral regions. These differences between the *color* temperature and the *effective* temperature indicate that the sun does not radiate as a perfect blackbody and in these conditions there is not a single value that may be called temperature.

Hence we assign to the solar disk the temperature of the best fitting of the curves of emission of a blackbody, i.e., 5762°K

ACKNOWLEDGMENTS

I am indebted to many friends and colleagues for constructive criticism and discussion that have led to the preparation of this chapter, among them are Professor V. Storelli and Mr. A. Polverini. Also, I am indebted to Professor G. De Maria, Director of the Istituto di Chimica Fisica of the University of Rome, for his continuous support, to Miss V. Goodier, for the language supervision, and to Mr. E. Gobbi for the graphics.

3

Solar Irradiance, Total and Spectral

*MATHEW P. THEKAEKARA**

NASA/GODDARD SPACE FLIGHT CENTER
GREENBELT, MARYLAND

3.1 INTRODUCTION

The earth atmosphere system receives energy from the sun at the prodigious rate of 5.445×10^{24} J or 1.5125×10^{18} kWh/year. The total output of all man-made energy producing systems in the world in 1970 was less than 0.004% of this, 2×10^{20} J. Of this energy, all but 4% was from a rapidly dwindling supply of fossil fuels. Solar energy is the only nonpolluting form of energy available to man. The supply is abundant, cost-free, and inexhaustible, however, it is thinly distributed over a wide area, and at ground level, the supply is highly variable. As the energy crisis becomes worse and conventional fuels become more expensive, more attention is paid to making practical use of the sun's energy. Solar energy conversion systems are expensive to build and maintain, they have to be made cost-effective. Hence, we need to know how much energy is available and at what wavelengths. This chapter will attempt to make a survey of the quantitative data on solar energy that are currently available.

3.2 THE EXTRATERRESTRIAL SOLAR ENERGY

The solar constant and the extraterrestrial solar spectrum may seem at first to be of little importance for solar energy conversion systems. Most of these systems are below the atmosphere and the energy that reaches the

* Deceased

ground is quite different from that available outside the atmosphere. However, a closer examination of the problem shows that the extraterrestrial values are of great practical significance.

The solar constant is the total energy that falls on unit area exposed normally to the rays of the sun at the average sun–earth distance. The distribution of this energy as a function of wavelength is the extraterrestrial solar spectrum.

These quantities are of importance primarily for spaceborne solar energy conversion systems. There are such systems on all satellites. Spaceborne power stations with very large solar cell panels capable of producing electrical power between 3000 and 15,000 MW have been shown to be technically feasible, and have elicited considerable interest from the news media and the public. Since solar cells are spectrally sensitive, the important parameter is not the solar constant itself but the distribution of solar energy in the visible and near infrared portions of the spectrum. The extraterrestrial values are also of significance, though indirectly, for ground-based systems. Precise measurement of the solar spectrum and total solar energy is a tedious and time-consuming process. In many cases solar energy parameters of sufficient accuracy can be obtained by computing the fractional loss due to the absorbing and scattering properties of the atmosphere. The extraterrestrial solar spectrum is the important input parameter for such computer programs.

3.3 THE SOLAR CONSTANT

The NASA/ASTM standard value of the solar constant is 1353 W m^{-2} (in the preferred SI units). Expressed in other units, which are perhaps more conventional or more familiar in engineering applications, the value is 1.940 cal cm^{-2} min^{-1}, or 429.2 Btu ft^{-2} h^{-1}, or 125.7 W ft^{-2}, or 1.81 hp m^{-2}. The estimated error in this value of the solar constant is $\pm 1.5\%$. The value is based on nine long series of measurements, all made from high altitude platforms, Convair 990, balloons, X-15 aircraft, and Mariner-Mars probe, during the period 1967–1970. Different types of instruments were used, cavity radiometers, normal incidence pyrheliometers, Ångström pyrheliometers, etc.; and the calibration of the instruments was referred to three scales of radiometry, the absolute electrical units scale, the international pyrheliometric scale IPS 56, and the thermodynamic Kelvin temperature scale. This value of the solar constant, 1353 W m^{-2}, is referred to as the NASA/ASTM standard since it has been adopted as the design criterion for NASA space vehicles (Anonymous, 1971c) and has been issued as the engineering standard by the American Society of Testing and Materials (Anonymous, 1974a). The NASA/ASTM standard includes also a detailed set of values of the spectral distribution of the sun's energy. These standards have been formally adopted by ANSI (American National Standards Institute) and have been proposed for the ISO (International Standards Organization).

3.4 SOLAR SPECTRAL IRRADIANCE FOR ZERO AIR MASS

The extraterrestrial solar spectrum in the wavelength range 0.2–2.6 μm is shown in Fig. 3.1. The spectral irradiance values are given in tabular form in Table 3.1. These data are based on detailed measurements made from a Convair 990 jet aircraft at an altitude of 11.6 km (Anonymous, 1971c, 1974a; Drummond and Thekaekara, 1973; Thekaekara, 1970, 1973). In the wavelength range 0.3–2.6 μm, which contains over 95% of the solar energy, four spectroradiometric instruments were used: a Perkin–Elmer monochromator with lithium fluoride prism (1 P28 photomultiplier tube and thermocouple detectors), a quartz double-prism monochromator (EMI 9558 QA and lead sulfide detectors), a filter radiometer, and a polarization-type interferometer. The range of the Perkin–Elmer monochromator extended to 4 μm. A Michelson-type infrared interferometer was used for the wavelength range 2.6–15 μm. The spectral irradiance curve obtained by these measurements was modified slightly in the visible and near IR in the light of the extensive filter radiometer data obtained by the Eppley–JPL team under the direction of A. J. Drummond (Drummond *et al.*, 1967, 1968; Drummond and Hickey, 1968). Data from other sources were used for the two extreme ends of the solar spectrum, below 0.3 μm and beyond 15 μm. The spectral irradiance values given in Table 3.1 are averages over 10 nm bandwidths for the range

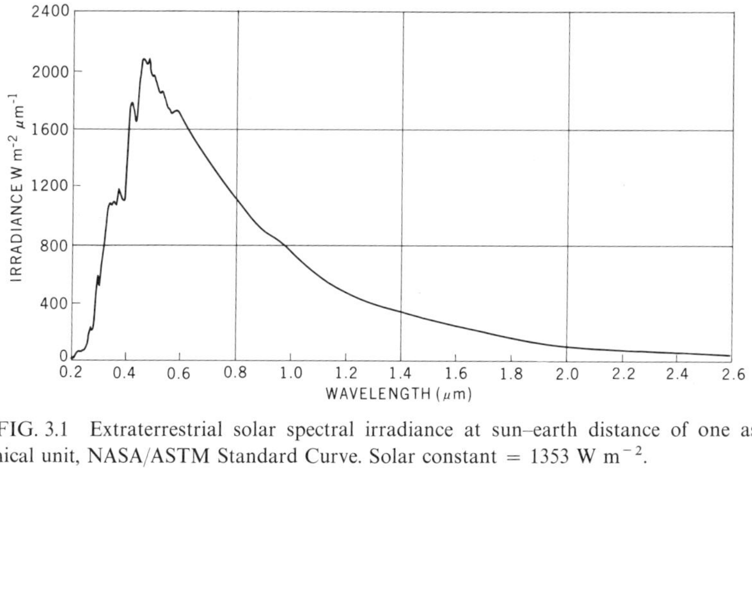

FIG. 3.1 Extraterrestrial solar spectral irradiance at sun–earth distance of one astronomical unit, NASA/ASTM Standard Curve. Solar constant = 1353 W m^{-2}.

TABLE 3.1

Solar Spectral Irradiance—Standard Curve[a]

λ	E_λ	$E_{o-\lambda}$	$D_{o-\lambda}$	λ	E_λ	$E_{o-\lambda}$	$D_{o-\lambda}$	λ	E_λ	$E_{o-\lambda}$	$D_{o-\lambda}$
.115	.007	.0025	.0001	.510	1882	324.926	24.015	1.55	267	1186.109	87.665
.120	.900	.0048	.0003	.515	1833	334.214	24.701	1.60	245	1198.909	88.611
.125	.007	.0070	.0005	.520	1833	343.379	25.379	1.65	223	1210.609	89.475
.130	.007	.0071	.0005	.525	1852	352.591	26.059	1.70	202	1221.234	90.261
.140	.030	.0073	.0005	.530	1842	361.826	26.742	1.75	180	1230.784	90.967
.150	.070	.0078	.0005	.535	1818	370.976	27.418	1.80	159	1239.259	91.593
.160	.230	.0093	.0006	.540	1783	379.979	28.084	1.85	142	1246.784	92.149
.170	.630	.0136	.0010	.545	1754	388.821	28.737	1.90	126	1253.484	92.644
.180	1.250	.0230	.0016	.550	1725	397.519	29.380	1.95	114	1259.484	93.088
.190	2.710	.0428	.0031	.555	1720	406.131	30.017	2.00	103	1264.909	93.489
.200	10.7	.1098	.0081	.560	1695	414.669	30.648	2.1	90	1274.559	94.2024
.210	22.9	.2778	.0205	.565	1705	423.169	31.276	2.2	79	1283.009	94.8269
.220	57.5	.6798	.0502	.570	1712	431.711	31.907	2.3	69	1290.409	95.3739
.225	64.9	.9858	.0728	.575	1719	440.289	32.541	2.4	62	1296.959	95.8580
.230	66.7	1.3148	.0971	.580	1715	448.874	33.176	2.5	55	1302.809	96.2903
.235	59.3	1.6298	.1204	.585	1712	457.441	33.809	2.6	48	1307.959	96.6710
.240	63.0	1.9356	.1430	.590	1700	465.971	34.439	2.7	43	1312.509	97.0073
.245	72.3	2.2738	.1680	.595	1682	474.426	35.064	2.8	39	1316.609	97.3103
.250	70.4	2.6306	.1944	.600	1666	482.796	35.683	2.9	35	1320.309	97.5838
.255	104.0	3.0666	.2266	.605	1647	491.079	36.295	3.0	31	1323.609	97.8277
.260	130	3.6516	.269	.61	1635	499.284	36.902	3.1	26.0	1326.459	98.0383
.265	185	4.4391	.328	.62	1602	515.469	38.098	3.2	22.6	1328.889	98.2179
.270	232	5.4816	.405	.63	1570	531.329	39.270	3.3	19.2	1330.979	98.3724
.275	204	6.5716	.485	.64	1544	546.899	40.421	3.4	16.6	1332.769	98.5047
.280	222	7.6366	.564	.65	1511	562.174	41.550	3.5	14.6	1334.329	98.6200
.285	315	8.9791	.663	.66	1486	577.159	42.657	3.6	13.5	1335.734	98.7238
.290	482	10.9716	.810	.67	1456	591.869	43.744	3.7	12.3	1337.024	98.8192
.295	584	13.6366	1.007	.68	1427	606.284	44.810	3.8	11.1	1338.194	98.9056
.300	514	16.3816	1.210	.69	1402	620.429	45.855	3.9	10.3	1339.264	98.9847
.305	603	19.1741	1.417	.70	1369	634.284	46.879	4.0	9.5	1340.254	99.0579
.310	689	22.4041	1.655	.71	1344	647.849	47.882	4.1	8.70	1341.1641	99.12521
.315	764	26.0366	1.924	.72	1314	661.139	48.864	4.2	7.80	1341.9891	99.18618
.320	830	30.0216	2.218	.73	1290	674.159	49.826	4.3	7.10	1342.7341	99.24124
.325	975	34.5341	2.552	.74	1260	686.909	50.769	4.4	6.50	1343.4141	99.29150

.330	1059	39.6191	2.928	.75	1235	699.384	51.691	4.5	5.92	1344.0351	99.33740
.335	1081	44.9691	3.323	.76	1211	711.614	52.595	4.6	5.35	1344.5986	99.37905
.340	1074	50.3566	3.721	.77	1185	723.594	53.480	4.7	4.86	1345.1091	99.41678
.345	1069	55.7141	4.117	.78	1159	735.314	54.346	4.8	4.47	1345.5757	99.45127
.350	1093	61.1191	4.517	.79	1134	746.779	55.194	4.9	4.11	1346.0049	99.48299
.355	1083	66.5591	4.919	.80	1109	757.994	56.023	5.0	3.79	1346.3999	99.51219
.360	1068	71.9366	5.316	.81	1085	768.966	56.834	6	1.8200	1349.2049	99.71950
.365	1132	77.4366	5.723	.82	1060	779.694	57.627	7	.9900	1350.6099	99.82335
.370	1181	83.2191	6.150	.83	1036	790.174	58.401	8	.5850	1351.3974	99.88155
.375	1157	89.0641	6.582	.84	1013	800.419	59.158	9	.3670	1351.8734	99.91673
.380	1120	94.7566	7.003	.85	990	810.434	59.899	10	.2410	1352.1774	99.93920
.385	1098	100.3016	7.413	.86	968	820.224	60.622	11	.1650	1352.3804	99.95420
.390	1098	105.7916	7.819	.87	947	829.799	61.330	12	.1170	1352.5214	99.96462
.395	1189	111.5091	8.241	.88	926	839.164	62.022	13	.0851	1352.6224	99.97209
.400	1429	118.0541	8.725	.89	908	848.334	62.700	14	.0634	1352.6967	99.97758
.405	1644	125.7366	9.293	.90	891	857.329	63.365	15	.0481	1352.7524	99.98170
.410	1751	134.224	9.920	.91	880	866.184	64.019	16	.037100	1352.7950	99.98485
.415	1774	143.036	10.571	.92	869	874.929	64.665	17	.029100	1352.8281	99.98730
.420	1747	151.839	11.222	.93	858	883.564	65.304	18	.023100	1352.8542	99.98923
.425	1693	160.439	11.858	.94	847	892.089	65.934	19	.018600	1352.8751	99.99077
.430	1639	168.769	12.473	.95	837	900.509	66.556	20	.015200	1352.8920	99.99202
.435	1663	177.024	13.083	.96	820	908.794	67.168	25	.006170	1352.9454	99.99596
.440	1810	185.706	13.725	.97	803	916.909	67.768	30	.002970	1352.9683	99.99765
.445	1922	195.036	14.415	.98	785	924.849	68.355	35	.001600	1352.9797	99.99850
.450	2006	204.856	15.140	.99	767	932.609	68.928	40	.000942	1352.9860	99.99897
.455	2057	215.014	15.891	1.00	748	940.184	69.488	50	.000391	1352.9927	99.99946
.460	2066	225.321	16.653	1.05	668	975.584	72.105	60	.00019000	1352.9956	99.99967
.465	2048	235.606	17.413	1.10	593	1007.109	74.435	80	.00006160	1352.9981	99.99986
.470	2033	245.809	18.167	1.15	535	1035.309	76.519	100	.00002570	1352.9990	99.99992
.475	2044	256.001	18.921	1.20	485	1060.809	78.404	120	.00001260	1352.9994	99.99995
.480	2074	266.296	19.681	1.25	438	1083.884	80.109	150	.00000523	1352.9997	99.99997
.485	1976	276.421	20.430	1.30	397	1104.759	81.652	200	.00000169	1352.9998	99.99999
.490	1950	286.236	21.155	1.35	358	1123.634	83.047	250	.00000070	1352.9999	99.99999
.495	1960	296.011	21.878	1.40	337	1141.009	84.331	300	.00000034	1352.9999	99.99999
.500	1942	305.766	22.599	1.45	312	1157.234	85.530	400	.00000011	1352.9999	99.99999
.505	1920	315.421	23.312	1.50	288	1172.234	86.639	1000	.00000000	1353.0000	100.00000

[a] λ, wavelength in micrometers; E_λ, solar spectral irradiance averaged over small bandwidth centered at λ, in W m^{-2} μm^{-1}; $E_{0-\lambda}$, integrated solar irradiance in the wavelength range 0 to λ, in W m^{-2}; $D_{0-\lambda}$, percentage of solar constant associated with wavelengths shorter than λ; solar constant, 1353 W m^{-2}. Note: lines indicate change in wavelength interval of integration.

0.3–0.75 μm. Wider bandwidths for averaging were used for longer wavelengths. The estimated accuracy of these values is ±5%.

A more detailed description of the solar spectrum giving irradiance over 1 Å intervals has been published for the wavelength range 0.3–0.61 μm (Thekaekara, 1974). While this is useful for some of the practical applications like filter radiometry, atmospheric and solar modeling, and meteorology, such detailed spectral information is not essential for solar energy conversion systems.

3.5 VARIATIONS WITH SUN–EARTH DISTANCE

The solar constant and the extraterrestrial solar spectrum are defined for the semimajor axis of the earth's orbit or one astronomical unit, equal to 1.496×10^{13} cm. According to the well-known Kepler's law, the earth like

TABLE 3.2

Variations of Sun–Earth Distance, Total Solar Irradiance outside the Atmosphere, the Solar Declination, and the Ephemeris Transit during the Year 1976

Date		Sun–earth distance in astronomical units	Irradiance due to the sun ($W\ m^{-2}$)	Solar declination			Ephemeris transit		
				deg	min	s	h	min	s
Jan	1	0.983 34	1399	−23	05	11	12	03	16
	4	0.983 32	1399	−22	49	45	12	04	41
Feb	1	0.985 31	1393	−17	24	08	12	13	32
	12	0.987 03	1389	−14	02	51	12	14	17
Mar	1	0.990 97	1378	−7	37	00	12	12	21
	20	0.995 94	1364	−0	11	42	12	07	27
Apr	1	0.999 41	1355	+4	30	26	12	03	50
	4	0.999 98	1353	+5	39	32	12	02	57
May	1	1.007 70	1332	+15	02	57	11	57	03
	14	1.010 72	1324	+18	36	18	11	56	17
June	1	1.014 13	1316	+22	02	27	11	57	47
	21	1.016 30	1310	+23	26	24	12	01	77
July	1	1.016 72	1309	+23	06	54	12	03	49
	3	1.016 73	1309	+22	58	07	12	04	12
	26	1.015 66	1312	+19	27	28	12	06	27
Aug	1	1.014 95	1313	+18	02	30	12	06	15
Sep	1	1.009 14	1329	+8	18	55	11	59	54
	23	1.003 37	1344	−0	02	08	11	52	17
Oct	1	1.001 07	1350	−3	09	00	11	49	36
	5	0.999 89	1353	−4	41	46	11	48	22
Nov	1	0.992 39	1347	−14	24	01	11	43	37
	3	0.991 87	1375	−15	02	01	11	43	36
Dec	1	0.985 98	1392	−21	47	19	11	49	09
	22	0.983 68	1398	−23	26	22	11	58	44

all other planets of the solar system moves along an elliptical orbit with the sun at one of the foci of the ellipse. Hence the sun–earth distance varies according to the seasons of the year, being a minimum around January 3 and a maximum around July 4. These seasonal variations are known with considerably higher precision than the absolute value of the solar constant. The values of the sun–earth distance in astronomical units and the total solar irradiance outside the atmosphere for a certain number of days in 1976 are given in columns 2 and 3, respectively, of Table 3.2. The total solar irradiance is the solar constant (1353 W m^{-2}) divided by the square of the sun–earth distance. Columns 4 and 5 give two other parameters, the solar declination and the ephemeris transit, respectively, which are required for computing the air mass. The dates chosen are the first day of each month and a few other dates for which one or the other of the variables has a flexion point. Variations from year to year on a given date are quite negligible. Seasonal variations of spectral irradiance are in the same ratio as those of total irradiance.

3.6 ACCURACY OF THE STANDARD VALUES

In all applications of the NASA/ASTM standard values, there are two questions of major importance, the absolute accuracy of these values and their intrinsic variability, apart from that due to the sun–earth distance.

The question of accuracy is closely related to problems of instrumentation and radiation scales, which will be discussed later. The nine values from which the solar constant was derived varied between a maximum of 1368 W m^{-2} and minimum of 1338 W m^{-2}. The standard value of 1353 W m^{-2} has a stated uncertainty of ± 21 W m^{-2}. For some of the applications in meteorology, atmospheric physics, solar physics, etc., an absolute accuracy of 0.1% or ± 1.4 W m^{-2} is desirable and as many authors believe, it is possible within the state-of-the-art of radiometric measurements. Variations of the solar constant with sporadic or cyclic variations in the sun itself cannot be detected unless the precision, if not the absolute accuracy, of the solar constant is 0.1% or better.

The uncertainties in the distribution of the solar energy as a function of wavelength are considerably greater than in the solar constant itself. The four different instruments used by the GSFC experimenters on board CV990 did not yield identical curves; there was a certain amount of scatter between them, though the variations were within the estimated error limits of the instruments. There were greater differences between the GSFC curve (weighted average of the four instruments) and the Eppley–JPL filter data, as shown by Drummond and Thekaekara (1973, p. 112).

The differences between the NASA/ASTM standard and the solar spectral curves derived from earlier ground-based measurements are considerably greater. Two of these which have received relatively wide circulation are those due to Johnson (1954) and to Labs and Neckel (1968, 1970). In the wavelength range 0.3–0.5 μm, Johnson values are higher by about 8% and Labs and Neckel values are lower by 10–20% compared to the standard values. It will be recalled that Johnson had scaled the experimental data of Dunkelman and Scolnik (1959) by 8.8%. The values of Labs and Neckel are probably low because of the difficulty of estimating the solar continuum in a wavelength range which is so rich in Fraunhofer lines. The solar spectral curve derived by Stair and Ellis (1968) on the basis of their measurements at Mauna Loa agrees closely with the NASA/ASTM curve. They used two instruments, a Leiss double-prism monochromator and a filter radiometer. The wavelength range was limited by the photomultiplier detector from 0.3 to 0.53 μm. The spectral irradiance lamps used for calibration by the GSFC experimenters were of the same type as those used by Stair and Ellis and were quite different from those used by Labs and Neckel and those of the measurements from which the Johnson curve was derived.

3.7 VARIATIONS IN SOLAR ENERGY OUTPUT

The possible variations in the energy output of the sun is a question which has been studied by many investigators. The most significant measurements are those made by C. G. Abbot and his co-workers of the Smithsonian Institution (Abbot, 1966, Aldrich and Hoover, 1964). The experiments extended for over half a century, and were made at many different locations, in Washington, D.C., on Mount Wilson, California, in Calama, Chile, in Algeria, and elsewhere. Abbot concluded that while the solar constant has a definite value, close to 1.94 cal cm^{-2} min^{-1}, there are small variations with different periodicities. The most clear periodicity is that of 273 months or 22 years and 9 months. This period is that of the solar magnetic cycle and twice the period of the sunspot cycle. The cyclic variations in solar energy output are about 2 or 3%, and occasionally greater changes occur as when a large sunspot crosses the solar disk. Among other authors who have examined in detail the Smithsonian data and Abbot's analysis, there are some who firmly believe that significant variations do occur and others who say that the experimental errors are too large to draw any firm conclusions. C. W. Allen (1958) concludes that the solar constant variations observed at the Smithsonian Institution are due to either data processing methods or changes in atmospheric transparency, and that the solar constant is constant to within 0.1%. On the other hand, Kondratyev and Nikolsky (1970) strongly support the Smithsonian findings. Based on his own measurements made from balloons at high altitude over a six-year period, Kondratyev finds that the solar constant is a maximum for sunspot numbers between 80 and 100 and that the solar constant value is lower by 2 to 2.5% at sunspot maxima and minima. Bossalasco and his group (1965) analyzed data from four widely separated stations and concluded that the maximum value of the solar constant occurs for sunspot number N about 160, that the solar constant is significantly lower for $N < 160$ and $N > 160$, and that the minimum at $270 < N < 330$ is about 15% lower. But these conclusions have been disputed by Kondratyev who ascribes the low value to increased atmospheric turbidity caused by the nuclear tests made during the same time period as the solar data. But a decrease as large as 15% in solar energy received on the ground, whether it is due to the sun itself or due to man-made atmospheric opacity, should be of great interest to those working on solar energy conversion systems.

There are several intriguing meteorological phenomena which seem to follow the changes in the sun. There are many cyclic and noncyclic changes occurring in the sun which have been long well known to solar physicists, and there are also many terrestrial phenomena which seem to bear a correlation to these solar changes. Among such phenomena are the number of days when Etesian winds blow over Greece, the wintriness index of the northern hemisphere sea level pattern, the annual march of temperature in many European cities, changes in Meridional sea level pressure, growth rings of trees, ozone density, geomagnetism, glacier movement, etc.

While a great deal has been published about changes in the solar constant and their effects on weather, there is hardly any mention of the changes in spectral distribution in the visible and near infrared where the energy output is greatest. The reason is not that changes do not exist, but that they are totally unknown and unexplored. The cyclic changes which occur in the UV of $\lambda < 0.3$ μm and in the radio wave range of $\lambda > 1$ cm are well established. But these two extreme ends of the solar spectrum contribute, respectively, 1 and 10^{-8}% to the total solar energy output. To establish on firm experimental basis the changes in the solar constant and the solar spectrum, measurements are needed with considerably greater accuracy and precision than have hitherto been possible.

3.8 SOLAR ENERGY VARIATION AT GROUND LEVEL

The problem of extraterrestrial solar irradiance and radiation scales, their possible variations and margin of uncertainty are all quite relevant to the problem of more immediate interest, the irradiance, total and spectral, at ground level. Solar energy conversion systems are mostly on the ground. The variability of energy incident on a collector surface on the ground is considerably greater than that of the extraterrestrial solar energy. On days of clear sunshine the energy increases from zero at sunrise to a maximum

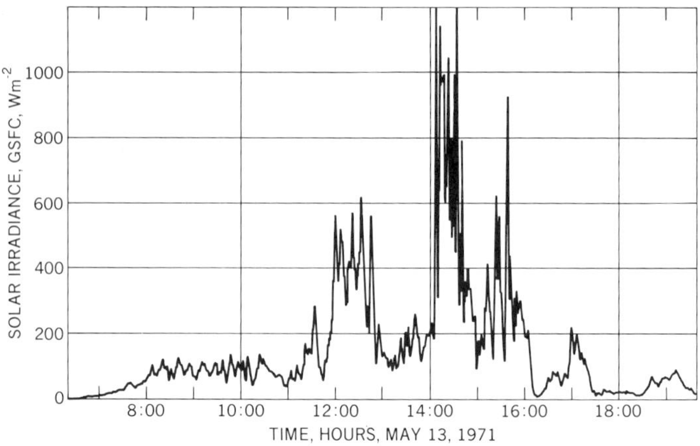

FIG 3.2 Global irradiance due to the sun and sky on a horizontal surface, measured at GSFC on May 13, 1971. Total energy received during the day, 175 cal cm^{-2} (732 J cm^{-2}).

at solar noon and decreases to zero at sunset. At any moment clouds may intercept the sun and decrease the energy to a low value due to the diffuse sky radiation. Figures 3.2 and 3.3, based on measurements made on two consecutive days, illustrate the wide variations that can be expected. These measurements were made as part of the GSFC international comparison of working standard pyranometers. The instrument was an Eppley pyrano-

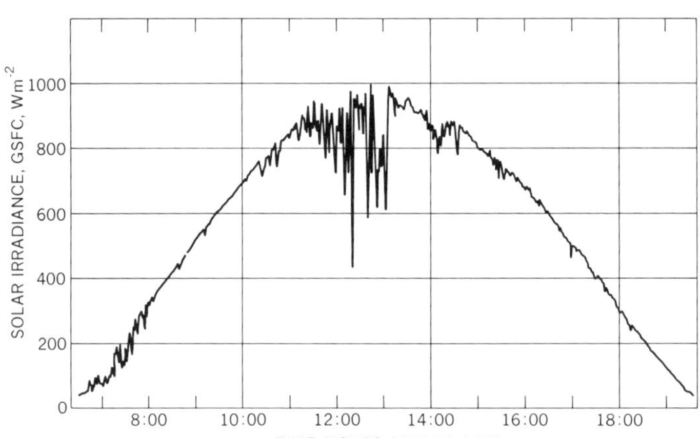

FIG. 3.3 Global irradiance due to the sun and sky on a horizontal surface, measured at GSFC on May 14, 1971.

meter, model 2, mounted on a roof top. The readings were taken every 4 s. The x axis shows hours in Eastern daylight saving time and the y axis shows total irradiance. May 13 (Fig. 3.2) was heavily overcast during most of the morning. A short interval of sunshine in the afternoon was followed by a heavy cloudburst at 18:00. The next day was one of relatively clear sunshine with a few passing clouds. The high value, near 1200 W m^{-2}, after 14:00 on May 13, is 30% higher than might be expected on a clear day for the given solar elevation; this is obviously due to reflection from the clouds. Such abnormally high values are of short duration.

3.9 ATMOSPHERIC ATTENUATION OF SOLAR ENERGY

Through most of the solar spectrum, the absorption of a monochromatic beam of light is governed by the logarithmic decrement law known also as Beer's law or Bouguer's law.

$$E_\lambda = E_\lambda{}^0 e^{-c_\lambda m} \tag{3.1}$$

where $E_\lambda{}^0$ and E_λ are irradiance at a given wavelength λ outside the atmosphere and after transmittance through air mass m, respectively. c_λ is an attenuation factor, often referred to as the extinction optical thickness. The coefficient c_λ is the sum of three terms, c_1 due to Rayleigh scattering, c_2 to ozone, and c_3 to aerosols or atmospheric turbidity. In the infrared ($\lambda > 0.69\ \mu$m), there is also selective absorption by the polyatomic gaseous constituents of the atmosphere, mainly, water vapor and carbon dioxide, and continuum attenuation due to scattering and absorption by particulate matter and water droplets. The selective absorption is characterized by many thousands of lines of the vibration–rotation spectrum of the molecules, as will be shown by a high dispersion ($\lambda/\Delta\lambda > 10^5$) spectrograph. The total effect over finite bandwidths is not simple enough to be expressed by Eq. (3.1); nor is it correct to assume that the total energy at ground level can be expressed by an integral over all wavelengths of the right-hand side of Eq. (3.1).

There is a considerable amount of literature on atmospheric attenuation. Some of the attempts to solve the problem require elaborate mathematical tools and a great deal of computer time. A simpler approach is presented here. Solar irradiance spectra at ground level is computed on the basis of the standard NASA/ASTM curve, assuming atmospheric attenuation parameters which are considered to be highly reliable.

The Rayleigh attenuation coefficient c_1 and the ozone attenuation coefficient c_2 are based on the data developed by Elterman (1968). They are valid for the US standard atmosphere. The total amount of ozone in a vertical path is assumed to be 0.34 cm (at NTP). Other values of ozone density can be introduced if needed. In Elterman's notation these constants c_1 and c_2 are, respectively, τ_r', Rayleigh optical thickness (h–∞) and τ_3', ozone optical thickness (h–∞). The values for $h = 0$ (i.e., at sea level) were used for the present computation. Elterman's tables list the values for 22 discrete wavelengths. For other wavelengths of the standard table a linear interpolation for c_1 and c_2 was found to be sufficiently accurate.

For atmospheric turbidity, instead of using Elterman's coefficients, an equation of the form developed by A. Ångström is used:

$$c_3 = \beta/\lambda^\alpha \tag{3.2}$$

where β is the Ångström β coefficient, α the wavelength exponent, and λ the wavelength in microns. This permits a greater flexibility in choosing α and β parameters corresponding to different levels of atmospheric pollution.

In the infrared, as stated earlier, a fourth parameter to account for the molecular absorption bands is required; here no single expression applies to all the absorption bands. The experimental results of Gates and Harrop (1963) and Gates (1960) seem to be most appropriate for handling this complex problem. The expressions are relatively simple; the accuracy is adequate; and the parameters are based on actual measurements of solar irradiance. The experimental results were interpreted in the light of the random model theory (disordered distribution of many lines within a band) developed by Goody (1958) for water vapor and the regular model theory (regular distribution of absorption lines) developed by Elsasser (1942) for carbon dioxide. Thus, in the infrared, Eq. (3.1) has to be modified as

$$E_\lambda = E_\lambda{}^0 e^{-c_\lambda m} T_{\lambda i} \tag{3.3}$$

where $T_{\lambda i}$ is a transmittance factor which can have one of three values,

$$T_{\lambda 1} = e^{-c_4\sqrt{wm}}, \qquad T_{\lambda 2} = e^{-c_5\sqrt{wm}}, \qquad \text{or} \qquad T_{\lambda 3} = (1 - c_6\sqrt{m}) \tag{3.4}$$

c_4, c_5, and c_6 are, respectively, the coefficients $-c_1$, $-c_2$, and c_5 of Gates and Harrop (1963), m is the air mass, and w the amount of precipitable water vapor (the height in centimeter of the liquid layer if all the water vapor in a vertical column of unit cross section were condensed into liquid). For the present computation, w was assumed to be 2 cm which is a global annual average for midlatitudes. Other values can be introduced if needed.

The values of atmospheric extinction coefficients c_1, c_2, and c_3 are given in Table 3.3. The values of c_1 and c_2 from Elterman (1968) are valid for the US standard atmosphere from ground level to outside the atmosphere and for 0.34 cm of ozone. The c_3 values are based on Eq. (3.2) with $\alpha = 0.66$ and $\beta = 1.70$.

TABLE 3.3

Atmospheric Extinction Optical Thickness due to Rayleigh Scattering, Ozone Absorption, and Turbidity

Wavelength λ (μm)	Extinction optical thickness c_1	c_2	c_3 ($\alpha = 0.66$; $\beta = 0.17$)
0.270	1.928	70.956	0.4034
0.280	1.645	35.816	0.3938
0.300	1.222	3.413	0.3763
0.320	0.927	0.303	0.3606
0.340	0.717	0.022	0.3465
0.360	0.564	0.001	0.3336
0.380	0.450	0.000	0.3220
0.400	0.364	0.000	0.3112
0.450	0.223	0.001	0.2880
0.500	0.145	0.012	0.2686
0.550	0.098	0.031	0.2522
0.600	0.069	0.045	0.2382
0.650	0.050	0.021	0.2259
0.700	0.037	0.008	0.2151
0.800	0.021	0.003	0.1970
0.900	0.013	0.000	0.1822
1.026	0.007	0.000	0.1671
1.060	0.003	0.000	0.1636
1.670	0.001	0.000	0.1212
2.170	0.000	0.000	0.1020
3.500	0.000	0.000	0.0744
4.000	0.000	0.000	0.0681

In Table 3.4 are given the values of c_4, c_5, and c_6 (Gates and Harrop, 1963), where it is also shown which subscript is applicable at each of the wavelengths. The value of w (precipitable water vapor) should be in millimeter when these constants are applied to Eq. (3.4).

3.10 TOTAL AND SPECTRAL SOLAR IRRADIANCE AT GROUND LEVEL

Computations of solar spectral irradiance at ground level have been made for different sets of parameters of ozone density, precipitable water vapor, turbidity coefficients, and air mass. The spectral irradiance outside the atmosphere, $E_\lambda{}^0$, is assumed to be that given the NASA/ASTM standard.

TABLE 3.4

Molecular Absorption Coefficients[a]

λ	i	c_i	λ	i	c_i	λ	i	c_i	λ	i	c_i
0.6983	4	0.0345	1.0940	4	0.03960	1.7320	5	0.00457	3.2140	4	0.41000
0.7000	4	0.0000	1.0980	5	0,00427	1.7820	5	0.00693	3.2450	4	0.36600
0.7100	4	0.0000	1.1010	6	0.07350	1.8620	4	0.78000	3.2600	4	0.37600
0.7200	4	0.0000	1.1280	4	0.30000	1.9550	4	0.20500	3.2850	6	0.24900
0.7277	4	0.0250	1.1310	4	0.27200	2.0080	6	0.28200	3.3170	6	0.28500
0.7300	4	0.0000	1.1370	4	0.28300	2.0140	6	0.22100	3.3440	4	0.31700
0.7400	4	0.0000	1.1440	4	0.21500	2.0570	6	0.23300	3.4030	6	0.22600
0.7500	4	0.0000	1.1470	4	0.23400	2.1240	5	0.00852	3.4500	6	0.16800
0.7621	6	0.2472	1.1780	5	0.00799	2.1560	5	0.00937	3.5070	6	0.10600
0.7700	4	0.0000	1.1890	5	0.00656	2.2010	6	0.11900	3.5380	6	0.14000
0.8000	4	0.0000	1.1930	6	0.06810	2.2660	6	0.10600	3.5730	5	0.01000
0.8059	4	0.0232	1.2220	5	0.00465	2.3200	6	0.11200	3.6330	4	0.03540
0.8250	4	0.0000	1.2360	5	0.00336	2.3380	6	0.13500	3.6730	4	0.06480
0.8350	4	0.0000	1.2640	6	0.16800	2.3560	6	0.16200	3.6960	4	0.03000
0.8465	4	0.1400	1.2760	6	0.11400	2.3880	4	0.11400	3.7120	6	0.07280
0.8600	4	0.1200	1.2880	5	0.00420	2.4150	4	0.13000	3.7650	4	0.03630
0.8700	4	0.1400	1.3140	5	0.00924	2.4530	4	0.14100	3.8120	4	0.04010
0.8750	4	0.1400	1.3350	4	0.13200	2.4940	4	0.21400	3.8880	6	0.19300
0.9000	4	0.1300	1.3840	4	0.90000	2.5370	4	0.53300	3.9230	6	0.18000
0.9150	4	0.1200	1.4320	4	0.42600	2.9000	4	0.54900	3.9480	6	0.18300
0.9250	4	0.2300	1.4570	4	0.27200	2.9410	4	0.37500	4.0450	6	0.24200
0.9300	4	0.2800	1.4720	4	0.28900	2.9540	4	0.38200			
0.9400	4	0.2000	1.5420	5	0.00284	2.9730	4	0.28200			
0.9500	4	0.2100	1.5720	6	0.07730	3.0050	4	0.29700			
0.9550	4	0.1900	1.5990	6	0.06200	3.0450	4	0.39800			
0.9650	4	0.1700	1.6080	6	0.08020	3.0560	4	0.38100			
0.9750	4	0.0500	1.6260	6	0.05770	3.0970	4	0.46200			
0.9850	4	0.0580	1.6440	6	0.06670	3.1320	4	0.28100			
1.0180	5	0.0030	1.6500	6	0.06620	3.1560	6	0.19400			
1.082	5	0.0053	1.6760	5	0.00460	3.2040	4	0.52000			

[a] $c_i = c_4$, c_5, or c_6 [see Eq. (3.4)].

Some of the results are presented here in graphical and tabular form. Figures 3.4 and 3.5 give the spectral curves for air mass 1, 4, 7, and 10. These correspond, respectively, to solar zenith angles 0°, 75.5°, 81.7°, and 84.3°. The continuous curve on top is the extraterrestrial spectrum, same as Fig. 3.1. Figure 3.3 is for $\alpha = 1.3$ and $\beta = 0.02$, and corresponds to a relatively clear atmosphere. A higher value of $\beta = 0.04$, that is, a more turbid atmosphere, is assumed for the curves of Fig. 3.5. The integral values of irradiance are given in the inset.

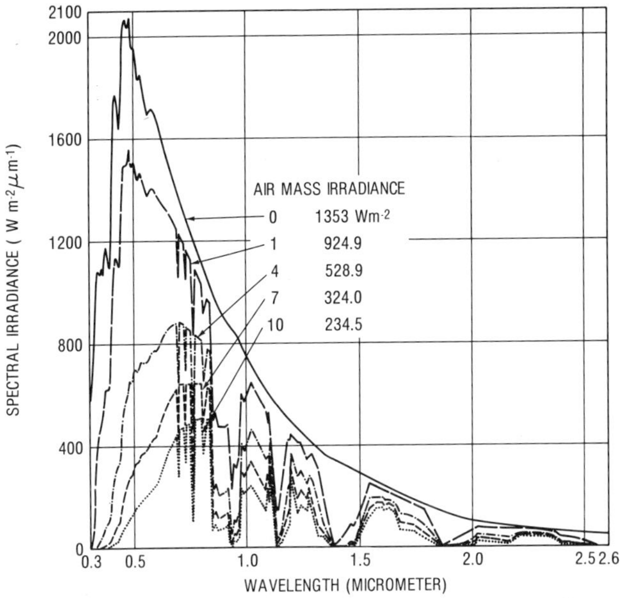

FIG. 3.4 Solar spectral irradiance for different air mass values, assuming US standard atmosphere, 20 mm of precipitable water vapor, 3.4 mm of ozone, very clear air ($\alpha = 1.3$, $\beta = 0.02$).

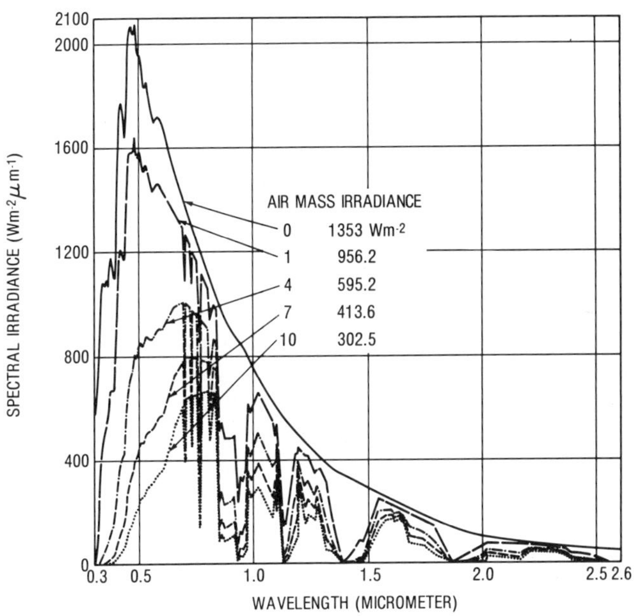

FIG. 3.5 Solar spectral irradiance for different air mass values, assuming US standard atmosphere, 20 mm of precipitable water vapor, 3.4 mm of ozone, relatively clear air ($\alpha = 1.3$, $\beta = 0.04$).

TABLE 3.5

WAVE LENGTH \ AIR MASS	0	α 0.66; β 0.085				α 0.66; β 0.170			
		1	4	7	10	1	4	7	10
0.540	1783.0	1371.7	624.5	284.4	129.5	1207.3	374.8	116.4	36.1
0.545	1754.0	1354.2	623.2	286.8	132.0	1192.8	375.2	118.0	37.1
0.550	1725.0	1336.6	621.7	289.2	134.5	1178.2	375.4	119.6	38.1
0.555	1720.0	1335.7	625.5	293.0	137.2	1178.3	378.8	121.8	39.2
0.560	1695.0	1319.2	622.0	293.3	138.3	1164.7	377.8	122.6	39.8
0.565	1705.0	1330.0	631.3	299.6	142.2	1175.0	384.6	125.9	41.2
0.570	1712.0	1338.4	639.5	305.6	146.0	1183.3	390.7	129.0	42.6
0.575	1719.0	1346.9	647.8	311.6	149.9	1191.6	396.9	132.2	44.0
0.580	1715.0	1346.7	652.0	315.7	152.8	1192.3	400.6	134.6	45.2
0.585	1712.0	1347.3	656.6	320.0	156.0	1193.6	404.5	137.1	46.5
0.590	1700.0	1340.7	657.7	322.6	158.3	1188.6	406.3	138.9	47.5
0.595	1682.0	1329.4	656.4	324.1	160.0	1179.4	406.6	140.2	48.3
0.600	1666.0	1319.6	655.8	325.9	162.0	1171.5	407.3	141.6	49.2
0.605	1647.0	1311.0	661.3	333.6	168.2	1164.6	411.8	145.6	51.5
0.610	1635.0	1307.9	669.6	342.8	175.5	1162.6	418.0	150.3	54.0
0.620	1602.0	1294.2	682.4	359.9	189.7	1151.9	428.2	159.2	59.2
0.630	1570.0	1280.9	695.6	377.8	205.2	1141.4	438.6	168.5	64.8
0.640	1544.0	1272.1	711.4	397.9	222.5	1134.9	450.7	179.0	71.1
0.650	1511.0	1257.1	723.9	416.9	240.1	1122.8	460.8	189.1	77.6
0.660	1486.0	1244.2	730.2	428.6	251.6	1112.5	466.9	195.9	82.2
0.670	1456.0	1226.8	733.8	438.9	262.5	1098.2	471.2	202.2	86.8
0.680	1427.0	1209.9	737.4	449.5	273.9	1084.3	475.6	208.6	91.5
0.690	1402.0	1196.2	742.9	461.3	286.5	1073.1	481.1	215.7	96.7
0.698	1374.6	1010.3	546.1	311.8	181.6	907.1	354.9	146.7	61.8
0.700	1369.0	1175.3	743.7	470.6	297.7	1055.4	483.6	221.6	101.6
0.710	1344.0	1157.4	739.2	472.1	301.5	1040.4	482.7	223.9	103.9
0.720	1314.0	1135.1	731.7	471.6	304.0	1021.3	479.6	225.2	105.8
0.728	1295.5	1003.1	582.3	351.7	215.5	903.3	382.8	168.8	75.5
0.730	1290.0	1117.8	727.1	473.0	307.7	1006.7	478.5	227.4	108.1
0.740	1260.0	1095.1	718.9	471.9	309.8	987.2	474.8	228.4	109.9
0.750	1235.0	1076.6	713.2	472.4	313.0	971.4	472.8	230.1	112.0
0.762	1205.5	794.0	357.1	163.6	69.1	717.2	237.8	80.3	25.0
0.770	1185.0	1039.2	700.8	472.7	318.8	939.4	467.9	233.1	116.1
0.780	1159.0	1019.4	693.6	472.0	321.1	922.3	464.7	234.1	118.0
0.790	1134.0	1000.3	686.7	471.4	323.6	905.8	461.6	235.2	119.9
0.800	1109.0	981.2	679.4	470.5	325.8	889.1	458.2	236.1	121.7
0.806	1095.1	874.4	547.7	355.9	234.4	792.7	370.0	179.2	88.0
0.825	1048.0	931.6	654.3	459.6	322.8	845.9	444.8	233.9	123.0
0.830	1036.0	921.8	649.3	457.3	322.1	837.3	442.0	233.3	123.2
0.835	1024.5	912.4	644.4	455.2	321.5	829.1	439.4	232.9	123.4
0.846	998.1	476.2	181.0	85.9	44.2	433.1	123.8	44.2	17.1
0.860	968.0	506.4	212.0	107.4	58.3	461.0	145.6	55.6	22.8
0.870	947.0	453.8	174.7	84.0	43.8	413.4	120.4	43.7	17.2
0.875	936.5	449.2	173.4	83.6	43.7	409.4	119.6	43.6	17.3
0.887	912.5	448.6	178.3	87.7	46.7	409.2	123.4	46.1	18.6
0.900	891.0	448.9	183.7	92.3	50.0	409.8	127.6	48.8	20.1
0.907	882.8	455.2	190.9	97.6	53.7	415.8	132.9	51.8	21.7
0.915	874.5	461.5	198.5	103.2	57.5	421.8	138.4	54.9	23.4
0.925	863.5	279.0	73.6	28.0	12.1	255.1	51.5	14.9	5.0
0.930	858.0	221.8	46.9	15.4	6.0	202.9	32.8	8.3	2.4

[a] Expressed in W m^{-2} μm^{-1}; H_2O, 20 mm; O_3, 3.4 mm.

Spectral irradiance at ground level in tabular form is given in Table 3.5 for the range 0.29–0.93 μm and in Table 3.6 for the range 0.94–4.045 μm. These values correspond to a considerably higher level of atmospheric pollution, as in large cities and industrial centers. Column 2 gives the extra-terrestial values; columns 3–6 give values at ground level for $\alpha = 0.66$ and $\beta = 0.085$; and columns 7–9 give values at ground level for $\alpha = 0.66$ and $\beta = 0.17$.

Several of the solar spectral irradiance tables (λ and E_λ) are available on punched cards for computer-aided applications. These include irradiance at ground level for the four sets of α and β, and for a large number of values of air mass, in the wavelength range 0.29–4.045 μm, extraterrestrial solar

Solar Irradiance for Different Air Masses[a]

AIR MASS / WAVE LENGTH	0	α 0.66; β 0.085				α 0.66; β 0.170			
		1	4	7	10	1	4	7	10
0.290	482.0	0.0	0.0	0.0	0.0	0.0	0.0	0.0	0.0
0.295	584.0	0.0	0.0	0.0	0.0	0.0	0.0	0.0	0.0
0.300	514.0	4.1	0.0	0.0	0.0	3.4	0.0	0.0	0.0
0.305	603.0	11.4	0.0	0.0	0.0	9.4	0.0	0.0	0.0
0.310	689.0	30.5	0.0	0.0	0.0	25.4	0.0	0.0	0.0
0.315	764.0	79.4	0.1	0.0	0.0	66.2	0.0	0.0	0.0
0.320	830.0	202.6	2.9	0.0	0.0	169.2	1.4	0.0	0.0
0.325	975.0	269.5	5.7	0.1	0.0	225.5	2.8	0.0	0.0
0.330	1059.0	331.6	10.2	0.3	0.0	277.9	5.0	0.1	0.0
0.335	1081.0	383.4	17.1	0.8	0.0	321.8	8.5	0.2	0.0
0.340	1074.0	431.3	27.9	1.8	0.1	362.7	14.0	0.5	0.0
0.345	1069.0	449.2	33.3	2.5	0.2	378.4	16.8	0.7	0.0
0.350	1093.0	480.5	40.8	3.5	0.3	405.4	20.7	1.1	0.1
0.355	1083.0	498.0	48.4	4.7	0.5	420.8	24.7	1.4	0.1
0.360	1068.0	513.7	57.2	6.4	0.7	434.8	29.3	2.0	0.1
0.365	1132.0	561.3	68.4	8.3	1.0	475.7	35.3	2.6	0.2
0.370	1181.0	603.5	80.5	10.7	1.4	512.3	41.8	3.4	0.3
0.375	1157.0	609.4	89.0	13.0	1.9	518.0	46.5	4.2	0.4
0.380	1120.0	608.0	97.2	15.6	2.5	517.6	51.1	5.0	0.5
0.385	1098.0	609.8	104.5	17.9	3.1	519.9	55.2	5.9	0.6
0.390	1098.0	623.9	114.5	21.0	3.9	532.6	60.8	6.9	0.8
0.395	1189.0	691.2	135.8	26.7	5.2	590.8	72.5	8.9	1.1
0.400	1429.0	849.9	178.8	37.6	7.9	727.4	95.9	12.7	1.7
0.405	1644.0	992.8	218.7	48.2	10.6	850.8	117.9	16.3	2.3
0.410	1751.0	1073.7	247.5	57.1	13.2	921.3	134.2	19.5	2.8
0.415	1774.0	1104.5	266.5	64.3	15.5	948.8	145.2	22.2	3.4
0.420	1747.0	1104.3	278.9	70.4	17.8	949.8	152.7	24.5	3.9
0.425	1693.0	1086.5	287.2	75.9	20.1	935.6	157.9	26.7	4.5
0.430	1639.0	1067.9	295.4	81.7	22.6	920.7	163.2	28.9	5.1
0.435	1663.0	1100.1	318.4	92.2	26.7	949.5	176.7	32.9	6.1
0.440	1810.0	1215.5	368.2	111.5	33.8	1050.3	205.2	40.1	7.8
0.445	1922.0	1310.4	415.3	131.6	41.7	1133.5	232.5	47.7	9.8
0.450	2006.0	1388.4	460.3	152.6	50.6	1202.2	258.8	55.7	12.0
0.455	2057.0	1434.8	486.9	165.2	56.1	1243.7	274.9	60.8	13.4
0.460	2066.0	1452.2	504.4	175.2	60.8	1260.1	285.9	64.9	14.7
0.465	2048.0	1450.7	515.7	183.3	65.1	1260.1	293.5	68.4	15.9
0.470	2033.0	1451.2	527.9	192.0	69.8	1261.8	301.6	72.1	17.2
0.475	2044.0	1470.3	547.3	203.7	75.8	1279.6	313.9	77.0	18.9
0.480	2074.0	1503.4	572.6	218.1	83.1	1309.6	329.7	83.0	20.9
0.485	1976.0	1443.3	562.4	219.2	85.4	1258.5	325.1	84.0	21.7
0.490	1950.0	1435.2	572.2	228.2	91.0	1252.6	332.0	88.0	23.3
0.495	1960.0	1453.6	592.9	241.9	98.7	1269.8	345.3	93.9	25.5
0.500	1942.0	1451.2	605.6	252.7	105.5	1268.8	353.9	98.7	27.5
0.505	1920.0	1440.1	607.6	256.4	108.2	1260.2	356.3	100.8	28.5
0.510	1882.0	1416.8	604.4	257.8	110.0	1240.9	355.7	101.9	29.2
0.515	1833.0	1384.9	597.3	257.6	111.1	1214.0	352.7	102.5	29.8
0.520	1833.0	1390.0	606.1	264.3	115.2	1219.5	359.1	105.7	31.1
0.525	1852.0	1409.5	621.3	273.9	120.7	1237.6	369.3	110.2	32.9
0.530	1842.0	1406.9	626.9	279.4	124.5	1236.4	373.9	113.1	34.2
0.535	1818.0	1393.6	627.7	282.8	127.4	1225.6	375.5	115.1	35.3

spectrum in the range 0.115–1000 μm (standard table) and the spectrum at 1 Å intervals in the range 0.3–0.61 μm. User organizations that need these data decks may apply to National Space Science Data Center, NSA/GSFC, Code 601, Greenbelt, Maryland 20771.

Four curves related to solar spectral irradiance are shown in Fig. 3.6. This figure is a revision of a similar one that was first published by the AFCRL Handbook (Campen, 1960) and has since been frequently reprinted. The older curve was based on the data of F. S. Johnson and Parry Moon. Figure 3.6 is based on the NASA/ASTM curve and the present computations. The temperature of the equivalent blackbody curve has been chosen to be 5762°K and as explained by Anonymous (1971c, p. 24), it is the temperature,

TABLE 3.6

AIR MASS / WAVE LENGTH	0	α 0.66; β 0.085				α 0.66; β 0.170			
		1	4	7	10	1	4	7	10
2.338	66.3	54.7	39.9	30.4	23.4	52.1	32.8	21.6	14.4
2.356	65.1	52.0	36.3	26.5	19.6	49.5	29.9	18.9	12.1
2.388	62.8	36.0	18.7	11.7	7.8	34.3	15.5	8.3	4.8
2.415	61.0	32.5	15.8	9.4	6.0	31.0	13.0	6.7	3.7
2.453	58.3	29.6	13.7	7.9	5.0	28.2	11.3	5.7	3.1
2.494	55.4	20.3	6.8	3.2	1.7	19.4	5.6	2.3	1.1
2.537	52.4	4.6	0.4	0.1	0.0	4.4	0.3	0.1	0.0
2.900	35.0	2.9	0.2	0.0	0.0	2.8	0.2	0.0	0.0
2.941	33.4	6.0	1.0	0.3	0.1	5.7	0.8	0.2	0.1
2.954	32.8	5.7	0.9	0.3	0.1	5.5	0.8	0.2	0.1
2.973	32.1	8.7	2.2	0.9	0.4	8.4	1.8	0.6	0.3
3.005	30.8	7.8	1.8	0.7	0.3	7.5	1.6	0.5	0.2
3.045	28.8	4.7	0.7	0.2	0.1	4.5	0.6	0.1	0.0
3.056	28.2	4.9	0.8	0.2	0.1	4.7	0.7	0.2	0.1
3.097	26.2	3.2	0.4	0.1	0.0	3.1	0.3	0.1	0.0
3.132	24.9	6.8	1.7	0.7	0.3	6.5	1.5	0.5	0.2
3.156	24.1	18.7	12.6	8.9	6.3	17.9	10.7	6.7	4.2
3.204	22.5	2.1	0.2	0.0	0.0	2.0	0.2	0.0	0.0
3.214	22.1	3.4	0.5	0.1	0.0	3.3	0.4	0.1	0.0
3.245	21.1	3.9	0.7	0.2	0.1	3.8	0.6	0.2	0.1
3.260	20.6	3.7	0.6	0.2	0.1	3.5	0.5	0.1	0.0
3.285	19.7	14.2	8.5	5.1	2.8	13.7	7.3	3.9	1.9
3.317	18.8	12.9	6.9	3.5	1.3	12.4	5.9	2.7	0.9
3.344	18.1	4.2	0.9	0.3	0.1	4.1	0.8	0.2	0.1
3.403	16.5	12.3	7.8	5.1	3.2	11.9	6.7	3.9	2.2
3.450	15.6	12.5	8.9	6.7	5.0	12.0	7.7	5.1	3.5
3.507	14.5	12.5	9.9	8.1	6.7	12.1	8.5	6.2	4.6
3.538	14.2	11.8	8.8	6.9	5.5	11.3	7.6	5.3	3.8
3.573	13.8	10.9	5.4	2.6	1.3	10.5	4.6	2.0	0.9
3.633	13.1	10.8	8.3	6.7	5.5	10.4	7.1	5.2	3.8
3.673	12.6	9.1	6.1	4.6	3.5	8.8	5.3	3.5	2.5
3.696	12.3	10.4	8.2	6.7	5.6	10.1	7.1	5.2	3.9
3.712	12.2	10.9	9.0	7.6	6.5	10.5	7.8	5.9	4.6
3.765	11.5	9.5	7.2	5.9	4.8	9.1	6.3	4.6	3.4
3.812	11.0	8.9	6.7	5.4	4.4	8.6	5.8	4.2	3.1
3.888	10.4	8.1	5.6	4.0	2.9	7.8	4.8	3.1	2.0
3.923	10.1	8.0	5.6	4.2	3.1	7.7	4.9	3.3	2.2
3.948	9.9	7.8	5.5	4.0	3.0	7.6	4.8	3.2	2.1
4.045	9.1	6.7	4.1	2.6	1.5	6.5	3.6	2.0	1.1
TOTAL $W\ m^{-2}$	1353	889.2	448.7	255.2	153.8	800.2	303.1	133.3	63.4

[a] Expressed in $W\ m^{-2}\ \mu m^{-1}$; H_2O, 20 mm; O_3, 3.4 mm.

which, according to the Stefan–Boltzmann equation, a blackbody of the size of the sun should have in order to give an irradiance of 1353 $W\ m^{-2}$ at one astronomical unit. The lowest curve with many sharp dips in it is the solar spectral irradiance observed on the ground under "normal" conditions, for air mass two, that is, when the sun is at 60° from the vertical. This is the solar spectral distribution that has been adopted for theoretical calculations of solar cell efficiency by the Photovoltaic Workshop sponsored by ERDA and NASA (Brandhorst *et al.*, 1975). A table of values of spectral irradiance and of number of photons corresponding to this curve is available in Brandhorst *et al.* (1975).

Solar Irradiance for Different Air Masses[a]

AIR MASS / WAVE LENGTH	0	α 0.66; β 0.085				α 0.66; β 0.170			
		1	4	7	10	1	4	7	10
0.940	847.0	313.4	95.0	39.6	18.5	286.9	66.7	21.3	7.6
0.950	837.0	296.5	86.3	35.0	16.0	271.6	60.7	18.9	6.7
0.955	828.5	321.1	102.3	44.1	21.2	294.2	72.1	23.9	8.8
0.965	811.5	344.4	120.4	55.1	27.8	315.7	85.0	30.0	11.6
0.975	794.0	576.9	346.0	224.6	150.1	529.1	244.9	122.7	63.2
0.985	776.0	544.6	316.1	201.2	132.4	499.8	224.2	110.3	56.1
1.018	719.2	617.5	391.0	247.5	156.7	567.8	279.4	137.5	67.6
1.082	620.0	512.9	290.4	164.4	93.1	473.1	210.3	93.5	41.5
1.094	602.0	464.1	303.1	210.8	149.9	428.4	220.0	120.3	67.3
1.098	596.0	503.7	304.1	183.6	110.9	465.0	220.9	105.0	49.9
1.101	591.8	504.8	362.7	267.3	198.8	466.1	263.6	153.0	89.5
1.128	560.5	135.1	27.7	9.1	3.6	124.9	20.2	5.3	1.6
1.131	557.0	152.2	35.3	12.6	5.3	140.7	25.8	7.3	2.4
1.137	550.1	143.1	31.7	11.0	4.5	132.4	23.2	6.4	2.1
1.144	542.0	191.2	57.4	24.2	11.6	176.9	42.1	14.1	5.3
1.147	538.5	174.5	48.2	19.3	8.8	161.5	35.3	11.2	4.1
1.178	507.0	399.3	195.1	95.4	46.6	370.0	143.8	55.9	21.7
1.189	496.0	402.2	214.5	114.4	61.0	372.8	158.4	67.3	28.6
1.193	492.0	424.0	310.8	233.3	176.6	393.1	229.7	137.4	82.9
1.222	464.3	391.8	235.3	141.3	84.9	363.6	174.7	83.9	40.3
1.236	451.2	390.8	254.1	165.2	107.4	363.0	189.1	98.5	51.3
1.264	426.5	329.2	209.7	140.0	94.3	306.1	156.7	84.1	45.5
1.276	416.7	342.6	238.6	172.6	126.3	318.7	178.6	104.0	61.2
1.288	406.8	347.3	216.1	134.4	83.7	323.2	162.1	81.3	40.7
1.314	386.1	298.3	137.6	63.5	29.3	277.9	103.6	38.6	14.4
1.335	369.7	190.6	85.0	46.7	27.7	177.6	64.2	28.6	13.7
1.384	343.7	5.7	0.1	0.0	0.0	5.3	0.1	0.0	0.0
1.432	321.0	44.6	5.4	1.3	0.4	41.7	4.1	0.8	0.2
1.457	308.6	85.4	20.6	7.7	3.3	80.0	15.8	4.8	1.7
1.472	301.4	77.4	17.4	6.2	2.6	72.4	13.3	3.9	1.3
1.542	270.4	239.3	165.9	115.0	79.7	224.5	128.5	73.6	42.1
1.572	257.3	222.6	168.1	130.4	102.1	209.0	130.7	83.9	54.4
1.599	245.4	216.0	166.7	131.5	104.5	203.0	129.9	85.0	56.0
1.608	241.5	208.5	157.4	122.1	95.7	195.9	122.7	79.1	51.4
1.626	233.6	206.7	160.7	127.5	101.9	194.3	125.6	82.8	55.0
1.644	225.6	197.9	152.4	120.1	95.5	186.1	119.3	78.3	51.8
1.650	223.0	195.7	150.9	119.1	94.7	184.1	118.2	77.6	51.4
1.676	212.1	181.9	114.8	72.4	45.7	171.2	90.1	47.5	25.0
1.732	187.9	161.5	102.5	65.1	41.3	152.2	80.9	43.0	22.9
1.782	166.6	136.7	75.6	41.8	23.1	129.0	59.9	27.9	12.9
1.862	138.2	4.0	0.1	0.0	0.0	3.8	0.1	0.0	0.0
1.955	112.9	42.7	14.5	6.8	3.6	40.5	11.6	4.6	2.1
2.008	102.0	69.4	35.8	17.7	6.4	65.7	28.9	12.2	3.8
2.014	101.2	74.7	45.5	28.8	17.8	70.8	36.7	19.8	10.4
2.057	95.6	69.5	41.3	25.3	14.8	66.0	33.4	17.5	8.7
2.124	87.4	70.0	35.9	18.4	9.5	66.4	29.2	12.8	5.6
2.156	83.8	66.0	32.3	15.8	7.7	62.7	26.3	11.0	4.6
2.201	78.9	66.1	49.1	38.0	29.7	62.8	40.1	26.7	17.9
2.266	72.4	61.6	46.8	36.8	29.3	58.6	38.4	26.0	17.9
2.320	67.6	57.2	43.2	33.8	26.8	54.4	35.5	24.0	16.5

Table 3.7 gives data on total irradiance at ground level for the four air mass* values and the four sets of turbidity parameters. The partition of energy between the UV, visible, and IR ranges of the spectrum is also shown. It is significant that as the air mass increases or as the turbidity increases, the relative amount of energy in the infrared increases, and that in the visible

* The air mass is defined as the ratio of the pathlength of the radiation through the atmosphere at any given angle to the sea level path straight through the atmosphere (vertically). Air mass zero refers to the absence of atmospheric attenuation at one astronomical unit from the sun, α and β are coefficients used to calculate the atmospheric turbidity, and different values correspond to different levels of atmospheric pollution.

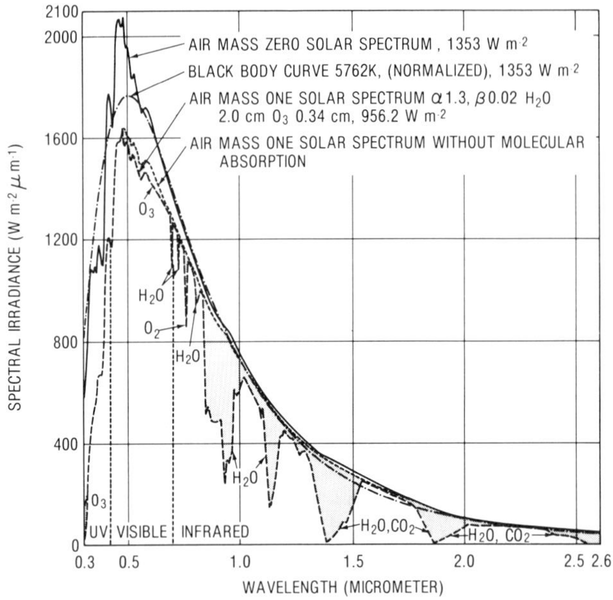

FIG. 3.6 Four curves related to solar spectral irradiance.

and UV, decreases. The variation of total irradiance as a function of air mass is shown in Fig. 3.7 for the four sets of α and β. The y axis is on the log scale. For monochromatic radiation, where Eq. (3.1) holds, the plot of $\log E_\lambda$ versus air mass m is a straight line and the slope of the line gives the coefficient c_λ; but the linear relation does not at all hold for total irradiance. Even for monochromatic radiation, the plot of $\log E_\lambda$ versus air mass can often lead to large uncertainties in extrapolation to zero air mass, since both the turbidity and amount of water vapor can change significantly during the time required for a wide range of air mass. The ozone density is also subject to change.

Table 3.7 and Fig. 3.7 show the effect of turbidity and air mass on the total irradiance on the ground. Two other atmospheric parameters which affect ground insolation are the amount of precipitable water vapor and of ozone. Table 3.8 gives samples of the results obtained when these parameters are also changed within the limits that occur naturally. Twelve sets of parameters as shown in each row of the first four columns have been chosen; the next five columns give the total irradiance on a surface at sea level exposed normally to the sun's rays. The corresponding air mass and solar

TABLE 3.7

Solar Irradiance for Different Air Mass Values[a]

AIR MASS	SOLAR ZENITH ANGLE (DEGREES)	TURBIDITY FACTORS		TOTAL IRRADIANCE $W\ m^{-2}$	RATIO OF TOTAL IRRADIANCE TO SOLAR CONSTANT %	FRACTION OF THE TOTAL ENERGY IN THE		
		α	β			UV, $\lambda<0.4\mu m$ %	VISIBLE $0.4\ \mu m<\lambda<0.72\mu m$ %	INFRARED $\lambda>0.72\mu m$ %
0	0			1353.0	100.0	8.7	40.1	51.1
1	0	1.30	0.02	956.2	70.7	4.8	46.9	48.3
4	75.5	1.30	0.02	595.2	44.0	1.23	44.2	54.5
7	81.8	1.30	0.02	413.6	30.6	0.35	39.4	60.3
10	84.3	1.30	0.02	302.5	22.4	0.102	34.7	65.2
1	0	1.30	0.04	924.9	68.4	4.6	46.4	49.0
4	75.5	1.30	0.04	528.9	39.1	1.04	42.1	56.9
7	81.8	1.30	0.04	342.0	25.3	0.26	35.9	63.8
10	84.3	1.30	0.04	234.5	17.3	0.065	30.3	69.6
1	0	0.66	0.085	889.2	65.7	4.7	46.4	48.9
4	75.5	0.66	0.085	448.7	33.2	1.14	42.4	56.5
7	81.8	0.66	0.085	255.2	18.9	0.30	36.3	63.4
10	84.3	0.66	0.085	153.8	11.4	0.08	30.7	69.2
1	0	0.66	0.17	800.2	59.1	4.5	45.4	50.1
4	75.5	0.66	0.17	303.1	22.4	0.88	38.3	60.8
7	81.8	0.66	0.17	133.3	9.85	0.14	30.0	69.8
10	84.3	0.66	0.17	63.4	4.69	0.039	22.9	77.1

[a] US standard atmosphere: H_2O, 20 mm; O_3, 3.4 mm.

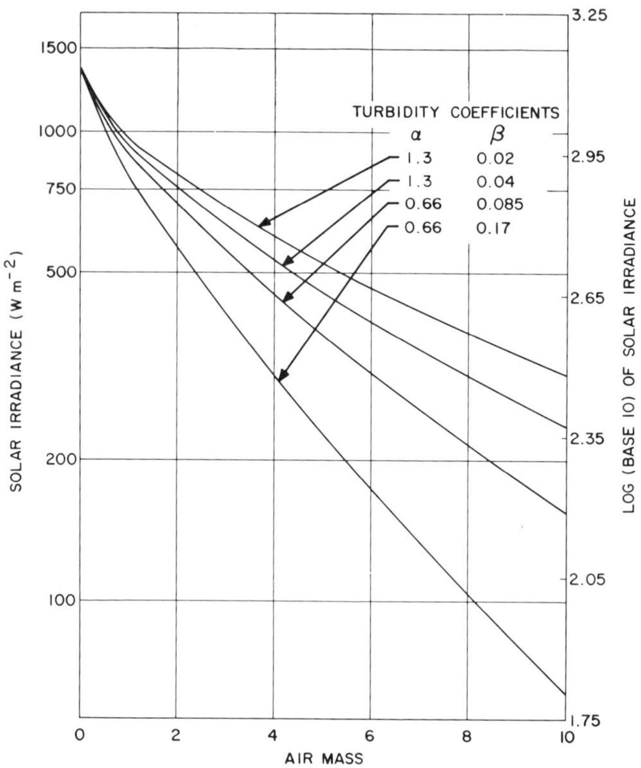

FIG. 3.7 Semilog plots of solar irradiance versus air mass. Curves show the variation of total energy received directly from sun by unit area exposed normally to the sun's rays for different values of solar zenith angle and atmospheric turbidity. H_2O: 20 mm; O_3: 3.4 mm.

zenith angles are shown at the top of each column, five values of air mass have been considered, 1, 1.5, 2, 3, and 5. Table 3.8 is an abridged version of a more detailed table that gives total and spectral irradiance for 32 sets of atmospheric parameters and 10 values of air mass.

As stated earlier, these values are for normal incidence. The direct radiation from the sun on a horizontal surface is obtained by multiplying these values by the cosine of the zenith angle or dividing them by the air mass.

3.11 CONCLUSION

With the increased interest in solar energy conversion, the quantitative data on available solar energy will become more accurate and abundant. Research and development programs in heating and cooling, photovoltaic

TABLE 3.8

Direct Solar Irradiance as a Function of Air Mass for Varying Atmospheric Parameters[a]

				Solar Irradiance				
				For air mass				
				1	1.5	2	3	5
				For solar zenith angles				
Atmospheric parameters								
H_2O (cm)	O_3 (cm)	β	α	0	48.2°	60°	70.5°	78.5°
2.0	0.20	0.02	1.3	961	876	805	693	534
2.0	0.20	0.17	0.66	804	672	568	414	232
2.0	0.34	0.02	1.3	956	870	798	684	523
2.0	0.34	0.17	0.66	800	668	563	409	228
2.0	0.55	0.02	1.3	949	861	788	672	508
2.0	0.55	0.17	0.66	795	661	556	402	222
0.5	0.34	0.02	1.3	1024	943	875	767	612
0.5	0.34	0.17	0.66	859	727	622	464	273
1.0	0.34	0.02	1.3	992	909	840	729	571
1.0	0.34	0.17	0.66	832	700	595	439	252
5.0	0.34	0.02	1.3	898	807	732	615	454
5.0	0.34	0.17	0.66	750	616	513	363	193

[a] Expressed in W m^{-2}.

conversion, etc., need such data for determining the efficiency of the system. Spectral distribution data are as important as total irradiance data for most applications. Theoretical computations of ground level irradiance have become easier through a better understanding of atmospheric processes and greater availability of computers. Current research efforts in improving solar energy data are in many directions: greater accuracy in extraterrestrial measurements, improvements of instruments, definition of radiation scales and standards, spectral and total irradiance monitoring at a larger number of stations, and studies of atmospheric processes and pollution. These research efforts are of interest not only for solar energy applications but for many related fields.

4

Solar Energy Availability Prediction from Climatological Data

A. A. M. SAYIGH

MECHANICAL ENGINEERING DEPARTMENT
COLLEGE OF ENGINEERING
RIYADH UNIVERSITY
RIYADH, SAUDI ARABIA

4.1 INTRODUCTION

The amount of solar radiation received on the earth's surface is attenuated by the terrestrial atmosphere. On a clear day it may reach the ground level after having been depleted of up to 15% of its original value outside the atmosphere. This attenuation is caused by the following effects:

scattering by molecules much smaller than the wavelength of the radiation (Rayleigh scattering);
selective absorption by gases present in the atmosphere and particularly by O_3, O_2, H_2O, CO_2;
scattering by aerosols (dust, smoke, pollen, etc.) of size comparable to or larger than the wavelength of the radiation; and
scattering and absorption by cloud masses.

For further details about these effects, see Chapters 2 and 3.

Therefore the solar radiation reaching the earth is made of *direct* and *diffuse* radiation, also known as short and long wavelength radiation, respectively. Practically all the energy of wavelength shorter than 0.3 μm and larger than 2.5 μm does not reach us and the selective absorption is limited to O_3, O_2, H_2O, and CO_2 with the absorption bands increasing with the thickness of the layers of these gases according to the Lambert–Beer law.

Owing to the importance of knowing accurately the total solar radiation on the earth's surface and because of the unreliability in the past of solar radiation measuring equipment and the present high cost of such equipment (Moon, 1940; Haurwitz, 1948), several authors (Bonhomme and Ometto, 1975; Cuomo, 1974; Morgan and Craig, 1975; Brook and Miller, 1963), attempted to estimate the solar radiation using well-established climatological data. Climatological data such as temperature, humidity, sunshine hours, rainfall, and, to a less extent, cloudiness, have been measured almost all over the world in the last 50 years. The instruments have also been perfected and cheaply produced. In this chapter several formulas for estimating total direct and diffuse solar radiation which have been proposed by numerous authors will be discussed and their closeness to the measured radiation in some part of the world are stated.

4.2 ESTIMATION OF TOTAL SOLAR RADIATION

A. The Use of Sunshine Hours

Masson (1966) plotted the total solar radiation H on a horizontal plane in langleys (ly) per day against the hours of sunshine per day, S and fitted a hyperbolic segment to the result. The segment tends to be a straight line if the hours of sunshine are equal or greater than 7 h. The results were expressed in mathematical forms as

$$(S + 2.64)^2/(2.64)^2 - (H - 60)^2/(99)^2 = 1 \tag{4.1}$$

or

$$H = 60 + (1406.2S^2 + 7426.6S)^{1/2} \tag{4.2}$$

Sabbagh *et al.* (1973) used the average daily sunshine hours S per month with arbitrary curve similar to the radiation curve and two arbitrary constants. In another formula, they used the average hours of the length of the day Z per month:

$$H = (A + B \ln n)S \tag{4.3}$$

and

$$H = (A' + B' \ln n)Z \tag{4.4}$$

where $A' = 35$, $B' = 11$ and $A = 42$, $B = 12$, and $n = 1, 2, 3, 4, 5, 6, 6, 5, 4, 3, 2, 1$ for the months 1, 2, 3, 4, 5, 6, 7, 8, 9, 10, 11, 12, respectively (month 1 = January, and so on). Formulas (4.3) and (4.4) will be compared with formulas presented later on in this chapter.

B. The Use of the Ratio of Sunshine Hours to Length of Day

One of the earliest expressions was Ångström's regression (1924), which is

$$H = H_0'(A' + B'S/Z) \tag{4.5}$$

where A' and B' are arbitrary constants (Fritz, 1951, suggested that $A' = 0.35$ and $B' = 0.61$, H is the monthly average horizontal solar radiation in W m^{-2}, and H_0' the monthly average horizontal solar radiation for a clear day. H_0' can be obtained from charts (Duffie and Beckman, 1974, Fig. 4.1), while Z can be calculated from the equation

$$Z = (2/15) \cos^{-1}(-\tan \phi \tan \delta)$$

where δ is the declination angle of the place and ϕ the latitude of the place. An easier way of calculating Z is from a nomogram that was developed by Whillier (1965) and is shown in Fig. 4.2 (see the example in the figure).

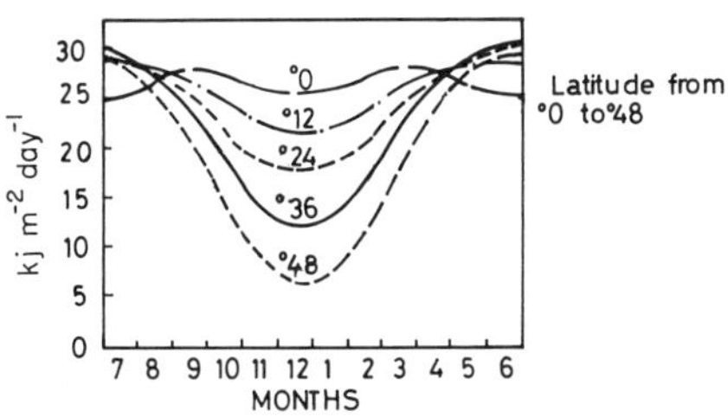

FIG. 4.1 Solar radiation on a horizontal plane for a clear day at various latitudes.

A better form of Eq. (4.5) is suggested by Page (1964):

$$H = H_0(A + BS/Z) \tag{4.6}$$

where A and B are arbitrary constants, and H_0 is the average monthly insolation at the top of the atmosphere. Duffie and Beckman (1974) used Eq. (4.6) for the town of Madison, Wisconsin with H_0 varying from 41.4×10^3 $\text{kJ}/m^2 \cdot \text{day}$, in June to $11.6 \times 10^3 \text{ kJ}/m^2 \cdot \text{day}$, in December; $A = 0.30$ and $B = 0.34$.

C. The Use of Angström's Equation with an Extra One or Two Parameters

Bennett (1967) derives two sets of equations. One set for Canadian stations for June and December are, respectively,

$$H = H_0(311.6 + 4.94S_r) \tag{4.7}$$

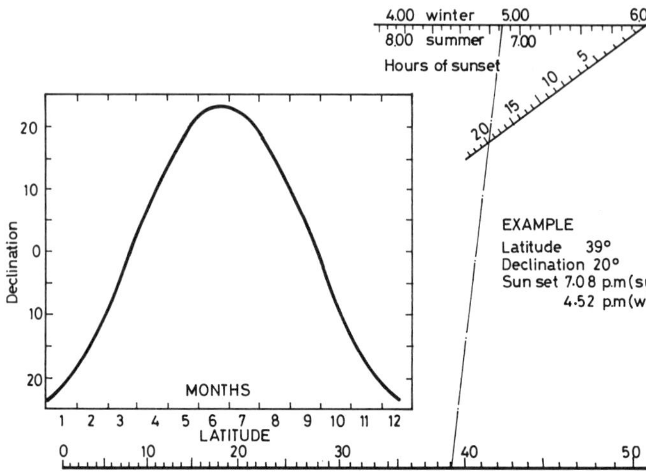

FIG. 4.2 A nomogram for calculating the length of the day (Z) in hours. Day length: winter = 9.72; summer = 14.28.

and

$$H = H_0(194.8 + 7.85S_r) \tag{4.8}$$

where H_0 equals 1.98 lg/min and S_r is the monthly mean daily percentage of possible sunshine.

The second set for North America in which he introduces the station elevation h in feet, i.e., for June and December are, respectively,

$$H = H_0(314.0 + 4.22S_r + 0.0144h) \tag{4.9}$$

and

$$H = H_0(128.1 + 7.20S_r + 0.0144h) \tag{4.10}$$

In another article, Bennett (1965) writes Eqs. (4.9) and (4.10) as

$$H/H_0 = A + BS_r + Ch \tag{4.11}$$

and for United States, A varies from 188 in April to 291 in August, B varies from 3.768 in August to 5.574 in April, and C varies from 0.0130 in July to 0.0226 in December. Swartman and Ogunlade (1966) use the relative humidity R in three different equations:

$$H = 490D'^{0.357}R^{-0.262} \tag{4.12}$$

$$H = 460e^{0.607(D'-R)} \tag{4.13}$$

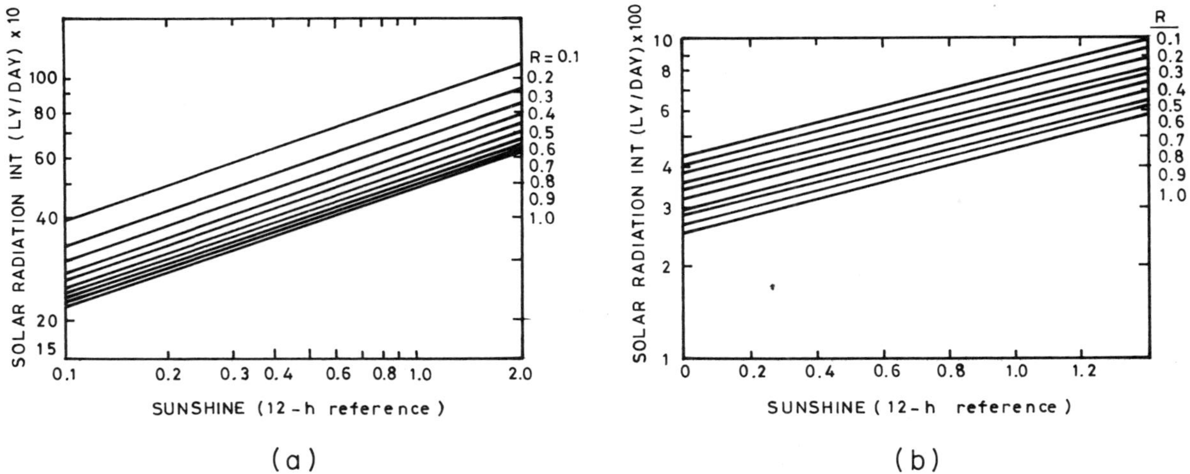

FIG. 4.3 (a) Graph for correlation $H = 490\, D'^{0.357}\, R^{-0.262}$. (b) Graph for correlation $H = 460 \exp[0.607(D' - R)]$.

and

$$H = 464 + 265D' - 248R \tag{4.14}$$

where D' is the ratio of sunshine hours to 12. Swartman and Ogunlade claim that Eqs. (4.12)–(4.14) give a better estimation of solar radiation than Eqs. (4.7) and (4.8) by Ångström. They plotted Eqs. (4.12) and (4.13) as shown in Fig. 4.3. This is a convenient way of reading off the solar radiation if D' and R are known. Equations (4.12)–(4.14) are tested later on and compared with the recent formula of the author and his colleagues which will be discussed in detail.

D. The Use of Cloud Cover and Other Parameters

Bennett (1965) used the following equation in estimating average solar radiation by using the monthly mean daily cloud cover Cd, i.e.,

$$H/H_0 = A + B((1 - Cd + S_r)/2) \tag{4.15}$$

where A and B are arbitrary constants and the rest of the terms are as before. However, in another article Bennett (1969) correlated the daily insolation with the total sky cover or opaque cloud cover and the mean daily sunshine percentage for 562 stations, and he claimed that for opaque cloud cover the correlation is much better than for total sky cover.

Majumdar *et al.* (1972) derived their formula for clear sky and minimum smoke and dust. The formula relates solar radiation to the air mass and to the transmission coefficient of water vapor, i.e.,

$$H = H_0 A^m (T_w)^{(Wm_r)^2}$$

where H_0 is the effective radiation intensity above the troposphere (=1.91 lg/min), T_w the transmission coefficient of water vapor (equal to the fractional transmission through 1 cm optical path of water vapor, i.e., $Wm_r = 1$), m_r the relative air mass (=ratio of the actual air mass along the path of the beam to the vertical air mass at the same place), m the absolute air mass (=$(p/1000)m_r$, where p is the local atmospheric pressure in millibars). Normal pressure at sea level is 1000 mbar. W is the water vapor content in the atmosphere (=$0.16p_w$ cm; in most places in India where this research, was carried out), p_w the surface vapor pressure in millibars, q is a coefficient (=0.25), and A the coefficient independent of altitude or humidity. Therefore,

$$H = 1.91(0.8644)^{\frac{pm_r}{1000}}(0.8507)^{(Wm_r)^{0.25}} \tag{4.17}$$

or

$$H = 1.91(0.8644)^{\frac{pm_r}{1000}}(0.9028)^{(p_w m_r)^{0.25}} \tag{4.18}$$

in ly/min. Reddy (1971) proposed his complex formula which estimates the daily total solar radiation at the earth surface in g·cal/cm²/day:

$$H = K(1 + 0.8S_r)(1 - 0.2J)/R \tag{4.19}$$

where $K = (\lambda z + \psi_{i,j} \cos \phi)\, 10^2$ g·cal/cm²/day, ϕ is the latitude of the place in degrees, $\lambda = 0.2/(1 + 0.1\phi)$ the latitude factor, Z and S_r as previously defined, J the ratio of number of rainy days during the month to the number of days of the same month, R the mean relative humidity per day in the month, and $\psi_{i,j}$ the seasonal factor ($i = 1,2,3$ as inland station, coastal stations, and hilly stations, respectively, while $j = 1,2,3,\ldots,12$ stand for January to December). For hill stations, K' was used instead of K; $K' = K + \varepsilon$, where $\varepsilon = 0.5\, h\, (12 - Z) \times 10^{-2}$ where h is the height of the station above sea level in meters. Table 4.1 gives the values of the different seasonal factors. In another formula Reddy (1971) uses the air temperature and humidity and derives an equation slightly simpler than Eq. (4.19):

$$H = K(0.6 + 0.02t_m S_r - 0.04\sqrt{R}) - R(4.3 - \sqrt{t_m}) \tag{4.20}$$

where H, K, S_r, and R are as previously defined and t_m is the mean daily temperature. Kettani and Lam (1974) used Eq. (4.19) and attempted to draw the isoradiation lines for the Arabian Peninsula.

TABLE 4.1

	Months of the year											
ψ	1	2	3	4	5	6	7	8	9	10	11	12
ψ_1	1.28	1.38	1.54	1.77	2.05	2.30	2.48	2.41	2.36	1.73	1.38	1.17
ψ_2	1.46	1.77	2.05	2.15	2.05	2.05	2.10	2.17	2.14	1.96	1.60	1.43
ψ_3	1.60	1.81	2.00	2.17	2.25	2.26	2.24	2.20	2.10	1.92	1.74	1.60

4.3 ESTIMATION OF THE TOTAL SOLAR RADIATION FROM METEOROLOGICAL DATA

The fact that there is a lack of solar radiation measurement in many countries should not act as a deterrent to solar energy research because it is possible to obtain solar radiation data either by using (a) the known solar radiation data for a given place and applying them to places having the same climate, latitude, and topography, or (b) the meteorological data, latitude, and topography of the place in estimating their solar radiation.

Reviewing Eqs. (4.1)–(4.20), one can deduce that various parameters such as sunshine hours, latitude, altitude, temperature, and relative humidity are more related to solar radiation, some more than others. Nevertheless in order

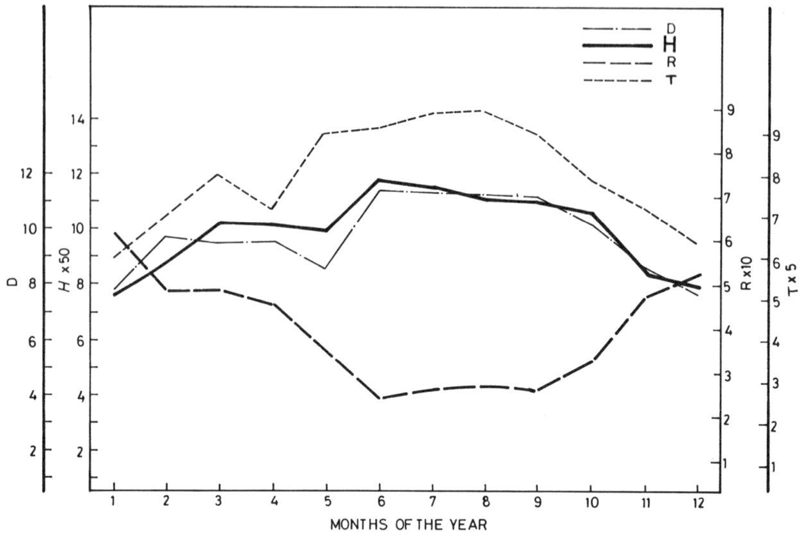

FIG. 4.4 Sunshine hours, radiation, relative humidity, and maximum temperature for Riyadh (1970–1975).

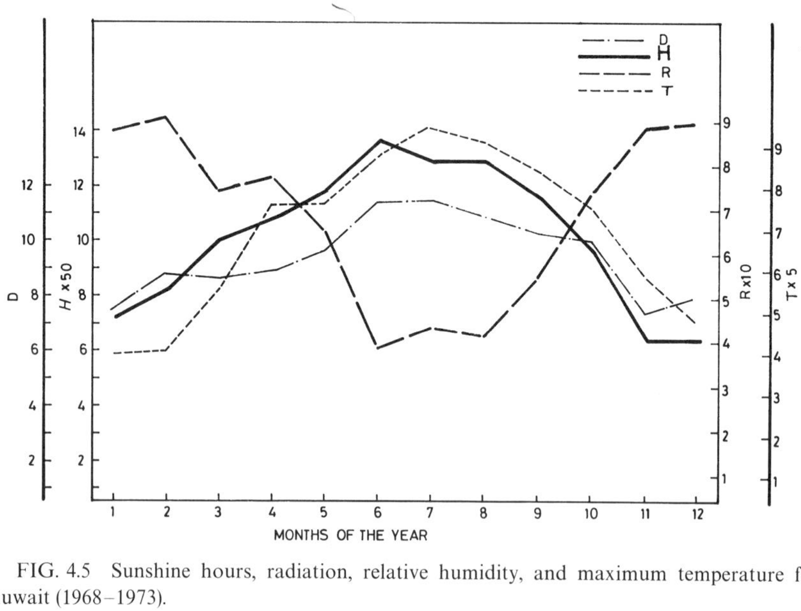

FIG. 4.5 Sunshine hours, radiation, relative humidity, and maximum temperature for Kuwait (1968–1973).

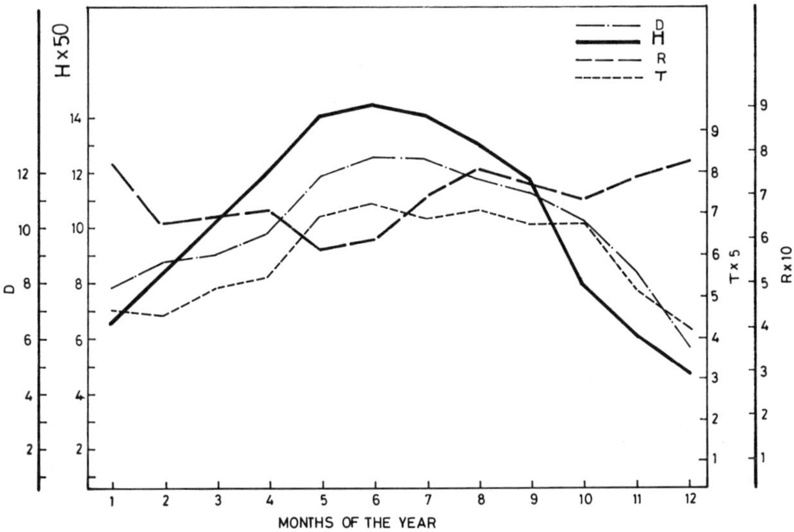

FIG. 4.6 Sunshine hours, radiation, relative humidity, and maximum temperature for Bahtim (1970–1974).

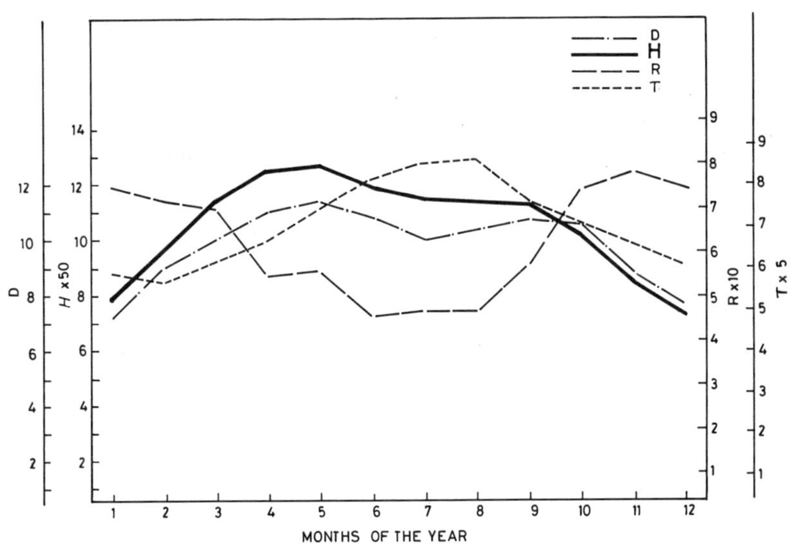

FIG. 4.7 Sunshine hours, radiation, relative humidity, and maximum temperature for Port Sudan (1941–1970).

to estimate solar radiation, one has to consider all these parameters and to delete some of them is bound to result in error as will be shown later. Using these parameters, the author and two of his colleagues deduced a formula (Sabbagh *et al.*, 1975), that can predict total solar radiation on a horizontal plane with much greater accuracy than any of the previously mentioned formulas. However its validity was tested only in arid and semiarid zones due to the lack of comparable data from other localities. In the following paragraphs, the proposed formula will be fully discussed.

4.4 SOME FACTORS WHICH INFLUENCE TOTAL SOLAR RADIATION

Several factors that contribute toward the reduction of solar radiation on the ground level have been mentioned earlier. Other factors are indicative of the amount of total solar radiation, large or small, but are not directly proportional to the amount of solar radiation as will be seen later on. These factors are

the amount of sunshine available per day;

the air temperature at the ground level, whether mean daily temperature, maximum or minimum temperature per day;

the altitude and latitude of the place. Both these factors influence the solar zenith angle which is the angle between the vertical at a point on the earth and the line joining that point to the sun.

This line or path is used in defining the air mass. The air mass is the ratio of the mass of air in a column of unit cross section along the path of the solar rays to the mass of air in a vertical column of unit cross section.

In a first approximation,

$$m = \sec \bar{Z} \tag{4.21}$$

where m is the air mass and $\bar{Z}$ is the zenith angle. But $\bar{Z}$ is a function of latitude of the place, solar declination for the day, and the hour angle of the sun as shown in Chapter 2. As for the altitude of the place, it does influence the solar radiation intensity as shown earlier in the K factor of Eq. (4.19) or it affects the air mass.

Figures 4.4–4.7 show the variation of daily mean sunshine hours per month, the mean monthly solar radiation, the mean monthly relative humidity, and the maximum monthly average air temperature in Riyadh, Saudi Arabia, Kuwait Airport, Kuwait, Bahtim, Egypt, and Port Sudan, Sudan (from Sabbagh *et al.*, 1975).

4.5 A GENERAL EMPIRICAL FORMULA FOR ESTIMATING TOTAL SOLAR RADIATION

A general formula in two forms was proposed by Sabbagh *et al.* (1975), and tested in various locations. These locations are shown in Table 4.2. The formula is

$$\begin{aligned} H &= 1.53K \exp[\Phi(D - R^{1/3}/100 - 1/T_{\max})] \quad \text{(I)} \\ &= 1.53K \exp[\Phi(D' - R^{1/3}/100 - 1/T_{\max})] \quad \text{(II)} \end{aligned} \tag{4.22}$$

TABLE 4.2

Locations Where Formula (4.22) Was Tested

Country	Station	Station coordinates: Latitude N	Longitude E	Altitude M	Period
Egypt[a]	Kharga	25°27′	30°32′	72.8	1970–1974
	Bahtim	30°08′	31°34′	74.5	
	Mersa Matruh	31°20′	27°18′	30.0	
	Tahrir	30°47′	31°00′	14.0	
Kuwait[b]	Kuwait	29°14′	47°59′	45.0	1968–1973
Lebanon[c]	Beirut	33°55′	36°45′	24.0	1973–1974
Sudan[d]	Khartoum	15°36′	32°33′	380.0	1941–1970
	Port Sudan	19°35′	37°13′	5.0	
Saudi Arabia[e]	Qatif	26°30′	50°00′	7	1970–1975
	Taif	21°14′	40°21′	1740	
	Hofuf	25°30′	49°34′	150	
	Kharj	24°10′	42°25′	430	
	Riyadh	24°42′	46°43′	600	

[a] The Monthly Weather Report, publication of the meteorological authority of the Arab Republic of Egypt.

[b] Annual Climatological Report, State of Kuwait.

[c] Meteorological data of the AUB Agricultural Research and Education Center, House Sneid, Lebanon.

[d] Climatological data, Sudan Meteorological Department.

[e] Climatological data, Ministry of Agriculture Department of Water Exploration and Research Division.

where Φ is the latitude angle in radians, D the ratio of the actual mean sunshine hours to 12 h, D' the ratio of the actual sunshine hours to the length of the day, R the mean relative humidity in percentage, T_{max} the mean maximum temperature, and K the latitude factor in $g \cdot cal \cdot cm^{-2}\ day^{-1}$. Table 4.3 shows the comparison between the two forms of Eq. (4.22) and the measured total solar radiation in various locations, while Table 4.4 shows the comparison between the estimated and the measured values of monthly mean total solar radiation in the same locations. In the previous two tables the percentage difference is the percentage of the estimated value minus the measured value over the measured value.

Formula (4.22) was compared with eight different empirical formulas regarding the calculation of total solar radiation. Each one of these eight formulas was computed separately to give the best fit as far as the measured values of solar radiation were concerned. Table 4.5 shows the comparison among five of these formulas in calculating total yearly mean solar radiation at various locations and it is clear that formula (4.22) is the best one among them all.

Using formula (4.22) and Figs. 4.8–4.10, the isoradiation maps for the Arabian Peninsula and North Africa were drawn (see Figs. 4.11–4.13, Sayigh and El-Salam, 1976).

TABLE 4.3

Estimation of Average Yearly Total Solar Radiation by Using Formulas (4.22) and Its Comparison with the Observed Radiation in Various Locations

Location	H_{obs} ($g \cdot cal \cdot cm^{-2}\ day^{-1}$)	I		II	
		H_{est}	% Difference	H_{est}	% Difference
Kharga	496.27	511.66	+3.09	506.33	+2.03
Bahtim	480.36	494.47	+2.93	486.12	+1.20
Mersa Matruh	429.46	432.35	+0.62	428.51	−0.22
Tahrir	476.27	499.72	+4.93	490.32	+2.95
Kuwait	479.05	472.78	−1.31	481.06	−0.42
Beirut	466.33	452.90	−2.88	490.76	+5.24
Khartoum	547.07	540.78	−1.15	538.89	−1.49
Port Sudan	495.83	506.23	+2.09	502.84	+1.41
Qatif	485.25	483.55	−0.35	480.02	−1.08
Taif	448.69	413.32	−7.88	395.52	−11.85
Hofuf	482.88	483.93	+0.22	479.45	−0.71
Kharj	466.11	451.62	−3.11	446.57	−4.19
Riyadh	468.75	458.65	−2.15	454.03	−3.14

TABLE 4.4

Comparison between Estimated and Measured Monthly Mean Total Solar Radiation in Various Locations

	Month	H_{exp}	H_{est}	% Difference		H_{exp}	H_{est}	% Difference
Bahtim,	Jan	312.3	316.97	−1.49	Riyadh,	351.0	330.79	5.76
Egypt	Feb	461.8	385.73	16.42	Saudi	411.0	399.52	2.79
	Mar	423.8	356.19	15.95	Arabia	479.0	482.45	−0.72
	Apr	533.9	467.66	12.41		480.0	432.63	9.87
	May	625.4	560.94	10.31		470.0	478.76	−1.86
	June	555.5	678.72	−22.18		560.0	605.47	−8.11
	July	658.4	701.54	−6.55		550.0	618.72	−12.49
	Aug	602.2	660.22	−9.64		530.0	526.29	0.69
	Sept	517.3	617.09	−19.29		522.0	491.34	5.87
	Oct	431.9	478.08	−10.69		504.0	436.51	13.39
	Nov	303.0	333.21	−9.97		397.0	371.01	6.54
	Dec	285.1	307.44	−7.84		371.0	330.25	10.98
	Year	475.88	488.65 (av)	−2.68	Year	468.75	458.650 (av)	+2.15
Port	Jan	363.4	379.31	−4.37	Kuwait,	330.9	344.66	−4.15
Sudan,	Feb	452.4	456.91	−0.99	Kuwait	383.1	421.81	−10.10
Sudan	Mar	534.9	525.84	2.24		475.1	495.34	−4.26
	Apr	594.9	572.86	3.71		514.4	518.29	−0.75
	May	603.6	562.36	6.83		562.0	542.90	3.39
	June	567.4	581.63	−2.51		651.6	598.75	8.11
	July	543.6	560.11	−3.04		617.1	581.74	5.73
	Aug	537.0	573.62	−6.82		615.7	586.22	4.79
	Sept	539.4	551.36	−3.17		552.6	561.23	−1.56
	Oct	479.4	514.14	−7.25		454.4	499.31	−9.88
	Nov	396.2	423.48	−6.88		296.5	363.72	−22.67
	Dec	339.8	373.09	−9.79		295.2	344.06	−16.55
	Year	495.83	506.23 (av)	−2.09	Year	479.05	488.17 (av)	−1.90

TABLE 4.5

Comparison between the Various Formulas in Calculating Total Mean Yearly Solar Radiation

Location	Measured radiation ($g \cdot cal \cdot cm^{-2} day^{-1}$)	Calculated radiation[a] I		II		III		IV		V	
		H	% Difference	H	% Difference	H	% Difference	H	% Difference	H	% Difference
Kharga	496.27	590.58	+19.0	608.28	+22.70	586.70	+8.22	601.84	+21.77	511.66	+3.09
Bahtim	480.36	460.48	−4.14	473.19	−1.49	475.89	−0.93	527.10	+9.73	494.74	+2.93
Matruh	429.46	418.39	−2.57	410.54	−4.41	440.20	+2.50	512.29	+19.29	432.35	+0.67
Tahrir	476.27	506.31	+6.31	493.88	+3.69	475.15	−0.24	542.84	+13.98	499.72	+4.93
Kuwait	479.05	440.35	−8.08	435.94	−8.99	429.57	−10.33	497.67	+3.89	472.78	−1.31
Beirut	466.33	456.22	−2.17	494.31	+6.00	436.73	−6.35	524.26	+12.30	452.90	−2.88
Khartoum	547.07	537.42	−1.76	426.94	−3.68	583.57	+6.57	522.01	+4.58	540.78	−1.15
Port Sudan	495.83	459.76	−7.27	451.21	−9.01	461.33	−6.96	517.40	+4.35	506.23	+2.09
Qatif	485.25	432.21	−10.93	446.43	−8.00	458.87	−5.45	480.77	−0.92	483.55	−0.35
Taif	448.69	514.71	+14.71	523.11	+16.58	524.51	+16.09	524.22	+16.83	413.32	−7.88
Hofuf	482.88	469.59	−2.75	497.37	+3.00	478.38	−0.93	495.42	+2.60	483.93	+0.22
Kharj	466.11	502.93	+7.89	536.03	+15.00	513.81	+10.23	516.94	+10.91	451.62	−3.11
Riyadh	468.75	503.05	+7.32	525.00	+12.00	521.62	+11.28	514.00	+9.65	458.65	−2.15

[a] I: $H = 523.753D^{0.872}R^{-0.174}$

II: $H = 389.022 \exp[0.711(D - R)]$

III: $H = 419.391D - 292.730R + 330.571$

IV: $H = (42 + 12 \ln n)D$

V: $H = 1.53K \exp\left[\phi\left(D - \frac{R^{1/3}}{100} - \frac{1}{T_{max}}\right)\right]$

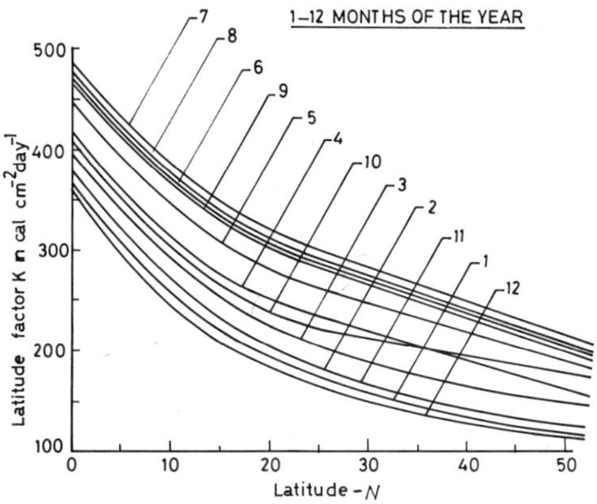

FIG. 4.8 The *K* factor for inland stations.

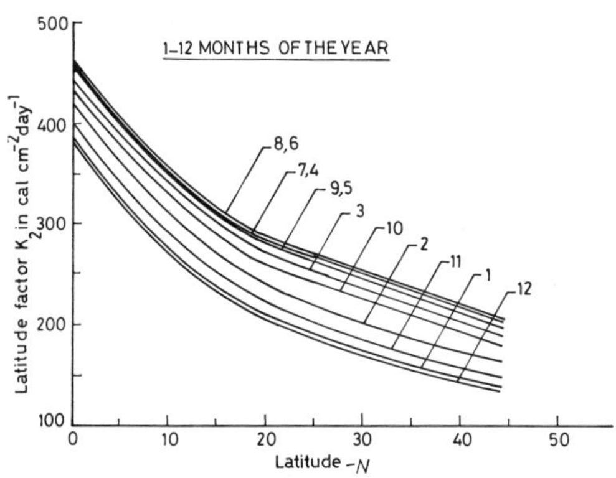

FIG. 4.9 The *K* factor for coastal stations.

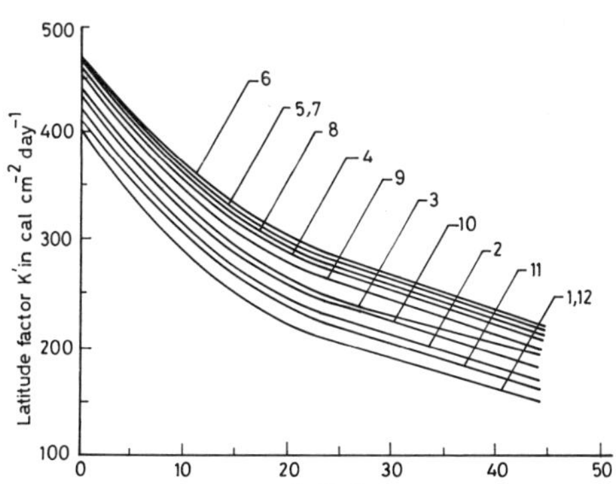

FIG. 4.10 The *K* factor for hill stations.

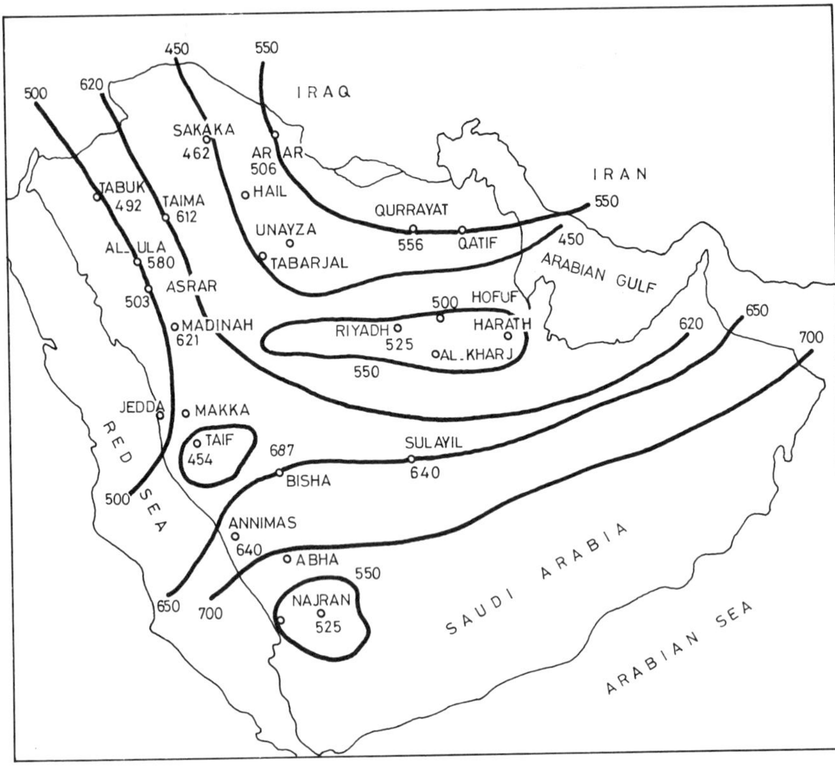

FIG. 4.11 Isoradiation map for the Kingdom of Saudi Arabia (summer months).

4.6 ESTIMATION OF DIRECT AND DIFFUSED RADIATION

A. During Days with No Clouds

When solar radiation enters the atmosphere some of it is absorbed, some is scattered, and the rest penetrates the atmosphere. The penetrated rays are called direct radiation while that part of the scattered rays that reach the earth are called diffused radiation. The term diffused radiation means the relatively short wavelength radiation coming from the sky and not from atmospheric thermal radiation which has much longer wavelengths than the scattered sky radiation. Both types of radiation are affected by the atmospheric ozone content, water vapor content, dust content, and the solar altitude, other radiation-depleting agents also affect them. Liu and Jordan found in 1960, that in dust-free localities and with minimum water vapor content in the atmosphere, the diffuse and the direct radiation varied with

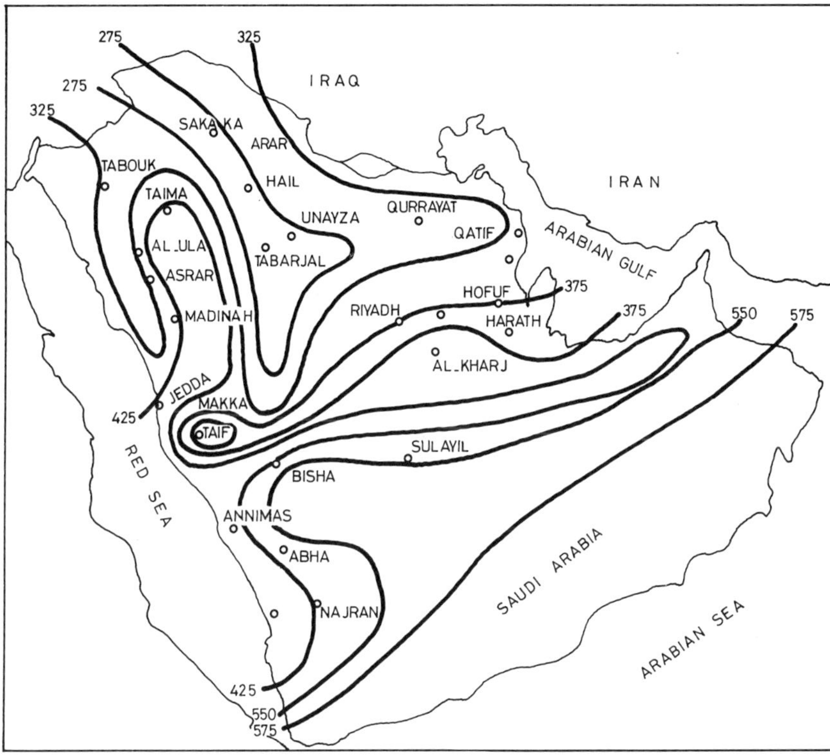

FIG. 4.12 Isoradiation map for the Kingdom of Saudi Arabia (winter months).

each other in a linear way and both of them are functions of the air mass. The same authors proposed the following formulas, for cloudless and dust-free localities:

$$\tau_d = 0.2710 - 0.2939\tau_D \tag{4.23}$$

and

$$\tau_d = 0.3840 - 0.4160\tau_T \tag{4.24}$$

where τ_d is the transmission coefficient for diffuse radiation on horizontal surface, and H_f/H_0 the ratio of diffuse radiation to extraterrestrial intensity of solar radiation. The extraterrestrial solar radiation used by Liu and Jordan was 0.1396 W/cm^2. Also, $\tau_D = H_d/H_0$ is the ratio of direct radiation to extraterrestrial intensity of solar radiation, and $\tau_T = H/H_0$ the ratio of solar radiation to extraterrestrial intensity of solar radiation.

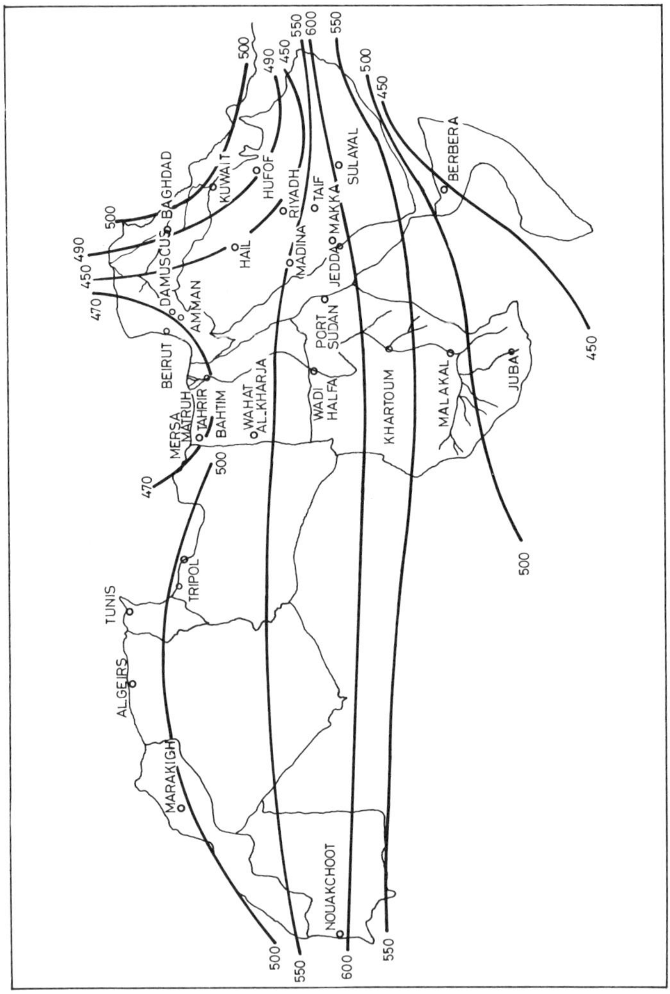

FIG. 4.13 Yearly average Isoradiation lines for the Arab countries.

B. During Cloudy Days

Liu and Jordan (1960) show that the extraterrestial radiation per day on a horizontal surface can be calculated by using the formula

$$H_0 = (24/\pi) rB(\cos\phi \cos\delta \sin\omega + \omega \sin\phi \sin\delta) \tag{4.25}$$

where $B = 442\ \mathrm{Btu/h/ft^2} = 2\ \mathrm{ly/min} = 0.1396\ \mathrm{W/cm^2}$, r is the intensity of normal solar radiation to the solar constant which is $0.1396\ \mathrm{W/cm^2}$, and varies from 0.9666 on the first of July to 1.0335 on the first of January. Also ω is the sunset hour angle in radians, ϕ the latitude in degrees, and δ the solar declination in degrees; the hour angle can be determined from the equation

$$\cos\omega = -\tan L \tan\delta \tag{4.26}$$

From this, it is claimed by the same authors that the diffuse solar radiation on a horizontal surface and during cloudy days can be deduced to an accuracy of $\pm 5\%$ from Fig. 4.14. Another expression worth mentioning here is the ratio between hourly to daily diffuse radiation

$$r_f = \frac{\pi}{24}\left(\frac{\cos h - \cos\omega}{\sin\omega - \omega\cos\omega}\right) \tag{4.27}$$

This equation was derived by Whillier (1956). For a standard tropical sky (15 mm precipitable water vapor, 300 dust particles/$\mathrm{cm^3}$ and 2.5 mm ozone at 760 mm atmospheric pressure), Numan (1975) proposed two formulas, one to measure total radiation on a horizontal surface, and the second to measure diffuse radiation on a horizontal surface in $\mathrm{W/m^2}$, i.e.,

$$H = 494 \sin\beta + 0.6187 H_d \tag{4.28}$$

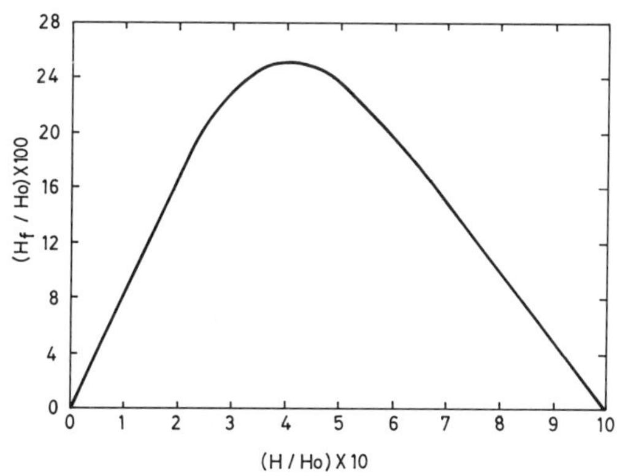

FIG. 4.14 The relationship between the daily total radiation and daily diffuse radiation on a horizontal surface.

and

$$H_f = 494 \sin\beta - 0.3813 H_d \tag{4.29}$$

where β is the solar altitude (degrees) and H_d the direct horizontal radiation. Numan (1975) also included several tables showing the ratio R of diffuse radiation on inclined surface to that on horizontal surface, at various sun altitude angles and solar azimuth angles.

Sharma and Pal (1965) proposed the following formula for the calculation of direct and diffuse solar radiation on horizontal surface in India:

$$H_d/H = [CN/(CN + K)][A \sin^2\beta/(\sin^2\beta + C)] \tag{4.30}$$

$$H_f = H - H_d \tag{4.31}$$

where

$$\begin{aligned} CN &= \text{clearance numbers} \\ &= 0.0 \text{ to } 0.5 \text{ for cloudy and overcast atmosphere} \\ &= 0.5 \text{ to } 0.7 \text{ for hazy atmosphere} \\ &= 0.7 \text{ to } 1.1 \text{ for clear atmosphere} \\ &= 1.1 \text{ to } 1.3 \text{ very clear atmosphere} \end{aligned}$$

K and A are constants which vary slightly with solar altitude as shown in Table 4.6 and

$$\sin\beta = \sin\lambda \sin\delta + \cos h \cos\delta \cos\lambda \tag{4.32}$$

where λ is the latitude of the place (degrees), δ the declination angle of the sun (degrees), h the hour angle of the sun (15° for every hour from noon), and β the solar altitude angle with the horizontal (degrees).

Stanhill (1966) used a linear relationship between the ratio of diffuse radiation to the total radiation (H_f/H) and the ratio of total radiation to the extraterrestrial radiation (H/H_0), for the semiarid land of Gilat. The relationship is

$$H_f/H = -1.755(H/H_0) + 1.539 \tag{4.33}$$

Stanhill (1966) showed that his results for diffuse radiation were in close agreement with those of New Delhi which were reported by Choudhary (1963).

Another equation suggested by Stanhill (1966) was

$$H_f/H = -0.786D' + 0.964 \tag{4.34}$$

where D' is the ratio of hours of sunshine per day to the astronomical length of the day.

TABLE 4.6

The Values of K and A Used in Eq. (4.30)[a]

Solar altitude (degree)	A	K
10	1.7500	0.882
15	1.8250	0.972
20	1.8885	1.075
25	1.9563	1.175
30	2.0290	1.245
35	2.0747	1.300
40	2.1159	1.335
45	2.1489	1.365
50	2.1703	1.390
55	2.1910	1.410
60	2.2074	1.428
65	2.2177	1.440
70	2.2279	1.453
75	2.2346	1.458
80	2.2423	1.463
85	2.2500	1.470
90	2.2542	1.475

[a] From Sharma and Pal (1965).

Another formula similar to (4.34) was derived by Sayigh for the town of Hofuf in the eastern province of Saudi Arabia:

$$H_f/H = -0.025D' + 0.30 \tag{4.35}$$

This gives an average annual value for H_f/H of 0.27.

4.7 CONCLUSION

Numerous authors have proposed various formulas to calculate total or global solar radiation. This interest dates back to the early twenties when Ångstrom (1924) suggested his formula. Nevertheless because of the numerous factors that affect solar radiation reaching the ground, there will continue to be more writers suggesting more formulas. A general formula that could be used at any season and at any location, with very high accuracy is very difficult to obtain. But several formulas with reasonable accuracy (± 5 to $\pm 10\%$) have been shown in this chapter. As for the extraterrestrial radiation again various authors have used different values, mainly between 2 and

1.8 ly/min (0.1395 and 0.1256 W/cm^2). The most up-to-date value is 0.1353 W/cm^2 which is mentioned in Chapter 3.

As for diffuse radiation to total radiation ratio, this certainly varies greatly from place to place. The variation is due to the variations in atmospheric conditions, dust and smoke, water vapor, air molecules, and other suspended matter. Liu and Jordan (1960) suggested the figure 0.12 as a minimum and 0.25 as a maximum, while Stanhil (1966) showed that this ratio varies from 0.3 to 0.36 with a mean value of 0.33. Drummond (1956) suggested the value of 0.30 in tropical locations. In some stations in Britain (Kew and Eskdalmuir) the annual ratio during cloudless days was found to be 0.24. Mani and Chacko (1963) and Choudhary (1963) have found in two different parts of India that this annual ratio for cloudless condition is 0.18. The author found this ratio to be about 0.27 for the town of Hofuf where a sizable amount of smoke and dust exist all year round.

5

Heat Transfer for Solar Energy Utilization

J. A. SABBAGH

MECHANICAL ENGINEERING DEPARTMENT
COLLEGE OF ENGINEERING
RIYADH UNIVERSITY
RIYADH, SAUDI ARABIA

5.1 INTRODUCTION

Progress in the utilization of solar energy depends, to a large extent, on progress in the field of heat transfer. A vast knowledge of the heat transfer process is required to build efficient solar energy collectors for the solar energy that reaches the earth's surface by radiation, as well as to build heat exchangers to transfer the collected energy to the energy conversion devices or to where it will be utilized. The problem facing solar energy utilization with regard to heat transfer is that in spite of the knowledge we have gained on the subject, new solutions utilizing more suitable and realistic boundary conditions for the solar energy problems are needed. The low temperature of solar energy applications necessitates large, costly solar heat exchangers that require more exact heat transfer analysis so that more efficient and compact heat exchangers can be built. This chapter contains a general survey of the available heat transfer knowledge that is relevant to solar energy utilization. No detailed solutions will be presented since this is beyond the scope and size of this chapter, but the sources for the detailed work will be indicated. Whenever possible, an idea of the difference between the presented solution and the actual problem will be given.

Heat transfer occurs mainly by three mechanisms. The first is by conduction through solid materials in the presence of a temperature difference. The second mechanism is radiation, in which energy moves in space by electromagnetic waves. In a moving fluid, the fluid molecules gain heat or lose it by conduction or radiation and carry it by their movement from one place to another. This process, the third mechanism, is called convection. The heat transfer may be accompanied by other physical phenomena such as heat generation within the medium, vapor condensation, liquid evaporation, etc.

5.2 CONDUCTION

The heat flow by conduction can be represented by the equation

$$q = -k_n \frac{\partial T}{\partial n} A \tag{5.1}$$

where q is the rate of heat transfer, k_n the thermal conductivity of the material in the direction n, and A the area normal to the direction of heat flow. The general three-dimensional equation of conduction can be derived by heat balance for an element (Gebhart, 1971), and the result in Cartesian tensor form is

$$\frac{\partial}{\partial x_i}\left(k_{ij}\frac{\partial T}{\partial x_j}\right) + q''' = \rho c \frac{\partial T}{\partial t} \tag{5.2}$$

where q''' is the rate of heat generation inside the conductor and ρ and c are its density and specific heat, respectively. The above equation reduces to

$$\nabla^2 T = 0 \tag{5.3}$$

for steady state, constant conductivity, and zero heat generation. The two basic conduction problems usually encountered in solar energy utilization are conduction through a flat plate and through the wall of a cylinder (Fig. 5.1). The solution of the above Laplace equation for an infinite flat plate is

$$T - T_1 = (x/L)(T_2 - T_1) \tag{5.4}$$

and the rate of heat transfer per unit area is

$$q'' = (k/L)(T_1 - T_2)$$

The solution of the Laplace equation for a long circular cylinder is

$$T = [1/\ln(r_2/r_1)][(T_2 - T_1)\ln r + T_1 \ln r_2 - T_2 \ln r_1] \tag{5.5}$$

The rate of heat transfer per unit area is

$$q'' = -k(T_2 - T_1)/r \ln(r_2/r_1) \tag{5.6}$$

For detailed treatment of the problems of conduction and the solutions for the general equation, see Carslaw and Jaeger (1959) and Jakob (1949a).

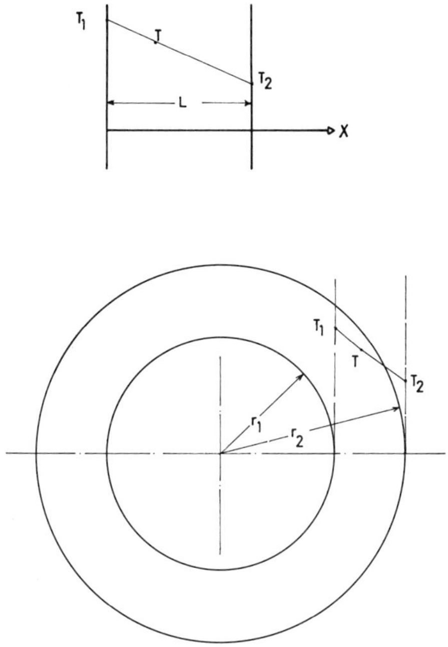

FIG. 5.1 Steady-state temperature distribution in flat plate and cylindrical walls.

Conduction in Extended Surfaces

The use of fins to increase the heat transfer surface is utilized in some solar collectors. The problem described here is the case of a longitudinal fin. If a flux q'' is applied and transferred to the tube through the thickness a of the fin (see Fig. 5.2), then the following relation can be written (Bliss, 1959; Kraus, 1964):

$$q''l\,dx = \frac{d}{dx}\left(-kal\,\frac{dT}{dx}\right)dx \tag{5.7}$$

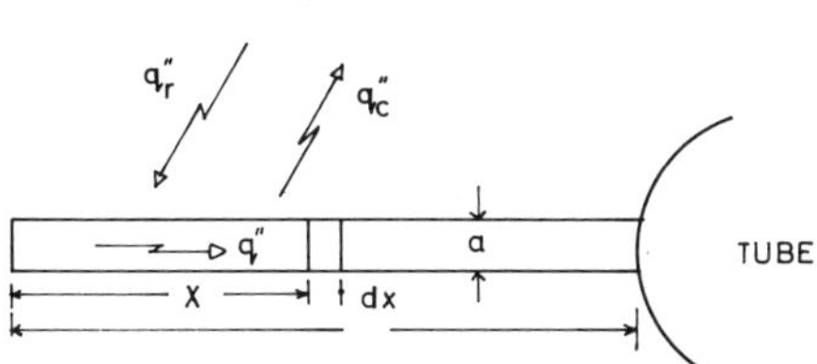

FIG. 5.2 Heat flow in a uniform rectangular fin.

In order to solve this differential equation, the boundary conditions can be taken as

$$T = T_0 = \text{const} \quad \text{at} \quad x = L, \qquad \text{and} \qquad dT/dx = 0 \quad \text{at} \quad x = 0$$

and the heat flux q'' can be assumed equal to $q_r'' - q_c''$ where q_r'' is the net radiation transferred by conduction through the fin and q_c'' is the heat transferred from the fin to the atmosphere by convection, which is proportional to the difference in temperature between the fin surface and the atmosphere. Therefore, the differential equation will be

$$q_r'' - h(T - T_a) = -ka\, d^2T/dx^2 \tag{5.8}$$

The solution for this equation is

$$T = c_1 e^{bx} + c_2 e^{-bx} + T_a + (q_r''/h) \tag{5.9}$$

where $b^2 = h/ka$. Application of the boundary condition gives

$$T = T_0 \frac{\cosh bx}{\cosh bL} + \left(T_a + \frac{q_r''}{h}\right)\left(1 - \frac{\cosh bx}{\cosh bL}\right) \tag{5.10}$$

and

$$q'' = (\tanh bL/bL)[q_r'' - h(T_0 - T_a)]$$

The value of $(\tanh bL)/bL$ can be taken as the efficiency factor for the fin. For a detailed study of the subject of extended surfaces, see Jakob (1949a) and Kraus (1964).

5.3 CONVECTION

Convection is a term used to describe the method of heat transfer when it involves the movement of the material, gas or liquid. However, the actual heat transfer process from or to the moving molecule of the material may be caused by conduction or radiation. The general equation which describes the energy exchange by convection for the case of an incompressible flow is

$$\rho c \frac{DT}{Dt} = k\, \nabla^2 T + \frac{DP}{Dt} + q''' + \mu\phi \tag{5.11}$$

where

$$\frac{D}{Dt} = \frac{\partial}{\partial t} + u\frac{\partial}{\partial x} + v\frac{\partial}{\partial y} + w\frac{\partial}{\partial z} \tag{5.12}$$

and q''' is the heat generation and ϕ the viscous dissipation. This equation must be solved taking into account the momentum equation

$$\rho \frac{DV}{Dt} = -\nabla P + g\rho\beta(T - T_\infty)i + \mu \nabla^2 v \tag{5.13}$$

and the continuity equation

$$\nabla \cdot V = 0 \tag{5.14}$$

The term $g\rho\beta(T - T_\infty)$ appears only if the effect of buoyancy is considered. If the flow is caused by buoyancy effect, then the convection is called natural convection to distinguish it from the forced convection where the motion is imposed on the fluid by external means. In many cases, natural and forced convection exist together and the convection is called mixed convection. In both forced and natural convection, the flow could be laminar or turbulent. Laminar flow has its streamlines run side by side and the transport of momentum and energy occurs by molecular diffusion. In turbulent flow, fluid particles fluctuate around the mean flow motion and the momentum, energy, and mass are diffused in the flow by the eddies which have different sizes and move in the stream. Solutions of the above nonlinear partial differential equations are extremely difficult even for the laminar flow case. In turbulent flow, theoretical analysis of only limited simple problems can be done, while empirical and semiempirical approaches are utilized for the rest of the problems.

The heat convection problems usually associated with solar energy utilization are internal, natural, and/or forced convection in pipes or between plates. External convection problems are convection from tubes or flat plates and convection to the atmosphere. The heat transfer process may be accompanied by evaporation and/or condensation as in the case of a solar still. In the following sections, solutions and discussions for these problems will be presented.

A. Forced Convection over a Flat Plate

In the flow over a surface at large Reynolds number, the effect of viscosity appears in the region close to the surface and vanishes at large distances from the surface. The region where the effect of viscosity appears is called the boundary layer. Similarly, a thermal boundary layer exists near the solid surface (Fig. 5.3) in which there is a large temperature gradient and heat exchange occurs between the solid surface and the flow. The momentum and energy equation can be approximated in the boundary layer. For a semi-infinite plate, the boundary layer equations are

$$u\frac{\partial u}{\partial x} + v\frac{\partial v}{\partial y} = \nu\frac{\partial^2 u}{\partial y^2} - \frac{1}{\rho}\frac{\partial P}{\partial y}, \quad u\frac{\partial T}{\partial x} + v\frac{\partial T}{\partial y} = \alpha\frac{\partial^2 T}{\partial y^2}, \quad \frac{\partial u}{\partial x} + \frac{\partial v}{\partial y} = 0 \tag{5.15}$$

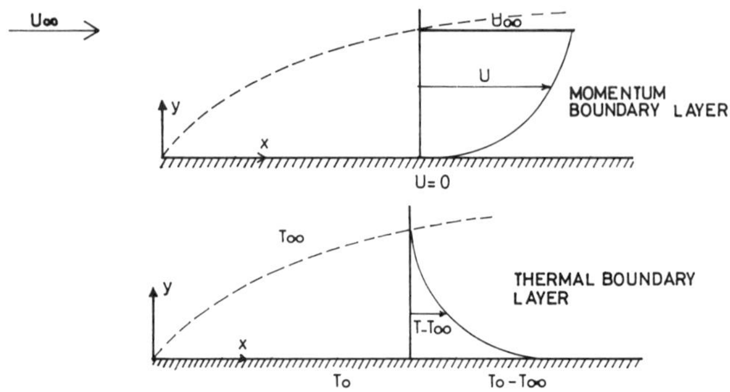

FIG. 5.3 Momentum and thermal boundary layer.

The solution of the boundary layer equations for an isothermal flat plate with zero-pressure gradient gives a local heat transfer coefficient (Schlichting and Kestin, 1968; Eckert, 1942; Pohlhausen and Angew, 1921)

$$h_x = 0.332 \frac{k}{x} \sqrt{\mathrm{Re}_x} \sqrt[3]{P_L} \tag{5.16}$$

or in dimensionless form

$$\mathrm{Nu}_x = 0.332 \sqrt{\mathrm{Re}_x} \sqrt[3]{\mathrm{Pr}} \tag{5.17}$$

The total heat transfer coefficient is

$$\mathrm{Nu} = 0.664 \sqrt{\mathrm{Re}} \sqrt[3]{\mathrm{Pr}} \tag{5.18}$$

For an inclined flat plate or wedge (see Fig. 5.4), the velocity outside the boundary layer can be written in the form cx^m where $m = \beta\pi/[2 - (\beta/\pi)]$. β is defined as an angle in Fig. 5.4 while β in Eq. 5.13 is the thermal coefficient of volumetric expansion.) The solution of the boundary layer equations gives a heat transfer coefficient (Eckert, 1942; Kays, 1966)

$$h_x = (\Psi k/\sqrt{v})\sqrt{x^{m-1}} \tag{5.19}$$

where

$$\Psi = \mathrm{Nu}_x/\sqrt{\mathrm{Re}_x} = \mathrm{const} \tag{5.20}$$

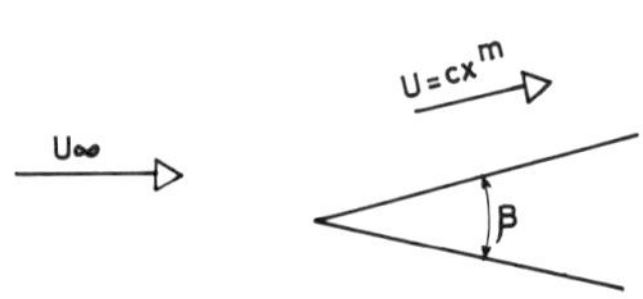

FIG. 5.4 Flow over a wedge.

for $\Pr = 0.7$, and $m = 0.333$, $\Psi = 0.384$; for $m = 1$, $\Psi = 0.496$; and for $m = 4$, $\Psi = 0.813$.

A semiempirical approach is used for turbulent boundary layers. From the definition of what is called a mixing length and the experimental data (Von Karman, 1921) the following velocity distribution for turbulent flow near a solid surface is suggested.

Very close to the surface the flow is laminar and the region is called a laminar sublayer	$u^+ = y^+$	$y^+ < 5$
After the laminar sublayer the flow is a mixture of laminar and turbulent	$u^+ = 5 \ln y^+ - 3.05$	$5 < y^+ < 30$
In the fully turbulent flow region	$u^+ = 2.5 \ln y^+ + 5.5$	$y^+ > 30$

This is called the universal velocity distribution. The variable u^+ is defined as $\bar{u}/u^*$, $y^+ = yu^*/\nu$, and the variable u^* is called the shear velocity and is equal to $\sqrt{\tau_0/\rho}$. The relationships between the momentum flux and the flow velocity gradient and between the thermal energy flux and the temperature gradient similar to the relationships for the laminar flow case were suggested by Gebhart (1971). Therefore,

$$\tau_t = K_m(\partial u/\partial y) \quad \text{and} \quad q_t''/\rho c = K_h(\partial T/\partial y) \tag{5.21}$$

where K_m and K_h are the eddy diffusivities. The total momentum and energy flux in the flow will be

$$\tau = \tau_m + \tau_t = (\nu + K_m)(\partial u/\partial y) \tag{5.22}$$

and

$$q''/\rho c = (1/\rho c)(q_m'' + q_t'') = (\alpha + K_h)(\partial u/\partial y) \tag{5.23}$$

where α and ν are the normal molecular diffusivities. From these relationships the heat transfer coefficient can be correlated with the shear stress and the fluid properties. The shear stress can be found empirically or semiempirically. Assuming $K_h = K_m$, Gebhart (1971) derived the following heat transfer coefficient:

$$\mathrm{Nu}_x = \frac{0.0288(\mathrm{Re}_x)^{0.8}\,\Pr}{1 + (0.85/10\sqrt{\mathrm{Re}_x})\{\Pr - 1 + \ln[1 + \frac{5}{6}(\Pr - 1)]\}}. \tag{5.24}$$

The above analysis is suitable for a smooth plate. A completely empirical relation for the heat transfer from a rough plate is

$$\mathrm{Nu} = 0.037(\mathrm{Re})^{0.8}\sqrt[3]{\mathrm{Pr}} \tag{5.25}$$

This relation is valid for $\mathrm{Pr} \geqslant 0.7$.

B. Forced Convection over a Circular Cylinder

The analysis used for the wedge can also be used for the laminar flow case and large Reynolds number to determine the heat transfer coefficient near the stagnation point (Kays, 1966). The equation is

$$\mathrm{Nu}_r = 0.81\sqrt{\mathrm{Re}_r}\,\mathrm{Pr}^{0.4} \tag{5.26}$$

where Nu_r, Re_r are the Nusselt and Reynolds numbers with r as the length. The total heat transfer coefficient can be determined from the following relationship, which is given by Hilpert (1933):

$$\mathrm{Nu} = 1.12B\sqrt[3]{\mathrm{Pr}}\,(\mathrm{Re})^n \tag{5.27}$$

where Re is the Reynolds number and is equal to DU_∞/ν, and B and n are constants to be selected depending on the Reynolds number. For Re = 4000 to 40,000, the values of n and B are 0.618 and 0.174, respectively; and for Re = 40,000 to 250,000 the values of n and B are 0.805 and 0.0239, respectively.

C. Natural Convection over a Flat Plate and a Circular Cylinder

The boundary layer equations with the buoyancy force term can be solved for a vertical isothermal plate and laminar flow (Ostrach, 1953) to give the velocity and temperature profile and the following relationship for the heat transfer coefficient:

$$\mathrm{Nu} = \tfrac{4}{3}F(\mathrm{Pr})\sqrt[4]{\mathrm{Gr}} \tag{5.28}$$

where

$$\mathrm{Gr} = g\beta L^3(T_0 - T_\infty)/\nu^2 \tag{5.29}$$

(ν is kinematic viscosity) and is called the Grashof number and F is a function of the Prandtl number; for $\mathrm{Pr} = 5$, $F = 0.675$; and for $\mathrm{Pr} = 1$, $F = 0.401$.

Other formulas are used for $\mathrm{Pr} \simeq 1$ to calculate the heat transfer from a vertical plate or cylinder with large diameter:

$$\begin{aligned} \mathrm{Nu} &= 0.59\sqrt[4]{\mathrm{Ra}}, \qquad 10^4 < \mathrm{Ra} < 10^8 \\ \mathrm{Nu} &= 0.13\sqrt[3]{\mathrm{Ra}}, \qquad 10^8 < \mathrm{Ra} \end{aligned} \tag{5.30}$$

Gebhart (1971) and Fishenden and Sounders (1950) suggested using for horizontal and inclined plates (heated surface facing upward or cooled surface facing downward),

$$\begin{aligned} \mathrm{Nu} &= 0.54 \sqrt[4]{\mathrm{Ra}}, \qquad 10^5 < \mathrm{Ra} < 10^7 \\ \mathrm{Nu} &= 0.14 \sqrt[4]{\mathrm{Ra}}, \qquad 10^7 < \mathrm{Ra} < 3 \times 10^{10} \end{aligned} \tag{5.31}$$

McAdams (1954) suggested for a heated surface facing downward or a cooled surface facing upward:

$$\mathrm{Nu} = 0.27 \sqrt[4]{\mathrm{Ra}}, \qquad 3 \times 10^5 < \mathrm{Ra} < 3 \times 10^{10} \tag{5.32}$$

D. Forced Convection through a Tube

1. Laminar Flow

The energy equation for laminar steady flow with fully developed velocity profiles can be reduced to

$$\frac{\partial^2 T}{\partial y^2} + \frac{\partial^2 T}{\partial x^2} = \frac{u}{\alpha}\frac{\partial T}{\partial x} \tag{5.33}$$

where $\partial T/\partial x$ is a transport term, and for negligible axial heat transfer, the equation becomes

$$\frac{\partial^2 T}{\partial y^2} = \frac{u}{\alpha}\frac{\partial T}{\partial x} \tag{5.34}$$

which, in polar coordinates, is

$$\frac{1}{r}\frac{\partial}{\partial r}\left(r\frac{\partial T}{\partial r}\right) = \frac{u}{\alpha}\frac{\partial T}{\partial x} \tag{5.35}$$

The following relation can be written for a fully developed temperature profile:

$$\frac{\partial}{\partial x}\left(\frac{T_0 - T}{T_0 - T_m}\right) = 0 \tag{5.36}$$

where T_m is the average temperature and can be defined from

$$T_m A u_m \rho c = \int_A u\rho c T \, dA \tag{5.37}$$

and A is the cross-sectional area of the flow and u_m is the average velocity. The constant heat rate per unit length on the surface of the tube is

$$\frac{\partial q''}{\partial x} = \frac{\partial}{\partial x}[h(T_0 - T_m)] = 0, \qquad \frac{\partial T_0}{\partial x} - \frac{dT_m}{dx} = 0 \tag{5.38}$$

for constant h. Therefore, Eq. (5.36) gives

$$\frac{dT}{dx} = \frac{dT_0}{dx} = \frac{dT_m}{dx} \tag{5.39}$$

and Eq. (5.35) becomes

$$\frac{1}{r}\frac{\partial}{\partial r}\left(r\frac{\partial T}{\partial r}\right) = \frac{u}{\alpha}\left(\frac{dT_m}{dx}\right)$$

The boundary conditions for this case are

$$\text{at} \quad r = r_0, \quad T = T_0$$

$$\text{and at} \quad r = 0, \quad \partial T/\partial r = 0$$

The solution of the above equation after substituting the value of u from the solution of the momentum equation is

$$T = T_0 - \left(\frac{2u_m}{\alpha}\right)\frac{dT_m}{dx}\left(\frac{3}{16}r_0^{\,2} + \frac{r^2}{16r_0^{\,2}} - \frac{r^2}{4}\right) \tag{5.40}$$

and

$$q_0'' = h(T_0 - T_m) = \frac{r_0 u_m \rho c}{2}\frac{dT_m}{dx} \tag{5.41}$$

which gives

$$h = (48/11)(k/D)$$

For the case of an isothermal tube, $\partial T_0/\partial x = 0$ and Eq. (5.36) gives

$$\frac{dT}{dx} = \frac{T_0 - T}{T_0 - T_m}\frac{dT_m}{dx} \tag{5.42}$$

and the differential equation will be

$$\frac{1}{r}\frac{\partial}{\partial r}\left(r\frac{\partial T}{\partial r}\right) = \frac{u}{\alpha}\frac{T_0 - T}{T_0 - T_m}\frac{dT_m}{dx} \tag{5.43}$$

The solution of this equation gives

$$\mathrm{Nu} = hD/k = 3.658 \tag{5.44}$$

The complete derivation and solution of this equation and the one for constant heat transfer is given by Kays (1966), who also gives the Nusselt number for the case of two parallel plates where one is insulated and the other has a constant heat flux, as

$$\mathrm{Nu} = 2hb/k = 5.385 \tag{5.45}$$

where b is the distance between the two plates. The Nusselt number when one plate is insulated and the other kept at a constant temperature is Nu = 4.86. The same reference gives a very good description of the solution of the problem of an underdeveloped temperature profile and the cases of variable surface temperature or variable surface heat flux. For more information about the solution to the problem of forced laminar flow and heat convection in a tube, see Jakob (1949a), Kays (1955), Sellars *et al.* (1956), and Lundberg *et al.* (1963).

2. *Turbulent Flow*

The same semianalytical approach described in the section on forced convection over a flat plate can be used for turbulent flow. Gebhart (1971) used this analysis and gave a formula for the heat transfer coefficient for a smooth pipe in the form

$$\mathrm{Nu} = \frac{0.0396(\mathrm{Re})^{3/4}\,\mathrm{Pr}}{1 + (0.991/\sqrt[8]{\mathrm{Ra}})\{\mathrm{Pr} - 1 + \ln[1 + \frac{5}{6}(\mathrm{Pr} - 1)]\}} \tag{5.46}$$

Gebhart used a friction factor of the form

$$f = 0.3164/\sqrt[4]{\mathrm{Re}} \tag{5.47}$$

Kays (1966) suggests the following formulas:

$$\begin{aligned} 0.5 < \mathrm{Pr} < 1, &\qquad \mathrm{Nu} = 0.022\mathrm{Pr}^{0.6}\,\mathrm{Re}^{0.8} \\ 0.1 < \mathrm{Pr} < 20, &\qquad \mathrm{Nu} = 0.0155\mathrm{Pr}^{0.5}\,\mathrm{Re}^{0.83} \end{aligned} \tag{5.48}$$

and Gebhart suggests a relation of the form

$$\mathrm{Nu}_m = 0.023\mathrm{Re}_m^{0.8}\,\mathrm{Pr}_m{}^n(\mu_m/\mu_s)^{0.14} \tag{5.49}$$

where n is 0.4 for heating and 0.3 for cooling, and m and s are mean and surface temperatures.

E. Natural and Mixed Convection Inside a Tube and between Two Parallel Plates

For fluid between two parallel plates, if the two plates are horizontal and the upper plate temperature is higher than that of the lower plate or if the lower plate is heated and the Grashof number based on gap b is less than 1700, the heat transfer will be essentially by conduction. The heat transfer will also be by conduction for two vertical plates with Grashof numbers based on b less than 2000. For higher Grashof numbers, heat transfer will be by convection and Globe and Dropkin (1959) suggest the following formulas for Prandtl number ranges from 0.02 to 8750 and Rayleigh numbers from

3×10^5 to 7×10^9:

$$\mathrm{Nu} = 0.069 \sqrt[3]{\mathrm{Gr}}\,\mathrm{Pr}^{0.407} \tag{5.50}$$

Jackson *et al.* (1958) derived the following equation for a vertical pipe:

$$\mathrm{Nu} = 1.128\left\{\mathrm{Re}\,\mathrm{Pr}\frac{D}{L_x} + \left[3.02\left(\mathrm{Gr}\,\mathrm{Pr}\frac{D}{L}\right)\mathrm{Pr}\right]^{0.4}\right\}^{1/2} \tag{5.51}$$

where Re Pr D/L_x is evaluated for the length of the heated section and the other term is for the length of the total heated section. Eckert *et al.* (1955) also studied forced, mixed, and natural convection inside pipes for the case of turbulent flow. He suggested that the limit between free and mixed flow can be determined by the following equation:

$$\mathrm{Re}_x = 8.25(\mathrm{Gr}_x\,\mathrm{Pr})^{0.4} \tag{5.52}$$

and the limit between forced and mixed flow can be determined by

$$\mathrm{Re}_x = 15(\mathrm{Gr}_x\,\mathrm{Pr})^{0.4} \tag{5.53}$$

The buoyancy effect for mixed flow in horizontal and inclined tubes produces a secondary motion normal to the axial motion. Morton (1959) used the perturbation method to solve the energy and momentum equations for the mixed laminar flow in a horizontal tube. He gives a Nusselt number in the form of

$$\begin{aligned}\mathrm{Nu} &= \frac{q''xD}{A(T_0 - T_m)} \\ &= 6[1 + 0.0586 - 0.0852\mathrm{Pr} + 0.2686\mathrm{Pr}^2]\left(\frac{\mathrm{Ra}\,\mathrm{Re}}{4608}\right)^2\end{aligned} \tag{5.54}$$

His result is accurate only for small Ra, where

$$\mathrm{Ra} = \beta g T a^4/k\nu, \qquad \mathrm{Re} = a^3\gamma \tag{5.55}$$

and it gives a 10% error for Ra Re = 3000. Iqbal and Stachiewicz (1966) and Chang and Hong (1972) studied the problem for inclined tubes. Iqbal used a perturbation method similar to Morton. His results are subject to the same limitations as Morton's work. Chang used a combination of the boundary vorticity method and the line iteration method, and obtained detailed numerical results to illustrate the effect of the tube inclination, Rayleigh, Reynolds, and Prandtl numbers on flow and heat transfer characteristics. Sabbagh *et al.* (1976) show experimental results for mixed flow in a tube; he shows the effect of the angle of inclination, Rayleigh and Reynolds numbers on the Nusselt number and the temperature and the velocity profile (see Figs. 5.5, 5.6).

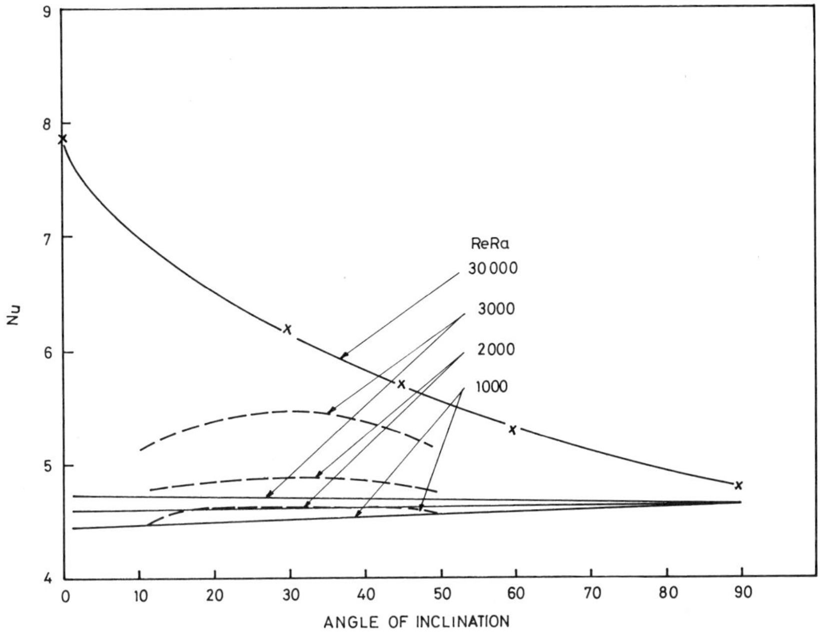

FIG. 5.5 Effect of angle of inclination on Nu.—×—×—from Sabbagh *et al.* (1976);— from Chang and Hong (1972); --- from Iqbal and Stachiewiez (1966).

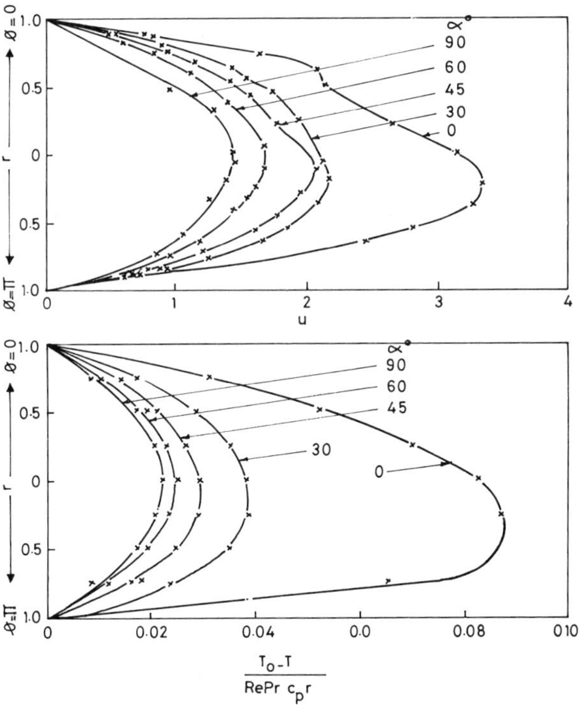

FIG. 5.6 Effect of inclination angle α on temperature and velocity profiles. Re = 746; Ra = 40.8; Pr = 0.72.

F. Convection Transfer to the Atmosphere

Convection transfer to the atmosphere occurs by wind motion and by natural convection. The atmospheric air flow is usually turbulent and the coefficient of heat transfer caused by this flow depends on the wind speed and the surface roughness. Natural convection occurs when the temperature of the surface is higher than the air temperature. This also causes unstable lapse conditions. A measurement of the instability caused by the lapse condition is the Monin–Obukhov length, defined as

$$L = -c_p u^{*2} T / K g q_0'' \tag{5.56}$$

where q_0'' is the surface convection heat flow (positive upward), K the Karman constant (equal to 0.4), and T the air temperature at any height z. Therefore $L < 0$ for the lapse condition and $L > 0$ for the inversion condition, where the temperature of the air is larger than the surface temperature and $L = \infty$ for the adiabatic condition. The flow is stable when $L > 0$ or $L = \infty$. The heat transfer rate q_0'' can be written as

$$q_0'' = c_p T \left(\frac{K_h}{K_m}\right) \frac{\partial T}{\partial z} \bigg/ \frac{\partial u}{\partial z} \tag{5.57}$$

Beckman *et al.* (1973) integrates Eq. (5.57) to a reference length z_r using a velocity profile for the neutral stable case of the form

$$\frac{u}{u^*} = \frac{1}{K} \ln\left(\frac{z}{z_0} + 1\right) \tag{5.58}$$

This velocity profile was given by Geiger (1965) and Sellers (1965). Beckman also used a value for the ratio $K_h/K_m = 1.6$, as suggested by Geiger and Sellers. The result of the integration gives

$$\mathrm{St}_n = \frac{q_0''}{\rho c_p u^*(T_0 - T_r)} = \frac{1.6K}{\ln[(z_r/z_0) + 1]} \tag{5.59}$$

Beckman *et al.* (1973) also suggested considering the effect of the surface roughness by adding its resistance. He suggested the use of a Stanton number for the sublayer derived by Owen and Thomson (1963) for fully developed flow of the form

$$\mathrm{St}_s = 1.6(u^* z_0/\nu)^{0.45} \tag{5.60}$$

Therefore, the total Stanton number is

$$\mathrm{St} = \frac{1}{(1/\mathrm{St}_n) + (1/\mathrm{St}_s)} \tag{5.61}$$

For the lapse period, Beckman *et al.* (1973) suggested the use of an empirical formula of the form

$$\mathrm{St}_l = \left(\frac{1}{1 + 0.1z_r/L}\right)\mathrm{St}_n \tag{5.62}$$

and therefore,

$$\mathrm{St} = \frac{1}{(1/\mathrm{St}_l) + (1/\mathrm{St}_s)} \tag{5.63}$$

Equation (5.57) was also integrated by using the lapse period velocity profile (Beckman *et al.*, 1973, and Swinbank, 1964).

$$u = \frac{u^*}{K}\ln\frac{e^{(z+z_0)/L} - 1}{e^{z_0/L} - 1} \tag{5.64}$$

and a ratio for the eddy diffusivity

$$K_h/K_m = 3 - 1.4e^{1.5z/L} \tag{5.65}$$

The Stanton number equals

$$K\int_0^{z_r/L} [3 - 1.4e^{1.5z/L}] - 1\left[\frac{e^{(z+z_0)/L}}{e^{(z+z_0)/L} - 1}\right] d\left(\frac{z}{L}\right) \tag{5.66}$$

The value of the surface roughness height z_0 is usually related to, but different from, the actual size of the surface roughness element. It is usually derived from the experimental velocity profile. Average values of z_0 are tabulated by Priestly (1959). For smooth mud flats, $z_0 = 3.3 \times 10^{-5}$ ft, and for deserts $z_0 = 1 \times 10^{-4}$ ft. The value of z_0 for grass depends on the wind velocity.

5.4. HEAT TRANSFER WITH VAPOR CONDENSATION

Vapor condenses in two forms. In the first kind of condensation the condensate forms drops. The heat transfer in this type of condensation is large because a large part of the heat transfer surface continues to be exposed to the vapor. Many investigations have studied dropwise condensation (Drew *et al.*, 1935; Griffiths and Lee, 1967; Westwater, 1968, and Tunner *et al.*, 1965). The second kind of condensation formed is a film. This kind of condensation is easy to analyze theoretically. A simplified analysis for inclined flat plates and cylinders was developed by Nusselt (Gebhart, 1971). The result of this analysis for a flat plate gives a value for the heat transfer coefficient:

$$\mathrm{Nu} = hL/K = 0.943[(\rho - \rho_v)gc_pL^3 \sin\phi/\nu K]^{1/4}[h_{Fg}/c_p(T_v - T_0)]^{1/4} \tag{5.67}$$

and for a horizontal cylinder, the value of the heat transfer coefficient is

$$h = 0.725[K^3\rho(\rho - \rho_v)h_{Fg}g/D\mu(T_v - T_0)]^{1/4} \tag{5.68}$$

Recently Sparrow and Gregg (1959) did a boundary layer analysis and reported their results:

$$\frac{\mathrm{Nu}_x}{[(\rho - \rho_v)c_p x^3 g h_{fg}/4\nu K c_p(T_v - T_0)]^{1/4}} = F\left[\mathrm{Pr}, c_p\frac{(T_v - T_0)}{h_{fg}}\right] \tag{5.69}$$

In a turbulent film condensation where the values of $\mathrm{Re} = 4m/\mu > 1800$, the following relation can be used for a vertical surface:

$$h = 0.0076[K^3\rho(\rho - \rho_v)g/\mu^2]^{1/3}\,\mathrm{Re}^{0.4} \tag{5.70}$$

and for a horizontal tube, the film becomes turbulent at $\mathrm{Re} > 3600$, where m is the mass rate of condensate flow at the lower edge of the surface per unit width.

5.5 LIQUID VAPORIZATION

When a warm solid surface submerged in a liquid has a free surface, the heat transfers from the solid surface to the liquid by natural convection and vaporization occurs at the free surface. Jakob (1949a) suggested that the natural convection equation may be used to estimate the rate of heat transfer for $\mathrm{Gr}\,\mathrm{Pr} < 10^9$. However, for a higher solid surface temperature than in the previous case, small vapor bubbles are formed on the hot surface. The bubbles grow and rise to the liquid-free surface by buoyant forces. The heat transfer in this case is different from the previous case. Jakob (1949a) suggested using the following empirical relation for the heat flux $q'' = c(\Delta T)^n$, where $n = 4$ and $c = 0.168$ for a horizontal surface submerged in water under atmospheric pressure. The temperature difference is in degrees Fahrenheit and q'' is in Btu/ft^2 h.

5.6 THERMAL RADIATION

Radiation energy is transferred through space in the form of photons which travel with the speed of light. These photons show electromagnetic wave characteristics and can be treated as waves. Thermal radiation is emitted characteristics and can be treated as waves. Thermal radiation is emitted from the body molecules or atoms as a result of their internal motion. The wavelength of the emitted radiation usually covers a large range with a very small band in the visible region.

A. Radiation Exchange between Surfaces

There are too many variables involved in the calculation of the energy exchange between surfaces, and the problem is extremely difficult to solve if all the variables are taken into account. Simple but inaccurate methods may be sufficient for many engineering applications. In the following, an approximate method will be described and assumptions will be made. Radiation is diffuse and follows the Lambert law—that is, the intensity of radiation over the surface of a hemisphere erected over a small radiating flat surface in its center varies with the cosine of the angle between the normal on the radiating surface and the line joining the center with the point at the hemisphere. All surfaces are considered black or gray and the values of the emissivity ε, the absorption α, and the reflectivity ρ are uniform over every surface. The value of ε is always equal to α and constant. The medium between the surfaces is transparent and does not radiate. Consider two elements dA_1 and dA_2 in the two surfaces A_1 and A_2 such that dA_1 emits radiation and dA_2 receives it. From the Lambert law the radiation intensity at any point on the hemisphere over the element dA_1 is

$$H = H_1 \cos\theta$$

where H_1 is the intensity at the intersection of the normal on dA_1 and the hemisphere surface. The energy received from the element dA_1 by the element $2\pi r^2\, d\theta$ on the surface of the hemisphere (see Fig. 5.7) is equal to

$$dq_{dA_1} = H_1 \cos\theta(2\pi r^2\, d\theta \sin\theta)$$

and therefore the energy received by the hemisphere is equal to

$$q_{dA_1} = \int_0^{\pi/2} H_1 2\pi r^2 \sin\theta \cos\theta\, d\theta = H_1 \pi r^2$$

This is equal to the energy emitted from the element dA_1, which is equal to $q_1''\, dA_1$. Therefore

$$H_1 = q_1''/\pi r^2 \qquad \text{and} \qquad H = (q_1''/\pi r^2)\cos\theta$$

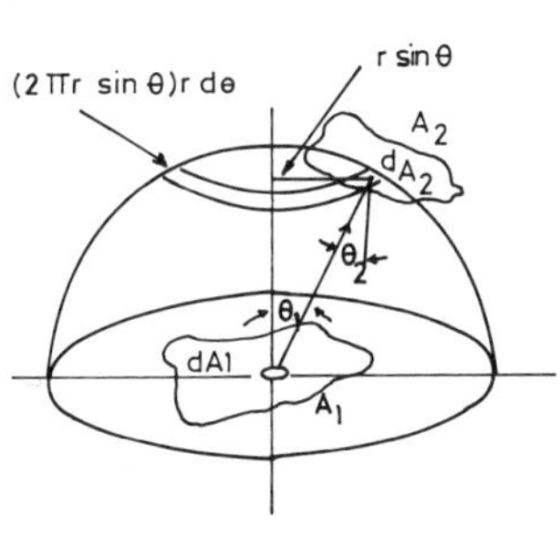

FIG. 5.7 Radiation from surface A_1 to A_2.

The energy received by the area projection of the element dA_2 is equal to

$$(q_1''/\pi r^2)\cos\theta_1\,(\cos\theta_2\,dA_2)$$

and therefore the total energy received by the area A_2 from A_1 is

$$\int_{A_1}\int_{A_2}(q_1''/\pi r^2)\cos\theta_1\cos\theta_2\,dA_2\,dA_1$$

If the energy received by A_2 from A_1 is equal to $q''_{1-2}F_{12}A_1$, then F_{12} will be equal to

$$(1/A)\int_{A_1}\int_{A_2}(1/\pi r^2)\cos\theta_1\cos\theta_2\,dA_2\,dA_1$$

The factor F_{12} is called the angle factor. Similarly, the energy emitted from A_2 and received by A_1 is equal to $q''_{2-1}F_{21}A_2$ and therefore it is clear that $F_{12}A_1 = F_{21}A_2$. For surfaces $A_1, A_2, \ldots, A_n$, the energy received by the surfaces A_1 to A_n from the surface A_1 is $q''_{12}A_1\sum_1^n F_{1n}$. The angle factor between all the surfaces from 1 to n can be represented by the following matrices:

$$F_{ij} = \begin{pmatrix} F_{11} & \cdots & F_{1n} \\ \vdots & & \vdots \\ F_{n1} & \cdots & F_{nn} \end{pmatrix}$$

The multireflection from each surface of an enclosure composed of a large number of gray surfaces will be treated as new emission by other surfaces, which tends to make the problem very complex. Gebhart (1971) suggested the use of the absorption factor B_{ij} which sums the energy absorbed by A_j from the emission of A_i. The absorption factor B_{ij} can be represented by matrix

$$\begin{pmatrix} B_{11} & \cdots & B_{1n} \\ \vdots & & \vdots \\ B_{n1} & \cdots & B_{nn} \end{pmatrix}$$

and for the enclosure $\sum_{j=1}^n B_{ij} = 1$. The net rate of energy loss from A_j is

$$q_j = q_j''A_j - \sum_i^n B_{ij}q_i''A_i$$

The total fraction of $q_1''A_1$ absorbed by A_j for surfaces A_1 to A_n with absorptivity equal to emissivity equal to ε_n and reflectivity equal to ρ_n is

$$B_{ij} = F_{ij}\varepsilon_j + F_{11}\rho_1 B_{1j} + F_{12}\rho_2 B_{2j} + \cdots + F_{1n}\rho_n B_{nj}$$

For each surface of the enclosure, a similar equation can be written and the solution of the n equations gives the value of B_{ij}. The total energy loss and energy absorbed by each surface can be calculated.

B. Thermal Radiation Exchange with the Atmosphere

Figure 5.8 shows the energy exchange for a surface exposed to the atmosphere. H_d is the direct solar radiation reaching the earth's surface, H_f the diffuse solar radiation that exists because of the scattering of part of the incoming solar radiation in the atmosphere, q_e'' the emitted radiation from the surface to the atmosphere, q_a'' the radiation emitted by the atmosphere, and q_r'' the direct, diffuse, and atmospheric radiation reflected by the surface. Chapter 4 deals with the estimation of various solar radiation intensities.

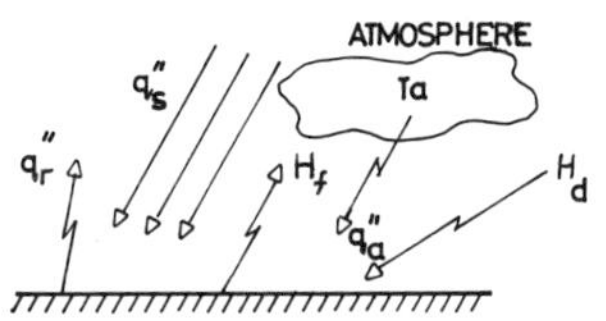

FIG. 5.8 Thermal radiation exchange with the atmosphere.

5.7 SOME HEAT TRANSFER PROBLEMS RELATED TO HEAT EXCHANGERS

A. Heat Transfer through an Insulated Wall or Pipe

For the wall shown in Fig. 5.9, the heat transfer from fluid one to the wall surface is equal to

$$q_{F1} = A_1 h_F (T_F - T_1)$$

and the heat transfer from surface one to the interface of the wall insulator is equal to

$$q_{12} = A_2 (k_w/b_w)(T_1 - T_2)$$

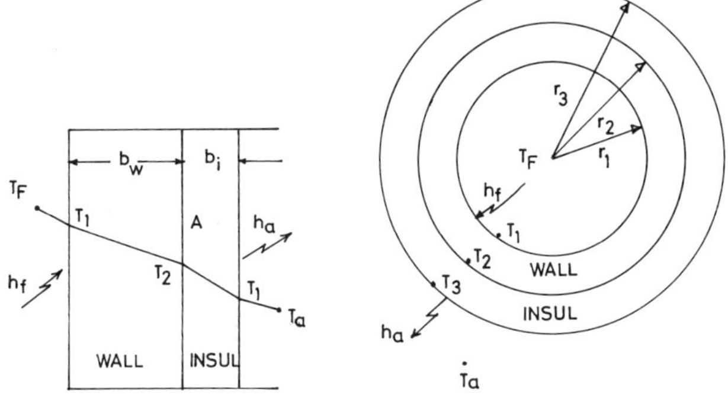

FIG. 5.9 Heat transfer from insulated wall and pipe.

The heat transfer from the wall-insulation interface to the second surface of the insulator is

$$q_{23} = A_3(k_i/b_i)(T_2 - T_3)$$

and the heat transfer from the insulation surface to the ambient is

$$q_{3-a} = A_3 h_a(T_3 - T_a)$$

where $A_1 = A_2 = A_3 = A$, the area of the wall, and h_F and h_a are the convective heat transfer coefficients from the fluid to the wall and from the insulation to the ambient, respectively. For steady heat transfer the value is

$$q_{F1} = q_{12} = q_{23} = q_{3a} = q = AU(T_F - T_a)$$

where U is the total heat transfer coefficient. The temperature difference is

$$T_F - T_a = (T_F - T_1) + (T_1 - T_2) + (T_2 - T_3) + (T_3 - T_a)$$

$$= \frac{q}{AU} = \frac{q}{Ah_F} + \frac{q}{A(k_w/b_w)} + \frac{q}{A(k_i/b_i)} + \frac{q}{Ah_a}$$

and from the above equation,

$$U = \frac{1}{(1/h_F) + [1/(k_w/b_w)] + [1/(k_i/b_i)] + (1/h_a)}$$

The heat transfer in a circular pipe from the fluid to the inside surface is equal to

$$q = 2\pi r_1 L h_F(T_F - T_1)$$

and through the solid wall

$$q = \frac{2\pi r_1 L(T_1 - T_2)}{\ln(r_2/r_1)}$$

Similar to the previous case:

$$(T_F - T_a) = \frac{q}{LU_0} = \frac{2\pi}{(1/h_F r_1) + [\ln(r_2/r_1)/k_w] + [\ln(r_3/r_2)/k_i] + (1/h_a r_3)}$$

where U_0 is the total heat transfer coefficient. The total heat transfer coefficient is defined in this case in terms of unit length rather than unit area.

B. Mean Temperature Difference in Heat Exchanger

As the fluids travel down the tube, the temperature of fluid a usually increases and the temperature of fluid b decreases. A mean temperature difference

$$\Delta T = \tfrac{1}{2}[(T_{a1} - T_{b1}) + (T_{a2} - T_{b2})]$$

can be taken to calculate the total rate of heat transfer with constant average heat transfer coefficient. A logarithmic mean temperature difference for large temperature changes expressed as

$$\Delta T = \frac{(T_{a2} - T_{b2}) - (T_{a1} - T_{b1})}{\ln[(T_{a2} - T_{b2})/(T_{a1} - T_{b1})]}$$

can be used to calculate the rate of heat transfer.

C. Friction in the Heat Exchangers

The heat transfer coefficient by convection depends on the flow velocity as in the previous section. The increase in the flow velocity causes an increase in the pressure loss by friction. This will tend to increase the power required for pumping the flow through the heat exchanger. In the case of natural circulations, the friction loss will be balanced by the buoyant forces that cause the circulations. The friction loss is caused by the viscous forces and is proportional to $\nu\, du/dy$ in the laminar flow case and to $(\nu + K_m)\, du/dy$ in the turbulent flow case as mentioned earlier. The value of the friction loss can be taken as $fL\rho u_m{}^2/2gd$, where L is the length of the circulation, d the diameter of the pipe, and for noncircular pipe, d can be taken as the hydraulic diameter that is equal to 4 × flow area/perimeter. The variable f is a friction factor that can be calculated from the analytical and empirical formulas shown in Table 5.1.

TABLE 5.1

Friction Factor f Calculated from the Analytical and Empirical Formulas

Type of flow	Formula[a]	Note
Laminar-developed flow between two flat plates	$f = 24/\text{Re}$	$\text{Re} = u_m b/\nu < 2000$
Laminar-developed flow in a circular pipe	$f = 64/\text{Re}$	$\text{Re} < 2000$
For transition in pipe	$1/\sqrt{F} = -0.87 \ln[(z_0/3.7d) + (2.51/\text{Re}\sqrt{F})]$	Colebrook (1939)
For completely turbulent flow in pipe	$1/\sqrt{F} = -0.87 \ln[(z_0/d) + 1.14]$	Colebrook (1939)

[a] z_0 is the roughness of the surface in feet. Moody (1944) draws the relationships between F, z_0/d, and Re in a graph which is available in most standard fluid mechanics textbooks.

6

Solar Energy Utilization—Liquid Flat Plate Collectors

W. W. S. CHARTERS

MECHANICAL ENGINEERING DEPARTMENT
UNIVERSITY OF MELBOURNE
PARKVILLE, VICTORIA, AUSTRALIA

6.1 INTRODUCTION

Although flat plate collectors have been in use for over thirty years now, it is only in recent times that efforts have been made to study their operating characteristics in any detail. Essentially all collectors consist of several basic elements as detailed in Fig. 6.1.

(1) A flat plate upon which the short wave solar radiation falls and is absorbed and transformed into thermal energy.

FIG. 6.1 A photograph of cross sections of flat plate collectors, Melbourne University.

(2) An insulated container to eliminate or substantially reduce thermal losses from the collection system.

(3) One or more plates of glass or transparent plastic to reduce the upwards heat loss from the hot collector plate; Figs. 6.2 and 6.3 show the solar properties of various glass covers.

(4) Tubes or channels as illustrated on the photograph in Fig. 6.4 to circulate the liquid required to remove the thermal energy from the plate to the storage tank.

These collectors are potentially useful in supplying low-grade thermal energy at temperatures generally less than 90°C and may be used in systems for the supply of domestic or industrial hot water or for space heating and/or cooling applications. All the models used to predict the thermal performance of such collectors stem from the original work of Hottel and Woertz (1942),

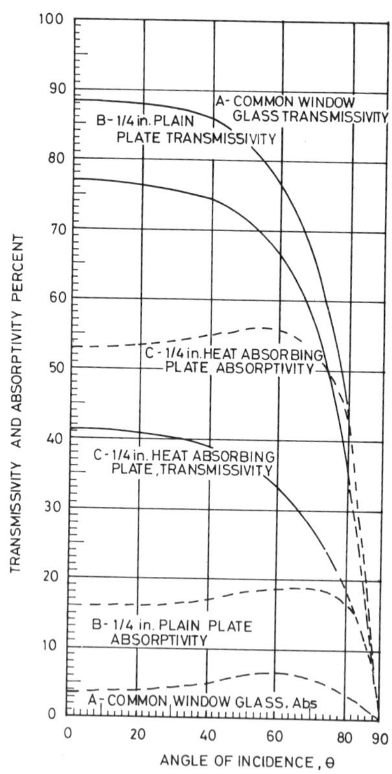

FIG. 6.2 Transmissivity and absorptivity versus angle of incidence for three types of glass (from Yellott, 1961).

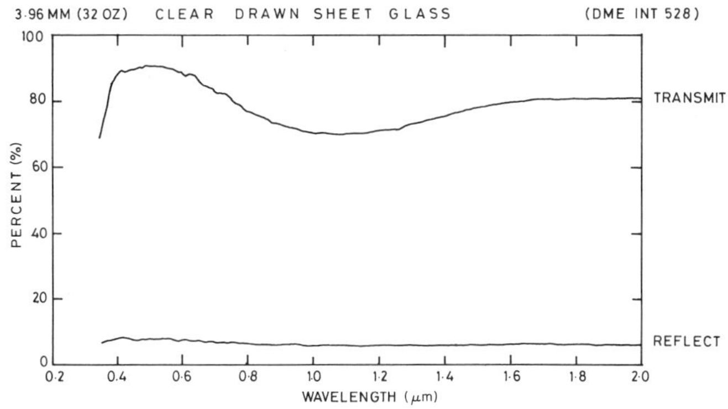

FIG. 6.3 Spectral normal transmissivity and reflectivity versus wavelength for locally produced glass (from CSIRO Mechanical Engineering Division, Highett, Victoria, Australia).

FIG. 6.4 A photograph of different collector plates, Melbourne University.

Hottel and Whillier (1958), and Bliss (1959). Since that time significant improvements have been made on these original theoretical models by authors such as Tabor (1958a) and Klein (1973). It is remarkable that at this stage little experimental work is available to allow validation of the proposed theoretical models. One of the few experimental investigations available was recently completed by Watmuff (1975).

Although a variety of fluids may be used to perform the heat transfer in the collector flow channels, virtually all practical systems to date have used water or water ethylene glycol solutions, the latter additive being used primarily to alleviate problems due to the freezing of liquid in the tubes under subzero atmospheric conditions.

6.2 HEAT TRANSFER PROCESSES

The apparent simplicity of the flat plate device is unreal when one considers the practical operation of a solar collector under the range of input conditions encountered in transient operation. Major variations in solar insolation and wind velocity may occur in the course of a single short test run (Fig. 6.5). In general, a steady state model is assumed to be adequate for performance prediction. This involves the use of empirical correlations for all of the heat transfer mechanisms related to the energy collection system (see Fig. 6.6).

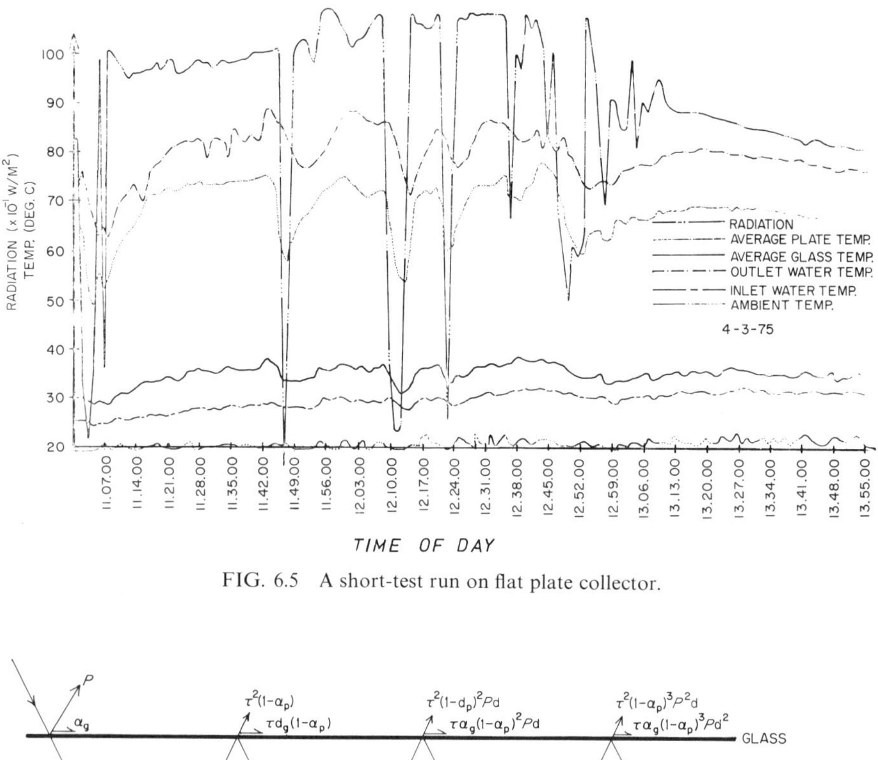

FIG. 6.5 A short-test run on flat plate collector.

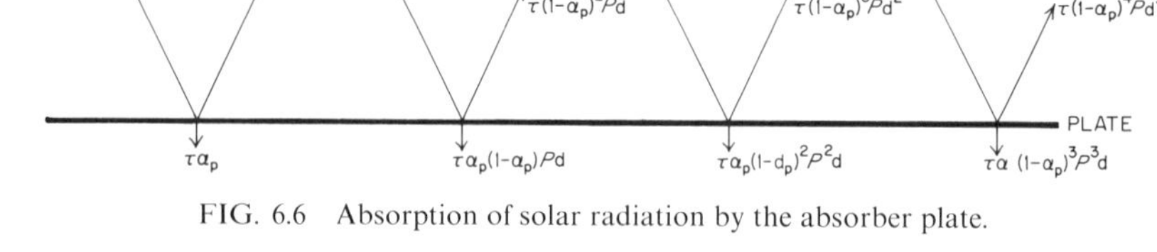

FIG. 6.6 Absorption of solar radiation by the absorber plate.

In general, it is easier to specify reasonable relationships to allow for radiative effects than it is to find corresponding empirical expressions for the convective components, but in the following section an attempt is made to cover both aspects. For the conduction loss through the sides and base of the insulated box, use of the one-dimensional Fourier conduction equation is quite adequate.

A. Radiative Effects

(i) To allow for the radiative interchange within the gap between the hot collector plate and the top transparent cover, the general assumptions made are that all surfaces are gray and diffuse reflectors and that the aspect ratio is sufficiently large to neglect edge effects. This leads to the equation

$$Q_{\text{rad}} = \frac{A_{\text{p,g}}(T_{\text{p}}^{4} - T_{\text{g}}^{4})}{(1/\varepsilon_{\text{pL}}) + (1/\varepsilon_{\text{gL}}) - 1}$$

(ii) The radiant interchange between the top cover and the sky may be formulated by assuming that the glass cover is a gray surface situated within a black enclosure at some specified sky temperature less than or equal to the ambient temperature. The deficit between ambient and sky temperature is discussed in more detail but is generally less than 12°C.

$$Q_{\text{rad}} = \sigma\varepsilon_{\text{gL}}A_{\text{g}}(T_{\text{g}}^{4} - T_{\text{sky}}^{4})$$

Several procedures have been proposed to estimate the effective sky temperature T_{sky}. Bliss (1961) suggests a procedure for estimating the intensity of atmospheric radiation upon horizontal surfaces, together with the results of calculations for typical atmospheric conditions. In his work, Bliss plotted the net radiation exchange R between an exposed horizontal thermally black surface of area 0.093 m^2 (1 ft^2) at ambient temperature for various relative humidities ϕ and dewpoint temperatures. From this plot it is possible to find an estimate of an effective sky temperature.

For example, if the ambient temperature is 37.8°C (100°F), the relative humidity is 10%, and the dewpoint temperature is 0°C (32°F), then R is found to be 104.115 W/m^2 (33 Btu/h ft^2).

Since $R = \sigma(T_{\text{a}}^{4} - T_{\text{sky}}^{4})$, then

$$T_{\text{sky}} = 21.28°\text{C} \quad (70.3°\text{F})$$

Values of T_{sky} have been calculated for various ambient temperatures, relative humidities, and dewpoint temperatures and the results are shown in Table 6.1.

The influence of clouds and the ground for a collector tilted so that it can see the ground are difficult to estimate, but it is likely that both the

TABLE 6.1

Values of T_{sky} for Various Ambient Temperatures, Relative Humidities, and Dewpoint Temperatures[a]

T_a (°C)	ϕ (%)	Dewpoint temperature (°C)	R (W/m^2)	T_{sky} (°C)	$T_a - T_{sky}$ (°C)
37.8	10	00.0	104.1	21.3	16.5
37.8	50	26.7	56.8	29.1	8.7
23.9	100	23.9	53.6	15.6	8.3
23.9	10	−9.4	104.1	4.7	19.2
10.0	100	10.0	56.8	−1.7	11.7
10.0	10	−17.8	101.0	−12.0	22.0

[a] From Watmuff (1975).

clouds and the ground will tend to increase the effective sky temperature over that for a clear sky. Values of effective sky temperature on clear days which have been obtained for February and March 1975, in Melbourne (Australia) indicated maximum values of $T_a - T_{sky}$ of approximately 12°C.

B. Convection Processes

The convection processes occurring within a flat plate collector are complicated by the inclined operating position. The three significant areas where heat transfer by convection occurs in a collector are

(1) the combined free and forced convection in the inclined riser tubes or channels of the collector plate,

(2) the free convection in the parallel gap between the inclined heated collector plate and the cooler cover glass, and

(3) the combined free and forced convection from the flat cover plate inclined to the horizontal and exposed to the surrounding atmosphere.

1. Forced Convection in an Inclined Tube

In general, the flow of the liquid through the riser tubes in solar collectors is so slow that it is laminar. In a typical collector, the tube length to diameter ratio ($L/D = 100/1$) is such that the entrance region is of considerable importance. In this developing flow region, the heat transfer coefficients in the entrance section are several times higher than those occurring for fully developed laminar flow.

It has been well established that the cross circulation promoted by buoyancy forces can have a marked effect on the heat transfer to fluid flowing in a tube, especially at low Reynolds numbers. This cross circulation when superimposed on the forced convection leads to a complex heat transfer situation for which many correlations have been developed.

The majority of cases reported deal with the situation where the tube wall is maintained at uniform temperature. Martinelli and Boelter (1942) suggested a possible means of accounting for the effects of free convection in vertical tubes. A review of earlier investigations of these cases (horizontal and vertical inclinations) has been made by McAdams (1954). A more recent study was made by Oliver (1962) in which he gives an excellent review of past work, and suggested the following correlations based on a study of previous data and his own experiments.

$$\overline{\mathrm{Nu}} = \left[\frac{\mu_\mathrm{b}}{\mu_\mathrm{w}}\right]^{0.14} 1.75[\overline{\mathrm{Gz}} + 0.0083(\overline{\mathrm{Gr}}\ \overline{\mathrm{Pr}})^{0.75}]^{1/3}$$

for tubes with L/d ratio greater than 70, and

$$\overline{\mathrm{Nu}} = \left[\frac{\mu_\mathrm{b}}{\mu_\mathrm{w}}\right]^{0.14} 1.75[\overline{\mathrm{Gz}} + 0.00056(\overline{\mathrm{Gr}}\ \overline{\mathrm{Pr}}\ L/d)^{0.7}]^{1/3}$$

for tubes with L/d less than 70, provided $\overline{\mathrm{Gz}}$ is not less than $\pi\,\overline{\mathrm{Nu}}$. Previously, analysts of solar collectors have used a variety of expressions to predict the film heat transfer coefficient. Whillier (1967) quoted the following expression without reference,

$$\overline{\mathrm{Nu}} = 4.36 + \frac{0.067(d/L)\mathrm{Re\ Pr}}{1 + 0.04[(d/L)\mathrm{Re\ Pr}]^{2/3}}$$

This expression is identical to the one proposed by Hausen (1943) except for the substitution of 4.36 in Hausen's relation. Note that the average Nusselt number approaches a constant value of 4.36 (or 3.66) when the tube is very long corresponding to those for fully developed laminar flow. Hottel and Whillier (1967) assumed a constant value of the heat transfer coefficient of 363.4 W/m^2 °C (64 Btu/h ft^2 °F), while Whillier and Saluja (1965) proposed a constant value of 227.12 W/m^2 °C (40 Btu/h ft^2 °F). Duffie and Beckman (1974) proposed the use of the relation quoted in Kreith (1968).

$$\overline{\mathrm{Nu}} = \frac{\mathrm{Gz}}{4} \ln\left[\frac{1}{1 - (2.654/\mathrm{Pr}^{0.167}\ \mathrm{Gz}^{0.5})}\right] \qquad \text{for} \quad L/d < 0.0048\ \mathrm{Re}$$

However, investigation reveals that this relation has been derived from Polhausen's solution for heat transfer with laminar flow over a flat plate at uniform temperature. Since the conditions in collectors approach that of uniform heat flux in tube flow, this expression may well prove unsuitable.

Kupper *et al.* (1969) performed a study of the combined free and forced convection with water as the working medium in a horizontal tube receiving uniform heat flux. Reynolds numbers ranged from 100 to 2000, Grashof numbers from 300 to 300,000, and Prandtl numbers from 4 to 9 and as such, the work was evidently attempting to model the conditions usually encountered in solar collectors. A tentative correlation was proposed for the fully developed region of the following form,

$$\overline{\mathrm{Nu}} = 4.36 + 0.048\ \mathrm{Pr}^{1/3}(\mathrm{Re\ Ra})^{1/5}$$

Baker (1967) reported results obtained from an experimental investigation into the laminar heat transfer characteristics of a tube-in-strip, similar to the type used in solar collectors. For such a horizontal tube, he reported that there could be a significant variation in the circumferential wall temperature distribution at any given section. He found that a higher heat transfer coefficient (approximately, 567.83 $\mathrm{W/m^2\ ^\circ C}$ or 100 $\mathrm{Btu/h\ ft^2\ ^\circ F}$) could be used over that generally recommended for solar collector design (40–60 $\mathrm{Btu/h\ ft^2\ ^\circ F}$). This study revealed that unsteadiness of the flow, indicated by temperature fluctuations, existed at Reynolds numbers well below the usual lower limit of transition to turbulent flow.

Iqbal (1966) presented both an experimental and a theoretical study of the effects of free convection superimposed on forced flow in inclined tubes. The theoretical analysis considered fully developed laminar flow under uniform heat flux and a constant pressure gradient. The solutions for the velocity and temperature equations were obtained by a perturbation analysis in terms of a power series of Rayleigh number. The experimental study was performed for 45° and 90° tube inclinations.

Cheng and Hong (1972) reported results from a theoretical study of the free convection effects of fully developed upward laminar flow in a uniformly heated inclined tube. The numerical results showed that no maximum value of Nusselt number exists for any inclination angle. The authors concluded that the determination of a suitable heat transfer coefficient in solar collectors based on theoretical and experimental data for horizontal tubes is unrealistic, and that the results of their numerical simulation for the limiting case of the thermally and hydrodynamically fully developed situation would provide a guide for the selection of a suitable film heat transfer coefficient for use in solar collectors.

The range of values used in the literature for similar situations in collector design indicates an incomplete knowledge in the area. An experimental investigation similar to the type performed by Baker but for inclined collector tubes should reveal more about this particularly complicated heat transfer problem.

2. Convection Heat Transfer between the Collector Plate and the Glass Cover

The problem of predicting the natural convection heat transfer in a fluid enclosed in a gap and heated from below is fundamental to the study of a wide range of engineering situations as well as to those occurring in solar collectors. Consequently, this particular problem has been subjected to many investigations. In 1900 Bénard reported experimental observations which indicated that a critical temperature gradient was exceeded before convective motion became established for a horizontal layer of fluid heated from below. He also noted that the initial motion had a steady cellular form. Rayleigh laid the foundations for a theoretical explanation of this cellular convection first described in the Bénard experiment. He found that a certain dimensionless grouping (later known as the Rayleigh number) fixed the temperature difference at which convection first occurs. A number of analytical treatments of this problem were made, but it was Mull and Reiher (1930) who were the first to make a thorough experimental investigation. Since then the major works of interest are those by

(a) Jakob (1946), who showed that the results of the experiments of Mull and Reiher could be simplified, improved, and generalized.

(b) De Graaf and Van der Held (1953) also questioned the use of the term $(h/L)^{-1/9}$ as proposed by Jakob where h is height of a collector gap. Their measurements of the heat transfer were carried out in an air layer with inclinations ranging from 0° to 90° (in 10° increments).

(c) Wilkes and Peterson (1937), who used an area 2.44 m (8 ft) by 0.814 m (2 ft 8 in.) and a gap width of 9.2 cm ($3\frac{5}{8}$ in.) to measure the heat transfer characteristics of air gaps at inclinations of horizontal, 45° to horizontal and vertical. Their work was aimed at engineers and architects wishing to find an estimate of heat losses through frame constructions. Their results are expressed in dimensional form and cover only a limited range of temperature differences. Their recommendations are for air and expressed in British units. A basic limitation of this work is the small range of Rayleigh numbers covered.

(d) Dropkin and Somerscales (1965) presented the results of an experimental investigation of convective heat transfer in liquids confined by two parallel plates and inclined at various angles to the horizontal. The experiments covered a range of Rayleigh numbers between 5×10^4 and 7×10^8 and Prandtl numbers between 0.02 and 11,560. The liquids used were water, silicone oils, and mercury.

(e) Eckert and Carlson (1961) studied the temperature field in an air layer enclosed between two isothermal vertical plates of different temperatures using a Mach–Zehnder interferometer. Local heat transfer coefficients were derived from the temperature gradients in the air, normal to the plate surfaces. An expression for the total heat transfer coefficient was derived.

(f) Hollands (1975) studied the natural convection of air between two parallel isothermal copper plates each of area 56 × 61 cm and spaced 1.27 cm apart, with the lower plate at a higher temperature than the upper plate. The apparatus was inserted into a vacuum (or pressure) vessel in which the air pressure can be varied from 10 Pa to 700 kPa. This provision of changing the air pressure allowed a wide range of Rayleigh numbers to be studied without altering plate spacing or temperature difference. Runs were made at inclinations to the horizontal of 0°, 15°, 30°, 45°, 60°, and 75°. Based on his extensive experimental program, Hollands recommends the relation,

$$\mathrm{Nu} = 1 + 1.44\left[1 - \frac{1708}{\mathrm{Ra}\cos\phi}\right]^{*}\left[1 - \frac{1708(\sin 1.8\phi)1.6}{\mathrm{Ra}\cos\phi}\right] + \left[\left(\frac{\mathrm{Ra}\cos\phi}{5830}\right)^{1/3} - 1\right]^{*}$$

for $10^3 < \mathrm{Ra} < 10^5$, where $[\Delta]^*$ indicates that if the argument inside the brackets is negative, the quantity is zero. It should be noted that for a solar collector, the Rayleigh number will often exceed 100,000.

(g) Tabor (1958) examined the results of a number of investigations. In addition to the work reviewed so far in this section, he also studied a report by the Housing and Home Finance Agency in Washington, Report No. 32, published in 1954. The results are reported in dimensional form for builders and architects. Tabor nondimensionalized the results to obtain more generalized forms.

For horizontal planes, and lower plate at higher temperature,

$$\mathrm{Nu} = 0.168\,\mathrm{Ra}^{0.281}, \qquad 10^4 < \mathrm{Gr} < 10^7$$

For planes inclined at 45°, with heat flow upwards,

$$\mathrm{Nu} = 0.102\,\mathrm{Ra}^{0.310}, \qquad 10^4 < \mathrm{Gr} < 10^7$$

For vertical planes,

$$\mathrm{Nu} = \begin{cases} 0.0685\,\mathrm{Ra}^{0.327}, & 1.5 \times 10^5 < \mathrm{Gr} < 10^7 \\ 0.0369\,\mathrm{Ra}^{0.381}, & 1.5 \times 10^4 < \mathrm{Gr} < 1.5 \times 10^5 \end{cases}$$

Tabor strongly recommends the use of the expressions derived from this Washington report when using large plates. This recommendation stems from the fact that some 146 tests were carried out with air gaps varying from 1.27 cm ($\frac{1}{2}$ in.) to 8.26 cm ($3\frac{1}{4}$ in.) (i.e., h/L from $18 \rightarrow 96$) and for both low and high emissivity surfaces.

(h) The exact solutions of the hydrodynamic equations governing the free convection in enclosed spaces have proved to be a difficult task. In recent years there have been a number of attempts to solve these equations using approximate numerical methods. A review of the approximations, the theory involved, and the methods used in placing the equations in a suitable form can be found in a thesis by Peterson (1971). Most researchers for simplicity have assumed two-dimensional flow. The upper Rayleigh numbers studied have been much lower than those occurring in a collector operating under typical conditions and as such, cannot really be directly applied. An extension of the numerical methods to analyze situations with higher Ra in two-dimensional inclined enclosures is currently being performed at Melbourne University and it is expected that the results will be published soon in thesis form (Koutsoheras, 1976).

3. *Combined Free and Forced Convection Heat Transfer from a Collector Cover Exposed to the Atmosphere*

The nature of combined free and forced convection heat transfer from a plate inclined to the horizontal and exposed to the atmosphere is perhaps the most difficult heat transfer problem in the analysis of collector performance.

Due to the inclination of the collector, the direction of the wind velocity with respect to the surface has a profound effect on the nature of the flow over the plate. The magnitude of the wind velocity may vary over a wide range of frequencies and amplitudes. Moreover, the eddies caused by flow around surrounding structures may introduce gross nonuniformities into what could normally be considered the free stream velocity for the plate.

(a) The exact solution of the simplified Navier–Stokes equations for laminar flow (i.e., the laminar boundary layer equations) of a constant property fluid over a flat plate, assuming negligible frictional heating, pressure gradients, and buoyancy forces, was found by Polhausen and Angew (1921). It can be shown that the theoretical mean Nusselt number of this situation is given by

$$\overline{\mathrm{Nu}} = 0.664\mathrm{Pr}^{1/3}\,\mathrm{Re}^{1/3}, \qquad 0.6 < \mathrm{Pr} < 10$$

However, this expression is not directly applicable to the case being considered because of the following reasons:

(1) The contribution of buoyancy forces is significant at low velocities and for high temperature differences between the plate and the free stream.

(2) The inclination of a collector introduces significant pressure gradients in the flow.

(3) Conditions will often be turbulent.

In the absence of wind velocity, the convective heat transfer reduces to natural convection. McAdams (1954) has reviewed the literature for the relations predicting free convection heat transfer from horizontal and vertical plates.

For angles of inclination between 0° and 40° from the vertical, Rich (1953) has shown that the Grashof number can be modified by the angle of inclination to give good agreement for free convection from inclined plates using horizontal data.

(b) *Empirical Expressions* Hottel and Woertz (1942), Yellott (1961), Whillier (1967), and Duffie and Beckman (1974) have used a linear relation first quoted in McAdams (1954) which was suggested by Jurges (1924)

$$h_w = 1.0 + 0.21V, \qquad V < 16$$

where h_w is the convection coefficient (Btu/h ft^2 °F) and V is the bulk velocity of the air (ft/s). Various forms of this expression have been quoted for different systems of units.

Jurges' data were obtained for the flow of air at room temperature and pressure parallel to a vertical copper plate 0.153 m^2 or 1.64 ft^2. It is not at all clear what experimental techniques were used and what allowance was made for the radiation component of the heat transfer. Hence, the expression is subject to considerable uncertainty.

Rowley *et al.* (1930) investigated the heat transfer from horizontal test surfaces, 30.48 cm^2 or 12 in.^{2}, for different air velocities parallel to the surface.

Results are in the form of plots of

(i) surface conductance versus air velocity, for each surface for a number of mean temperature differences, and

(ii) surface conductance versus mean temperature difference, for each surface for a number of air velocities.

Rowley and Eckley (1932) extended the work of Rowley *et al.* (1930) to look at the situation where the direction of the wind velocity was not parallel to the test surface. The authors reported that "for wind velocities up to 24 km/h (15 mph), the coefficients were substantially the same for all angles 15° to 90°, all being less than for parallel flow. Above a 15 mph velocity, the coefficients were reduced as the angle was increased." Hence, in the real situation, the average convection coefficient for a collector cover would probably be less than or equal to the value obtained by Rowley *et al.*

(1930) for flow parallel to the surface, provided that atmospheric turbulence was negligible and that wind speed was small enough so that turbulent flow did not develop on the cover.

The similarity of the values of surface conductance as obtained by Rowley *et al.* and the wind convection coefficient as used by Hottel and Woertz (1942), Yellott (1961), Whillier (1967), and Duffie and Beckman (1974) is cause for concern because it appears plausible that Jurges's data include a radiation component of the heat transfer to the surroundings. Another cause of doubt is that Jurges's expression implies that the convection coefficient is independent of the plate temperature or is valid only for a limited range of plate temperatures that have not been specified. The fact that analysts of collector models may have doubly accounted for radiation heat transfer from the collector cover is disturbing. This problem is one of the least understood of the heat transfer processes within flat plate collectors. More realistic empirical information is required to improve our knowledge in this area.

6.3 STEADY STATE MODELS OF COLLECTOR PERFORMANCE

The most commonly used models of the performance of flat plate collectors are those developed by Hottel and Woertz (1942), Hottel and Whillier (1958), and Bliss (1959). The major assumption that has been made in each of these works is that the thermal capacitance effects of the collector components are negligible.

A typical collector plate is shown in Fig. 6.7. The riser tubes are joined to two header tubes at each end of the plate. The effect of the headers on collector performance has been assumed to be negligible as they form a relatively small portion of the collector surface area and contain a disproportionally large volume of water.

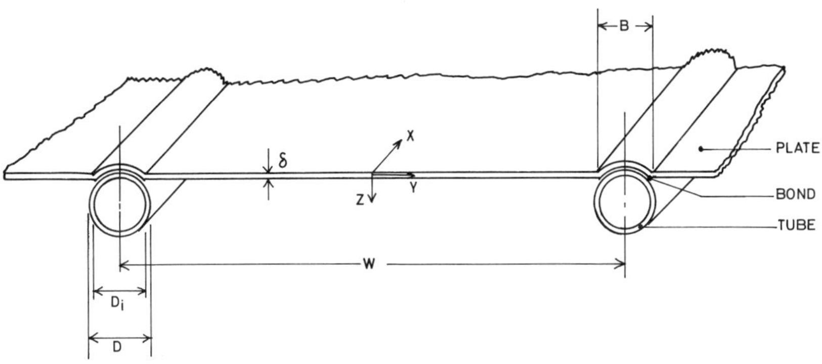

FIG. 6.7 A typical collector plate.

A. Temperature Distribution across the Plate

The steady state heat conduction in a solid having uniform physical properties with internal heat generation is described by the equation

$$(\partial^2 T/\partial x^2) + (\partial^2 T/\partial y^2) + (\partial^2 T/\partial z^2) + (\dot{q}/K) = 0 \qquad (6.1)$$

where $\dot{q}$ is the heat generated by internal sources per unit volume.

If it is assumed that for a thin collector plate the temperature in the z direction is uniform, then the net flow of heat to an element of plate surface of area dA will be evenly distributed throughout the depth, i.e.,

$$\dot{q}_{\text{net}} = \dot{q}\delta \qquad (\text{for } \delta = \text{plate thickness})$$

where $\dot{q}_{\text{net}}$ consists of the heat input to the plate from absorption of incident solar radiation less the heat losses due to conduction through the insulation at the underside of the plate and due to convection and radiation to the glass cover and the surrounding walls. So

$$(\partial^2 T/\partial x^2) + (\partial^2 T/\partial y^2) + (\dot{q}_{\text{net}}/K\delta) = 0 \qquad (6.2)$$

($\partial^2 T/\partial z^2 = 0$, since $T(z)$ is assumed constant). If, as has been generally assumed, $\partial^2 T/\partial x^2$ can be neglected, the equation will reduce to

$$(d^2 T/dy^2) + (\dot{q}_{\text{net}}/K\delta) = 0 \qquad (6.3)$$

Now,

$$\dot{q}_{\text{net}} = \dot{q}_{\text{in}} - \dot{q}_{\text{lost}} \qquad (6.4)$$

where $\dot{q}_{\text{in}} = f_1(H) = (\alpha\tau)_{\text{eff}} H$. Given a collector geometry for which the heat transfer relations and the material properties are known

$$\dot{q}_{\text{lost}} = f_2(T_{\text{p}}, T_{\text{g}}, T_{\text{a}}, T_{\text{e}}, V_{\text{wind}}, \dot{m}) \qquad (6.5)$$

Now,

$$T_{\text{g}} = f_3(T_{\text{p}}, T_{\text{e}}, V_{\text{wind}}, \dot{m}, T_{\text{a}}) \qquad (6.6)$$

and if it is assumed that $(T_{\text{g}}/T_{\text{e}})$ is constant, then

$$\dot{q}_{\text{lost}} = f_4(T_{\text{p}}, T_{\text{a}}, V_{\text{wind}}, \dot{m}) \qquad (6.7)$$

So that for a given steady mass flowrate and a steady wind velocity,

$$\dot{q}_{\text{lost}} = f_5(T_{\text{p}} - T_{\text{a}}) \qquad (6.8)$$

This has generally been written as

$$\dot{q}_{\text{lost}} = U_{\text{L}}(T_{\text{p}} - T_{\text{a}}) \qquad (6.9)$$

where U_{L} is a constant, most often known as the overall loss coefficient for the collector. By manipulation of the above equations, a second-order differential equation is obtained that has been used to describe the temperature distribution across a collector plate, i.e.,

$$\frac{d^2T}{dy^2} + \frac{(\alpha\tau)_{\text{eff}}H - U_L(T_p - T_a)}{K\delta} = 0 \tag{6.10}$$

The solution of this equation for T is given by the expression

$$T = A_1 e^{by} + A_2 e^{-by} + (H/U_L) + T_a \tag{6.11}$$

where $b = \sqrt{(U_L/K\delta)}$ and A_1 and A_2 are constants. The first boundary condition to be applied is that the temperature gradient at the center of the region between tubes will be zero at all locations, i.e.,

$$\left[\frac{dT}{dy}\right]_{y=0,x} = 0 \tag{6.12}$$

The second boundary condition is that the temperature of the plate at the base of the *fin* [$y = (W - B)/2$ is equal to $T_{b,x}$, i.e.,

$$T_{y=(W-B)/2} = T_{b,x} \tag{6.13}$$

The application of these two boundary conditions yields the result

$$T = T_{b,x}\frac{\cosh by}{\cosh b([W - B]/2)} + \left(T_a + \frac{H}{U_L}\right)\left[1 - \frac{\cosh by}{\cosh b([W - B]/2)}\right] \tag{6.14}$$

B. Useful Heat Gained by the Fluid—Q_u

The energy conducted to the region of the tube from both sides of the plate per unit length in the flow direction ($q_{\text{fin-tube}}$), has been determined as

$$q_{\text{fin-tube}} = -2K\delta\left[\frac{dT}{dy}\right]_{y=(W-B)/2} \tag{6.15}$$

Differentiation of the equation derived for T and substitution into the above equation gives

$$q_{\text{fin-tube}} = (W - B)F[H - U_L(T_{b,x} - T_a)] \tag{6.16}$$

where $F = [\tanh b(W - B)/2]/[b(W - B)/2]$, and is usually known as the fin efficiency. The energy gain of the section of plate directly above the tubes due to absorption of solar radiation is

$$q_{\text{tube-sect}} = B[H - U_L(T_{b,x} - T_a)] \tag{6.17}$$

Hence, the total energy gain of the collector tubes per unit length in the flow direction may be expressed as

$$q_u = q_{\text{fin-tube}} + q_{\text{tube-sect}} \tag{6.18}$$

or

$$q_u = [B + F(W - B)][H - U_L(T_{b,x} - T_a)] \tag{6.19}$$

The resistance to heat flow to the tube from the plate may be considered to consist of three components: (i) the resistance due to the bonding material between the plate and the tube; (ii) the resistance due to the wall thickness of the tube; and (iii) the resistance due to the temperature gradient in the fluid at the tube wall. Hence,

$$q_u = \frac{T_{b,x} - T_{f,x}}{(1/C_b) + (1/C_W) + (1/\pi D_i h_{f,x})} \tag{6.20}$$

where C_b is the conductance of the bond, C_W the conductance of the tube wall, and $h_{f,x}$ the local film heat transfer coefficient.

The useful energy gain of the fluid can then be expressed in terms of the known dimensions, the physical parameters, and the local fluid temperature by solving the equation for T_b and substituting to obtain q_u from

$$q_u = WF'[H - U_L(T_{f,x} - T_a)] \tag{6.21}$$

for

$$F' = \frac{1/U_L}{W[1/U_L(B + (W - B)F) + (1/C_b) + (1/C_W) + (1/\pi D_i h_{f,x})]} \tag{6.22}$$

where F' has been called the "collector efficiency factor." Bliss (1959) calculated F' for a variety of collector geometries.

Consider the energy balance on a fluid element flowing through a pipe of length dx which is receiving a uniform heat flux q_u, so that

$$\dot{m}C_p T_{f,x} - \dot{m}C_p T_{f,x+dx} + q_u\, dx = 0 \tag{6.23}$$

or

$$\dot{m}C_p(dT_{f,x}/dx) - WF'[H - U_L(T - T_a)] = 0 \tag{6.24}$$

If the assumption is made that F' and U_L are constant (and independent of x), then the solution of the differential equation is

$$T = T_a + (H/U_L) - [(H/U_L) - (T_{in} - T_a)] \exp(-U_L W_c F' x/\dot{m}C_p) \tag{6.25}$$

The outlet temperature may then be found by putting $x = L_c$ in the above equation to give

$$T_{\text{out}} = T_a + (H/U_L) - [(H/U_L) - (T_{\text{in}} - T_a)] \exp(-U_L F' A_c/\dot{m}C_p) \tag{6.26}$$

where $A_c = W_c L_c$, the area of collector. The total useful energy collection rate Q_u may be expressed as

$$Q_u = \dot{m}C_p(T_{\text{out}} - T_{\text{in}}) \tag{6.27}$$

Substitution of T_{out} already derived gives

$$Q_u = AF_R[H - U_L(T_{\text{in}} - T_a)] \tag{6.28}$$

where

$$F_R = \dot{m}C_p/AU_L[1 - \exp(-F'U_L A_c/\dot{m}C_p)] \tag{6.29}$$

F_R has been termed the "heat removal factor" of the collector.

C. Evaluation of Overall Loss Coefficient

The energy loss from the collector plate consists of radiation and convection to the cover and the edges, and conduction through the box insulation. Hottel and Woertz (1942) have shown that the energy loss rate through the N glass covers of a collector Q_t can be found by solving a nonlinear system of $N + 1$ equations which are listed below:

$$\begin{aligned}
Q_t &= \frac{\sigma A(T_p{}^4 - T_{g_1}^4)}{(1/\varepsilon_p) + (1/\varepsilon_g) - 1} + h_{C_1}A(T_m - T_g) \\
&\vdots \\
Q_t &= \frac{\sigma A(T_{g_{i-1}}^4 - T_{g_i}^4)}{(2/\varepsilon_g) - 1} + h_{C_i}(T_{g_{i-1}} - T_{g_i}) \\
&\vdots \\
Q_t &= \varepsilon_g \sigma A(T_{g_N}^4 - T_{\text{sky}}^4) + h_w(T_{g_N} - T_a)
\end{aligned} \tag{6.30}$$

where T_p is the mean absolute temperature of the collector plate, T_{p_i} the mean absolute temperature of the ith glass cover, ε_p, ε_g the infrared emissivities of the plate and glass, h_{C_i} the convection coefficient between surfaces forming the ith air space, and h_w the convection coefficient between the top cover and the atmosphere. Note that here it has been assumed that the glass does not absorb solar radiation, so that the energy loss rate from the collector plate to the first glass cover is the same as the loss rate between subsequent glass covers and the loss rate between the top glass cover and the atmosphere.

Absorption of solar radiation by the glass covers will in fact occur, but it will tend to raise the glass temperatures and reduce the losses from the plate. Hottel and Woertz (1942) argued that this effect justifies the use of the expressions given. Hottel and Whillier (1958) later attempted to account for the absorption of solar radiation in the glass by defining an effective transmittance of the glass cover system that allows for the recovery of part of the energy absorbed by the glass.

It has also been assumed that there is a negligible temperature drop through each glass cover, i.e., that the thermal resistance within a cover may be neglected when compared to the thermal resistance between the surfaces, and so a mean temperature may be assigned to each cover.

If Q_t and the N mean glass temperatures are considered as unknowns in the equations, then the mean collector plate temperature must be known for a solution. The mean collector plate temperature can be found by integrating the equations describing the plate temperature distributions as

$$T_{\mathrm{p}} = \frac{1}{A}\int_0^{(W-B)/2}\int_0^L T\,dx\,dy. \tag{6.31}$$

Once T_{p} has been found, the $N + 1$ equations given may be solved iteratively so that Q_t and the N mean glass temperatures are obtained. Repeatedly determining Q_t as a function of the operating conditions is a tedious task. Hottel and Woertz (1942) realized this fact and developed an approximate relation to enable quick calculation for Q_t.

$$Q_t = \frac{A(T_{\mathrm{p}} - T_{\mathrm{a}})}{[N/C\sqrt[4]{(T_m - T_{\mathrm{a}})/(N+f)}] + (1/h_w)} + \frac{\sigma A(T_{\mathrm{p}}^4 - T_{\mathrm{a}}^4)}{(1/\varepsilon_{\mathrm{p}}) + [(2N + f + 1)/\varepsilon_{\mathrm{g}}] - N} \tag{6.32}$$

where C is a coefficient whose value is dependent on collector tilt, and f has values of 0.76, 0.36, and 0.24 for wind speeds of 0, 10, and 20 mph, respectively.

Klein (1973) developed a more accurate empirical relation for Q_t using the values of convection coefficients and glass emissivity recommended by Tabor (1958a). For a collector tilted 45° from the horizontal, Klein suggests the relation

$$Q_t = \frac{A(T_{\mathrm{p}} - T_{\mathrm{a}})}{(N^*/0.165[(T_{\mathrm{p}} - T_{\mathrm{a}})/(N^* + f^*)]^{0.31}) + (1/h_w)} + \frac{A\sigma(T_{\mathrm{p}}^4 - T_{\mathrm{a}}^4)}{[1/\varepsilon_{\mathrm{p}} + N^*(1-\varepsilon_{\mathrm{p}})0.05] + [(2N^* + f - 1)/\varepsilon_{\mathrm{g}}] - N^*} \tag{6.33}$$

where $N^* = 1.00$, 1.85, and 2.65 for 1-glass, 2-glass, and 3-glass cover collector, respectively, and

$$f^* = 0.9556 - 0.211h_w + \left[1.0 + \frac{N^* - 1}{N^* + 3}\right] h_w^{\,2}$$

where Q_t is in Btu/h, temperatures are in °R, and area is in ft^2.

In this work he uses relations recommended by Tabor (1958) to estimate the energy loss rate from the collector plate through the bottom insulation and by Whillier (1967) to find the edge losses.

$$Q_{\text{bottom}} = \frac{A(T_p - T_a)}{(\delta_I/K_I) + (1/h_{\text{bottom}})} \tag{6.34}$$

where δ_I and K_I are the thickness and conductivity of the insulation and h_{bottom} is the convection coefficient between the bottom of the collector and the atmosphere.

$$Q_{\text{edge}} = h_{\text{edge}} A_p (T_p - T_a) \tag{6.35}$$

where A_p is the outer perimeter area of the collector and h_{edge} is the heat transfer coefficient. The overall loss coefficient U_L is then defined by

$$U_L = \frac{Q_t + Q_{\text{bottom}} + Q_{\text{edge}}}{A(T_p - T_a)} \tag{6.36}$$

Duffie and Beckman (1974) plotted the top loss coefficient U_t versus T_p for a number of different ambient temperatures, glass temperatures, wind speeds, and plate emissivities. It is interesting to note from this plot that U_t is essentially a linear function of $T_p - T_a$ for black painted collectors with plate emissivities of 0.95. However, for collectors with selective plate surfaces of longwave emissivity 0.10, the top loss coefficient U_t is shown to be a nonlinear function of $T_p - T_a$.

6.4 SOME PERFORMANCE CHARACTERISTICS OF CONVENTIONAL TUBE AND PLATE COLLECTORS

A. Temperature Distributions during Steady State Periods

Many thermocouples were used in various places in a collector and Figs. 6.8–6.10 show the location of the thermocouples used.

1. Temperature Distribution in the Flow Direction

Typical periods of steady solar radiation flux chosen for study were those for 24-2-1975 and 5-3-1975. Measurements of the plate temperatures and the "local" water temperatures made at 13.38.00 and 11.47.20 on each

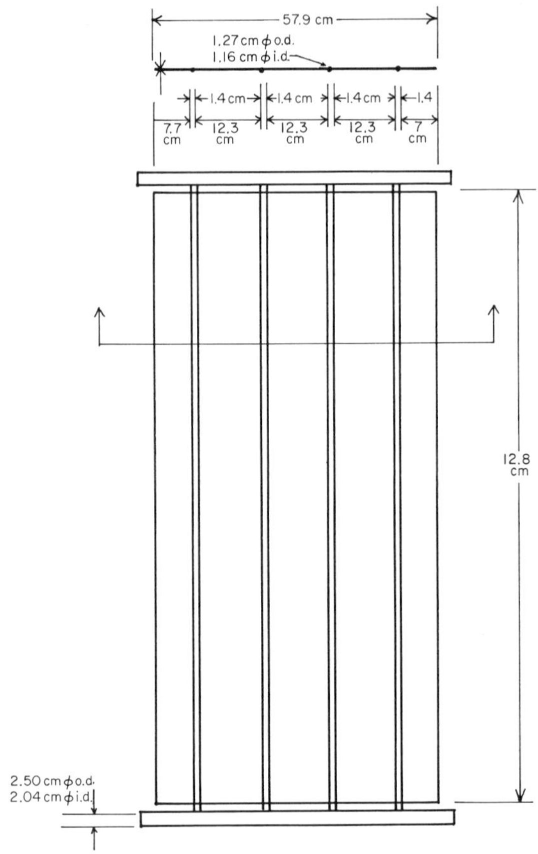

FIG. 6.8 Dimensions of the absorber plate.

of these days, respectively, are shown in Fig. 6.11. The first two curves show the temperatures along the center line of the region between the tubes and the water temperatures as measured along the plate and the bulk outlet temperature of the water on each of the days, respectively. The second two curves show the temperature distribution as measured across the plate. The curves have been drawn to indicate the general shape of the temperature profiles. The values of the steady solar radiation flux measured on each of these days were 900 and 930 W/m^2, respectively.

It is interesting to note that use of the theoretical solution predicts the temperature distribution of the water along the tube to a reasonable degree of accuracy for conditions on 5-3-1975. The fluid temperature distribution

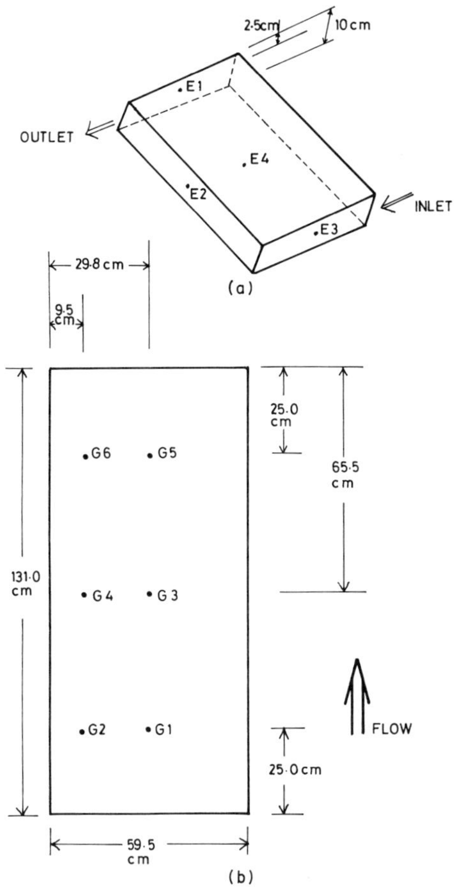

FIG. 6.9 (a) Location of the thermocouples on the collector box; (b) location of the thermocouples on the glass cover.

predicted by Eq. (6.25) is shown as a dotted line. Although the estimated experimental temperature at the outlet of the tube and the theoretical prediction of this temperature are within 1°C, it should be noted that the theoretical curve predicts a lower water temperature (by approximately 4°C) at other regions along the tube. This is mainly because an average value of the film heat transfer coefficient was used to evaluate the temperature distribution, while, as described earlier, the convection coefficient in the developing region of the flow is higher than that for fully developed flow. Sample calculations are required for the theoretical curve.

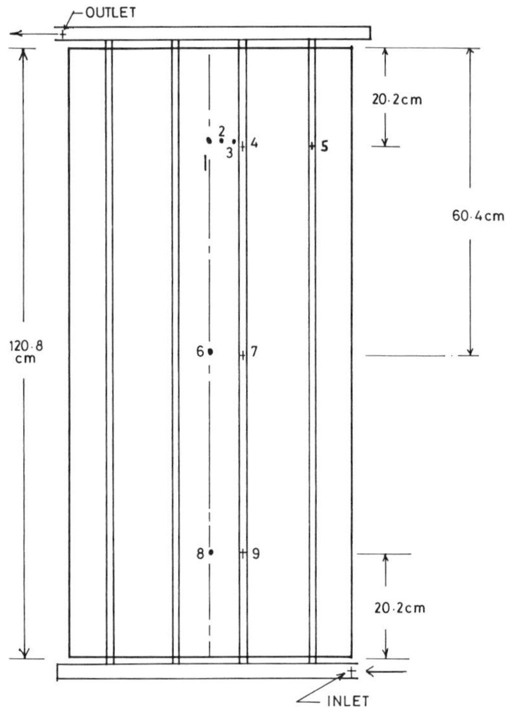

FIG. 6.10 Location of the thermocouples on the collector plate.

The theoretical curve was obtained in the following manner. The heat lost from the collector plate (Q_L) was calculated from the known absorbed solar radiation $(\alpha\tau)_{eff}AH$ and the energy gained by the fluid flowing through the collector $\dot{m}C_p\,\Delta T$. Knowing Q_L and representative values of plate and glass temperatures, the overall loss coefficient was calculated. On 5-3-1975, between 11.45.00 and 11.50.00, the average value of the overall loss coefficient was found to be $U_L = 5.0$ W/m^2 °C. The fin efficiency factor was calculated to be $F = 0.97$, and the collector efficiency factor (F') was then calculated. A value of bond conductance (C_b) of 40 W/m°C was used as recommended by Whillier and Saluja (1965). The thermal conductance of the wall was calculated from

$$C_W = \frac{2\pi K(D_o^2 - D_i^2)}{D_o^2 \ln(D_o/D_i) - [(D_o^2 - D_i^2)/2]} \tag{6.37}$$

where D_o is the outside diameter of the tube, and D_i the inside diameter of the tube. The value used for C_W was calculated as 80 W/m°C. The value

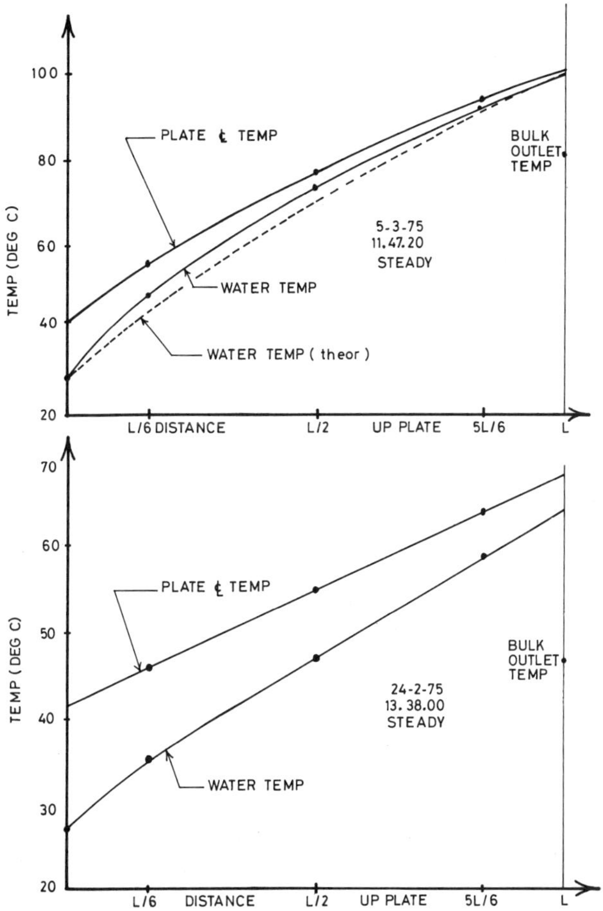

FIG. 6.11 Temperature distributions in the flow direction during steady conditions.

used for the film heat transfer coefficient (h_f) was 567.83 W/m^2 °C as recommended by Baker (1967) so that,

$$\pi D_i h_f = 20 \text{ W/m °C}$$

Use of all these parameters yielded a value for F' of 0.92 and hence the temperature of the water could then be estimated from

$$T_{f,x} = 209 - 179 \exp(-0.404x)(°\text{C})$$

where x is the distance along the plate in meters.

2. Temperature Distributions across the Plate

The temperature distributions across the "fin" and the local fluid temperature (as measured by thermocouples) are illustrated on Fig. 6.12. Curves have been drawn through the three points representing the plate temperatures. The fact that there is a temperature gradient at the center line of the plate for both sets of measurements is evidence of a nonuniform heat flux flowing to the water in the tubes. These results are of special significance since Dunkle and Davey (1970) predict a nonuniform but symmetrical flow distribution in the tubes. The temperature gradients that existed at the center of the plate are an indication of a possible asymmetrical flow distribution.

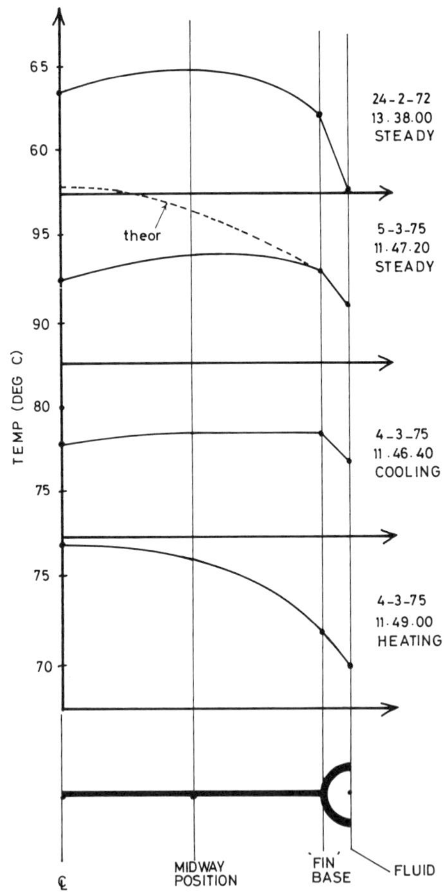

FIG. 6.12 Temperature distributions across the plate.

The theoretical temperature profile for 5-3-1975 is shown on the same figure as a dashed line. It should be noted that the largest temperature drop exists between the tube wall and the water.

B. Temperature Distributions during a Transient Period

The measurements reported here were recorded on 4-3-1975. The incident solar radiation had been steady (between 970 and 1000 W/m^2) for approximately 15 min before 11.46.40 on that day. At that time, the value of the solar radiation flux had dropped to a value of approximately 160 W/m^2 and by 11.49.00, it had returned to a value of 1000 W/m^2. The flowrate was maintained at a constant value of 0.097 l/min.

1. Temperature Distributions in the Flow Direction

The temperature distributions during the cooling and heating processes are shown on Fig. 6.13 as marked. As expected, the water temperature distribution is closer to the plate temperature distribution during the cooling process at 11.46.40 than for the heating process at 11.49.00. Klein (1973) proposed that the transient temperature distribution in a collector tube will be expressed as

$$T_{f,x} = T_{f,x}^{\text{ss}} - [(T_{\text{out}}^{\text{ss}} - T_{\text{out}})/L]x$$

where $T_{f,x}$ is the fluid temperature at a distance x up the plate and the superscript "ss" refers to steady state temperatures that will be obtained if enough time is allowed for equilibrium to occur. A careful examination of Fig. 6.13 will show that this proposal does not accurately describe the transient temperature profiles. Instead, during the cooling process the flow of cool water from the bottom header tube into the tube in the plate has the effect of *kinking* the temperature profile. The *kink* in the profile can be seen to extend up the tube during the following heating stage. *Note* that the estimated outlet temperature of the collector tube is close to the bulk outlet temperature of the collector in both cases.

2. Temperature Distributions across the Plate

The temperature distributions across the plate are also shown on Fig. 6.12 During the cooling process the temperature distribution was nearly isothermal. The temperature difference between the water and the fin base was only 1.7°C. During the heating process the temperature distribution across the plate was similar to the type of distribution expected during steady state operation, as there was a negative temperature gradient extending from the center of the region between the tubes to the *fin* base. Whether or not a

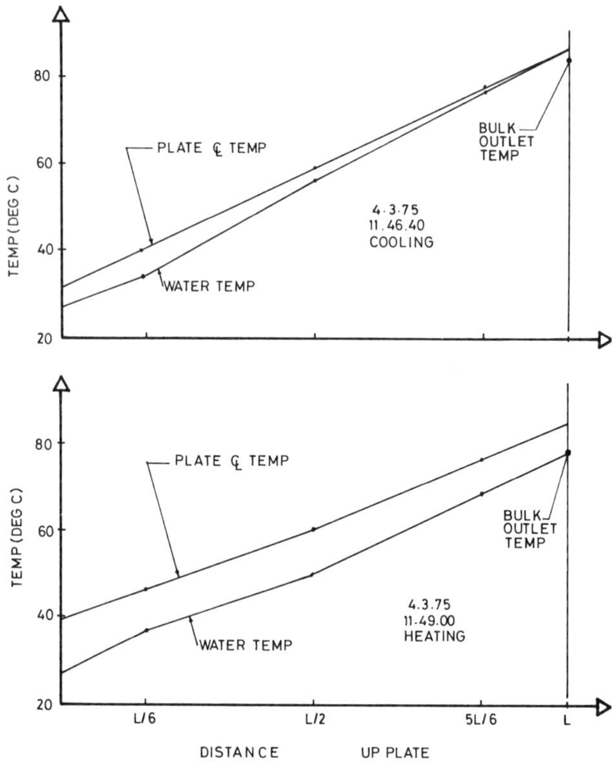

FIG. 6.13 Temperature distributions in the flow direction during transient conditions.

zero temperature gradient existed at the central region could not be determined accurately from these tests but it appeared that the temperature gradient was very small at this section. The temperature difference between the water and the fin base was 2.7°C.

3. Temperature Distribution in the Glass Cover

From a test run on 5-3-1975 at time 11.47.20, the following cover glass temperatures have been extracted to illustrate the cover temperature distribution

$$T_{G1} = 34.8°C, \qquad T_{G3} = 36.9°C, \qquad T_{G5} = 39.0°C$$

$$T_{G2} = 32.5°C, \qquad T_{G4} = 35.1°C, \qquad T_{G6} = 37.6°C$$

The temperature of the glass surface was very insensitive to variations in solar radiation flux of a moderate frequency and also appeared to be insensitive to fluctuations in wind speed. In Fig. 6.14 the incident solar radiation flux, mean glass temperature, and the wind speed are plotted against

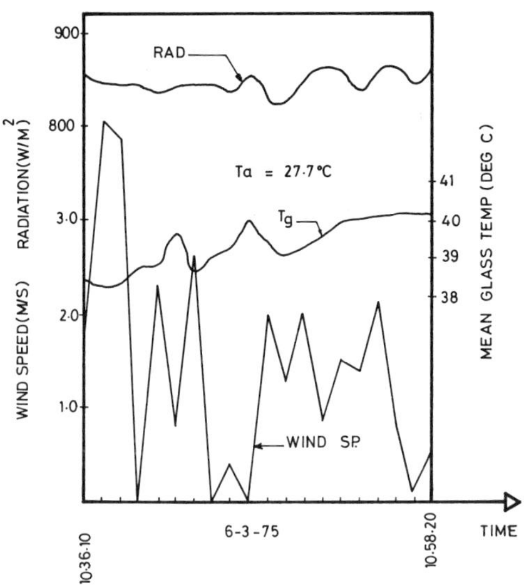

FIG. 6.14 Mean glass temperature, wind speed, and solar radiation flux versus time, for conditions between 10.36.10 and 10.58.20 on March 6, 1975.

time for conditions between 10.36.10 and 10.58.20 on 6-3-1975. The wind speed can be seen to have been varying considerably while the incident solar radiation flux remained relatively steady. It can be seen, however, that the mean glass temperature remains relatively steady throughout this period of steady solar radiation and varying wind speed.

C. The Prediction of Collector Performance from Hourly Average Meteorological Data

Much of the meteorological data that are required for the analysis of solar devices is most often recorded in the form of hourly integrated averages. It has been argued that the use of hourly integrated values of the relevant meteorological variables (e.g., solar radiation) can be used to adequately predict collector performance. This implies that the effect of thermal capacitance of a collector is of little importance. Klein (1973) used a numerical model to study the effects of transient conditions on collector performance and concluded that a transient effect will result from fluctuations in wind speed, but not from fluctuations in radiation intensity. The effect was shown to be of small influence since the numerical simulation utilized what were considered to be the most intermittent of conditions expected. However, the experimental results recorded in the tests described in this section show that these conditions could often occur. Duffie and Beckman (1974) plotted calculated values of U_L against plate temperature for a number of different

TABLE 6.2

Comparison of Theoretical and Experimental Values of Total Heat Loss Rate—Hourly Averages[a]

	T_g (°C)	H (W/m^2)	V_{wind} (m/s)	Q_{in} (W)	Q_u (W)	Q_L (exp)
	T_p (°C)	T_a (°C)	Q_t (Klein)	Q_{edge} (W)	Q_{bottom} (W)	Q_L (Klein)
	Q_{t_1} (W)	$Q_{L_{g \to a}}$ (W)	Q_{in_g} (W)	Q_{t_2} (W)	$Q_{t_{av}}$ (W)	Q_L (theor)
4.3.75	32.9	911	0.7	458	343	115 ± 33
Steady radiation	67.6	19.0	117	8	25	150 ± 12
conditions	104	151	78	73	88	121 ± 15
Transient	34.5	895	0.8	450	343	107 ± 32
radiation	67.6	19.9	117	8	25	150 ± 12
conditions	99	165	77	88	93	126 ± 15
5.3.75	35.9	830	0.3	417	314	103 ± 30
Steady radiation	68.9	21.0	112	8	25	145 ± 12
conditions	99	160	71	89	94	127 ± 15
Steady radiation	37.2	917	0.4	461	354	107 ± 33
conditions	73.9	21.5	126	9	27	162 ± 13
	113	171	79	92	102	138 ± 16
Steady radiation	38.0	907	0.6	456	337	119 ± 32
conditions	75.8	22.1	133	9	28	170 ± 13
	117	179	78	102	109	146 ± 16
Steady radiation	36.5	808	1.0	406	318	88 ± 29
conditions	70.3	22.5	113	8	25	146 ± 12
	102	162	69	93	97	130 ± 15
6.3.75	38.3	801	1.3	403	304	99 ± 29
	70.1	28.5	105	7	24	134 ± 12
Radiation	95	133	69	64	79	110 ± 15
reasonably						
steady over	41.0	862	1.0	434	351	83 ± 32
entire test	74.2	29.0	113	8	22	145 ± 12
(±50W/m^2	100	149	74	75	88	117 ± 15
every						
60–100 s.)	41.2	794	1.1	399	319	80 ± 29
	72.6	30.4	106	7	21	135 ± 12
	86	140	68	72	79	107 ± 15
10.3.75	33.0	593	0.1	298	215	83 ± 22
Rapid rise	52.4	23.9	60	5	15	80 ± 10
in rad.	51	84	51	33	42	62 ± 10
intensity—						
steady for						
15 min—						
rapid fall						

TABLE 6.2 (*Cont.*)

	T_g (°C) T_p (°C) Q_{t_1} (W)	H (W/m^2) T_a (°C) $Q_{L_{g\to a}}$ (W)	V_{wind} (m/s) Q_t (Klein) Q_{in_g} (W)	Q_{in} (W) Q_{edge} (W) Q_{t_2} (W)	Q_u (W) Q_{bottom} (W) Qt_{av}(W)	Q_L (exp) Q_L (Klein) Q_L (theor)
11.3.75	36.1	726	0.2	365	316	49 ± 27
Radiation	62.2	23.9	86	7	20	113 ± 11
steady	74	128	62	66	70	97 ± 14
Transient	37.2	797	0.3	400	317	83 ± 29
radiation	68.2	24.4	102	7	23	132 ± 12
	92	136	68	68	80	110 ± 15
Steady	38.3	830	0.4	417	354	63 ± 31
radiation,	72.9	24.7	114	8	25	147 ± 12
possible error	104	146	71	76	90	123 ± 16
in flow rate measurements	36.4	705	0.5	354	313	41 ± 27
	66.2	25.5	95	7	21	123 ± 12
	87	126	61	65	76	104 ± 16
14.3.75	42.7	759	0.5	382	288	94 ± 27
Rapidly varying	73.3	31.9	99	7	22	138 ± 12
radiation	91	124	65	58	74	103 ± 15
intensity	43.8	772	1.0	388	319	69 ± 28
throughout	74.9	34.6	100	7	21	128 ± 12
entire period.	93	122	66	56	79	107 ± 15
	39.6	480	0.9	241	197	44 ± 19
	60.8	34.9	60	5	14	79 ± 10
	58	72	41	30	44	63 ± 14
	40.2	505	1.0	254	230	24 ± 20
	64.2	34.6	70	5	15	90 ± 10
	67	83	43	39	53	73 ± 12
	35.8	401	0.9	202	156	46 ± 16
	53.6	32.8	46	4	11	61 ± 9
	47	53	34	19	33	48 ± 11

[a] From Watmuff (1965).

plate emissivities, ambient temperatures, and wind speeds. The effect of wind speed on U_L is marked and is an explanation of different cooling and heating rates leading to the transient effect as found by Klein. For high emissivity collector plates, U_L is shown to be a weak, almost linear function of plate temperature. Hence, one would expect that for a high emissivity collector plate, the heating and cooling rates would be similar and therefore the transient effect would be negligible. However, for low emissivity collector plates, U_L is quite a strong nonlinear function of plate temperature and one would expect to find substantial differences between the heating and cooling rates and, consequently, a significant transient effect.

Another factor that has been neglected in the analysis of transient conditions on collector performance is that convection processes are inherently subject to transient behavior since energy must be transported by the convecting fluid. While perhaps, energy transport occurs rapidly for forced convection phenomena, it is well known that considerable time is required for equilibrium to occur when free convection is a major mode of heat transfer (as is the case between a collector plate and its cover). In this section, hourly averages have been made of the solar radiation flux, glass and plate temperatures, inlet and outlet water temperatures, ambient temperature, and wind speed. It has been assumed that the means of the 52 sets of recording will be equivalent to the integrated means. The hourly average energy input rate to the collector and the hourly useful energy collection rate have been calculated in order to obtain the hourly average heat loss rates. Both methods of calculating the average energy loss from the collector have been based on the hourly averages of plate temperature, glass temperature, ambient temperature, and wind speed. The results of the calculations are listed in Table 6.2 with a brief description of the steadiness of the solar radiation during the period.

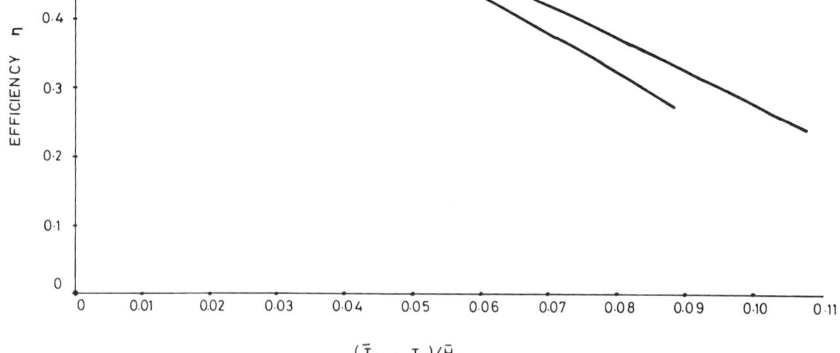

FIG. 6.15 Typical experimental steady state characteristic. Collector: single glazed selective surface.

In all but three of the 19 hourly periods studied, the total heat loss rate from the collector that was calculated by knowing the glass temperature and allowing for absorption of solar radiation, agreed with the experimentally determined value within the range of possible errors. In comparison, 12 of the 19 hourly average heat loss rates that were calculated using Klein's expression, lay outside the experimental range. The use of hourly integrated meteorological data and steady state heat transfer relations appear to be justified in predicting long term collector performance based on the existing available collector models. Finally, Fig. 6.15 shows an experimental steady state collector characteristic of a typical single glazed selective surface collector.

6.5 AREAS OF RESEARCH INTEREST FOR FUTURE WORK

1. Wind tunnel investigations of wind effects on the values assigned to U_L for various cover plate configurations and wind regimes.
2. Infrared measurements to improve the existing knowledge of effective sky temperatures under various meteorological conditions.
3. Comparison of test results from collectors on the basis of indoor simulator and outdoor test.
4. Fundamental studies of convective effects in inclined tube flow and inclined air gaps. These should include the effects of nonisothermal plates and transient heat exchange processes.
5. High temperature testing and extended life testing to assess degradation effects on selective surfaces and deterioration of overall collector performance.
6. Testing of adequacy of existing models of collector performance under a wide range of transient operating conditions and for differing collector configurations.
7. Performance of high temperature collectors utilizing honeycomb convection suppression devices or vacuum-sealed units.

7

Convective Heat Transfer Effects within Honeycomb Structures for Flat Plate Solar Collectors

W. W. S. CHARTERS

MECHANICAL ENGINEERING DEPARTMENT
UNIVERSITY OF MELBOURNE
PARKVILLE, VICTORIA, AUSTRALIA

7.1 INTRODUCTION

There has been a strong interest for many years now in the use of honeycomb cellular structures of various types because of their inherent structural properties. With their excellent strength to weight ratios, they have found countless applications in the aircraft and aerospace industries; and for these purposes they have been fabricated in many different materials such as aluminium, titanium, and even kraft paper. The range of cell shapes now available is quite extensive and only a few of the more commonly encountered forms are illustrated in Fig. 7.1. In more recent times there has been a renewed interest in these materials, not primarily for their structural properties but because they can be used to modify the heat transfer processes that occur in the vicinity of a heated body. For these purposes the honeycombs have been fabricated from other materials such as glass, or from one of the vast range of available plastics, in order to take account of the particular radiation properties of these substances. As the information is sometimes difficult to obtain, a brief tabulation is given in Table 7.1 of some of the conduction and radiation properties of the materials commonly used in the construction of honeycombs. These data should only be used as a guide when assessing the selection of a particular material

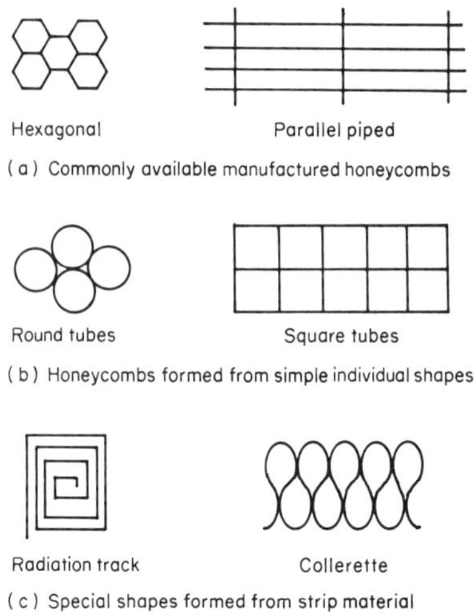

FIG. 7.1 Some types of cellular structures.

TABLE 7.1

Thermal Properties for Commonly Used Materials, Order of Magnitude Figures

Material	Aluminum foil	Titanium	Kraft paper	Glass	Plastic
Thermal conductivity, k (kcal/J m °C)[a]	180	18	0.2–0.3	0.7–0.9	0.15–0.25
Thermal capacity, c (J/kg °C)	920	—	—	750–950	1400–2300
Solar absorptivity, α_s	0.55	0.6–0.8	—	0.04	—
Infrared emissivity, ε_L[b]	0.2–0.4	0.5	0.7–0.9	0.91–0.94	—

[a] Conductivite thermique (A. Missenard. Original text in Library Department Heliophysique, Marseille, France.

[b] Infrared radiation (Bramson, 1958).

for some distinct purpose and not taken to be exact because of the large variations often encountered in practice with relatively small changes in a material specification.

Generally, for all practical applications the process of heat transfer is a multimode phenomenon and account must be taken of all three modes, conduction, convection, and radiation, when considering the overall performance of any one particular structure. In terms of conduction effects, the cellular structure can be envisaged as an extended surface used to increase the heat transfer area, and hence the cooling rate of a heated surface; in terms of convection effects, the structures have been proposed for the inhibition of, or modification to, the natural convection transport from a heated body; and in terms of the radiation effects, they have been proposed for the creation of artificial black bodies and special forms of thermal traps. It should be realized that at times the purpose of the additional surface structure may be to promote the heat transfer from the surface and at other times to diminish this transport as shown in Fig. 7.2. In this chapter attention is concentrated on the convective aspects of the general heat transfer problem and only superficial attention is paid to the radiation and conduction effects.

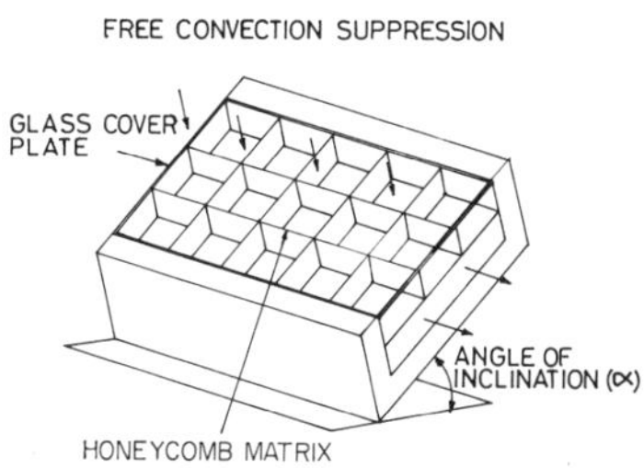

FIG. 7.2 Inclined flat plate solar absorber with honeycomb structure.

7.2 THE GENERAL PROBLEM OF HEAT TRANSFER WITHIN CELLULAR STRUCTURES

In nearly all cases of heat transfer that occur in practice, it is fair to say that there will be present all three fundamental modes of transport: conduction, convection (free and/or forced), and radiation. Each of these modes may be present to a greater or a lesser extent in any one particular problem being considered. For the case of heat transfer within cellular structures, the two predominant modes in practice are free convection and radiation; and, in general, the conduction component can be considered simply as a modifying factor on the other two predominant modes. As indicated earlier, the cellular structures are superimposed upon a heated body surface in order to influence the heat transfer effects that take place in the immediate vicinity of that body. To further illustrate this point it may be interesting to consider the modifications arising from the application of a honeycomb structure to a heated flat plate. Because of the heating of the flat plate, the final equilibrium temperature T_b attained by the surface will be determined both by the energy supplied to the plate and by the heat transfer losses from the plate in the form of free convection and radiation. The superposition of the cellular structure as shown in Fig. 7.3 can significantly modify these heat loss terms; and it is these modifying effects that will be considered in detail in the later sections of this chapter. At this stage it is perhaps sufficient to say that the presence of the structure is intended to suppress entirely the free convection component for small values of temperature difference $(T_b - T_a)$ for the open structure or $(T_b - T_c)$ for the closed structure, and to modify the emission and absorption properties of the baseplate in the case of the radiation component.

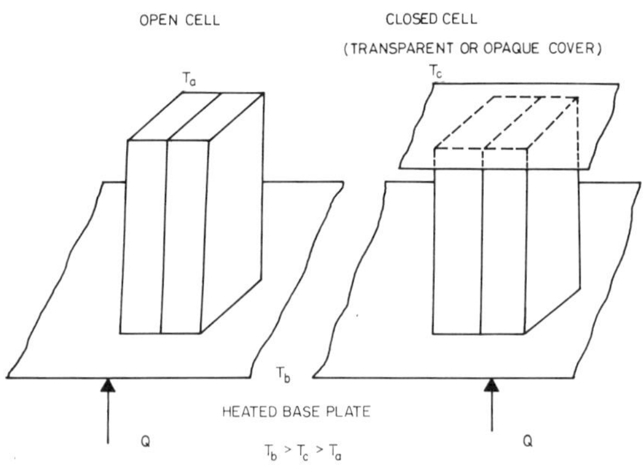

FIG. 7.3 A cellular structure to modify heat losses.

For the radiation terms, however, the imposed structure will also significantly modify the capture process for any incoming radiation to the surface and this is one of the aspects that has intrigued workers in the field of solar energy utilization in recent years since the publication by Francia (1961) of his initial work on honeycomb absorbers. In this work the author concentrated attention on the radiative aspects of the new collector and ignored the convective effects. This approach is of course entirely justified for space applications where there will be no atmospheric convection effects, but for all terrestrial applications care should be taken to consider both of these equally important heat transfer modes.

The subsequent work on heat transfer effects in cellular structures has virtually split off into two separate streams as indicated earlier in this chapter; and for this reason it has been decided to consider the convection and the radiation to be independent effects and to discuss later the possible interaction between these two modes. In the first of these streams, attention has been focused on the radiation process and this has involved the study of surface effects for black, gray, and selective absorbing surfaces and the exchange processes within long tubes of various cross section (Perrot *et al.*, 1962; Gallet, 1973). In most of this work the interaction between the conduction and the radiation has not been considered; and any conduction effects have been simply superimposed on the radiative effects. A second discrete area of research which has opened out from the original work has been the consideration of the natural convection within the cells of the honeycomb structure both in terms of the criteria for the onset of convective motion and in terms of the subsequently established convection. The basis for this work lies in the early studies of Rayleigh (1916) and Bénard (1901) on the stability of fluid layers and the nature of the convection patterns in infinite fluid layers. These fundamental works were extended, in 1940 by Pellew and Southwell who were perhaps the first authors to include in their analysis the effect of internal constraining walls within the fluid layer.

In the following sections a more detailed treatment of the subject is presented which includes under convection a natural division between the theoretical mathematical studies and those of a more experimental nature. This division is not so clear cut for the radiation field where in general a more integrated approach to the subject has been taken, but an attempt is made to differentiate between those studies that concentrate heavily on the theory and those that also include practical and/or experimental considerations. For all the radiation studies, the effect of the angle of inclination of the cells is, of course, unimportant, but in terms of the natural convection effects this is one of the important parameters that determine the heat transfer across the cell by modifying the movement of the contained fluid. The general case is that of the inclined cell with the heated baseplate inclined at some angle to the horizontal; but there are also two particular cases which have been treated at some length in the available published literature, i.e., the horizontal cell and the vertical cell. These three possible orientations of the cell structure are indicated in Fig. 7.4.

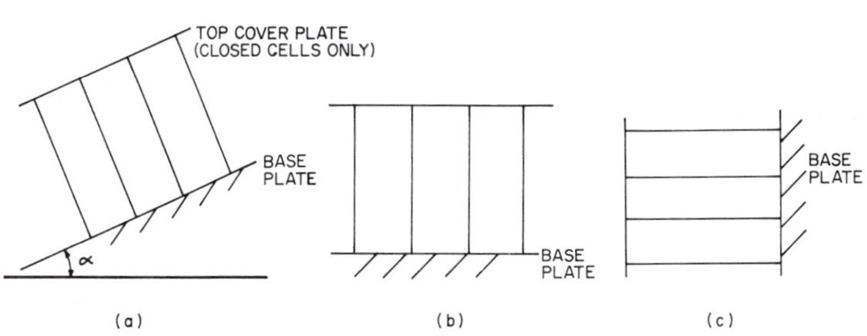

FIG. 7.4 Possible cell and fluid layer orientations. (a) Inclined cell, inclined fluid layer, $0 < \alpha < 90°$. (b) Vertical cell, horizontal fluid layer, $\alpha = 0$. (c) Horizontal cell, vertical fluid layer, $\alpha = 90°$.

7.3 SOME THEORETICAL ASPECTS OF THE FREE CONVECTION PROCESS FOR FLUID LAYERS

A. Unconstrained Fluid Layers

For a general understanding of the stability problem, it is instructive to consider first a horizontal layer of fluid of infinite width but finite depth contained between two rigid isothermal bounding surfaces. When the lower surface is gradually heated to a temperature in excess of the upper surface, an initial condition of stable conduction across the fluid layer is generated. This is typified by linear gradients of temperature and density across the layer as a stratified layer situation develops. This stagnant state, although potentially unstable, may be sustained until at some critical value of the temperature difference across the layer an instability occurs and there is an onset of free convection with a resultant increase in heat transfer rate across the gap. This critical condition is most often characterized by the critical Rayleigh number, for a given fluid layer

$$\mathrm{Ra_c} = (gB/K\nu)\cdot \Delta T_c L^3 = \Phi(\Delta T, L)$$

where Ra (Rayleigh number) = Gr × Pr (Grashof number × Prandtl number) which may be thought of as relating the buoyancy forces (caused by density differences) and the viscous shear forces that are brought into play as a result of fluid motion. This situation forms the basis of the classical Bénard problem that has been extensively studied in the literature on natural convection, i.e.,

$$\mathrm{Ra_{c_1}} = 1707 \text{ for enclosed layer with solid isothermal boundaries}$$

and

$$\mathrm{Ra_{c_1}} = 1100 \text{ for fluid layer with free top surface}$$

As mentioned earlier, the reasoning behind the introduction of the cellular matrix into the fluid layer is to provide extra internal slip surfaces for the fluid, and hence to increase the viscous shear forces. In this way it is possible to delay the onset of convection until some higher value of $\mathrm{Ra_c}$ has been attained so that for a given fluid layer, it is possible to increase the adverse temperature gradient while maintaining the heat transfer in the conduction regime. This is of considerable importance when one utilizes a fluid layer as an insulating gap.

A general treatment of the stability problem is possible, and some of the major attempts at providing a theoretical determination of the stability criteria may be found in the works of Chandrasekhar (1961), Kuo and Platzman (1961), Malkus and Veronis (1958), and Nakagawa (1960). For many of the problems that one encounters, it is sufficiently accurate to utilize an approximation first used by Boussinesq (1903) of treating the density as a constant in all of the defining equations except in so far as this parameter modifies the operation of the gravity force. This approximation is used generally in the formulation of the stability problem as it is apparent that the variations in density that lead to the buoyancy force terms in the momentum equations must be taken into account in the mathematical formulation of the problem. The relevant governing hydrodynamic equations of momentum, continuity, and energy with an appropriate equation of state for the working fluid are briefly summarized below.

1. Stagnant Fluid Solution Based on a Three-Dimensional Model

The basic hydrodynamic equations of momentum and continuity are used in conjunction with a thermal energy equation and an equation of state for the working fluid.

Momentum equations

$$\begin{aligned} \rho\, Du/Dt &= -(\partial p/\partial x) + \rho\nu\, \nabla^2 u \\ \rho\, Dv/Dt &= -(\partial p/\partial y) + \rho\nu\, \nabla^2 v - \rho g \\ \rho\, Dw/Dt &= -(\partial p/\partial z) + \rho\nu\, \nabla^2 w \end{aligned} \tag{7.1}$$

Continuity equation

$$(\partial u/\partial x) + (\partial v/\partial y) + (\partial w/\partial z) = 0 \tag{7.2}$$

Thermal energy equation

$$K\, \nabla^2 T = DT/Dt \tag{7.3}$$

State equation

$$-(\rho - \rho_0)/\rho_0 \cong (T - T_0)/T_0 \tag{7.4}$$

where the symbols ∇^2 and D/Dt are defined as

$$\nabla^2 = \left(\frac{\partial^2}{\partial x^2} + \frac{\partial^2}{\partial y^2} + \frac{\partial^2}{\partial z^2}\right) \qquad \text{and} \qquad D/Dt = (\partial/\partial t) + \mathbf{u}\cdot\nabla$$

It is evident that for the stagnant fluid solution there can be no fluid motion, which can be mathematically expressed as

$$u = v = w = 0 = \partial/\partial t \tag{7.5}$$

Substitution of these conditions into the basic equations results in the three expressions

$$-\partial p/\partial x = -\partial p/\partial z = 0 = \nabla^2 T; \qquad \text{and} \qquad -\partial p/\partial y = \rho g \tag{7.6}$$

from which it may be concluded that the resulting isobars and isotherms must be horizontal. As discussed earlier, this result indicates the reason for the essential instability of the inclined cell and the absence of a stability criterion such as that postulated for the vertical cell (horizontal fluid layer).

2. *General Case of Inclined Cell Based on a Two-Dimensional Model*

From extensive flow visualization tests, it is evident that the motion of the fluid in the inclined cell tends to be of a two-dimensional character once the cell is inclined at an angle greater than those for the transition region. Although the critical angle at which the change occurs from fully three-dimensional flow to two-dimensional flow has not been accurately measured, it would appear to lie between 10° and 20°, and at the higher angles of inclination, the two-dimensional roll form of the convective motion is quite apparent. For this reason and for simplicity the governing equations presented here are based on a two-dimensional model of the fluid flow following the original treatment by Batchelor (1954).

For a uniform incompressible fluid, the mass conservation equation can be written in the standard form as

$$(\partial u/\partial x) + (\partial v/\partial y) = 0 \tag{7.7}$$

To specify the equation of state for the working fluid in a convenient form, it is legitimate to write for both gases and liquids,

$$-(\rho - \rho_0)/\rho_0 \cong (T - T_0)/T_0 \tag{7.8}$$

where $1/T_0 = \beta =$ coefficient of thermal expansion. The equation is legitimate in this form provided we are dealing with small temperature changes across the fluid layer that is consistent with the normal Boussinesq (1903) approximation of constant properties except for the density term in the buoyancy force.

For a slow moving fluid such as would be generated after the onset of convection at low Ra, the heat transfer by modes other than conduction may be neglected as a first approximation as can the temperature changes due to fluid compression and viscous dissipation. It should be noted at this stage, however, that terms of the form $u\, \partial u/\partial x$ should not be neglected in any such analysis.

Using a nondimensional temperature perturbation as proposed by Batchelor (1954),

$$\theta = (T - T_0)/(T_i - T_0) \tag{7.9}$$

the thermal energy equation may be written as

$$(\partial\theta/\partial t) + u(\partial\theta/\partial x) + v(\partial\theta/\partial y) = K[(\partial^2\theta/\partial x^2) + (\partial^2\theta/\partial y^2)] \tag{7.10}$$

where $K = k/\rho c_p$ for gases and $K = k/\rho c_v$ for liquids. Essentially, in the analysis K may be regarded as constant as explained above for all small values of $(T - T_0)/T_0$. Allowing for the buoyancy effects, which are in this instance the basic driving force for the convection, the momentum equations become

$$\frac{\partial u}{\partial t} + u\frac{\partial u}{\partial x} + v\frac{\partial u}{\partial y} = \frac{-1}{\rho_0}\frac{\partial p}{\partial x} + \nu\left(\frac{\partial^2 u}{\partial x^2} + \frac{\partial^2 u}{\partial y^2}\right) \tag{7.11}$$

$$\frac{\partial v}{\partial t} + u\frac{\partial v}{\partial x} + v\frac{\partial v}{\partial y} = \frac{-1}{\rho_0}\frac{\partial p}{\partial y} + \nu\left(\frac{\partial^2 v}{\partial x^2} + \frac{\partial^2 v}{\partial y^2}\right) - g\left(\frac{\rho - \rho_0}{\rho_0}\right) \tag{7.12}$$

If we further assume that the time scale of the temperature change is large compared with the time scale of the convective motion and we consider only steady flows, then

$$\partial T/\partial t = 0$$

By defining a stream function such that

$$u = (K/L)(\partial\psi/\partial y), \qquad v = (-K/L)(\partial\psi/\partial x)$$

the energy equation may be rewritten as

$$(\partial\theta/\partial x)(\partial\psi/\partial y) - (\partial\theta/\partial y)(\partial\psi/\partial x) = \partial(\theta, \psi)/\partial(x, y) = \nabla^2\theta \tag{7.13}$$

By combining the force equations with the equation of state, we see further that

$$(1/\mathrm{Pr}[\partial(w, \psi)/\partial(x, y)] = -\mathrm{Ra}(\partial\theta/\partial y) + \nabla^2 w \tag{7.14}$$

where now w is the dimensionless vorticity $= -\nabla^2\psi$. The boundary conditions for the inclined cell as shown in Fig. 7.5 are

$$X = 0 \qquad 0 \leqslant Y \leqslant 1.0 \qquad \psi = \partial\psi/\partial X = 0$$
$$X = 1.0 \qquad 0 \leqslant Y \leqslant 1.0 \qquad \psi = \partial\psi/\partial X = 0$$
$$Y = 0 \qquad 0 \leqslant X \leqslant 1.0 \qquad \psi = \partial\psi/\partial Y = 0, \quad \theta = 1.0$$
$$Y = 1.0 \qquad 0 \leqslant X \leqslant 1.0 \qquad \psi = \partial\psi/\partial Y = 0, \quad \theta = 0$$

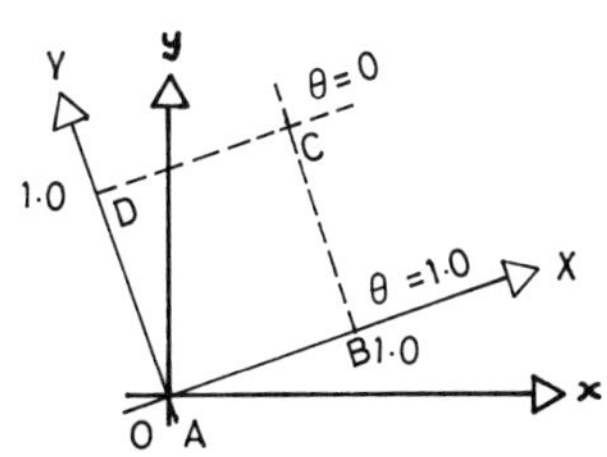

FIG. 7.5 Boundary conditions for an inclined cell.

For the case of thick highly conducting transverse walls, i.e., linear temperature gradient,

$$X = 0 \qquad 0 \leqslant Y \leqslant 1.0 \qquad \partial\theta/\partial X \quad \text{cannot be specified}$$
$$X = 1.0 \qquad \theta = 1 - Y$$

and for the case of perfectly insulating walls, i.e.,

$$X = 0, \qquad 0 \leqslant Y \leqslant 1.0 \qquad \partial\theta/\partial X = 0 \quad \text{no heat transfer from fluid to wall}$$
$$X = 1.0 \qquad \partial\theta/\partial Y \quad \text{cannot be specified}$$

Note: For walls of finite thickness and finite thermal conductivity, the temperature and heat flux distributions of the solid/fluid interface have to be matched for the fluid and the wall. This may be accepted by solving the two-dimensional conduction equation in the wall region (Koutsoheras and Charters, 1975).

B. Effect of Constraining Walls on Horizontal Fluid Layer (Vertical Cell)

When one considers the complications which ensue from the positioning of a cellular matrix structure within a horizontal fluid layer as shown in Fig. 7.6, the following points emerge for consideration:

1. Physical shape of the cell structure, e.g., hexagonal, circular, square, etc.
2. Honeycomb cell dimensions, expressed generally in terms of the aspect ratio L/D.
3. Thermal properties of the wall material, e.g., insulating or conducting.

In general, the nondimensional perturbation equations may be derived for a variety of cell configurations and solved for various boundary condi-

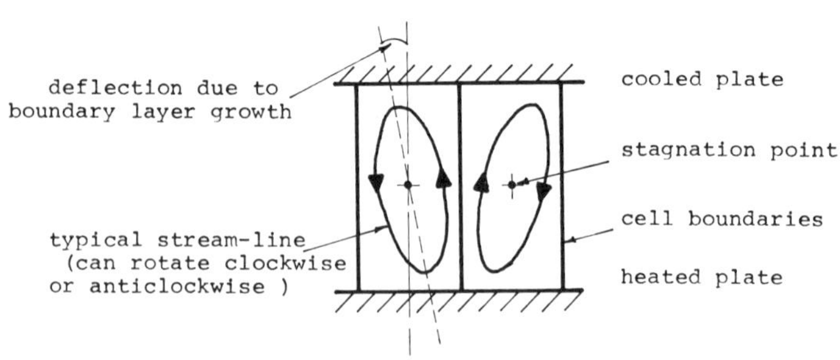

FIG. 7.6 Horizontal fluid layer. Three-dimensional motion, a definite critical Rayleigh number exists for this case.

tions; theoretical treatments of this type have been provided by Catton (1970), Edwards (1969), Edwards and Sun (1970), and Hollands (1966).

The effect of the thermally conducting walls has been allowed for by defining parameters describing the relationship of the thermal conductances around and across the cells; and an attempt has been made to determine the extent of the interaction between cells. In some more recent work the concept of conducting and nonconducting walls has been extended by Hollands (1972) to include the class of matrix cell with very thin walls that are virtually nonexistent for vertical and horizontal heat transfer but that still exist to restrain the fluid motion. The material chosen especially for this study (a transparent polyethylene film) was selected on the basis of its thermal properties to minimize any possible interaction between the convective and the radiative components of the heat transfer.

In general, it has been shown theoretically that the onset of convection is delayed by the matrix structure; the extent of the delay being a function of the cell material and the aspect ratio L/D; but that the convectional heat transport across the honeycomb tends toward the values obtained for unconstrained fluid layers as the higher values of Ra are approached.

$$\mathrm{Ra_c} = \Phi(\Delta T, L, L/D) \qquad \text{for a given fluid layer}$$

C. Unconstrained Inclined Fluid Layer

Predictions of the instability of free inclined fluid layers have been published by Kurzweg (1970) and Liang and Acrivos (1970). The work of Kurzweg indicates that a simple modification to $\mathrm{Ra_c}$ for a horizontal layer will suffice to modify this criterion for use with inclined fluid layers. There are, however, serious doubts as to the applicability of this work to the inclined cell problem because of the inherent assumption of the longitudinal roll form of the convective disturbance. In addition, there is implicit in the analysis the assumption that the principle of the exchange of stabilities that has been shown to be valid for the horizontal layer is equally applicable to the inclined layer. In the extensive work of Liang and Acrivos (1970), the solutions to the perturbation equations are provided in terms of wave numbers and a growth rate for the disturbance is postulated which reduces the equations to the Orr–Sommerfeld equation coupled with the energy equation. An interesting conclusion drawn in their work is that of the transition from stationary longitudinal rolls to stationary transverse rolls as the angle of inclination is increased from 0° to 90°. This concept of a *transitional region* dependent on the angle of inclination is reinforced by the observations in some of the experimental work presented in the following section. It appears that the type of convective flow established may well depend on the angle of inclination of the cell with a distinct transition from one type to another at some critical angle of inclination.

D. Unconstrained Vertical Fluid Layer (Horizontal Cell)

The special case of the unconstrained vertical fluid layer formed by two vertical isothermal and parallel plates which are maintained at different temperatures has been fully analyzed by Batchelor (1954). An obvious practical application of such a layer is in the formation of cavity walls for thermal insulation in building construction and also the practice of double window glazing for combined acoustic and thermal insulation (see Fig. 7.7). Because of the assumption of infinite width vertical bounding surfaces, a two-dimensional analysis is possible from the start and this is directly employed in place of the more complex three-dimensional analysis for the vertical cell. One of the more interesting features of this analysis is that it clearly shows that the stagnant fluid solution developed in Section 7.3A cannot exist for this cell orientation. For any finite temperature difference across the cell, it is evident that convective motion must ensue. This very complete analysis contains information on the solutions for very small Ra, the solutions for finite Ra and infinite cavity aspect ratio, and the solutions for very large Ra with finite cavity aspect ratios. An essential assumption in this work is that the boundary conditions on the top and bottom horizontal walls are of negligible importance to the analysis which is obviously not true for the finite cell problem.

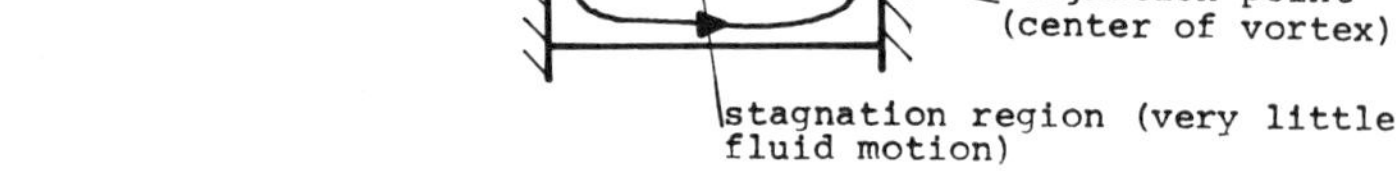

FIG. 7.7 Vertical fluid layer. Two-dimensional motion, no critical Rayleigh number.

E. Effect of Constraining Walls on Inclined Fluid Layers

From the stagnant fluid solutions, derived and presented in Section 7.3A, it is quite evident that the isobars and the isotherms must both be horizontal for the fluid layer. While this result is compatible with the configuration for the horizontal layer, an immediate inconsistency is apparent for the inclined fluid layer with constraining walls as shown in Fig. 7.8. As the isothermal bounding surfaces are no longer horizontal, the resulting distortion of the isotherms in the vicinity of the heating and the cooling surfaces

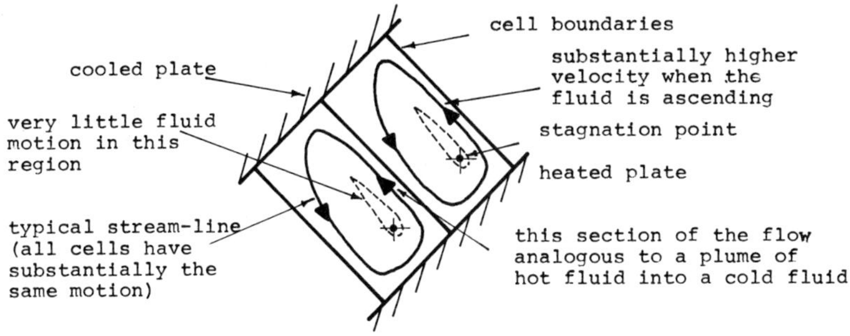

FIG. 7.8 Inclined fluid layer. Two-dimensional motion, no critical Rayleigh number.

can only be provided by motion of the fluid in these regions. This point is illustrated in Fig. 7.9. This indicates that there must be convective components included in the formulation of the general energy equation and that further, the concept of an initial stability criterion (Ra_{c_1}) is not directly applicable to the case of the inclined cell. In this respect it can be seen that the inclined cell problem is analogous to that of the unconstrained vertical fluid layer previously discussed. This result was first indicated by Charters and Peterson (1972) and the derivation of the equations governing the convective motion in the inclined cell (Section 7.3A) is due to Peterson (1971). The analysis has been based on a two-dimensional model as flow visualization studies indicated conclusively that a two-dimensional roll motion exists within the inclined cells for all angles of inclination greater than those for the so-called *transitional region.* Although no formal solution of these equations is yet available, numerical solutions using finite difference techniques have been carried out by Koutsoheras. The preliminary results of this work were presented by Koutsoheras and Charters (1975) and includes the effect of a real finite wall of finite conductivity. The cell temperature profile is generated by solving simultaneously the two-dimensional conduction equation in the wall and the convection hydrodynamic equations.

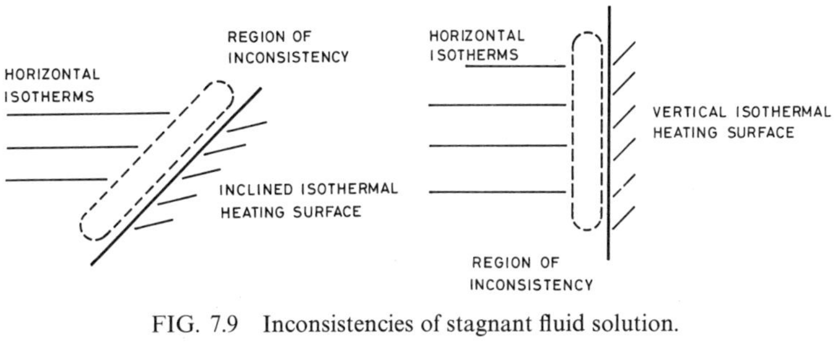

FIG. 7.9 Inconsistencies of stagnant fluid solution.

7.4 EXPERIMENTAL WORK ON FREE CONVECTION EFFECTS

A. Unconstrained Fluid Layer

There is a vast amount of literature dealing with the measurement of heat transfer in free convection for fluid layers whether they be horizontal, inclined, or vertical. In general, the experimental works do not deal with the stability problem but with the magnitude of the heat transfer across the fluid gap once the convection has been initiated. By reference to these works it is generally possible to obtain empirical correlations of Nusselt number and Rayleigh number for the different convection regimes from laminar to fully turbulent for most angles of inclination for the constrained fluid layer. In addition to the experimental works concerned with direct measurement, there are others that have provided a valuable insight into the actual physical phenomena in the convective transport process by using various techniques of flow visualization ranging from particle suspension techniques for viscous fluids to interferometric, schieleren, and other photographic methods. Among the more important of the references that provide data allowing the calculation of heat transfer across a fluid layer are the following:

Horizontal layers only: Schmidt and Silveston 1959; Globe and Dropkin (1959).

Horizontal and inclined layers: De Graaf and Van Der Held (1959).

Vertical layers: Batchelor (1954); Elder 1965; Jannot and Mordchelles-Regnier (1967).

B. Fluid Layers with Internal Constraints

The published experimental work on constrained fluid layers has been virtually restricted to extensive treatments of the horizontal layer (vertical cell). Good correlation has been obtained between the theoretical predictions and the measured experimental results for both low conductivity (phenolic) and high conductivity (aluminum) walls of the matrix used by Edwards (1969); and also for the thin-walled plastic matrices used by Hollands (1966). Some discrepancies have been indicated in the literature between the results obtained with small cells of low aspect ratio filled with air and the same cells filled with liquid, although in general the trends have been the same. Some prominent contributors to this section include Catton (1970), Catton and Edwards (1967), Hollands (1966), and Kent and Bowyer (1972).

There is not much suitable work on which to base empirical correlations for the heat transfer in inclined cells. In general, the tendency appears to have been to regard the horizontal layer (vertical cell) as the worst case for the free convection loss term and to utilize the data obtained for horizontal layers directly for cases involving inclined layers. In view of the uncertainties involving the transition region as the inclination angle is increased from zero, and with the distinct possibility that the form of the convective pattern changes radically, e.g., from a fully three-dimensional flow to a two-dimensional roll, this would appear to be an essentially hazardous pastime.

The most recent work published by Hollands (1975) gives empirical correlations for the case of the inclined cell for a restricted range of Ra.

In general, the studies involving free convection which have been quoted in the preceding sections of this chapter deal with fluid layers that are constrained between solid isothermal top and bottom surfaces although the equations derived in Section 7.3A are perfectly general within the limitations of the Boussinesq approximation, and the difference between the solid isothermal top surface and a free top surface only becomes apparent in the specification of the boundary conditions.

7.5 EFFECT OF RADIATION ON CONVECTIVE EFFECTS IN CELLS

An earlier work has dealt exclusively with the problems of convective heat transfer effects within cellular structures neglecting any interaction with the radiation component of the heat transfer. However, as the problem of predicting the magnitude of the radiation loss term entails a knowledge of the temperature distribution along the wall of the cell, and because this distribution may well be influenced by the convective heat transfer within the cell, it is necessary in practice to know the extent of any interaction between these two heat transfer modes. If it is possible to treat the two modes entirely separately, and finally to recombine the individual results by simple superposition, the prediction of the total heat transfer term across any given cell will obviously be considerably simplified.

For the above reasons a series of experiments was carried out to determine the extent of the dependence of the wall temperature distribution on the interaction between the convection and the radiation within a single cell. For this comparison, a glass cell of aspect ratio $L/D = 1.2/1$ was chosen to ensure a large convective component with relatively low temperature differences across the cell, the base temperature being in the order of 110°C for an ambient air temperature of about 20°C. The cell was tested both in the open cell and the closed cell configuration (Fig. 7.3) under vacuum conditions (to eliminate all free convection) and under free-air conditions to give the required comparisons.

The curves on Fig. 7.10 are plotted to show the differences obtained in this experimental series. It can be seen that there is a discrepancy between

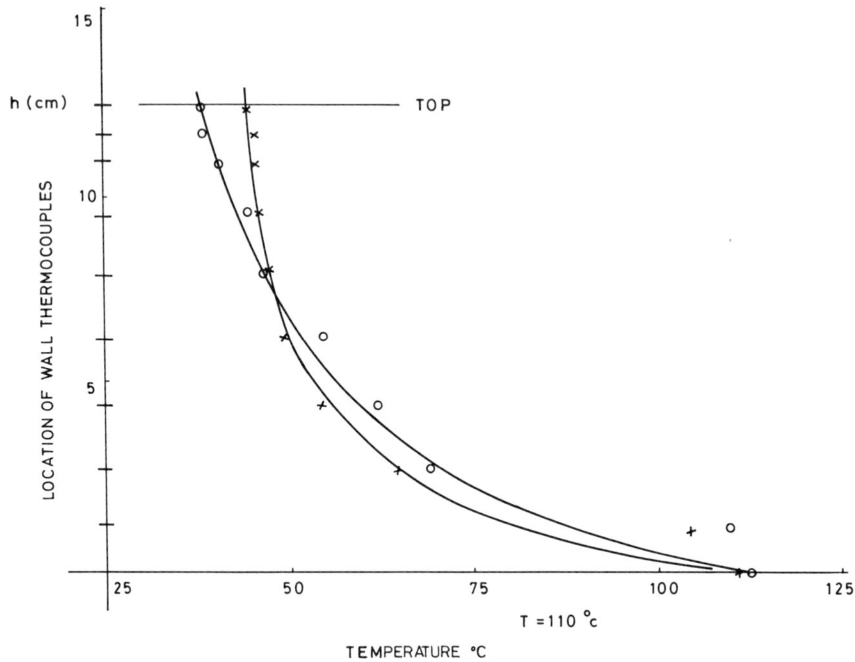

FIG. 7.10 Wall temperature profiles in single closed cell. —o— vacuum test, 10^{-5} mm Hg; —×— air test, atmospheric.

the wall temperature profiles, but that the order of magnitude is such that the prediction of the radiative heat transfer would not be seriously affected in the theoretical model for the range of temperature conditions used in the test series. As a good first approximation, in the range of temperature differences likely to be encountered in low to medium temperature solar apparatus, it appears that it is valid to separate out the convective and the radiative transfer terms in the prediction of the radiation heat transfer across the cell.

Because these initial test results were obtained using a single cell of rather low aspect ratio, it was decided to repeat the tests using a matrix of glass tubes to simulate more closely the practical case of a tube insulated within a tube bundle. The tubes selected to construct the matrix were 4.5 cm long, and approximately 1.5 cm in diameter giving an effective aspect ratio of 3/1. To minimize the effects of conduction in the walls of the tubes, the thickness of the tube wall chosen was the minimum obtainable, i.e., 0.2 mm. The central tube in the matrix of about 38 tubes was instrumented with eight thermocouples of the fine wire variety on the outside of the tube wall

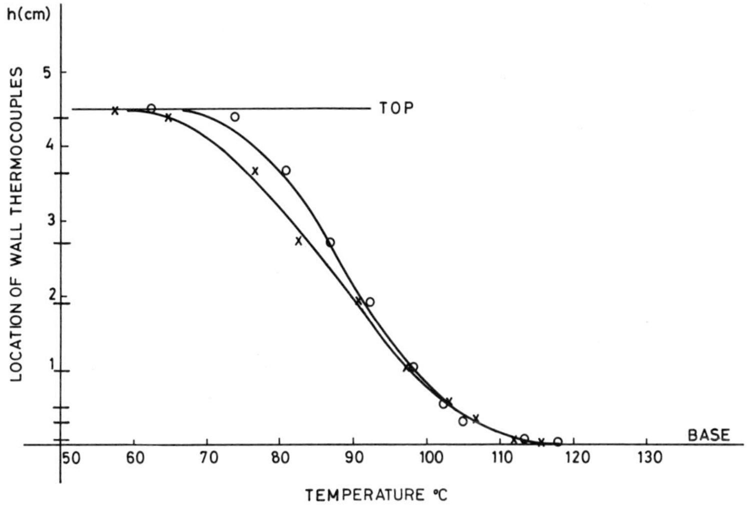

FIG. 7.11 Wall temperature profiles in central cell of cellular matrix. —o— vacuum test, 10^{-5} mm Hg; —x— air test, atmospheric.

to monitor the changes in the wall temperature profile under the varying conditions to be imposed in the test series. As for the single cell, a series of tests were performed under vacuum and under atmospheric conditions to identify the importance of the free convection term in the development of the wall temperature profile. As for the initial tests, the results from this second test series confirm the legitimacy of ignoring the free convection effects when determining the wall temperature profile to be used in the determination of the radiation heat transfer term (see Fig. 7.11).

8

Solar Air Heaters and Their Applications

M. KUDRET SELÇUK

JET PROPULSION LABORATORY
CALIFORNIA INSTITUTE OF TECHNOLOGY
PASADENA, CALIFORNIA

8.1 INTRODUCTION

The solar air heater has an important place among solar heat collectors. Its role has been mostly misunderstood or ignored. With the recent trend of providing both heating and cooling with solar energy, liquid heaters for high temperature operation have gained more popularity. As far as the ultimate application of heating air to maintain a comfortable environment is concerned, the solar air heater is the most logical choice. Heating a fluid by the sun, then transferring heat to the air results in loss of temperature potential available. If the limitations of solar energy applications in terms of attainable temperatures and related efficiencies as well as the low cost requirements are considered, it would be wiser to eliminate any heat transfer operation. Direct use of the air circulated through the solar air heater as the working substance also reduces the number of components required in the system. Solar-heated air could be used more effectively for drying under controlled conditions. Solar heaters supplying hot air to a conventional drier or special designs combining the air heater and the drying cabinet in one package have cost and efficiency advantages for drying applications. Solar air heaters also eliminate corrosion and leakage problems which may be difficult and costly to overcome. The cost of the air heater could be substantially lower than the liquid systems. Higher pressures experienced in liquid heaters necessitate the use of heavy-gauge sheet metal or tubes.

The air heater could be designed using less material, even some scrap of no commercial value.

The limitation of the heat transfer from the absorber plate to the working fluid for air heaters could be a major problem for improperly designed collectors. There are several ways to improve the heat transfer coefficients and/or increase surfaces (which will be discussed later) such that efficiencies would be comparable with liquid heaters. One disadvantage of solar air heaters is the need for handling larger volumes of air than liquids due to the low density of air as a working substance. Another disadvantage is the low thermal capacity of air. In cases where thermal storage is needed, water is superior to air. Water may be used both as a heat transfer and a heat storage substance. In many localities, however, an antifreeze solution has to be used to overcome the frost problem. If an antifreeze solution is used, the cost of the thermal storage fluid may become prohibitively high due to the high cost of antifreeze solutions. The suggested use of an antifreeze to water heat exchangers (water being the storage substance) reduces the storage system cost, but the slight loss in temperature available and the expense of an extra heat exchanger must be justified.

The final choice of the working fluid for solar energy is a complicated matter that necessitates a careful comparative analysis of thermal performance, operation, maintenance, and economic parameters.

In fact, the ultimate goal in solar energy—or any form of energy—utilization is to decrease the cost of energy collected in cents per kilowatt hour. The cost effective design may not necessarily be the one using the most efficient solar collector, it is rather a combination of high efficiency, low fabrication, installation, and operation costs, and other practical aspects that are related to the specific application in mind.

A brief glance at the available literature on the subject gives an idea of the work done so far in this field. The invention of the liquid heating flat plate solar water heater is credited to H. B. Saussure, a Swiss scientist, during the second half of the seventeenth century as reported by Ackerman (1915). Sir John Herschel in 1837 and Tellier in 1885 experimented with solar water heaters. Solar air heaters have been recently introduced. One of the early designs, the overlapped glass plate air heater, was described by Miller (1943) and introduced by Lof *et al.* (1946). His collaborators have tested the overlapped glass air heater (Lof, 1950; Lof and Nevens, 1953; Lof, 1956; Lof *et al.*, 1961) and used it at the solar house at Denver, Colorado, built in the late 1950s. Selçuk (1971) reported an extensive mathematical analysis of the overlapped glass air heater and its experimental verification. Bliss (1955) used a solar air heater—wire mesh type porous absorber. Whillier (1963, 1964), Close (1963), Khanna and Singh (1967), and Gupta and Garg

(1967), analyzed and reported the performance of nonporous air heaters. The thermal performance of solar air heaters having nonporous absorber plates with flow behind the plate was first examined by Whillier (1964). A thorough boundary layer analysis and its verification of the duct behind the absorber plate was reported by Malik and Buelow (1973). Charters (1971) and Ambrose and Bandopadhayay (1970) have also examined the heat transfer phenomena in asymmetrically heated ducts representing the flat plate solar collector.

Various geometries and collector efficiency and pressure drop variations for different flow conditions have recently been examined by Kuzay *et al.* (1974). Among special designs that combine the roof structure and the absorber to build solar heated structures is a galvanized steel roof described by Buelow and Sobel (1963). A porous absorber-type solar air heater with a slit and expanded sheet aluminum foil matrix was introduced by Shoemaker (1962). Porous matrix-type air heaters were extensively analyzed and tested by Chiou *et al.* (1965), Beckman (1968), and Hamit and Beckman (1969). Porous matrix-type air heaters with slit and expanded sheet iron matrix or blackened wood shavings were used in connection with solar drying experiments by Selçuk (1966, 1967) and Akyurt *et al.* (1972). Another porous absorber, a transpired honeycomb, has been extensively analyzed by University of California, Los Angeles team. Substantial efficiency improvement over convential nonporous absorbers is reported by Lalude and Buchberg (1971) and Buchberg *et al.* (1968, 1971).

8.2 DESIGN VARIATIONS OF SOLAR AIR HEATERS

Basically, all solar air heaters can be classified under two categories. The first type has a nonporous absorber in which the air stream does not flow through the absorber plate. Air may flow above and/or behind the absorber plate. The second type has a porous absorber that includes slit and expanded metal, transpired honeycomb, and overlapped glass plate absorber. Figures 8.1 and 8.2, illustrate basic design features of these solar air heaters.

A. Collectors with Nonporous Absorber Plates

Although a nonporous absorber may be cooled by the air stream flowing over both sides of the plate, as shown in Fig. 8.1a, the most common design is the one with air flow behind the absorbing surface. Air flow above the upper surface increases the convection losses from the cover plate, and therefore is not recommended if the air inlet temperature and/or temperature rise at the collector are large (Fig. 8.1b).

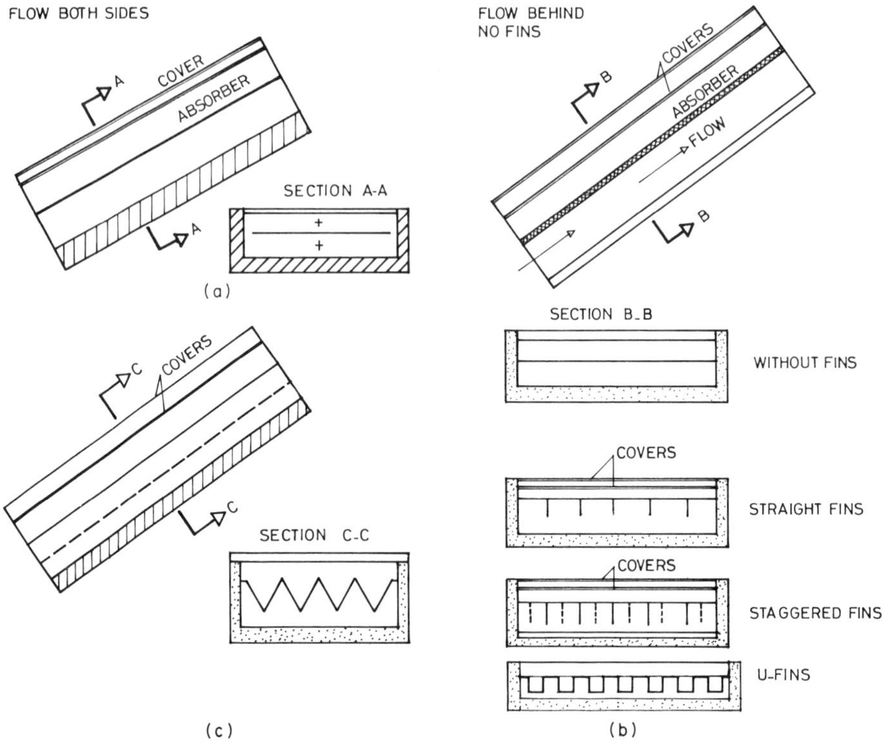

FIG. 8.1 Nonporous absorber-type air heater.

Transmission of the solar radiation through the transparent cover system and its absorption is identical to that of a liquid heater. Selective coating can be applied to improve collection efficiency provided that cost effectiveness criterion can be fulfilled. Low heat transfer rates result in efficiencies lower than liquid air heaters under the same radiation flux and temperature conditions. Performance improvement is possible either by

(a) roughening the rear of the plate to promote turbulence and improve the convective heat transfer coefficient, or

(b) adding fins to increase the heat transfer surface as shown in Fig. 8.1b. Usually turbulence is also increased which enhances the convective heat transfer.

Figure 8.1c illustrates a solar collector with Vee-corrugated copper foil as used by the CSIRO team (Close, 1963). Solar flux absorption is improved due to both surface radiative characteristics and the geometry of the corrugations, which help trap the reflected radiation. Detailed analyses and performance definition of three types of nonporous absorber plates with emphasis on the effect of adding fins with straight and staggered arrangements, as well as Vee-corrugated absorbers, will be discussed later.

B. Collectors with Porous Absorbers

A major drawback of the nonporous absorber plate is the necessity of absorbing all incoming radiation over the projected area from a thin layer over the surface, which is in the order of a few microns. Unless selective coatings are used, radiative losses from the absorber plate are excessive, therefore the collection efficiency cannot be improved. The pressure drop along the duct formed between the absorber plate and the rear insulation may also be prohibitive especially in the case of added fins to increase the heat transfer surface and turbulence rate.

A porous absorber shown in Fig. 8.2a overcomes these defects in two ways:

(a) The solar radiation penetrates to greater depths and is absorbed gradually depending on the matrix density. The cool air stream introduced from the upper surface of the matrix is first heated by the upper layers which are cooler than the bottom layers. The air stream warms up, while traversing the matrix layers. The lower matrix layers are hotter than the upper ones, therefore the air stream can effectively transfer heat from the matrix.

Improper selection of the matrix porosity and the thickness may result in reduced efficiencies since the additional matrix layers beyond an optimum may no longer absorb the solar radiation and heat the air stream further.

(b) The pressure drop for the porous matrix is usually lower than the nonporous absorber with flow behind the plate since flow per unit cross section would be much lower. Although the matrix hinders the flow, the pressure drops reported for porous matrix absorbers are still lower than for the nonporous absorbers. The solar air heating utilizing a transpired honeycomb is also very favorable from the pressure drop standpoint since the flow cross section is much larger (Fig. 8.2b).

A unique concept of using crushed glass layers to absorb solar radiation and heat the air as suggested by Whillier (1963) has not yet been examined. A porous bed made by forming layers of broken bottles (bottom dark top clear glass) may be readily used for agricultural drying purposes with minimal expenditure (Fig. 8.2c).

The overlapped glass plate air heater illustrated in Fig. 8.2d can be considered as a form of porous matrix, although overall flow direction is along the absorber glass plates instead of being across the matrix. Plate and air stream temperature increase gradually along the collector length and

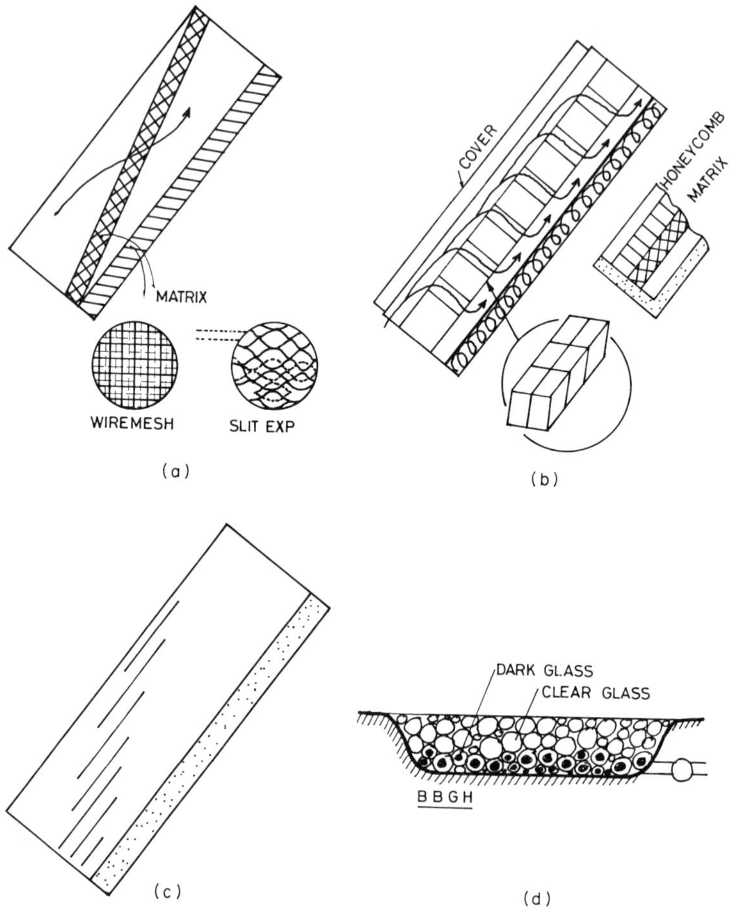

FIG. 8.2 Porous absorber-type air heaters.

across from top to bottom. Thus, thermal losses could be significantly reduced. The pressure drop is also significantly less than the nonporous flat plate absorber design.

8.3 CALCULATION OF THE EFFICIENCY OF A FLAT PLATE SOLAR AIR HEATER

The derivation of performance equations of the most popular solar air heater design, the flat plate with air flow on one side, based on the heat transfer model as introduced by Whillier (1964), will be presented. Equations describing the energy balances of various systems and control volumes are shown in Fig. 8.3.

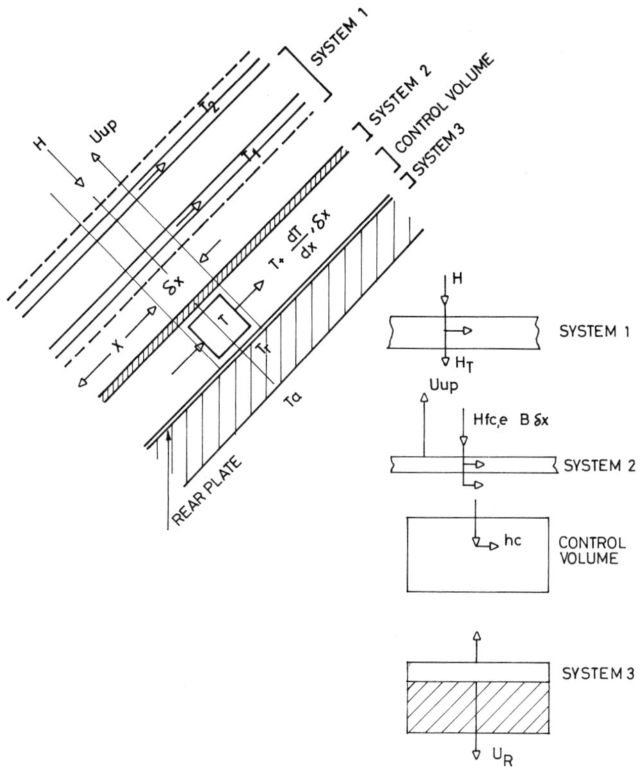

FIG. 8.3 The energy balance of various systems.

A. Transparent Cover (System 1)

The transparent glazing over the nonporous absorber plate could be made out of glass plates or plastic sheets or films or a combination of both. Determination of effective transmissivity for a given transparent cover combination is identical to that of a liquid air heater. A combination factor, effective absorptivity transmissivity product ($f_{c,e}$) defines the ratio of solar radiation absorbed by the absorber plate to the incident solar flux. $f_{c,e}$ can be calculated from

$$f_{c,e} = F_c + a(F_1 + F_2 + F_3 + \cdots) \tag{8.1}$$

where F_c is the fraction transmitted by cover plates and absorbed by the plate and F_1, F_2 the fraction absorbed by each glass cover. For two cover plates, for example, $F_c = 1.015 T_1 T_2 a$. For n number of cover plate arrangements, Table 8.1 lists (a) factors for use in Eq. (8.1).

Calculations of the short wave solar radiation transmitted through various transparent cover system arrangements are described in various

TABLE 8.1

Factor a for Various Covers in Eq. (8.1)

Number of cover plates	1	2	3
a	0.23	0.33	0.37

references: Whillier (1953), Duffie and Beckman (1974), as well as in this text (see Chapters 2 and 3). The exact analysis of the heat transfer model would require simultaneous solution of all energy balance equations from top glazing to rear insulation. In a simplified approach, the solar radiation incident upon the absorber plate is calculated by determining the incoming flux after being modified by the transparent cover system. The flux on the absorber can be computed by the term of $Hf_{c,e}$ and heat losses by $U_L(T_p - T_a)$. Thus, only two systems adjacent to the control volume identifying the air stream would suffice to compute the heat collected by the collector.

Heat balance on the absorber plate

$$Hf_{c,e}(B\,\delta x) = U_{up}(B\,\delta x)(T_p - T_a) + h_c(B\,\delta x)(T_p - T) + h_r E(B\,\delta x)(T_p - T_r) \tag{8.2}$$

Heat balance on the air stream (control volume)

$$WC_p(dT/dx)\,\delta x = h_c(B\,\delta x)(T_p - T) + h_c'(B\,\delta x)(T_r - T) \tag{8.3}$$

Heat balance on the rear plate

$$h_r E(B\,\delta x)(T_p - T_r) = h_c(B\,\delta x)(T_r - T) + U_r B\cdot\delta x(T_r - T_a) \tag{8.4}$$

Solution of these simultaneous equations yields the differential equation which defines the temperature variation of the air stream:

$$(WC_p/B)(dT/dx) = [1/(1 + U_L/h)][Hf_{c,e} - U_L(T - T_a)] \tag{8.5}$$

where

$$h = h_c + \frac{1}{1/h_c' + 1/Eh_r} \qquad U_r \ll U_{up}, \quad U_L \approx U_{up}$$

Letting $F' = 1/(1 + U_L/h)$ and solving the differential equation, the air temperature T at a distance x from the inlet end of the collector is found to be

$$T = Hf_{c,e}/U_L - (T_1 + T_a + Hf_{c,e}/U_L)\exp(-BF'U_L x/WC_p) + T_a \tag{8.6}$$

The air temperature at the collector exit at a distance L is

$$T_2 = T_1 + \{Hf_{c,e}/U_L - (T_1 - T_a)[1 - \exp(-F'U_L/GC_p)]\} \tag{8.7}$$

The overall heat transfer coefficient between the air stream and outside air, through the transparent cover U_0 is given by

$$U_0 = F'U_L, \qquad U_0 = 1/(1/U_L + 1/h) \tag{8.8}$$

Under ideal conditions where the heat transfer from the absorber plate to the air stream is perfect,

$$h = \infty, \quad F' = 1, \quad U_0 = U_L$$

Heat transfer texts refer to $(F'U_L/GC_p)$ as NTU, the number of transfer units. Substituting $F' = U_0/U_L$, the heat picked up by the air stream is found to be

$$\begin{aligned} q_u/A &= GC_p(T_2 - T_1) \\ &= \left(\frac{1}{1 + U_L/h}\right)\left(\frac{1 - \exp(-U_0/GC_p)}{U_0/GC_p}\right)(Hf_{c,e} - U_L(T_1 - T_a)) \end{aligned} \tag{8.9}$$

The terms in the second parenthesis is known as the flow factor F''. Using these definitions, the working equation is reduced to

$$q_u/A = F'F''[Hf_{c,e} - U_L(T_1 - T_a)] \tag{8.10}$$

A careful examination of Eq. (8.10) reveals possible means of increasing the amount of useful heat collected for given solar radiation input for a conventional nonporous flat plate collector. The efficiency is increased by increasing F', F'', $f_{c,e}$ and reducing U_L. U_L and $f_{c,e}$ are related with the characteristics of the upper surface of the absorber plate and the transparent cover system. F' and F'' can be increased by increasing the mass flowrate per unit area and h the convective heat transfer coefficient between the air stream and the absorber plate as well as increasing surfaces to transfer heat from the absorber plate to the air stream using fins.

Equation (8.10) is the same form used for water collectors. The solar air heater output can be determined once the radiation intensity, flowrate, ambient and inlet temperatures, and design features of the collector are given. For constant F', F'', U_L, and $f_{c,e}$, q_u/A is represented by a straight line if plotted against $(T_1 - T_2)$. The solar radiation intensity H, the main parameter identifying the performance of the air heater, must be above a threshold level for any useful heat collection.

The threshold level is defined by

$$H_c = U_L(T_1 - T_a)/f_{c,e}$$

where H_c denotes the critical radiation intensity. In actual practice, the critical radiation intensity (threshold intensity) must be increased by a factor P on the order of $P = 1.05$ to pay back the fan power consumed.

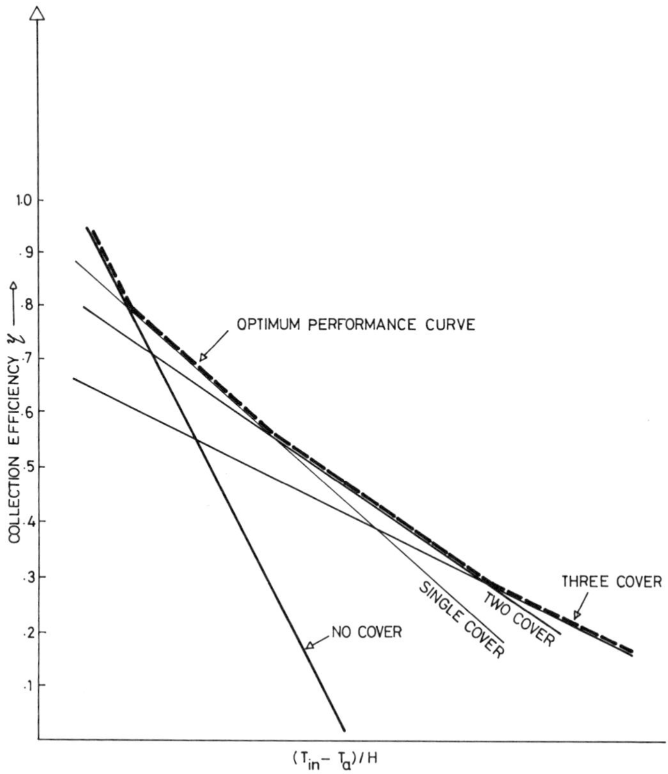

FIG. 8.4a

FIG. 8.4 (a) Generalized performance curves for typical flat plate nonporous collector. (b) flat plate solar air heater performance.---- optimum performance curve; ○ Vee-corrugated absorber, from Close (1963); □ finned absorber, from Kuzay *et al.* (1974); ▽ transpired honeycomb, from Buchberg *et al.* (1974). The theoretical performance covers are plotted from data by A. Whilier (1964).

The useful heat collected is, therefore, expressed with the linear relation of

$$q_u/A = CH - K(T_1 - T_a), \qquad H \geqslant PH_c \tag{8.11}$$

where $C = F'F''f_{c,e}$ and $K = F'F''U_L$.

A major difficulty in representing the performance using the above equation is the necessity of a separate line for each radiation intensity. The solar air heater efficiency could be calculated from

$$\eta = q_u/(A \cdot H) = C - K(T_1 - T_a)/H, \qquad H \geqslant PH_c \tag{8.12}$$

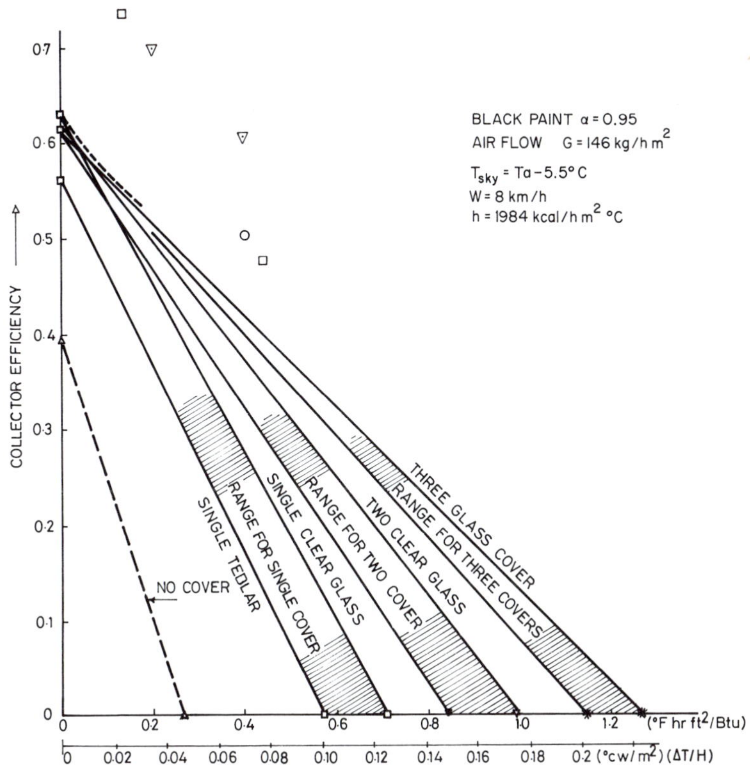

FIG. 8.4b

If the efficiency is plotted against $(T_1 - T_a)/H$, then a single generalized curve would represent the collector performance. Thus, various cover arrangements and/or selective coatings could be compared. Assuming U_L as a constant throughout the operation range is also misleading. For large variation of $(T_1 - T_a)/H$, U_L varies too. Therefore, the straight line assumption is not valid any more. Besides, for very low radiation intensities, $(T_1 - T_a)/H$ assumes very high values, which is also misleading. In this case it is necessary to use Eq. (8.10) instead of the generalized performance curve of Fig. 8.4 with utmost care in the proper evaluation of F', F'', and U_L.

The number of transparent cover layers glazing the absorber is the most important factor influencing the collector performance. For a low temperature rise, such as required for drying purposes, on the order of 5–10°C, no transparent covers are required. Increased transmission losses due to extra cover do not justify the heat loss reduction up to a point at which no cover and single cover efficiency curves intersect. Beyond this point the collector with a single cover collects more heat than the no cover collector. The best cover arrangement from the optimum performance curve is also shown in Fig. 8.4. Final selection of the number of covers has to be based on the economic performance criterion which will be discussed later.

8.4 PERFORMANCE OF VARIOUS COLLECTOR TYPES

In this section the thermal performance characteristics of various collector types will be discussed. The design variations of the conventional flat plate collector with flow behind the plate and collectors with porous absorbers will also be examined.

A. Nonporous Absorbers

Variations of the nonporous absorbers comprise:

absorbers with flow above the absorber plate only;
absorbers with flow on both sides of the absorber;
absorbers with air flow behind the absorber plate: without fins or with fins; and
Vee-corrugated plate.

The first two items will not be covered due to their limited applications in comparison with the flow behind the plate-type collector.

1. Flat Plate Collectors with Finned Absorbers

Adding fins to the rear side of the absorber improves the heat transfer, but also introduces extra pressure drop. Beyond some optimum point, increasing the number and depth of fins does not improve the performance since fan power requirements will also increase. Malik and Buelow (1973) have studied the heat transfer from the absorber plate to the air stream and reported correlations between the pressure drop and the fluid friction. Theoretical results were verified with the experiments.

A summary of the conclusions reached is as follows:

Tests with a flat sheet duct type A and corrugated sheet duct type B for Reynolds numbers from 10,000 to 50,000 were run to obtain pressure drop and heat transfer relations.

The Reynolds number is given by

$$\mathrm{Re} = (G \cdot d_{\mathrm{h}})/\mu$$

where G is the mass flowrate (g/cm^2 s), d_h the hydraulic diameter $= 4A/C$: area/wetted perimeter, and μ the dynamic viscosity (g/cm · s). Evaluating the properties at mean bulk temperature and correcting for viscosity effect due to bulk and wall temperature effect by $(\mu_B/\mu_W)^{0.14}$, the Nusselt number for heat transfer is

$$\mathrm{Nu} = [M_a \, \Delta h'/(T_W - T_B)]d_h/k$$

where T_W and T_B denote wall and bulk fluid temperatures, respectively, and $M_a \, \Delta h'$ is enthalpy rise of the air stream for a 0.9-meter long duct that gives a total heated wall area of 10.93 m^2.

$$M_a \, \Delta h' = M_a(t_{B_2} - t_{B_1})(0.24 + 0.45W)$$

Typical experimental values of Nusselt numbers for (duct A) and (duct B) adapted from Malik and Buelow (1973) are listed in Table 8.2. For rough and smooth ducts, the ratio of Nusselt numbers is given by

$$\mathrm{Nu}_{\mathrm{rough}}/\mathrm{Nu}_{\mathrm{smooth}} = (f/f_0)^{0.5}$$

where $f = -d_h(dp/dx)/(\rho v_m{}^2/2)$, the friction coefficient for fully developed flow, where dp/dx is the pressure gradient along the duct, ρ the density of air, and v_m the mean air velocity. The rough duct improves the heat transfer rate by a factor of $(f/f_0)^{0.5}$, whereas the pressure drop increases by (f/f_0). Values of $(f/f_0)^{0.5}$ corresponding to the flat and corrugated ducts are listed in Table 8.3.

TABLE 8.2

Nusselt Numbers for Solar Air Heater Ducts without Fins

Duct A (smooth plate)		Duct B (corrugated plate)	
Re	Nu	Re	Nu
9890	34.12	9740	38.65
14800	48.95	14580	55.80
20000	60.33	19600	63.30
24785	71.31	24580	88.02
29780	89.10	29390	88.93
		39220	121.81

Since improvement in heat transfer rates is lower than the rate of increase of pressure drop, increased fan power requirements must be carefully checked. Otherwise, the extra heat collected may not justify the increased power consumption.

TABLE 8.3

$(f/f_0)^{0.5}$ Factors for Air Heater Ducts

Duct A		Duct B	
Re	$(f/f_0)^{0.5}$	Re	$(f/f_0)^{0.5}$
10000	1.180	10000	1.435
20000	1.245	20000	1.461
30000	1.305	30000	1.542
40000	1.355	40000	1.601
50000	1.390	50000	1.656

Recent work by Kuzay *et al.* (1974) compares the performance of straight and staggered finned solar air heaters. The findings of this survey indicate substantial efficiency improvements.

Test runs using straight staggered galvanized fins 3.8 cm high and 49 cm long spaced about 5.1 cm and U-shaped staggered aluminium fins (2.5 cm high, 1.9 cm long), U section (top 1.0 cm, bottom 0.6 cm), show substantial friction factor increase. At a given Reynolds number for the straight finned ducts, the Nusselt number is reported to be 3.3 times that of a smooth duct. Similarly, a staggered U finned duct has a Nusselt number 33 times the smooth duct values. Corresponding friction factor increases are threefold for the straight fins and twentyfold for the U fins, respectively. Typical efficiencies for U finned absorbers at 44°C plate (18°C outside temperature) and 80°C plate (7°C outside temperature) are 74 and 49%, respectively. These figures are substantially higher than conventional no-finned absorbers which give collection efficiencies in the order of 45 and 35% for those two temperatures specified.

2. Vee-Corrugated Absorber

One way to increase both the heat transfer area available for the air stream and the effective absorption of the absorber plate is to use a Vee-corrugated absorber (VCA). If the Vee acceptance angle is kept below 60°, most of the solar radiation is absorbed after multiple reflections occurring on the sides of the VCA. For small acceptance angles, a cavity effect is obtained regardless of the specular absorptivity of the absorber surface. This effect alone is not sufficient since effective emissivity also increases. A selectively coated Vee groove of about 55° is suggested; thus, the heat transfer area is doubled and losses due to outgoing long wave radiation are kept to a minimum. The free convection between the Vee corrugation and the top glass cover is more than a plane absorber as reported by Charters and MacDonald (1973). The convective coefficient between the rear surface of the VCA and the air stream is also greater. Thus, the increased upper losses are offset by the increase of heat transfer to the air stream offering a net collection increase. Performance predictions of VCA were first given by Close (1963). A typical VCA has an efficiency of about 50% for 87.8°C air outlet temperature and 946.5 W/m^2 solar flux input. Conventional flat plate absorbers have an efficiency of about 10% for the same operating conditions. Hollands (1963) has examined the radiative properties of Vee-corrugated absorbers. Air heaters employing selectively coated Vee-corrugated copper foil absorbers were used in Australia for solar drying applications by Read (1974). The improvement of collection efficiency must be traded against the increase of pressure drop and absorber plate material usage for cost effective designs.

B. Porous Absorbers

The simplest form of a porous heater is a perforated absorber plate placed across the air stream. This design has limited heat transfer area, therefore, is not recommended. The most popular porous absorber types are wire mesh, slit and expanded metal, and transpired honeycomb absorbers.

1. Wire Mesh Absorber

Bliss (1955) first discussed the performance of a wire mesh absorber. The performance of a wire mesh or slit and expanded metal absorber are comparable, although heat transfer coefficients and available heat exchange surface per unit volume differ. An oversimplified heat transfer model of the porous matrix based on the relation of

$$\eta_c = F'F''[H_a - U_L(T_1 - T_a)]$$

used for nonporous absorbers is presented below: Referring to Fig. 8.5, the absorber efficiency factor F' for a matrix absorber is

$$F' = \frac{1}{1 + (U_c/h)}, \qquad U_L = U_c + U_a + U_r$$

where $U_a = 1/[(1/h_g) + (1/U_{ga})]$ is the heated air to ambient coefficient, $U_c = 1/[(1/hr_{cg}) + (1/U_{ga})]$ the collector to ambient coefficient, U_r the rear heat loss coefficient, and H_a the solar radiation absorbed by the absorber.

Among the heat transfer coefficients U_a and U_r could be closely predicted. However U_c and h must be experimentally determined.

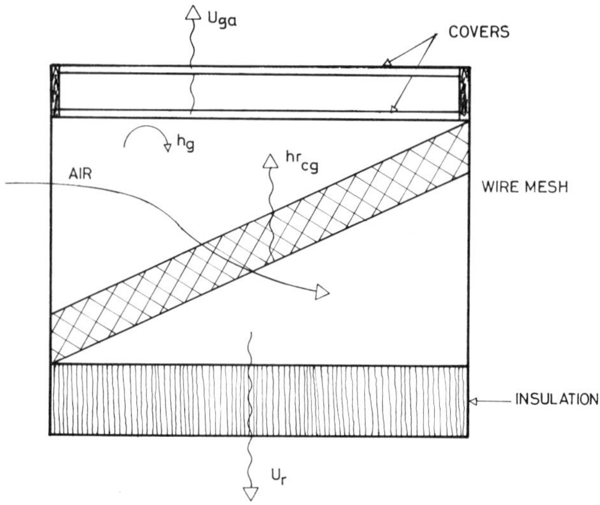

FIG. 8.5 A sketch of a wire mesh air heater.

2. *Slit and Expanded Metal Absorber*

Whether a wire mesh or slit and expanded matrix is used, the performance expected would be essentially the same. The slit and expanded foil matrix has, however, more potential in terms of cost reduction. Shoemaker (1962) first reported test data on the foil-type slit and expanded matrix and demonstrated performance improvements. These results give temperature flowrate relations, but without radiation data.

Duffie, Chiou, El-Wakil, Beckman, and Hamit, all of the University of Wisconsin, have further examined various types of slit and expanded matrices as solar heat absorbers and obtained correlations between the flow, pressure drop, heat transfer rates, and collection efficiency as referenced before. A typical collection efficiency around 75% at 45°C collection temperature is (1.3) times better than the nonporous absorber-type solar air heater under the same conditions.

The pressure drop ratio of a matrix type to a nonporous type is in the order of

$$\frac{1.3 \times 10^{-3}\ (\mathrm{lbf/ft^2})}{20.4 \times 10^{-2}\ (\mathrm{lbf/ft^2})} = \frac{1}{160}$$

Both heat transfer rates and pressure drops vary for different slit and expanded metal shapes and matrix thickness. The above data illustrate the potentialities in terms of efficiency improvement and pressure drop reduction.

The rigorous mathematical analysis developed by the University of Wisconsin team may be used for optimal matrix design. Analytical results, however, have to be verified by the test run of actual matrices to ascertain accurate data. Details of heat transfer analysis of the matrix is considered beyond the scope of this text. Without going into the mathematics, the following could be identified as factors influencing the matrix performance (Beckman, 1968; Hamit and Beckman, 1969, 1971):

(a) Increased flowrate improves the matrix efficiency. Figures above 95% are expected for $\Gamma = GC_pT_0/\sigma T_0^4 > 100$, where G is the mass velocity of air (lb/h-ft^2), C_p the specific heat of air (Btu/lb°F), T_0 the air inlet temperature, and σ the Stefan–Boltzmann constant.

(b) The increase of the dimensionless volumetric heat transfer coefficient improves the collector performance up to $\theta = (h_v/B_o\sigma T_0^3) > 50$, then does not have an effect. Here h_v denotes volumetric heat transfer coefficient and B_o is optical absorption coefficient.

(c) Increasing the bed conductivity has no effect on the performance of porous bed except at very low flowrates and rather high volumetric heat transfer coefficients in which case the maximum bed efficiency decreases.

(d) Fluid conduction has no effect on the overall performance of the collector.

(e) The bed efficiency is more sensitive to Γ and θ than dimensionless incident beam and diffuse fluxes and matrix thermal conductivity.

(f) Temperature profiles and overall matrix performance is insensitive to rear surface radiation effects.

3. *Overlapped Glass Plate Air Heater* (*OGPLAH*)

The overlapped glass plate air heater combines the advantages of low pressure drop and high efficiency for moderate temperature rises. The cost of the unit must, however, be carefully examined since the glass area requirements are four times the collection area, including the single cover glass. In case inlet air is at temperatures higher than the ambient, then double glass covers may be needed which increases the glass requirements five times as much as the absorber area.

The mathematical model of the OGPLAH is quite complicated. The reader is referred to Selcuk (1971) for its solutions. Figures 8.6 and 8.7a,b illustrate an OGPLAH and heat transfer model. Computer-plotted air temperature variations are shown in Fig. 8.8. Air temperature rise in the collector is plotted as a function of average air velocity in Fig. 8.9. As shown in Fig. 8.9 increased air velocities up to about 1 m/s increase the collection efficiency. Beyond this, efficiency levels off to about 75%, temperature gain reduces and pressure drop increases further. Variation of the spacing between

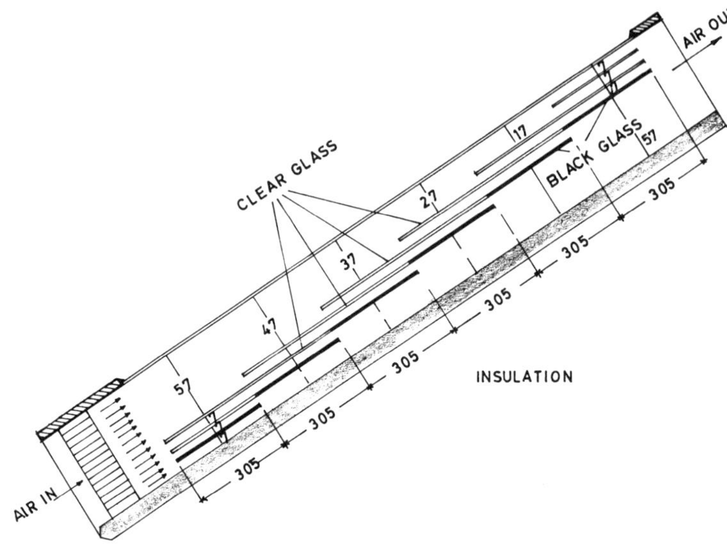

FIG. 8.6 Sketch of the overlapped-glass plate solar air heater. All dimensions, are in millimeter; all glass plates are 3 mm thick; width of the collector is 690 mm.

plates and number of plates slightly influences the collection efficiency. Optimum values lie about 5 to 7 mm spacing between glass plates, glass thickness less than 3 mm. The cost of the heat collected decreases with the increasing collector length. However, due to difficulties of fabrication and handling of the collector units, a reasonable size is about 1.8 to 2 m. Using clear glass as cover material improves collection efficiency, but the cost increase usually does not justify it. Using ordinary glass for intermediate staggered plates is in fact helpful due to increased absorption of the solar flux. If inexpensive configurated glass which has rough surfaces can be used, efficiency improvement is expected. Care must be exercised not to increase pressure drop excessively. The effect of semitransparent glass to rough surfaced glass has not yet been fully investigated.

4. *Transpired Honeycomb Absorber*

Transparent honeycomb structures made out of glass or clear plastics can be used to suppress convection between the absorber plate and cover glass for solar air heaters with nonporous absorber plates. A state of the art review paper by Buchberg *et al.* (1974) outlines their fundamental and applied research on honeycomb convection suppression as well as that of other investigators in France, (Perrot *et al.* 1967), Canada (Hollands, 1965, 1973), and Australia (Charters and Peterson 1972).

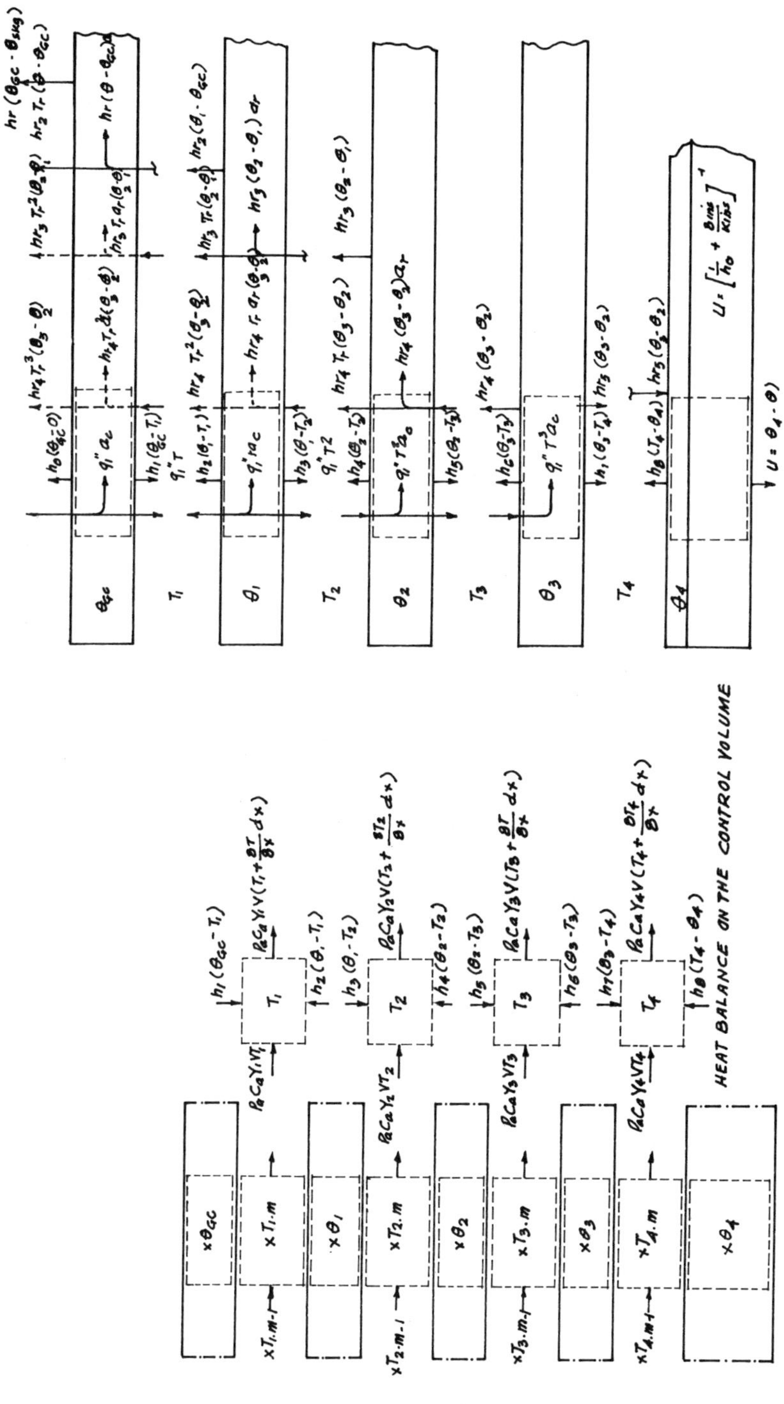

FIG. 8.7 (a) Coupled systems and control volumes (radiant exchange between glass plates taken into consideration); energy exchanges on the air streams shown only. (b) Energy balance of the glass plates (long-wave radiation).

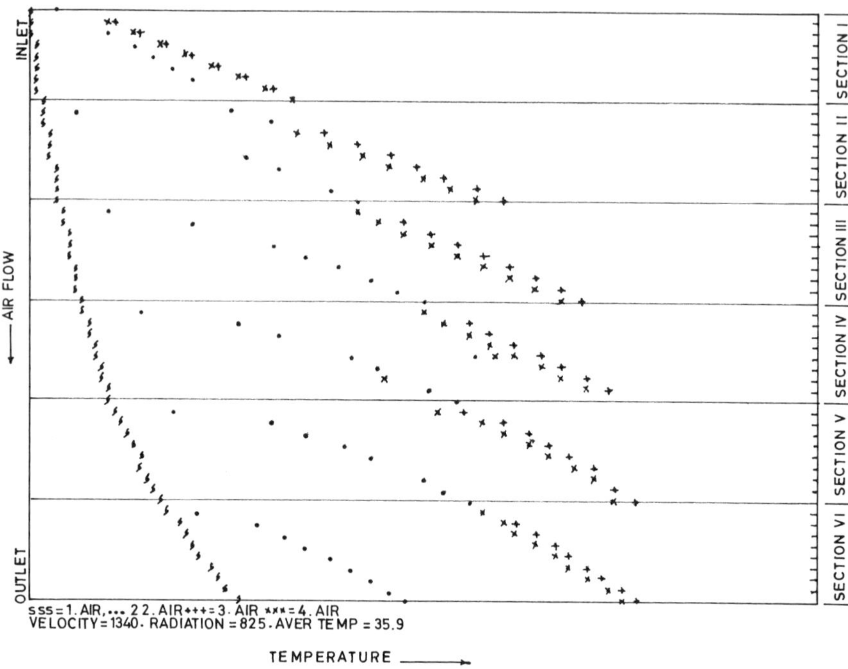

FIG. 8.8 Computer-plotted air temperatures. Each division represents 0.5°C temperature rise above ambient air inlet at ambient temperature.

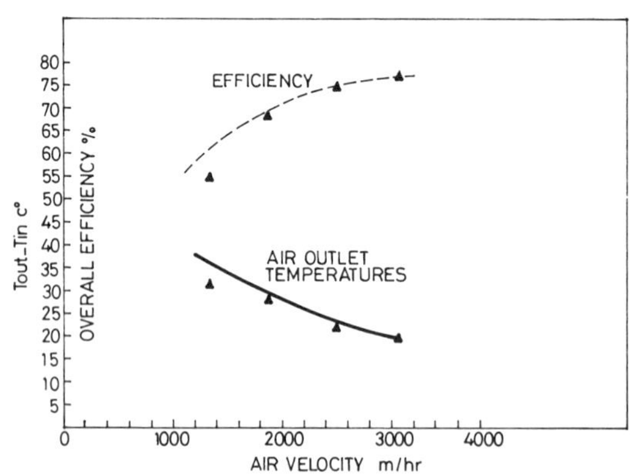

FIG. 8.9 Comparison of measured (▲) and calculated (---) efficiencies and air outlet temperatures for various air velocities.

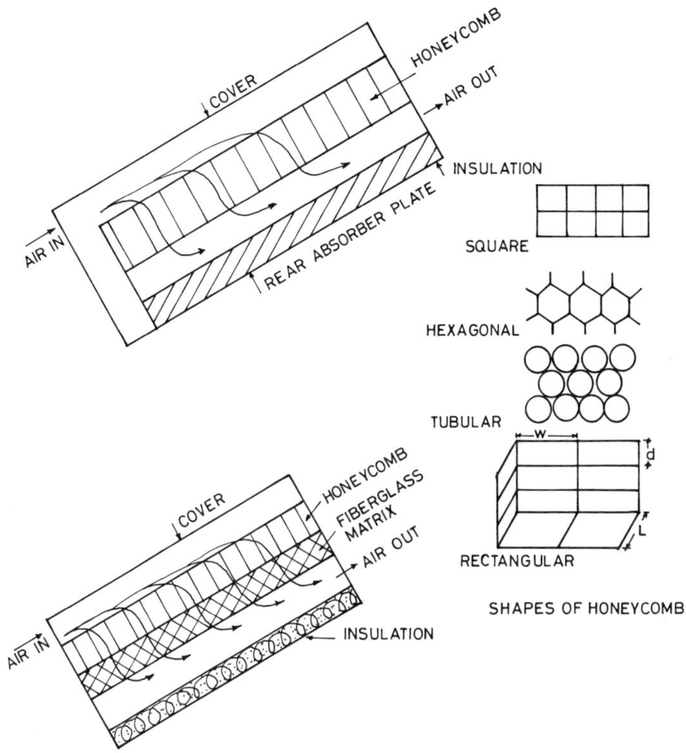

FIG. 8.10 Transpired honeycomb solar air heater design variations.

The major benefit of such an application is the reduction of upper heat losses from the absorber plate. The problem of heat removal from the absorber plate must still be solved by either roughened absorber plate design or added fins. Analysis of solar collectors with transparent honeycomb convection suppressors will not be covered here.* A solar air heater can be built passing air through a honeycomb structure. Such an arrangement is a kind of porous absorber. Various design possibilities exist; among them are transparent or opaque honeycombs, hexagonal or rectangular shapes, and combination of honeycomb structure with a porous matrix as shown in Fig. 8.10.

Buchberg and colleagues have developed a unique solar air heater employing transpired rectangular honeycombs having specularly reflecting walls

* Honeycomb collectors are examined in detail in Chapter 7.

coated with a thin lacquer film. The longer side of the rectangular is laid in the east–west direction. Optimum honeycomb dimensions: recommended length–width and depth ratios for the optimum performance are $w/d = 3.4$ and $L/d = 7.11$; an overcoat thickness of 0.0127 mm; and a mass flowrate of about 4.882 kg/h m^2. Figures 8.11 and 8.12, adapted from Buchberg *et al.* (1971), illustrate the variation of thermal efficiency with $(T_{av} - T_a)/H$ where T_{av} denotes average air temperature, T_a is ambient temperature, and H is solar flux. The availability factor β, which is defined as $\eta[1 - (T_a/T_p)]$,

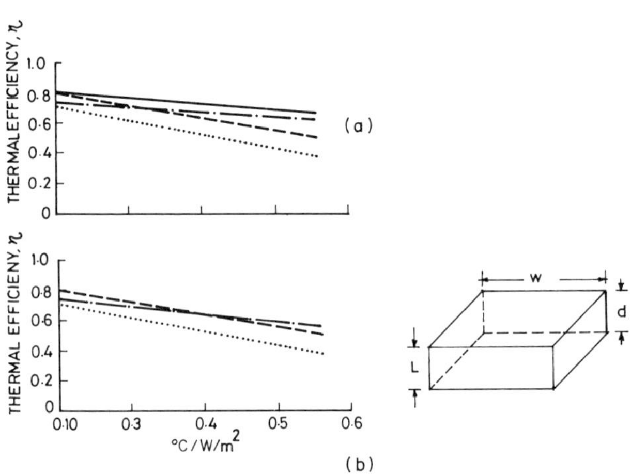

FIG. 8.11 Comparison of thermal efficiencies of experimental honeycombs with predictions and with efficiencies of heaters I and III after Gupta and Garg (1967). (a) $w/d = 3.4$; $L/d = 7.1$. (b) $w/d = 2.1$; $L/d = 4.17$.—experiment; –·–theory; ––I; ···· III.

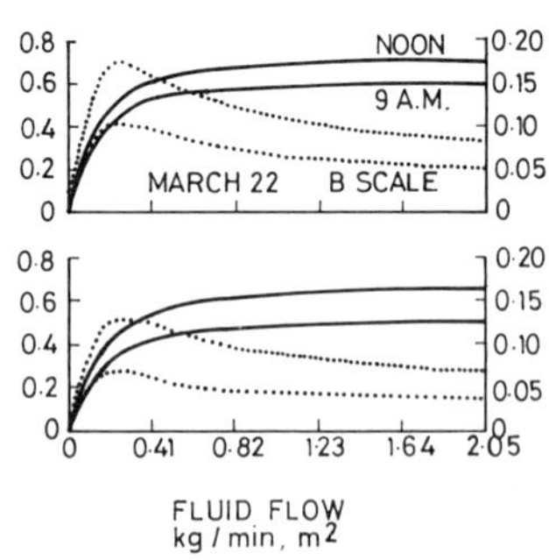

FIG. 8.12 Effect of fluid flowrate on thermal efficiencies for solar input of 413.3 W/m^2. $T_a = 23.75$°C; $T_i = 425$°C; $h_c = 5.19$ W/m^2 °C; $w/d = 5.0$; $L/d = 7.0$. Lacquer thickness, $\delta = 0.0127$ mm; $\gamma = \phi = 36$°N, tilt angle. From Buchberg *et al.* (1971).

peaks at about 0.976–1.465 kg/h m^2 mass flowrate (G), although the efficiency η increases with the increasing flowrate. Unnecessarily large flowrates are not recommended since the temperature rise of the air stream would be too low for any practical application.

8.5 APPLICATIONS OF SOLAR AIR HEATERS

Air heated by using one of the solar heat collectors described in the previous sections could be mainly used for the following processes:

(a) Heating buildings.
(b) Air conditioning buildings utilizing desiccant beds or an absorption refrigeration process.
(c) Drying agricultural produce and lumber.
(d) Heating greenhouses.
(e) Using air heaters as the heat source for a heat engine such as a Brayton or Stirling cycle. This specific application, however, requires temperatures in the order of hundreds of degrees which is beyond the practical range of operation of those collectors described. Recent proposals, suggesting the use of concentrating type collectors (Russell, 1974 and Jarvinen, 1975), are not yet fully developed and documented experimentally.

The following processes, which cannot be considered as solar heat collector applications, are also worth mentioning.

(a) Direct absorption of the solar radiation in the space to be heated such as rooms with thermopane windows. Such experiments were run in the early 1950s by Dietz and Czapek (1950) and recently by M. Telkes at the University of Delaware. The latter had a heat of fusion type thermal storage material, which acted as a thermal buffer in the room to be heated. A similar attempt was made by Trombe and Ducarroir (1972) in France, in which the concrete wall of the building acted as the absorber and the thermal storage medium.
(b) Drying agricultural produce in a package unit which combines the heater and container for the material to be dried. Examples are cabinet dryers or lumber stacks with weather-protecting plastic film covers.
(c) Direct absorption of the solar radiation in the greenhouse which acts as a solar air heater with no air circulation.

The term "greenhouse effect" usually refers to solar air or liquid heaters, employing transparent glazings. This phenomenon was believed to control the greenhouse temperature, namely, transmission of the short wave solar radiation through the greenhouse glazings by entrapment of the outgoing radiation. Recent investigations, however, indicate that radiation heat transfer is not the only factor influencing the greenhouse environment; instead, free convection is also significant (Damagnez, 1968 and Iqbal 1975).

The thermal environment of greenhouses and the effect of solar radiation was examined by various investigators, including Selçuk (1970), Oztoker and Selçuk (1971, Lawand *et al.* (1973), Takakura *et al.* (1969), Gac *et al.* (1968), Pozin (1971), and Seginer and Levav (1971).

A. Space Heating

Solar heated houses have been built and operated since the late 1940s (Telkes and Raymond, 1949; see also Chapter 17). The first house had ($Na_2SO_4 \times 10H_2O$) Glauber salt as the thermal storage medium. Lof (1963) used rockpile storage and overlapped glass plate solar heaters in the solar house illustrated in Fig. 8.13.

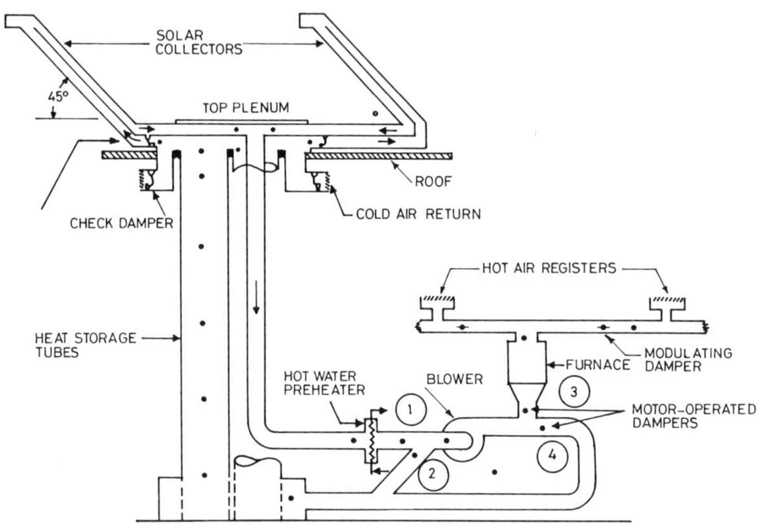

FIG. 8.13 Schematic of the Denver House air heating system after Lof *et al.* (1963).

A recent experiment employing solar heat collectors for space heating is the solar house of the University of Delaware (Boer *et al.*, 1973). The house has the feature of eutectic salt thermal storage and collectors that generate electricity via CdS cells as well as heating air. Among various air heater designs tested were the finned absorber-type design previously discussed. The solar cell panel–air heater combination is still being perfected.

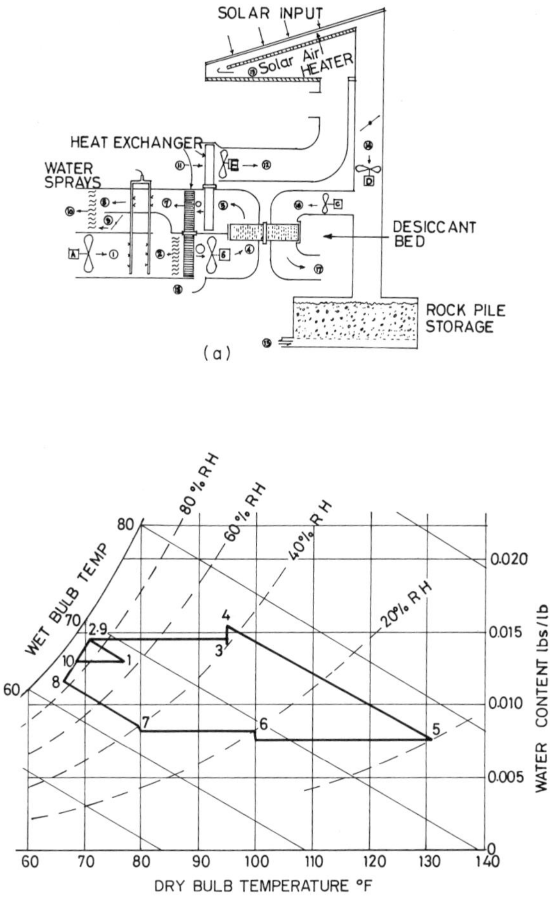

FIG. 8.14(a) Solar air conditioning system after Dunkle (1965). (b) The solar air conditioning cycle on psychometric chart.

Solar air conditioning is attainable using desiccant beds as suggested by Dunkle (1965), which is illustrated in Fig. 8.14a. The state points of the air processed are also marked in Fig. 8.14b.

Another approach to obtaining coolness using solar air heaters is to supply heat to the generator of an absorption air conditioner. A major restriction to be borne in mind is maintenance of generator temperatures above 86°C, which necessitates further higher air temperatures around 95°C due to poor heat transfer between the air stream and the solution in the generator.

A conceptual diagram illustrating the use of solar air heaters for a space heating air conditioning application is illustrated in Fig. 8.15. The proposed system has an air to water heat exchanger and suggests the use of water as the thermal storage medium.

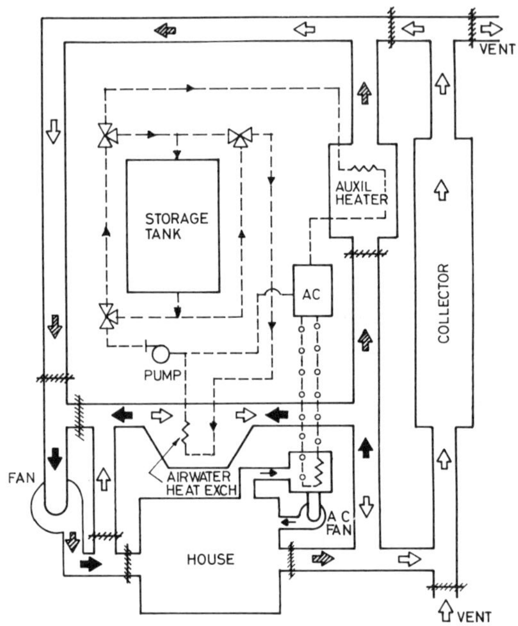

FIG. 8.15 Air system heating and air conditioning. ——— water line; –○– refrigerant line; ⇒ air flow.

B. Solar Driers

The advantages of solar drying over open sun exposure of agricultural produce have been accepted, although only limited applications can be observed.

Among driers proposed and constructed are a cabinet-type drier by Lawand (1963, 1966); a 1500 ft^2 plastic collector of Williams (1964), which aided corn and hay drying; and those slit and expanded sheet iron or metal chips matrix tested at the Middle East Technical University from 1965 until 1971 (Selçuk, 1966, 1967, Akyurt and Selçuk, 1973). Most of the early solar collector driers were constructed as a research tool, therefore, economy in construction had not been of primary importance.

Solar driers are expected to be used by farmers with limited technical skills and small capital; therefore, these proposed devices should be simple,

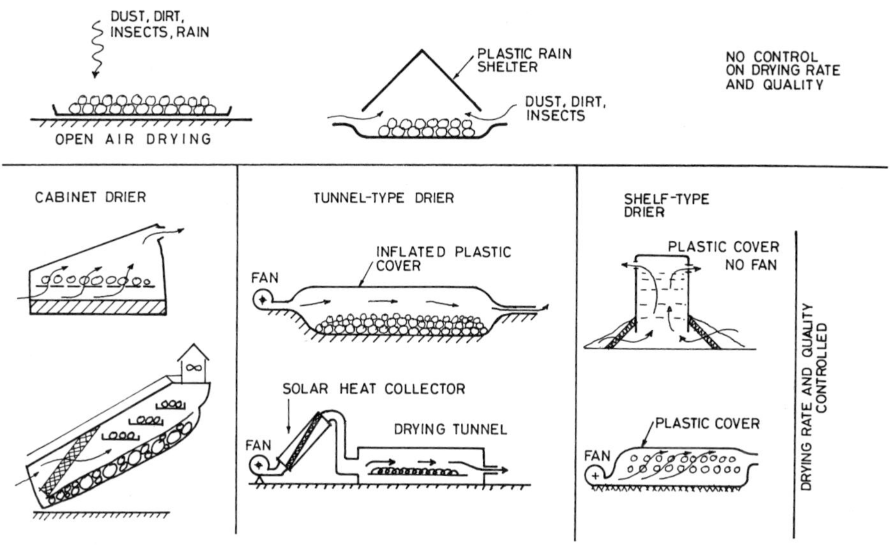

FIG. 8.16 Solar drying process and various drier designs.

inexpensive, and use least land area. Figure 8.16 illustrates basic solar drying processes and various designs tested or suggested. Protection from the rain by plastic sheets does not suffice to eliminate dust, dirt, and the entrance of insects. Totally enclosed driers provide both of these protections and enable control of the rate of drying. The cabinet-type driers may serve family needs while large-scale commercial applications necessitate tunnel or shelf-type driers. An inflated plastic drier is the simplest to build, but charging the wet produce, turning over during drying, and emptying could be a problem. During a recent experiment in Brazil, using a glass covered design, copra was successfully dried (Gosh, 1975). Shelf-type driers with plastic liners without forced air circulation are extensively used in Australia (Szulmayer, 1973).

Recently, the United States too, has emphasized various solar drying experiments (Cherne, 1974). Fruit and grain drying using solar radiation could provide substantial energy savings for those instances where bottled or natural gas is used to supply the heat.

A drying tunnel served by a battery of solar air heaters has to compete with the conventional heat sources since the investment required for the tunnel is not eliminated.

These driers must also be designed for large capacities and at a location other than the field itself. A field-type shelf-type drier, on the other hand, eliminates the problem of transporting the produce from the field to the tunnel site. Thus, the moisture, which has to be ultimately reduced to 20%, must also be carried. If the produce spoiled or lost during the transport is considered, there are major advantages to the field-type driers.

These driers do not offer long life, but are inexpensive and could well prove to be cost effective. Additional advantages are ease of construction by the farmer at the field where land value is of no importance and investment required is much lower.

9

Concentrating Collectors

A. B. MEINEL

OPTICAL SCIENCES CENTER
UNIVERSITY OF ARIZONA
TUCSON, ARIZONA

9.1 INTRODUCTION

The optics for solar concentrators generally can be classified as either *point focus* or *line focus* systems. Point focus systems have circular symmetry, and are generally used when high brightness concentration factors C are required, as in solar furnaces and central receiver power systems. Line focus systems have cylindrical symmetry, and are generally used when medium concentration is sufficient to reach the desired operating temperatures. A third group has occasionally been used that produces a line focus of limited length. These types are exemplified by the truncated cone (axicon) and the hemispherical bowl mirror.

In the following section we shall discuss the properties of these classes of optical systems, defining their optical performance and energy yields. Associated with these classes of optics, we have classes of mountings. The point focus systems require tracking of the sun during the day in two coordinates. Line focus systems require tracking only in one coordinate, and the choice of tracking mode is an important element in the choice of optical design to yield maximum output for the greatest simplification over a fully tracking system.

An important class of optical system that minimizes tracking to occasional season adjustments has also been recognized. The trade off between optical concentration factor and the angle over which the sun is accepted is also discussed in this section.

9.2 POINT FOCUS CONFIGURATIONS—PARABOLIC MIRRORS

The principal point focus optical system for solar collection is the parabolic mirror. The basic geometry of the parabolic mirror is shown in Fig. 9.1. This optical surface forms a point image on the optical axis and provides the maximum possible brightness of the solar image. A parabolic mirror, however, does have optical aberrations of significance. The first of these is coma. Coma means that even if the central image is essentially perfect, the images formed when the object is off-axis have coma, resulting in a degraded image. The second is scale error for the on-axis image. The origin of this scale error is shown in Fig. 9.2. Rays imaged from the central

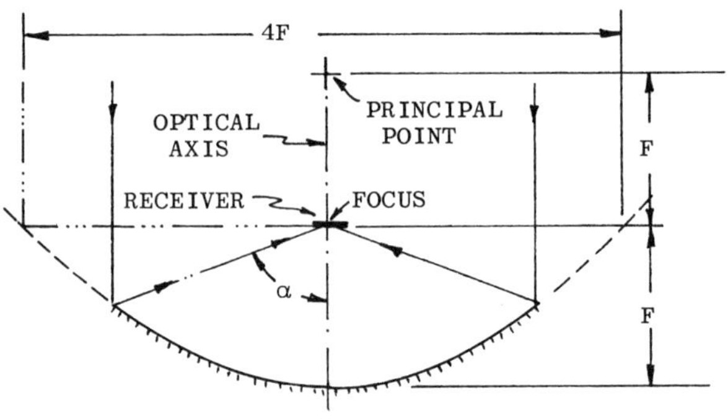

FIG. 9.1 Basic geometry of a paraboloid. The rays marked α are for a rim ray angle of 70°. The maximum aperture of a paraboloid is $4F$ for a plane receiver normal to the optical axis at the focus.

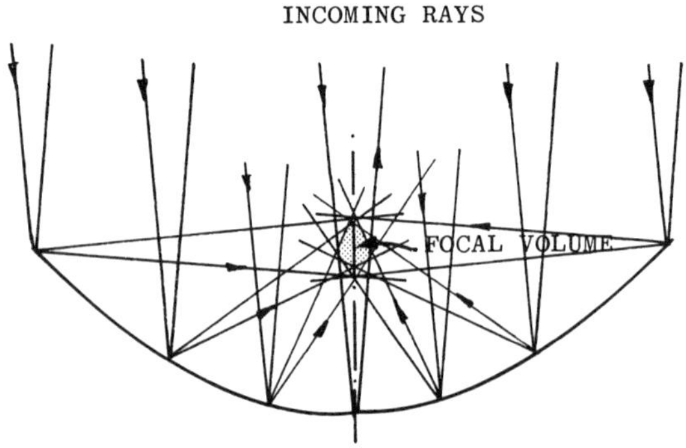

FIG. 9.2 Diagram illustrating the formation of the focal *ellipsoid* for a source having a finite angular size. All solar optical configurations where the distance from the mirror to the absorber varies result in a similar focal volume.

region of the paraboloid travel a short distance in arriving at the focus. Rays imaged from the edge of the mirror travel a significantly greater distance for "fast" paraboloids, resulting in a greater spread of the image. The result of this second aberration is that the three-dimensional image of the sun formed by a fast paraboloid is an ellipsoid, with major axis pointing toward the mirror. For an $F/0.5$ paraboloid, where the rim ray makes an angle α of 90° to the optical axis, the ratio of the major to both minor axes is 2:1. It should be noted that the boundaries of the focal ellipsoid are not sharply defined, as indicated by the various ray crossings in Fig. 9.2.

The comatic aberration of a fast paraboloid hardly looks like the traditional coma of an optical system due to its extreme magnitude. In Fig. 9.2 the ray pattern for the off-axis image aberration was shown by following the rays marked by the arrows.

The parabolic mirror has several variations in which it can be used. The case where the paraboloid is simulated by a number of individual flat or curved mirrors is one case of special importance, and will be discussed separately.

A variation of interest has been studied by Veinberg (1959). He shows that three distinct variants, each having the same absorber area are possible. They are shown in Fig. 9.3 as (A) absorber facing the mirror vertex, (B) absorber parallel to the optical axis, and (C) absorber facing the sun.

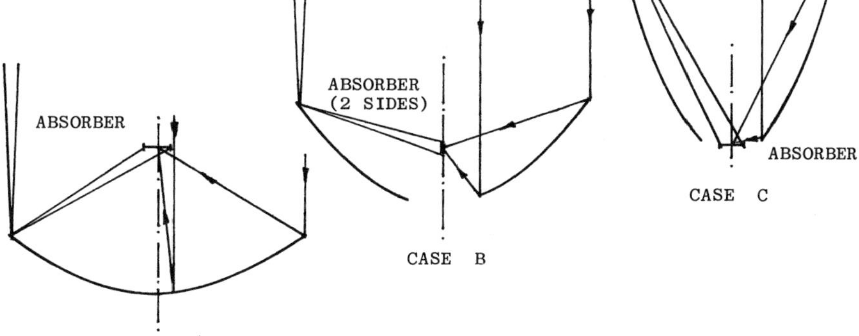

FIG. 9.3 Optical diagrams illustrating the three types of parabolic mirrors having (A) downward facing absorber, (B) coaxial absorber, (C) upward facing absorber.

Veinberg showed that the optimum aperture of each, weighing the angularity of the cones of sunlight approaching each absorber, was approximately equal, with the case (A) mirror being about 8% better than case (B) and 12% better than case (C). Case (C) is especially useful when the sunlight is to enter the top of the absorber cavity. A collector has been made of the type of case (C) by approximating the curve by three conical sections where the absorber was approximately the width of the opening at the bottom of the lowest cone.

9.3 CIRCULAR FRESNEL LENSES

Lenses are not used to a significant degree in current solar energy applications, except in the form of Fresnel lenses. The reason for standard lenses not being used is their weight and cost. Fresnel lenses are favorable from the weight and cost aspects and may find uses in current applications.

The principle of these lenses is shown in Fig. 9.4. A Fresnel lens is divided into a number of zones, the spacing of which can be as small as a few tenths of a millimeter to several centimeters. Within each aperture zone the surface of the lens is tilted so that it refracts light at the same angular amount as a normal spherical lens of the same focal length. In general, these lens facets are flat, but with tilted surfaces since the effect of curvature would be negligible in affecting the image size. If the facets were a significant portion of the aperture radius and larger than the desired focal image size, then curvature of the facets would be necessary.

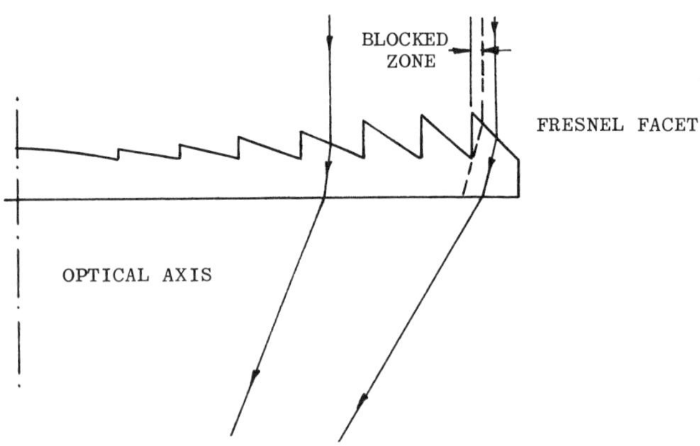

FIG. 9.4 Schematic diagram of a "fast" Fresnel lens. Note that the full aperture of the lens does not contribute to the brightness of the image due to a zone blocked by the adjacent facet. Faceted mirrors also suffer from blocked zones, as can be seen by imagining the facets to be made reflective on their top surfaces.

The equation for the tilt of the facet as a function of the aperture zone h and focal length F for a lens material having an index of refraction n is given by two transcendental equations:

$$\sin r_i = (1/n) \sin \phi, \qquad \sin \alpha = n \sin(\alpha - r_1) \tag{9.1}$$

where ϕ is the tilt of the plano wedge of the lens and α is the angle of convergence to the focus, where

$$\tan \alpha = h/F \tag{9.2}$$

Fresnel lenses can be made out of glass or plastic. Glass lenses are durable but expensive since the lifetime of a mold for the high temperatures required to press glass is short when the facet shape and polish must be maintained. Plastic lenses are economical and the molds last for many lenses, the major expense being that of making and polishing the mold. The groove shape with a plastic Fresnel lens can be precise, with sharp corners necessary for good energy concentration. Since a Fresnel lens has so many facets, slight rounding of the top and bottom corners of the facet can result in concentration efficiencies being in the range of 80–90% that of a standard lens of the same index material.

Fresnel lenses make good concentrating lenses for small solar energy applications, especially the circular ones. For example, the concentrators being tested with silicon and gallium arsenide solar cells are generally the circular Fresnel type. The choice is made in part for their low cost, but also since high concentrations can be obtained from a precision plastic lens, with brightness concentrations exceeding 2000. In these applications it is necessary to track the sun to keep the small solar image centered on the receptor element.

Cylindrical Fresnel lenses have often been considered as focusing elements where pipe-shaped absorbers are to be used. A lens has the distinct advantage that it can also serve as a window to keep weather and dust off the absorber and reduce heat losses. If the smooth side is placed outward, one has the problem that the facets must be shaped according to a different pair of transcendental equations, and the problem of increased off-axis image aberrations. The equations for a flat-face forward Fresnel lens are

$$\sin r = n \sin \phi, \qquad \alpha = r - \phi, \qquad \tan \alpha = h/F \tag{9.3}$$

The principal reason why Fresnel lenses are not as widely used as possible is that the focal length of the lens varies rapidly with the angle of incidence of the rays on the lens. This means that if one wished to use a cylindrical lens in an application where, for example, there is no tracking of the sun in the direction parallel to the length of the cylinder, the back focus distance between lens and focus shortens. A mirror Fresnel collector used in the same situation would have an invariable back focus distance with oblique angle, a significant engineering simplification.

Circular Fresnel lenses are potentially useful as optical concentrators for fully tracking mounting, but their maximum brightness concentration is considerably less than that of parabolic mirrors. This limitation is caused by the fact that the angle of deviation of the prismatic element of a lens is less than is possible by reflection by a mirror. A comparison with a paraboloid is shown in Fig. 9.8.

The optical aberrations of a circular Fresnel lens are both similar in nature to those affecting the image of the sun formed by a fast paraboloid, but slightly less in magnitude.

There is an interesting advantage of any transmission element in a point focus system over a reflective system. This advantage is that some wavelengths are absorbed in transmission through the plastic or glass enabling one to use glass windows near the focus with minimum thermal problems. The wavelengths that would produce serious heating by absorption in the high brightness beam near the focus are already absorbed out of the radiation.

9.4 THE CENTRAL RECEIVER

The central receiver concept is an old one, based in fact on the most ancient of solar energy events, the third century B.C. defense of Syracuse attributed to Archimedes. The story as repeated by Galen was that Archimedes stationed soldiers around the semicircular harbor walls of Syracuse each with a polished shield, reflecting sunlight to a central point in the harbor, where the solar flux was amplified to a point where it repelled the invading Roman fleet.

The central receiver concept was the subject of a patent issued to C. G. O. Barr in 1896. It was also the subject of several Soviet studies and experiments where a stationary boiler was surrounded by mirrors that directed sunlight to the boiler. The Soviet version used large mirrors on railway cars that traveled around a semicircular track during the day.

The concept of the central receiver is very simple. To avoid the cost and heat losses in transporting a working fluid to a central location, use sunlight itself as the transfer medium. To implement the concept, one needs a field of mirrors provided with the means of directing reflected sunlight to a central location, or a location at one edge of the field of mirrors.

In the typical central receiver, the *mirror* is composed of many smaller mirrors; each with its own heliostat to follow the sun. The heliostats are generally located in the horizontal plane, but, when the situation is favorable, can simply follow the existing terrain. The basic difference between a single mirror concentrator and the heliostat system is that the heliostat system has a *dilute mirror*. This means that the entire surface within the system is not covered with mirror surface. This diluteness is generally termed the *fill factor*. A central receiver with a fill factor of 40% means that 40% of the land area is covered by mirrors.

In Fig. 9.5 we show schematically a central receiver optical system where many small mirrors are separately mounted to act together like a dilute paraboloid. If for reasons of flux density at the focus one desires a diffuse focus, one can individually point the heliostats to provide any desired spot shape and flux density.

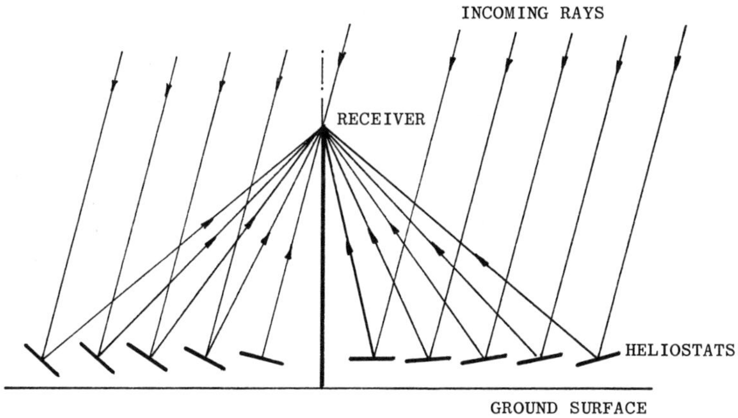

FIG. 9.5 Schematic arrangement of a central receiver–heliostat array, producing the optical effect of a dilute paraboloidal mirror. Note in this case the fill factor is large and the heliostat mirrors obscure adjacent heliostats.

The use of fill factors is important in regard to the efficiency of usage of each heliostat mirror. If the surface were fully occupied with mirrors when the sun is in the zenith, each mirror would partially obscure the neighbor when the sun is at an angle from the zenith. In the limit, a 100% fill-factor system would behave like a horizontal flat plate, having maximum collection at noon, falling off approximately as the cosine of the zenith distance during the day. The effect of a reduced fill factor is that each heliostat mirror acts like a fully tracking collector up to a certain zenith distance where mutual obscuration begins. Beyond this distance the cosine factor prevails. This effect is shown schematically in Fig. 9.6, where three fill factors are shown.

The principal system advantage of the dilute heliostat system is that the output during the day is *flat*, like the fully tracking system, yielding better system dynamics.

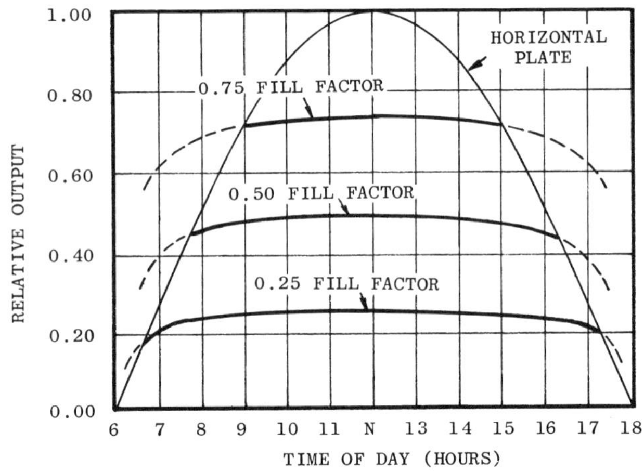

FIG. 9.6 Schematic graph of the diurnal variation of output from a central receiver heliostat as a function of fill factor. Each heliostat curve is approximately that of a fully tracking collector, but truncated by the angle where the field of mirrors appears to have fully filled the land surface.

The basic problem associated with the central receiver is that the heliostat mirrors require nonlinear drive rate in two coordinates to achieve the requirement of keeping the reflected image point on a fixed receiver. Along with this problem is the requirement that the heliostat be rugged enough to survive storms and operate successfully in a moderate wind, yet be inexpensive enough that the energy produced is competitive or is acceptable to the ultimate consumer. Lifetime of components is critical, as it is in all solar energy applications.

Since central receivers are a major part of the United States solar power program, and since the concept can have many smaller applications, it is important that the mathematical aspects of the heliostat system be available to the engineer.

9.5 CONCENTRATION LIMITS

Since parabolic point focus collectors are frequently used for solar furnaces, one is interested in the brightness concentration. There are two measures of concentration of importance in solar energy collectors, and the difference between them is often not appreciated, and hence not often clearly separated. The first is *brightness concentration C*, which refers to the maximum flux of energy per unit area within the image. This measure determines the maximum temperature at any one spot on the absorber. The second measure of concentration is the *radiation balance concentration* X, which refers to the average flux per unit area over the entire absorber. This measure determines the average temperature of the absorber based on the radiative balance between net incoming energy and net outgoing energy. In this section we shall refer to the brightness concentration C.

The source brightness of the sun per unit solid angle is conserved in any optical system, so $B'(T)$ at the focus is the same as $B(T)$ at the sun. The amount of energy in a solid angle of any optical system at its focus, $b'(T)$ is related to that emitted by the surface of the sun in 2π solid angle by the simple equation

$$b'(T) = b(T)[1/(4 \sin^2 \alpha)] \tag{9.4}$$

This equation means that if the optical system produces a focus having $\alpha = 90°$ (an $F/0.5$ system), the brightness of the image will equal the brightness of the surface of the sun (5750°K) less the energy lost in the traversal of sunlight through the earth's atmosphere. If the optical surface further has errors enlarging the solar image, or if the aperture is not filled, as can be the case for a quasiparabolic mirror made of many small heliostats, the maximum flux brightness and effective temperature will be lower than that given by Eq. (9.4). In Fig. 9.7 we show the curve of brightness concentration C as a function of α. The upper curve refers to a perfect mirror of 100% reflectance on the surface of the earth with the sun in the zenith. The shaded area shows the region occupied by actual solar furnaces constructed to date.

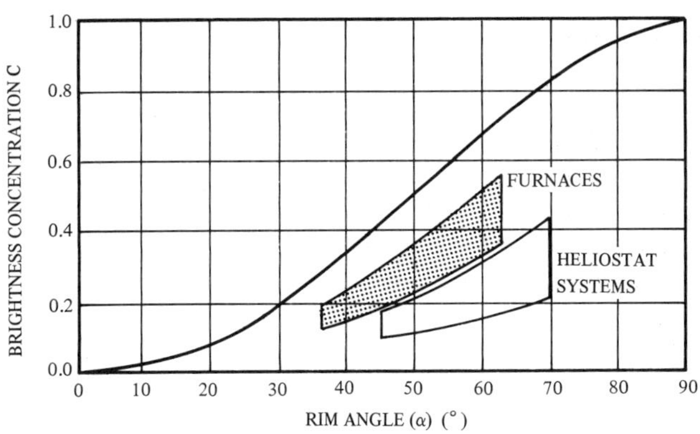

FIG. 9.7 Brightness concentration for a point focus concentrating collector as a function of rim angle α. For a perfect collector of $\alpha = 90°$ the flux through one standard atmosphere is 4365 W/cm^2.

The concentration limits and relative apertures for the maximum brightness paraboloid ($2\alpha = 180°$) is compared with a maximum brightness Fresnel lens ($2\alpha = 84°$). The Fresnel limit is set by the maximum inclination angle of the outermost glass facet, taken as 60°, producing a deviation of the ray of 42°. Note that at this angle the outer Fresnel groove is not performing efficiently, since the adjacent groove intercepts some of the rays from the facet. This loss is partially offset by the loss in the $F/0.5$ paraboloid due to the large angle of incidence of the edge rays on the receiver. By Eq. (9.4) the relative brightness concentration of the Fresnel lens is only 45% that of the paraboloid, confirmed by the relative apertures marked in Fig. 9.8.

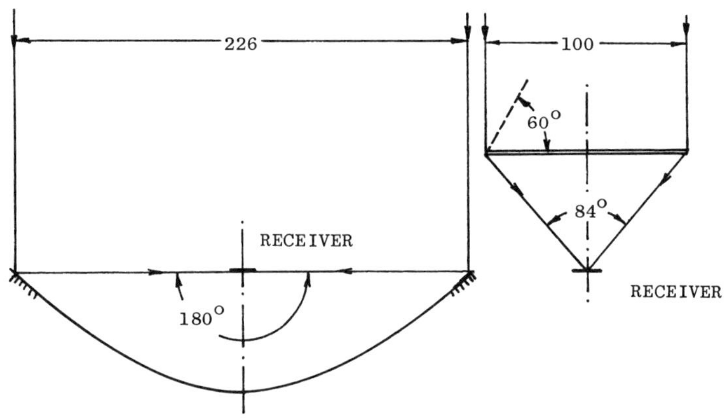

FIG. 9.8 Comparison of the relative aperture limits for a paraboloid and a Fresnel lens. The lens has facets inclined at 60° to the incoming sunlight.

9.6 POINT FOCUS MOUNTINGS

Point focus concentrating mirror systems are generally mounted in the fully tracking mode. This is because their field of good imagery is very small, and for maximum brightness concentration C and radiation balance concentration X, the absorber must be kept as small as the quality of the mirror surface permits. The tracking must therefore be accurate.

The maximum amount of available sunlight as a function of season and latitude for a fully tracking mounting is shown in Fig. 9.9. The maximum amount as a function of time of day is shown in Fig. 9.10. These two sets of curves are generally used as a point of reference for the curves for other types of mounting and tracking, as will be discussed in the next section.

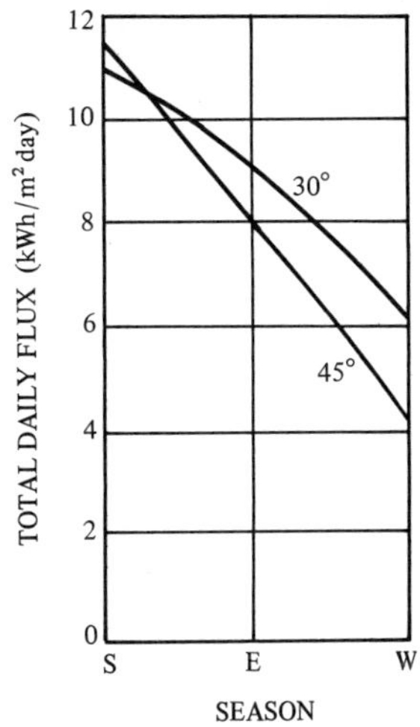

FIG. 9.9 Daily maximum collector yields in kilowatt hours per square meter, where 30° and 45° are latitudes.

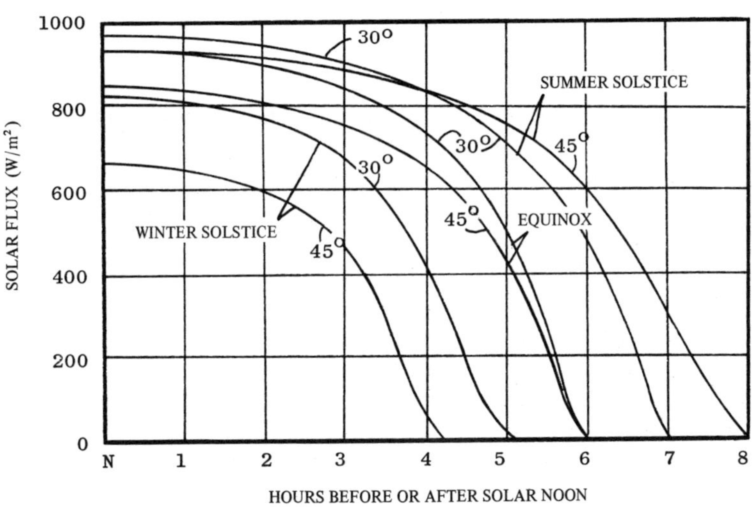

FIG. 9.10 Variation of direct solar flux for a fully tracking collector as a function of time of day, latitude, and season.

9.7 FULLY TRACKING MOUNTING CONFIGURATIONS

Fully tracking mountings can have several geometries. The simplest is that of a polar mounting, as shown in Fig. 9.11. In this case the one axis is placed parallel to the earth's axis and the second axis orthogonal. This means that the daily solar motion requires only tracking about one axis, the polar axis. Changes in declination are only required at infrequent times during the day, to accommodate the seasonal motion of the sun in declination.

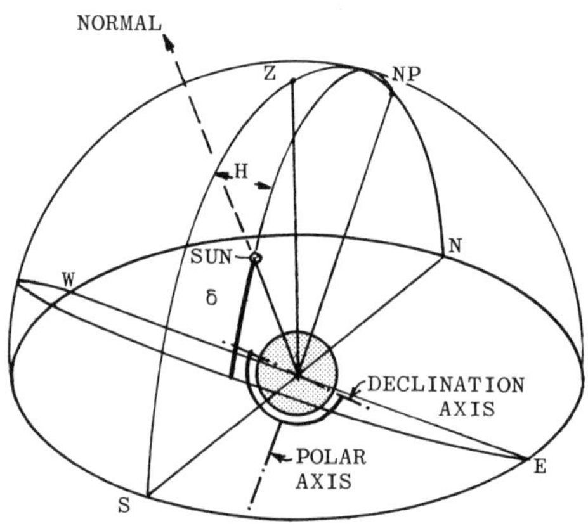

FIG. 9.11 Diagram of the polar mounting for a fully tracking collector.

The several possible configurations where the polar axis is not used require, in general, motion about two axes. These motions, moreover, are nonlinear motions. The additional complication of nonlinear motions is traded off for mechanical conveniences of these other mounting configurations.

The most common nonpolar mounting is the altitude–azimuth mounting. One axis is placed vertical and the second axis horizontal, as shown in Fig. 9.12. One particular advantage of this configuration is that the plane of gravity remains in the vertical plane through the mirror, minimizing the engineering complications in the design of the mirror and its support for minimizing gravity deflections of the mirror surface. Moments on the mirror axes and bearings are also minimized.

The mounting for heliostats results in the generalized case of nonlinear motions in two coordinates since the direction of the beam *after* reflection by the mirror is at an arbitrary angle with regard to the mounting since the absorber and the mirror are not rigidly attached. We shall now discuss the equations of motion of this generalized case.

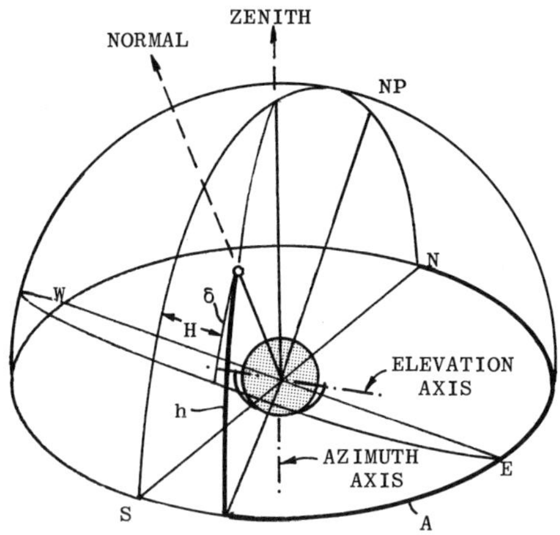

FIG. 9.12 Diagram of the altazimuth mounting for a fully tracking collector.

Creating or monitoring the nonlinear motions can be done either by means of a computer to instruct the appropriate drive motor, or by a closed-loop system that detects the error between the absorber and the solar image. The open-loop system, when used, is often supplemented by an error signal generator to modify the rates to keep the sun properly on the absorber. The closed-loop system seems to be preferred in recent systems, the error signal being the primary signal for the drive signal for the drive motors.

9.8 KINEMATICS OF FULLY TRACKING MOUNTINGS

The kinematics for several types of mountings are given below. The polar, altazimuth, alt–alt, and horizontal yoke mountings are given by Vasilevskis (1966) and for the generalized heliostat by Vartanyan *et al.* (1974). The equations, along with their geometries, are now summarized.

1. Polar Axis

The diagram of the geometry for a polar axis fully tracking mounting is shown in Fig. 9.11. The angle H is the hour angle, which is the solar right ascension less the time. It is measured from the meridian. The rate of change of H with time is the sidereal rate of motion of the sun. In the equations in this section the daily motion of the sun in declination is taken as zero. When the actual seasonal motion of the sun in declination is taken into account, the effect on the gross magnitude of the components of sidereal motion for the geometries herein discussed is small.

The equations for motion in a polar coordinate system are

$$d\alpha/dH = \text{const} = dH/dt, \qquad d\delta/dH = 0 \tag{9.5}$$

2. Altitude–Azimuth

The diagram for the altitude–azimuth geometry is shown in Fig. 9.12. In this case the fixed axis is vertical, with the variable axis being the elevation axis.

The equations for motion in an altazimuth system are

$$dz/dH = \cos\delta \sin p, \qquad dA/dH = -\cos\delta \cos p/\sin z \tag{9.6}$$

where

$$\cos z = \sin\phi \sin\delta + \cos\phi \cos\delta \cos H$$

$$\tan A = \frac{\sin H}{\cos\phi \tan\delta - \sin\phi \cos H} \tag{9.7}$$

$$\tan p = \frac{\sin H}{\tan\phi \cos\delta - \sin\delta \cos H}$$

A. NS Horizontal Axis (Altitude–Altitude)

The diagram for a horizontal axis geometry, where the fixed axis is north–south, is shown in Fig. 9.13.

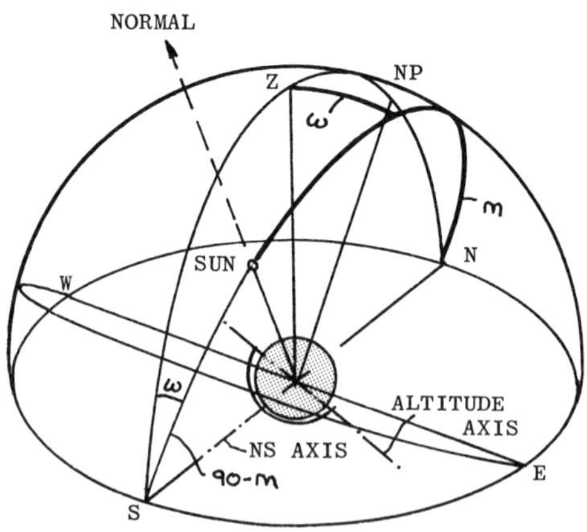

FIG. 9.13 Diagram of the NS horizontal axis (alt–alt) mounting for a fully tracking collector. Note that n is measured from the north horizon in the equations.

The equations for motion in a NS horizontal system are

$$dn/dH = -\cos\delta \sin q, \qquad d\omega/dH = \cos\delta \cos q/\sin q \tag{9.8}$$

where

$$\begin{aligned} \cos n &= \cos\phi \sin\delta - \sin\phi \cos\delta \cos H \\ \tan\omega &= \frac{\sin H}{\sin\phi \tan\delta + \cos\phi \cos H} \\ \tan q &= \frac{\sin H}{\cot\phi \cos\delta + \sin\delta \cos H} \end{aligned} \tag{9.9}$$

B. EW Horizontal Axis

The diagram for a horizontal axis geometry, where the fixed axis is east–west, is shown in Fig. 9.14. Note that in this case the elevation angle of the sun is measured from the north pole rather than the horizon or zenith. One could use either of these origins by appropriately modifying the equations, taken from Vaselivskis (1966).

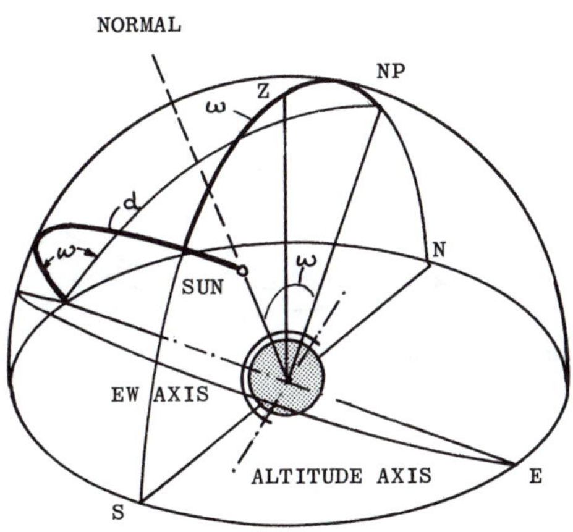

FIG. 9.14 Diagram of the EW horizontal axis mounting for a fully tracking collector. Note that the angle ω is measured south from the pole in the equations presented.

The equations for motion in an EW horizontal system are

$$dd/dH = -\cos\delta \sin q, \qquad d\omega/dH = -\cos\delta \cos d/\sin^2 d \tag{9.10}$$

where

$$\cos d = \cos\delta \sin H, \qquad \tan\omega = \cot\delta \cos H, \qquad \cot q = \sin\delta \tan H \tag{9.11}$$

9.9 GENERALIZED HELIOSTAT

The geometry is considerably more complicated for the heliostat than for the case where the absorber is attached to the optical axis of the mirror. The basic geometry is shown in Fig. 9.15. The point R defines the direction of the absorber as viewed from the heliostat. The heliostat axis is denoted in angle from the north pole since ζ and H are mutually combined in the equations. The normal to the heliostat mirror bisects the angle between the sun, S, and the ray from the heliostat to the absorber, R, the direction denoted by N. The kinematics in terms of the several possible mountings are now summarized.

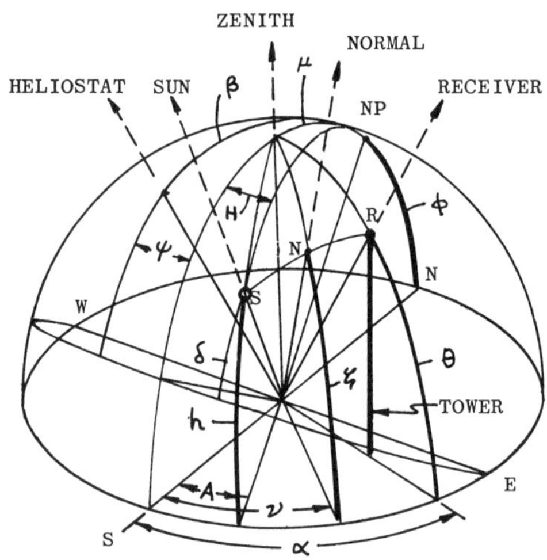

FIG. 9.15 Diagram of the basic geometry of the generalized heliostat system. Symbols are defined in the text.

The factors shown in Fig. 9.15 are:

- v the azimuthal angle of the heliostat normal N, measured from the south point, measured west;
- ζ the elevation angle of the heliostat normal N, measured from the horizon;
- α the azimuthal angle of the receiver R, from the south, measured west;
- θ the elevation angle of the receiver R, from the horizon, $\theta = 90 - \phi$,
- A azimuth of the sun S, measured from the south, west;

- h elevation angle of the sun S;
- δ declination of the sun S, measured from the equator;
- H hour angle of the sun S, measured from the meridian;
- ψ hour angle of the heliostat axis;
- β polar distance of the heliostat axis;
- ϕ latitude of the site;
- μ $90^\circ - \lambda$, where λ is the colatitude.

Vartanyan *et al.* (1974) presents the equations for the several mounting configurations in terms of intermediate quantities to simplify the equations. In the general case the equations for the motion of the normal to the heliostat N, are

$$\tan v = x/y, \qquad \tan \zeta = (z/y) \cos v \tag{9.12}$$

and

$$dv/dH = \frac{\dot{x}y - x\dot{y}}{x^2 + y^2}, \qquad d\zeta/dH = \frac{(x^2 + y^2)\dot{z} - xz\dot{x} - yz\dot{y}}{(x^2 + y^2 + z^2)\sqrt{x^2 + y^2}} \tag{9.13}$$

where the terms in x, y, and z and their derivatives with respect to time are given by

$$\begin{aligned} x &= \cos\delta \sin(H - \psi) - \sin\alpha \cos\theta \\ y &= \cos\delta \cos\beta \cos(H - \psi) - \sin\delta \sin\beta - \cos\alpha \cos\theta \\ z &= \cos\delta \sin\beta \cos(H - \psi) + \sin\delta \cos\beta - \sin\theta \end{aligned} \tag{9.14}$$

and

$$\begin{aligned} \dot{x} &= \cos\delta \cos(H - \psi) \\ \dot{y} &= -\cos\delta \cos\beta \sin(H - \psi) \\ \dot{z} &= -\cos\delta \sin\beta \sin(H - \psi) \end{aligned} \tag{9.15}$$

One could substitute these expressions for x, $\dot{x}$, etc., in Eq. (9.13), but the result would be quite cumbersome.

A. Altitude–Azimuth Heliostat

In the case of the altazimuth heliostat $\psi = 0$ and $\beta = 90 - \phi$, whence

$$\begin{aligned} x &= \cos\delta \sin H + \sin\alpha \cos\theta \\ y &= \cos\delta \sin\phi \cos H - \sin\delta \cos\phi + \cos\alpha \cos\theta \\ z &= \cos\delta \cos\phi \cos H + \sin\delta \sin\phi + \sin\theta \end{aligned} \tag{9.16}$$

and

$$\begin{aligned} \dot{x} &= \cos\delta\cos H \\ \dot{y} &= -\cos\delta\sin\phi\sin H \\ \dot{z} &= -\cos\delta\cos\phi\sin H \end{aligned} \tag{9.17}$$

and the angular velocities desired can then be obtained by inserting these expressions in Eq. (9.13).

When $\theta = 90^\circ$, we check these equations by noting that they reduce to the case for the normal altazimuth mounting, except $v = A$ and $\zeta = \frac{1}{2}(90^\circ + h)$, and that $\dot{v} = dA/dH$ and $\dot{\zeta} = \frac{1}{2}\, dh/dH$. The azimuth rate $\dot{\zeta}$ is therefore half the solar sidereal rate. It should be noted that each of the velocity terms must be multiplied by the solar rate to get the absolute drive rate in each coordinate.

B. Altitude–Altitude Heliostat

For the alt–alt heliostat, we have $\psi = 0^\circ$ and $\beta = 180^\circ - \phi$. This design is similar to that shown in Fig. 9.13, where the fixed axis is north–south, horizontal. In this case we have

$$\begin{aligned} x &= \cos\delta\sin H + \sin\alpha\cos\theta \\ y &= -\cos\delta\cos\phi\cos H - \sin\delta\sin\phi + \cos\alpha\cos\theta \\ z &= \cos\delta\sin\phi\cos H - \sin\delta\cos\phi + \sin\theta \end{aligned} \tag{9.18}$$

and

$$\begin{aligned} \dot{x} &= \cos\delta\cos H \\ \dot{y} &= \cos\delta\cos\phi\sin H \\ \dot{z} &= -\cos\delta\sin\phi\sin H \end{aligned} \tag{9.19}$$

C. Horizontal Yoke Heliostat

For the horizontal yoke heliostat mounting, with the horizontal fixed axis east–west, we have $\psi = 90^\circ$ and $\beta = 90^\circ$. In this case

$$\begin{aligned} x &= -\cos\delta\cos H + \sin\alpha\cos\theta \\ y &= -\sin\delta + \cos\alpha\cos\theta \\ z &= \cos\delta\sin H + \sin\theta \end{aligned} \tag{9.20}$$

and

$$\begin{aligned} \dot{x} &= \cos\delta\sin H \\ \dot{y} &= 0 \\ \dot{z} &= \cos\delta\cos H \end{aligned} \tag{9.21}$$

The reader is referred to the reference cited for additional cases of potential use; but care must be used since the English translation frequently mixes the Greek symbols for the angles involved.

D. Axis through Absorber

A geometry of particular interest is where the fixed axis of the heliostat is pointed at the absorber. This geometry enables the error sensor to look through the axis of the mounting at the absorber and pick up a solar reflection from the rear face of the mirror. If we examine Fig. 9.15, we see that β and ψ must be determined from a spherical triangle, as shown in Fig. 9.16. The expressions become cumbersome if the equations for β and ψ are inserted into the general expressions, so we shall leave them implicit.

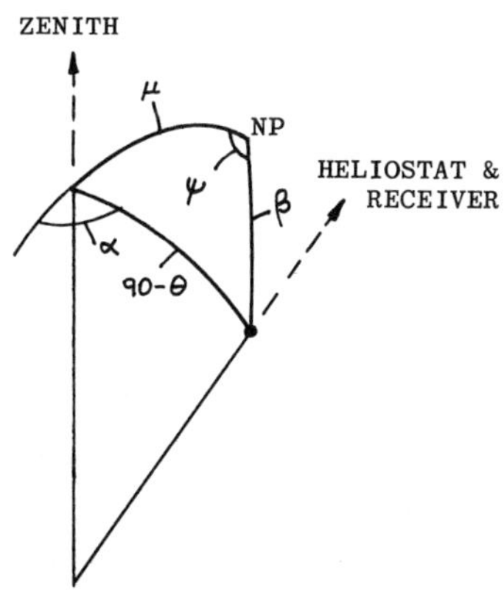

FIG. 9.16 Special geometry when the axis of the heliostat points toward the receiver.

We have

$$\cos\beta = \sin\theta\cos\mu + \cos\theta\sin\mu\sin\alpha \qquad \text{and} \qquad \sin\psi = \cos\theta\cos\alpha/\sin\beta \tag{9.22}$$

In Vartanyan *et al.* (1974) a closed expression is given, but this is for the special case where $\alpha = 0°$ and $\theta = 90°$, which means that the absorber is located in the zenith. The geometry then simply becomes that of the altazimuth mounting, a point which the authors state; but this is not the general case of interest when there is a field of heliostats under consideration and neither α, θ, β, or ψ coincide with the altazimuth geometry.

All two-axis mountings except the polar mounting have problems that develop when the sun passes close to the fixed axis. In this case the drive rates increase, reaching infinity as the sun passes through the angular position of a pole. For example, an altazimuth mounting located between $\pm 23.5^\circ$ latitude will have the sun cross the zenith of the heliostat during the day at some time of year, in general, four times a year. At these times the heliostat would need to be protected against rate and power excursions. The usual zone about the pole that is blocked out for radio telescopes is 5°–10°. This problem is termed *gimbal lock*.

9.10 LINE FOCUS CONFIGURATIONS

Line focus collectors have potential applications where medium brightness and radiation balance concentrations are sufficient. The receiver in these cases is an extended structure, and new questions that are not factors in the case of point focus systems arise with regard to the collection of the solar heat at the point of use. We shall begin with an examination of the optical configurations of the line focus type and end with a discussion of heat transfer in a distributed linear collector system.

9.11 PARABOLIC CYLINDER

The parabolic cylinder is the basic line focus optical configuration, the one-dimensional analog of the paraboloid. The optical surface produces a sharp line focus image whose width is set by the on-axis aberration of scale, as in the case of the on-axis image formed by a fast paraboloid. The basic geometry of the parabolic cylinder is shown in Fig. 9.17.

The parabolic cylinder and all cylindrical mirror systems have a useful property. The distance of the focal line from the mirror vertex is constant for all longitudinal field angles. Thus a parabolic cylinder oriented east–west will focus the sun on the absorbing cylinder at all times of day if the small north–south solar motion is guided out. This invariance is not present for cylindrical Fresnel lenses, a major complication reducing their applicability to simple solar collector configurations.

The maximum optical concentration of a cylindrical paraboloid having $\alpha = 90^\circ$ is $X = 73$ for a cylindrical absorber of the same diameter as the solar image formed by the rim rays. In practical applications radiation balance concentrations of $X = 30$–40 can be obtained. The imagery of a finite source is shown schematically in Fig. 9.18. The brightness concentration C can be as high as $C = 150$ for a perfect cylindrical parabolic mirror.

The optical equivalent of a parabolic cylinder can be made by slat or strip mirrors. It is the equivalent of the heliostat, but the mirror is long

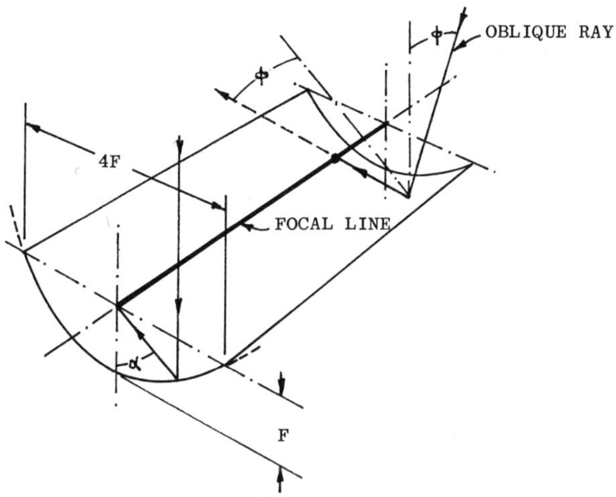

FIG. 9.17 Basic geometry of a parabolic cylinder mirror collector. Note that rays inclined parallel to the length of the mirror (ϕ) also come to a focus on the focal line.

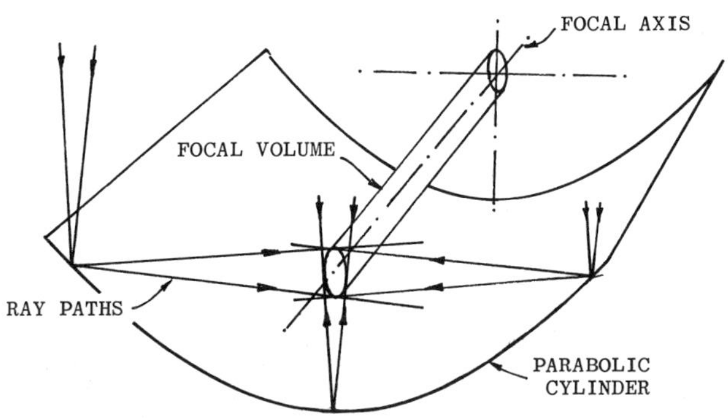

FIG. 9.18 Diagram illustrating the ray paths and focal volume of a parabolic cylinder, where the angle of the sun is enlarged $20\times$.

compared to its width. Tracking motion only about its long axis is required of its mounting. A typical slat-type cylindrical mirror is shown in Fig. 9.19. In this figure, if all six slats overlapped on the absorber, the brightness concentration would be approximately 6. If the slats were curved, then higher concentration would be possible. The limiting aberration of the concentration produced by cylindrical slats each having the same radius of curvature is astigmatism.

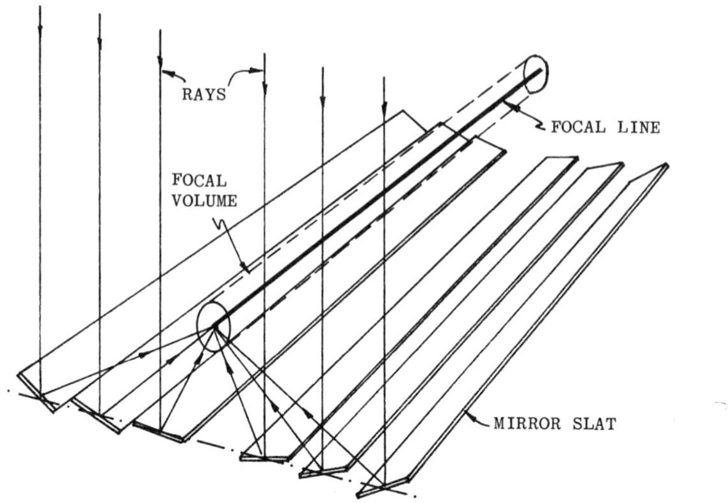

FIG. 9.19 A slat-mirror approximation to a parabolic cylinder. The size and shape of the focal volume is set by the width of the flat slats and the angular size of the sun.

9.12 CYLINDRICAL OPTICAL MOUNTINGS

Since cylindrical optical systems need only tracking capability in one coordinate, there are a number of possible mounting configurations. We shall discuss each in regard to relative ability for energy collection.

A. Polar Mounting

The basic geometry of the polar mounting is shown in Fig. 9.20. The axis of rotation of the system is pointed to the pole. The seasonal motion in declination then places the sun 23°.5 north and south of the normal to the noon position of the mirror. The resulting cosine projection factor is then only 0.92, and the difference between winter, summer, and the equinox is small, the winter suffering chiefly because of the increased air mass path length and increased absorption in winter. Differences between summer and winter are significant when one remembers that the sun spends most of each season near the solstice positions. If 4° increments of solar declination are used, one finds that the sun is within the summer and winter solstices 70 consecutive days, but only nine days in the same 4° zone at the equinox.

The variation of seasonal output for the polar mounting is shown in Fig. 9.21. The variation of output during the day as a function of time and season is shown in Fig. 9.22.

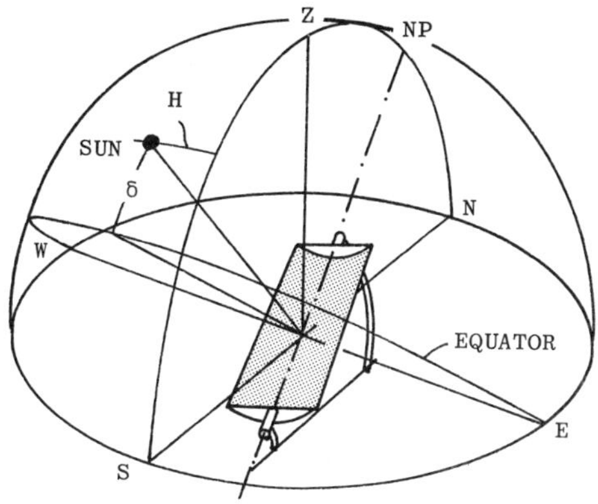

FIG. 9.20 Schematic diagram of a polar mounting of a cylindrical mirror.

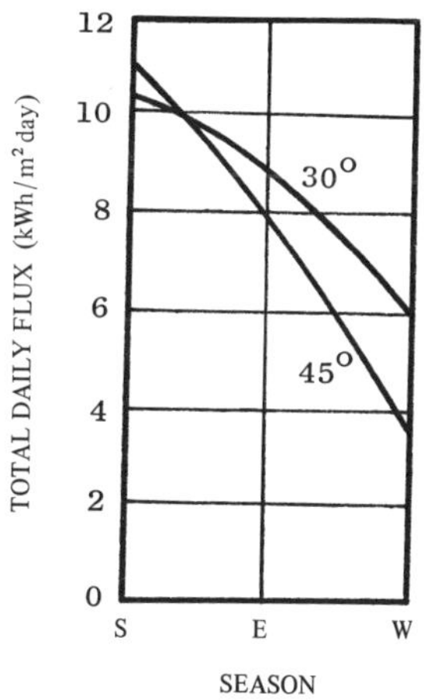

FIG. 9.21 Daily collector yield for a polar mounted collector, where 30° and 45° are latitudes.

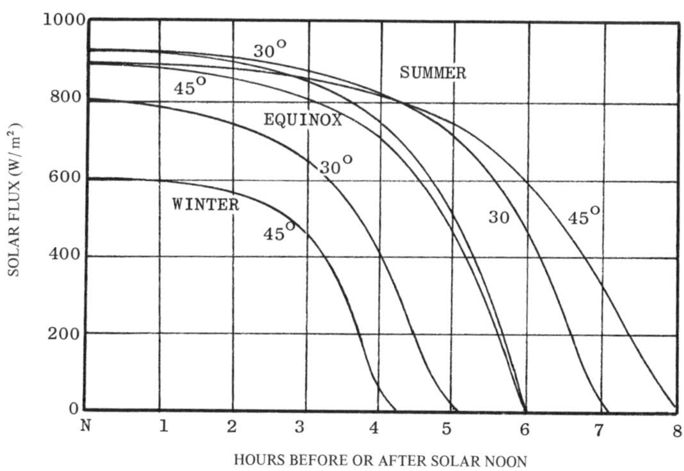

FIG. 9.22 Variation of flux collected by a polar mounting tracking the sun in the EW direction, where latitudes are given.

B. NS Horizontal: EW Tracking

The basic geometry of the north–south horizontal mounting having east–west diurnal motion is shown in Fig. 9.23. The seasonal performance is considerably altered from the polar mounting since the cosine factor for winter performance is small. The variation of seasonal output for the NS: EW tracking mounting is shown in Fig. 9.24. The variation of output during the day as a function of time and season is shown in Fig. 9.25.

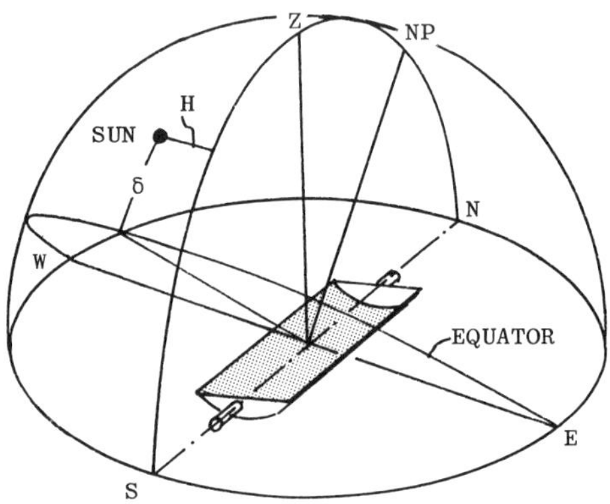

FIG. 9.23 Geometry of a NS horizontal: EW tracking collector.

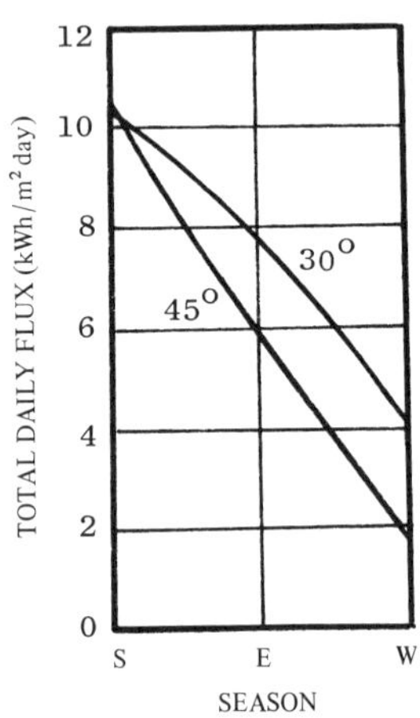

FIG. 9.24 Daily collector yield in kilowatt hours per square meter for a NS horizontal: EW tracking collector, where 30° and 45° are latitudes.

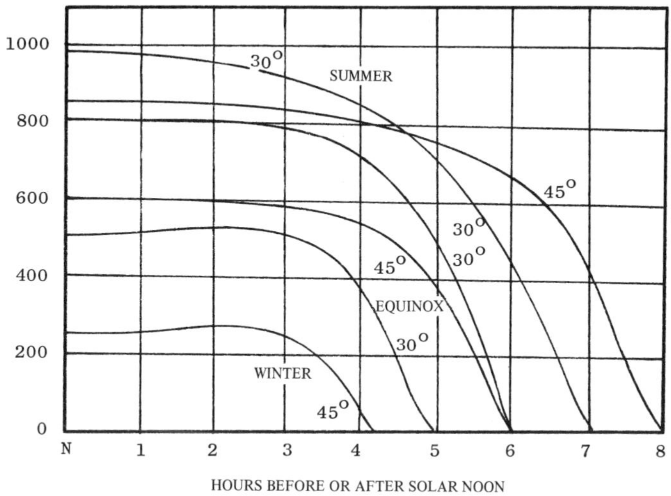

FIG. 9.25 Variation of solar flux for a collector with NS horizontal: EW tracking as a function of time of day and season for direct sun, where latitudes are given.

C. EW Horizontal: NS Tracking

The basic geometry of the east–west horizontal mounting having north–south tracking is shown in Fig. 9.26. The winter performance of this configuration is much better than the NS horizontal configuration, but the diurnal variation shows the peaked noontime performance of the horizontal

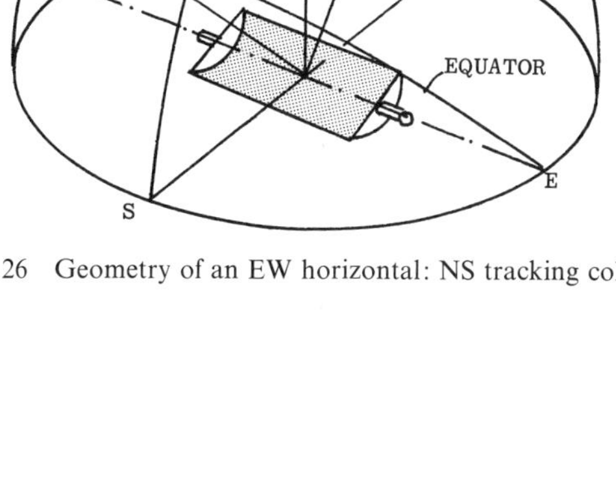

FIG. 9.26 Geometry of an EW horizontal: NS tracking collector.

flat plate rather than the flat diurnal performance of the fully tracking mounting. The variation of seasonal output for the EW horizontal configuration is shown in Fig. 9.27. The variation of output during the day as a function of time and season is shown in Fig. 9.28.

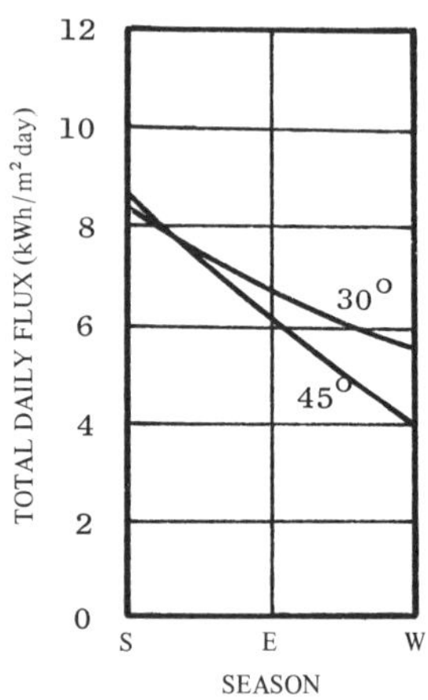

FIG. 9.27 Daily collector yield for an EW horizontal: NS tracking collector, where 30° and 45° are latitudes.

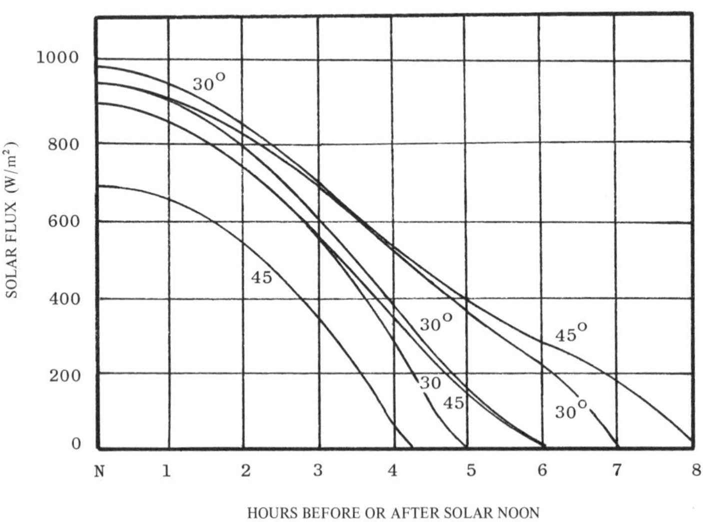

FIG. 9.28 Variation of solar flux for an EW horizontal: NS tracking collector, where latitudes are given.

9.13 FIXED MIRROR LINE FOCUS SYSTEMS

The fixed mirror line focus configuration is where the mirror *and* absorber are fixed, accepting the sun during a significant portion of the day and year. We shall discuss two variants of this general goal: (1) where the mirror is allowed seasonal incremental adjustment, and (2) where the absorber moves to track the solar image.

It is appropriate at this point to define what we mean when we use the two terms for concentration. By brightness concentration C, we refer to the maximum flux per unit surface mounting. The design approach they used is shown in Fig. 9.29, where an oversized absorber is used. During the day the solar image moves over the surface of the absorber, staying on the surface during the required time. It is clear that there is a major difference between the brightness concentration C of such a design and the radiation balance concentration X. C can remain on the order of 40 while X has been reduced to the range of 3–5.

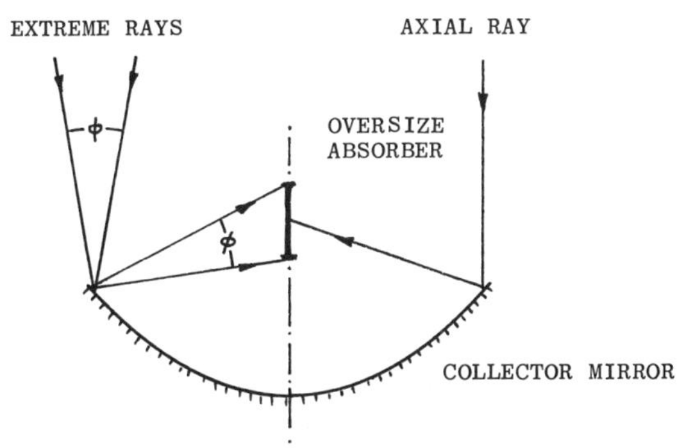

FIG. 9.29 Basic design of the type studied by Tabor and Zeimer for a semifixed mirror collector. Several mirror and absorber shapes were included in the study.

Two new approaches to fixed mirror designs were recently introduced, the Winston collector by Winston (1974) and the Trombe–Meinel (1974). The Winston design uses parabolic cusps having the geometry as shown in Fig. 9.30, approaching the problem from the side of high concentrations and small acceptance fields. The Trombe–Meinel approach is from the original Trombe design for a 180°, $X = 1$ optical configuration, but where the field angle is reduced to yield increased concentration through the extension of the *cusp* into quasiparabolic cusps of the Winston type, as shown in Fig. 9.31. Both types of design have acceptance fields narrow in the transverse direction and approaching 180° in the longitudinal direction.

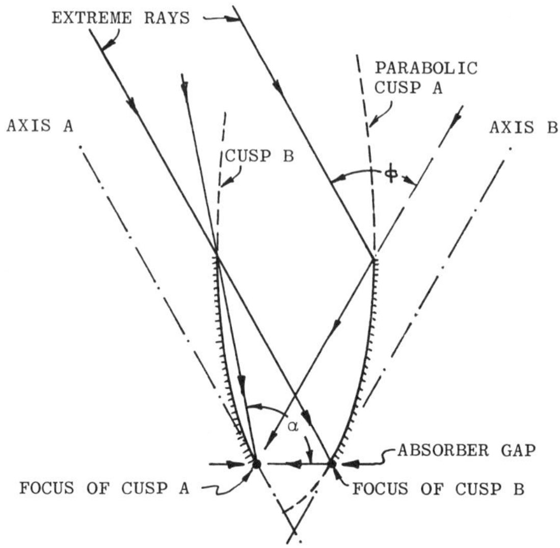

FIG. 9.30 Basic diagram for a Winston collector. Note that the foci of each cusp lie on the other at the absorber gap point. Length of a "full" Winston is where the sides of the parabolic cusps are parallel.

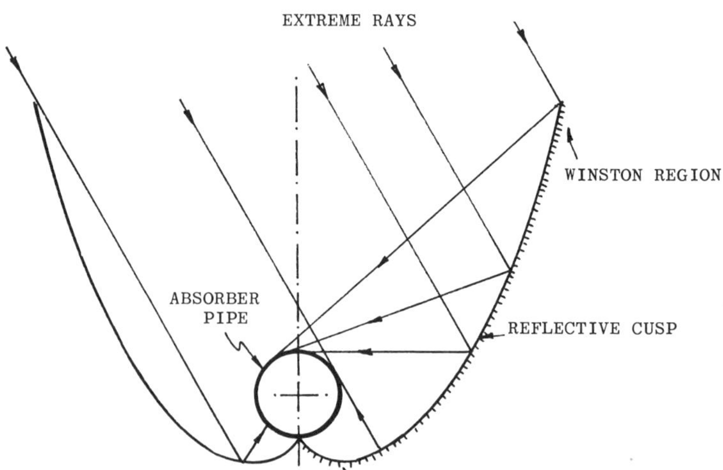

FIG. 9.31 Basic diagram of a Trombe–Meinel cusp collector. When the sides of the cusp are extended, they become parabolic; and the limit becomes identical in performance to the Winston, excepting that the surface of the absorber is larger than for a Winston.

The basic Trombe design, patented but never published, and rediscovered by Meinel, is for a field acceptance angle of 180°. The cusp shape is a simple geometric form in this case, being the locus of a string unwrapped from the pipe. The 180° cusp has unit concentration X, the aperture being equal to the area of the absorber arriving at some point on the absorber. This quantity determines the upper limit to heat transfer through the absorber and is important in engineering the absorber and heat transfer fluid. By radiation balance concentration X, we refer to the ratio of the cross section for collection of solar energy to the cross section of the absorber losing energy, which is important in arriving at the equilibrium temperature of the absorber and efficiency of extraction of energy by the absorber as a whole.

9.14 SEMIFIXED MIRROR SYSTEMS

A number of optical configurations have been explored where a compromise between radiation balance flux concentration X and field acceptance angle is made. If we examine the range of angle through which the sun travels during the day in the north–south direction for an east–west collector, we note that the sun is within a reasonable range of angles about a mean position for several hours each day, even at the solstices when the daily range is a maximum. The range of angles is illustrated graphically in Fig. 9.32, from which the relationship between duration and optical field angle can be determined.

The *fixed mirror* compromise then is to find optical configurations which eliminate daily tracking, using instead only seasonal adjustments. Tabor

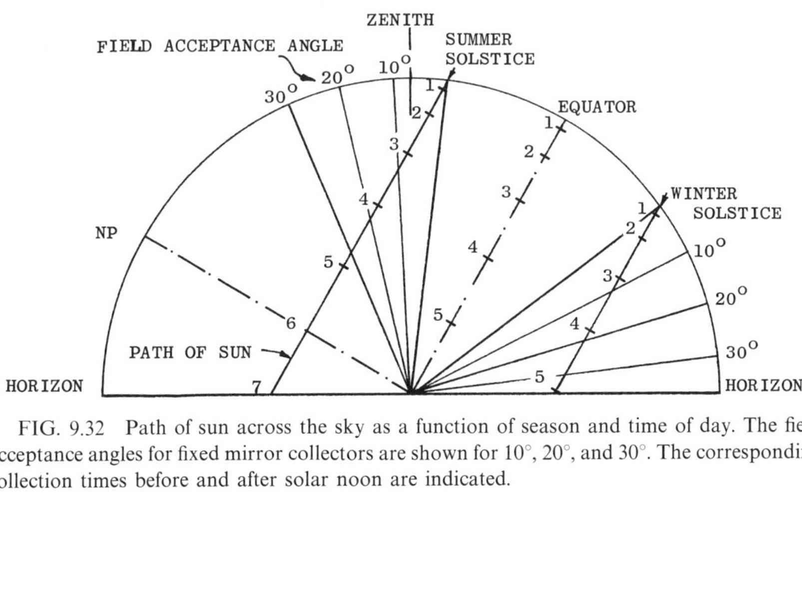

FIG. 9.32 Path of sun across the sky as a function of season and time of day. The field acceptance angles for fixed mirror collectors are shown for 10°, 20°, and 30°. The corresponding collection times before and after solar noon are indicated.

(1958) showed that a fixed mirror capable of avoiding seasonal repositioning was not possible with concentrations exceeding unity by any significant amount. Tabor and Zeimer (1962) explored other designs where seasonal motion was accepted as a reasonable simplification over a tracking circumference of the pipe, thus acting like a flat plate collector, with radiant transfer of solar flux to the absorber rather than conduction.

The Trombe–Meinel and Winston cusp designs flow into one family of fixed mirror collectors. As shown in Fig. 9.33, when the reflective cusp of the T–M collector are extended they approach the Winston cusp shape. The absorber, however, being a circular cylinder rather than a flat surface, has more reradiating surface and a lower net concentration X than a Winston design. It is similarly obvious that a cusp can be added to the absorber gap of the Winston design, in Fig. 9.31, to substitute for a pipe absorber.

When all of the fixed mirror designs are plotted in a graph of field acceptance angle (transverse) and radiation balance concentration, we note that the "full" Winston design delineates the limit to possible performance, as Winston recognizes in his 1974 paper. The Trombe–Meinel cusp designs

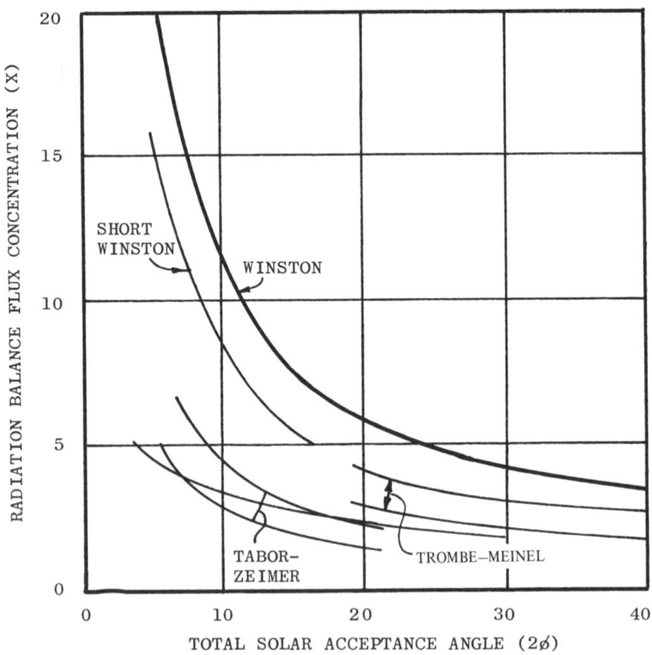

FIG. 9.33 Relationship between radiation balance concentration (X) and field acceptance angle for several fixed mirror systems.

lie lower, in the region where Winston has defined a class of *truncated* designs. The Tabor–Zeimer designs using parabolic cylinders lie lower than either of these two newer design classes.

9.15 FIXED-MIRROR MOVING-ABSORBER SYSTEMS

Two fixed-mirror moving-absorber configurations have been studied. They are: (1) the Russell cylindrical slat mirror and (2) the hemispherical bowl mirror, rediscovered by Steward and Krieth (1975) and Meinel (1974), but actually first built by Adams in 1878.

The Russell design is shown in schematic form in Fig. 9.34. The geometry is elegant, the optical image being stigmatic on a circle on which the absorber is moved. The mirror is made of thin strips of flat mirror, the width being compatible with the diameter of the absorber pipe to be used. It should be noted that these mirror strips have mutual obscuration when the sun angles become large in the transverse direction.

The Adams–Steward–Meinel design is not a true line focus system, but the absorber is elongated to accept the natural spherical aberration of the image formed by a deep spherical mirror as shown in Fig. 9.35. The absorber is linear and occupies the region between the paraxial focus at $R/2$, extending

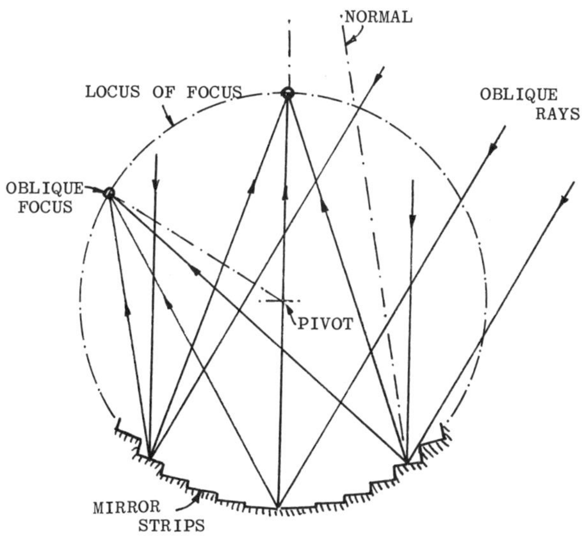

FIG. 9.34 Diagram of the Russell geometry for formation of a stigmatic line focus on a circular track. Radius of curvature of mirror strips (if curved ones are used) is approximately twice the on-axis focal length. For oblique rays the mirror steps cause shadowing of the mirror facets.

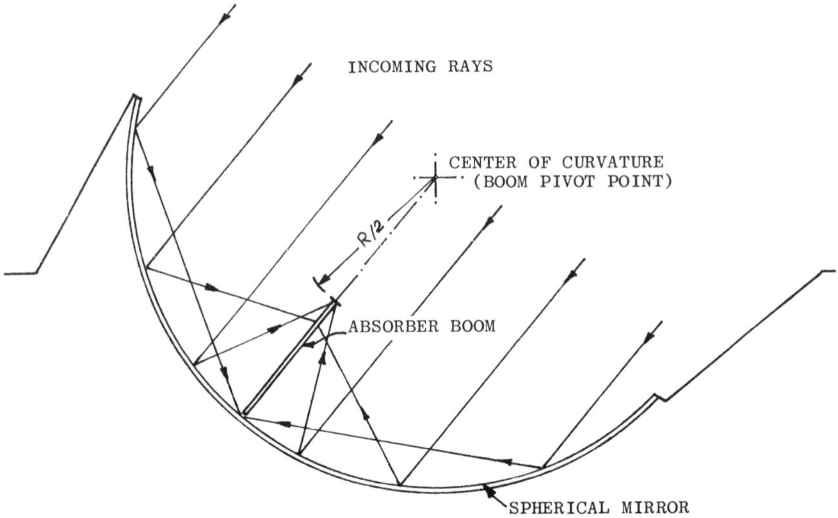

FIG. 9.35 Diagram of the Adams–Stewart–Meinel fixed hemispherical mirror movable focal boom design. Brightness concentration varies along the length of the boom, being brightest near the top, with a group of secondary maxima from multiple reflections at the bottom.

down to the mirror surface at R. When the absorber is pivoted at the center of curvature of the mirror, the motion of tracking the sun image during the day is as simple as driving a telescope. The basic configuration of this type of design is shown in Fig. 9.35. The flux brightness on the absorber of this type of collector is not uniform along its length, being highest near the paraxial focus, with an important secondary maximum near the mirror surface where multiply reflected rays converge.

Both of these fixed-mirror moving-absorber configurations are designed to avoid the engineering problems associated with heliostat mountings which are above ground and relatively vulnerable to severe weather. While the mirrors are thus more protected, the moving absorber boom adds the complication of requiring the heat to be transported through a mechanical joint. The hemispherical bowl concept is, moreover, limited to module size to the order of 100 m diameter, whereas the central receiver can be as large as a square kilometer in size.

9.16 MIRROR MATERIALS

The sensitivity of collectors to cost limits the materials that can be used for inexpensive or thin mirrors. Glass silvered on the rear or second surface is generally selected as most durable. Front surface mirrors can also be used, reducing the requirement for a transparent substrate as in the case of second-surface mirrors. Aluminum is the usual coating for first-surface mirrors, with an overcoating of aluminum oxide, magnesium flouride, or cerium oxide providing durability and enhanced reflection for the aluminum film. Silver is not satisfactory for front-surface mirror due to the tarnishing of the film. Overcoating lengthens lifetime, but diffusion of oxygen and sulfur dioxide through defects eventually leads to a reduction of reflectivity for sunlight. A tarnished silver film is still highly reflective in the thermal infrared.

Thin metal sheets can be used for mirror substrates, but a reflecting layer is required for good performance. Alzak is the trade name for aluminum sheets where a surface of high purity aluminum has been electropolished and overcoated with a thin layer of oxide.

Aluminized plastic films are attractive for their potentially low cost. Silver can also be used for the reflective film when properly protected. Plastic films have a diversity of physical properties and no single material has the ideal combination of properties. Sheldahl Corporation has developed a silvered plastic film in which two different plastics are combined to optimize properties. The substrate film is a type of Mylar that has good mechanical properties and a low thermal expansion coefficient compared to other plastics, matching aluminum rather well. Mylar is readily coated by continuous vacuum deposition. The silver layer is in turn protected by a layer of acrylic plastic flowed on and polymerized in place. The acrylic layer has excellent transparency and lifetime under solar flux, and is one of the hardest plastics adding lifetime under abrasive conditions compared to other plastics. Acrylic, however, has poor mechanical properties but this situation is compensated for by the excellent mechanical properties of the Mylar substrate. The acrylic also gives long lifetime to the reflectance, on the order of 90–92%, to the silver.

The low cost of metallized plastic films makes it attractive to examine ways of using these films for mirrors. A flat mirror, as would be used for heliostats, can be made by stretching the film over a rigid ring. Several designs are being evaluated as part of the central receiver power systems in the United States.

Concave mirrors of plastic film present difficulties. Preforming concave mirrors while retaining the surface smoothness of flat films is difficult. Several studies have been made of pressure deforming flat films to a desired radius of curvature. Babcock (1949) proposed using a procedure of evacuating the space behind an aluminized film to make mirrors of astronomical quality. More recently, Taylor (1975) has developed a design using air pressure between two plastic films to make a mirror concentrator for solar energy applications, as shown in Fig. 9.36.

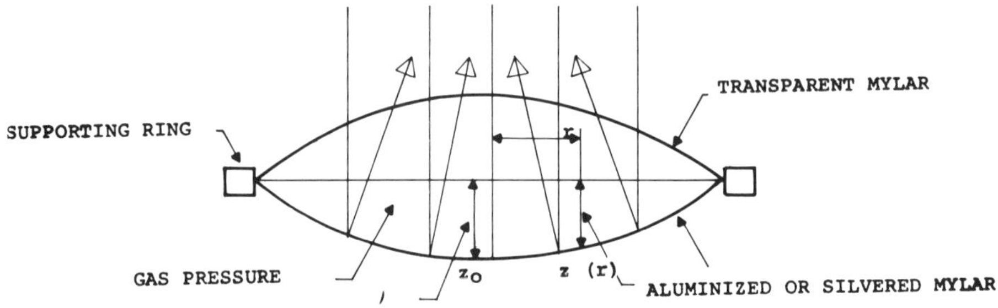

FIG. 9.36 Schematic arrangement for a double-layer plastic film mirror.

The deflection of a membrane surface of uniform cross section and elasticity is given by

$$z(r) = z_0[1 - 0.9(r/a)^2 - 0.1(r/a)^4 + \cdots] \tag{9.23}$$

where z_0 is the central deflection, r the radius of the zone where $z(r)$ is to be calculated, and a the semidiameter of the aperture. To examine the quality of the image formed by the deflected membrane, let us subtract a parabolic curve having the same central deflection. The residual between the two surfaces is

$$z(r) = z_0[0.1(r/a)^2 - 0.1(r/a)^4 + \cdots] \tag{9.24}$$

The image error is determined by the maximum slope error. In the above equation the maximum slope occurs at the edge, but the maximum error after refocusing is half as much. The slope of the above function is

$$\text{slope} = z_0[0.2(r/a) - 0.4(r/a)^3 + \cdots] \tag{9.25}$$

For the case where $a = R/2$ and the focal length is $R/2$, the image error normal to the rim ray is approximately 0.016 a. Thus for a mirror of 4 m aperture, the image enlargement over the perfect solar disk would be 3.2 cm.

The maximum deflection of such a membrane mirror is set by the tension stress and the modulus of rupture of the material. It is also set by the degradation of the metallized film due to stretching of the plastic film. This approach is entirely satisfactory for mirrors having small values of rim angle α, but experiments are needed if large values of α are required.

10

Solar Pond

S. B. SAVAGE

DEPARTMENT OF CIVIL ENGINEERING
AND APPLIED MECHANICS
MCGILL UNIVERSITY
MONTREAL, QUEBEC, CANADA

10.1 INTRODUCTION

Nonconvective solar ponds have been proposed as a simple, relatively inexpensive method of collecting and storing energy on a large scale. Typically, a solar pond is a shallow body of water about one meter deep containing dissolved salts to generate a stable density gradient (fresh water on top and denser salt water at the bottom). Part of the incident solar radiation entering the pond surface is absorbed throughout the depth, and the remainder which penetrates the pond is absorbed at the black bottom. If the pond were initially filled with fresh water the lower layers would heat up, expand, and rise to the surface. Because of the convective mixing and heat loss at the surface, only a small temperature rise in the water could be realized. On the other hand, convection can be eliminated by initially creating a sufficiently strong salt concentration gradient. In this case, thermal expansion in the hotter lower layers is insufficient to destabilize the pond. With convection suppressed, the heat is lost from the lower layers only by conduction. Because of its relatively low thermal conductivity, the water acts as an insulator and permits high temperatures (over 90°C) to develop in the bottom layers. Energy can be extracted from the pond by recycling the water in the hot layers of the pond through a heat exchanger. There are several possible applications for solar ponds ranging from the most ambitious one of power generation using, for example, organic fluid Rankine cycle engines (see Girardier *et al.*, 1973; Teagan and Sargent, 1973), which can operate with relatively low source temperatures to the most promising one of salt production (Hirschmann, 1970). Recently it has been suggested (Rabl and Nielsen, 1975) that solar ponds could be used for space heating at higher latitudes.

The history of the solar pond concept may be traced to the early part of this century when von Kaleczinsky (1902) suggested the use of stratified salt ponds as solar energy collectors. He had observed temperatures of 70°C at a depth of 1.3 m in Medve Lagoon, a naturally stratified salty pond in Hungary.

In 1948, Dr. Rudolf Bloch suggested to the Israel National Research Council that an artificially stratified pond be used to collect solar energy for practical utilization. Some ten years later, an extensive pioneering investigation was initiated at the National Physical Laboratory of Israel under the direction of Dr. Harry Tabor. Tabor (1959, 1961, 1963, 1964), Weinberger (1964), Elata and Levin (1965), and Tabor and Matz (1965) carried out theoretical investigations of the underlying physics of the solar pond and made laboratory and field tests to study the many factors affecting pond performance. The concept of the solar pond was well proven; temperatures of 95°C were developed in the pond bottom. Based upon the results of this work, economic analyses were made of very large scale ponds capable of producing electric power. At that time, in a country having a well-developed electricity grid, the solar pond could not compete economically with conventional power plants using petroleum, and the project was discontinued in 1966. Today, because of recent increases in oil prices, the economics will have to be reconsidered. However, large scale electric power production may not be the most suitable role for the solar pond, and other smaller scale applications in less well-developed areas for which solar ponds could readily supply the required low temperature heat might be more appropriate.

Since the beginning of Tabor's work, research on various aspects of solar ponds has been conducted in a number of other countries. In Chile, Hirschmann and Gaete (1962) made detailed studies of some physical characteristics associated with the operation of solar ponds. Hirschmann (1970) suggested the use of salt flats as solar collectors for industrial purposes. At MIT, Stolzenbach *et al.* (1968) developed numerical methods to predict temperature distributions within the solar pond. Dake (1973) has continued analytical and laboratory studies in Zambia. A group of Russian scientists at the Uzbek SSR Academy of Sciences (Eliseev *et al.*, 1971, 1973; Usmanov *et al.*, 1969, 1971a,b,c, 1973; Usmanov and Eliseev, 1973), has been actively engaged in very detailed and sophisticated research of the relevant physics. Dr. G. C. Jain (1973) has designed and is operating a solar pond (Fig. 10.1)

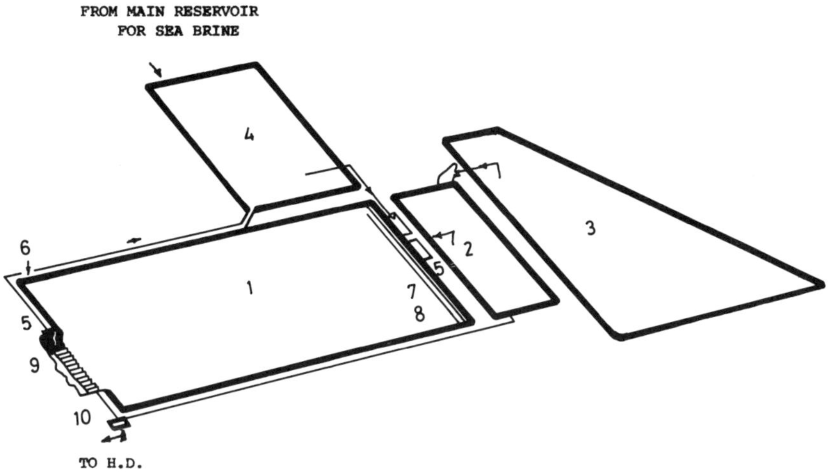

FIG. 10.1 Diagram of solar pond operated by Dr. G. C. Jain at the Central Salt and Marine Chemicals Research Institute, Bhavnagar, India. 1—Solar pond; 2—reservoir pond for concentrated solution; 3—bittern reservoir pond; 4—settling tank for sea brine; 5—inspection chamber; 6—overflow channel; 7—perforated pipe for surface flow; 8—perforated pipe for bottom feeding; 9—pipes for extracting heated solution; 10—collection and distribution tank for heated solution.

for use in the production of salt at the Central Salt and Marine Chemicals Research Institute in Bhavnagar, India. A research group at Lawrence Livermore Laboratory (Clark *et al.*, 1975) is developing solar ponds with the objective of producing electric power. Their ponds are different from the usual salt stratified ponds, but the overall concept is similar. An initial application of their pond module is to heat water for use in a uranium mine in New Mexico.

Compared with several other types of collectors and storage devices, solar ponds are at a rather rudimentary stage of their development. Bearing this in mind, this chapter will review previous work on solar ponds, concentrating on the physical principles governing their operation but at the same time pointing out the practical difficulties encountered.

10.2 FORMING AND MAINTAINING THE STABLE DENSITY GRADIENT

In order to generate the initial salt concentration gradient required to prevent convection, the usual practice is to fill the pond with several layers of salt solution with the successive layers changing stepwise in concentration from near saturation at the bottom to fresh water at the top. For a typical pond of a meter deep, one might use six or eight layers. Less dense layers can be floated successively on top of the existing ones, or successively, more dense layers can be put in at the pond bottom flowing in as under currents and gradually "lifting" the whole pond. Because of turbulent mixing and molecular diffusion, which occur during the filling, the staircase-type concentration profile evolves into a smooth nearly uniform concentration gradient. Various salts such as magnesium chloride ($MgCl_2$), sodium nitrate ($NaNO_3$), sodium chloride (NaCl), etc., have been suggested (Hirschmann, 1970) for use in the pond; naturally available salts are of a rather mixed chemical composition which varies from place to place. Magnesium chloride and sodium nitrate have higher saturation densities than sodium chloride and thus could provide more stability when the pond is hot.

In the absence of any flux of salt at the top and bottom of the pond, molecular diffusion will tend to erase the initially formed density gradient. Figure 10.2 shows the theoretical development of an initial stepwise (six layer) concentration profile toward the final asymptotic state where the concentration is uniform (Chepurniy and Savage, 1975). The evolution is quite slow; for a pond one meter deep and nearly saturated at the bottom, the time required for the difference in concentration between top and bottom of the pond to decay to half of its initial value is of the order of one year.

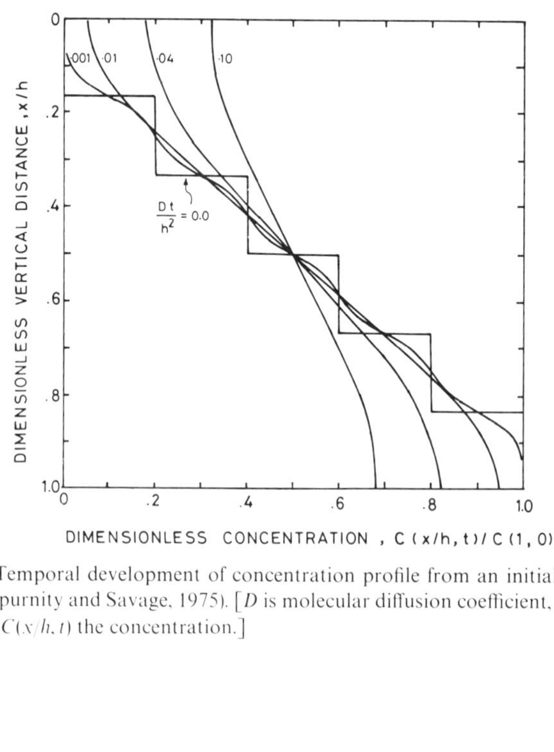

FIG. 10.2 Temporal development of concentration profile from an initial six layer staircase profile (Chepurnity and Savage, 1975). [D is molecular diffusion coefficient, t the time, h the total depth, and $C(x/h, t)$ the concentration.]

But to provide stability, the concentration gradient must be maintained and this entails a continuous upward salt transport of about 0.006 g/cm^2/day.

This diffusion loss can be restored and the concentration gradient maintained by periodically introducing concentrated salt solution at the bottom and washing the surface with fresh water (Tabor, 1963; Weinberger, 1964). Washing the surface with fresh water prevents the increase in concentration near the top which could occur, not only because of molecular diffusion but by evaporation and mixing due to wind, waves, etc. A more efficient approach, which does not require the continual addition of salt, is the "falling pond" concept. Liquid is withdrawn from the hot lower layer and put through a flash evaporator where heat is extracted. The solution, now having a higher concentration and smaller volume, is returned to the pond bottom. The difference between the rates at which fluid is extracted and returned causes a steady descent or "fall" of the pond, the rates being chosen so as to compensate for the molecular diffusion of salt upwards, and fresh water being added at the free surface to keep both the pond depth and surface concentration constant. The falling velocity estimated by Elata and Levin (1965) is approximately 0.2–0.5 mm/day. Usmanov and Eliseev (1973) suggested another way of preserving the density gradient which seemed to work well in laboratory model tests. The six layers of $MgCl_2$ solution filling the model pond were separated by thin transparent plastic films. The films had many small perforations to prevent the growth of air bubbles between the layers.

Wind-generated surface currents and waves can create an upper convection zone that decreases the maximum temperatures achievable in the pond bottom. Based upon their lab experiments, Elata and Levin (1965) suggested that such problems could be alleviated if the scale of convection was decreased by dividing the pond (near the surface) into a number of smaller areas using, for example, vertical transparent membranes.

It may be beneficial to cover the pond surface with an inflated clear plastic "mattress" similar to those that have been used to help heat swimming pools by solar energy. It would provide an additional insulating layer, reduce evaporative losses, stop dust and rain from entering the pond, and prevent wind-generated surface currents.

Another possible problem is the loss of salt and heat by percolation of the solution through the permeable pond bottom. To prevent this, the pond bottom can be covered with an impermeable liner of plastic or rubber (Dallaire, 1975), a technique which presently is used to prevent pollution from waste lagoons and ponds. Because of high temperatures developed at the solar pond bottom, there is the possibility that decaying organic matter may form a gas bubble under the pond. In one of the (unlined) ponds constructed by Tabor and his associates (1965), the formation of small bubbles beneath the pond destroyed the usefulness of the pond. Not only did the bubbles upset the density gradient during their rise, but they brought up mud from the pond bottom and prevented the original dirt from settling to the bottom. In waste lagoons such bubble problems have been avoided by placing a 15-cm sand blanket beneath the liner, so that the gas can escape up the pond slopes to the atmosphere. One-way check valves have been installed at those trouble spots where gas still collected.

Most of the papers mentioned above refer to quite strong concentration gradients that are presumably based upon a stability analysis of Weinberger (1964). Instability occurs in an overstable or oscillatory mode, the amplitude of an oscillatory motion increases in time. This type of dynamic instability can occur even when the pond is statically stable (i.e., density increases with depth). Stability analyses by others (see the review by Turner, 1973) indicate considerably smaller density gradients than that suggested by Weinberger (1964) for stability. The matter should receive more careful study to get an accurate estimate of the salt requirements for the pond. If one relies on Weinberger's analysis, then large quantities of salt are necessary, and to be economically viable the pond may have to be located near a ready source of concentrated waste brines or a natural salt supply.

10.3 ABSORPTION OF SOLAR RADIATION INCIDENT UPON THE POND

The solar radiation incident upon the earth's surface at given location and time for various atmospheric conditions has been discussed in Chapter 2, and the present section is concerned only with the subsequent behavior of the radiation that reaches the horizontal pond surface. Part of the radiation reaching the pond surface is reflected and the rest entering the pond is attenuated throughout its depth. The amount of radiation reflected depends upon the angle of incidence. Weinberger (1964) has calculated E, the ratio of radiation penetrating the pond surface and incident upon a horizontal surface to that incident on a horizontal plane just above the surface, by making use of Fresnel's equations (Williams and Becklund, 1972),

$$E = 2n(a^2 + b^2)\cos i \cos r$$
$$a = (\cos r + n\cos i)^{-1}, \qquad b = (\cos i + n\cos r)^{-1} \tag{10.1}$$

where n is the index of refraction, and i and r are the angles between the normal and the incident and refracted rays, respectively. The incident radiation is refracted according to Snell's law:

$$\sin i = n \sin r \tag{10.2}$$

For direct (beam) radiation, the angle of incidence (i.e., solar azimuthal angle) is given closely by

$$\cos i = \cos \phi \cos \delta \sin 2\pi(t - 6)/24 + \sin \phi \sin \delta \tag{10.3}$$

where ϕ is the latitude, t the time in hours (noon = 12), δ the declination of the sun given approximately by $\delta_0 \sin 2\pi D/365.25$, δ_0 the declination at the equinox (23.45°), and D the number of days measured from the spring equinox. For most locations it happens that when the radiation intensity is high, the incidence angle is small and the reflective losses are low (of the order of a few percent), and when the incidence angle is large, giving large reflective losses, the radiation incident upon a horizontal surface is small. The net effect is to give rise to relatively small total reflective losses for the direct radiation.

Weinberger (1964) assumed the diffuse radiation to be random and coming with equal intensity from all parts of the sky. Thus the fraction of diffuse radiation penetrating the surface was given by

$$2 \int_0^{\pi/2} E \cos i \sin i \, di \tag{10.4}$$

which Weinberger found to be approximately 0.93 for his pond.

Strictly speaking, the refractivity is wavelength dependent and Usmanov *et al.* (1971a) have presented data on the spectral refractivity of water and the index of refraction for several salt solutions of various concentrations. The values of n increase with concentration, but are not too different from that of 1.33 for fresh water which was assumed by Weinberger at the pond surface.

After entering the pond, the radiation is attenuated along its path by scattering and absorption. The absorption of radiation in a semitransparent medium may be described by Bouguer's law, which states that a thickness of the homogeneous absorber ds reduces the intensity of the incident radiation of wavelength λ an amount dq_λ proportional to both the intensity q_λ and ds, thus

$$dq_\lambda = -k_\lambda q_\lambda \, ds \tag{10.5}$$

where k_λ is the spectral absorption coefficient. Equation (10.5) may be readily integrated to yield for the transmitted radiation

$$q_\lambda^t = q_\lambda e^{-k_\lambda s} \tag{10.6}$$

The spectral transmittance is defined by

$$\tau_\lambda = q_\lambda^t / q_\lambda = e^{-k_\lambda s} \tag{10.7}$$

and the spectral absorption coefficient may be expressed as

$$k_\lambda = (1/s)\ln \tau_\lambda \tag{10.8}$$

Usmanov *et al.* (1971b) have presented spectral absorption coefficients k_λ for various uniform concentrations of $MgCl_2$ solutions. They also determined mean integrated absorption coefficients K defined such that

$$\int_0^\infty q_\lambda{}^t\,d\lambda = \int_0^\infty q_\lambda e^{-k_\lambda s}\,d\lambda = \left[\int_0^\infty q_\lambda\,d\lambda\right] e^{-Ks} \tag{10.9}$$

thus

$$K = -\frac{1}{s}\ln \frac{\int_0^\infty q_\lambda e^{-k_\lambda s}\,d\lambda}{\int_0^\infty q_\lambda\,d\lambda} \tag{10.10}$$

Their results showed a pronounced increase of K with concentration, but they have presented insufficient detail for one to readily make use of their results for the case of concentration varying with depth as occurs in the solar pond.

A simpler way of expressing the intensity of radiation H at any depth x is by the use of several exponential extinction coefficients, each corresponding to a particular spectral band (range of wavelengths). Thus

$$H = H_0 \sum_{j=1}^{m} \beta_j e^{-\eta_j x} \tag{10.11}$$

where H_0 is the intensity at $x = 0$, β_j a radiation component ($\sum_{j=1}^{m} \beta_j = 1$), and η_j the extinction coefficient for the jth spectral band. Note the analogy of this method of description with the mean integrated absorption coefficient K. Typical values for the β_j's and η_j's for water may be obtained from Defant (1961) and have been listed by Rabl and Nielsen (1975) as well as referenced by Anonymous (1972a). The values given by Rabl and Nielsen are listed in Table 10.1.

TABLE 10.1

Exponential Extinction Coefficients for Water[a]

j	β_j	η_j (cm^{-1})	Spectral band (μm)
1	0.237	3.2×10^{-4}	0.2–0.6
2	0.193	4.5×10^{-3}	0.6–0.75
3	0.167	0.03	0.75–0.9
4	0.179	0.35	0.9–1.2

[a] Quoted by Rabl and Nielsen (1975).

The short wavelengths can penetrate many meters with little absorption, while the near infrared is absorbed in the upper few centimeters. Water is virtually opaque to the longer wavelengths.

It should be noted that values in Table 10.1 are for water. To the writer's knowledge, comparable data for ponds having stratified salt solutions (which are more opaque) do not exist, but would be most useful.

Weinberger (1964) has presented measured absorptivity data for six samples of sea water (see also Sverdrup *et al.*, 1942) which show large variations between samples. For example, the radiation transmitted a length of one meter varied from 15 to 40% of the incident solar radiation. The differences disappear with filtration and thus they are due mostly to suspended matter. This does, however, point out the possibility of problems due to dust and other matter that fall on the solar pond. Because of the stable stratification, the pond is not turbid, as are some natural waters. Nevertheless, foreign matter may not necessarily settle to the bottom, but may seek the level corresponding to its own density.

Tabor and Matz (1965) have presented transmission measurements made in an actual stratified solar pond. The results are similar to that for one of the sea water samples described by Weinberger (1964), but show a higher absorption near the bottom where the salt concentration is higher, in accord with the results of Usmanov *et al.* (1971b).

The amount of radiation reaching a given depth of the pond is further complicated because of refraction which depends upon time of day and year. To reach a unit depth, the radiation must travel along an oblique path a distance

$$1/\cos r = [1 - (1/n^2) \sin^2 i]^{-1/2} \tag{10.12}$$

assuming a constant value for the index of refraction n. For example, taking an average value of $n = 1.35$, $\sec r$ equals 1.00, 1.17, and 1.49 for incidence angles i equal to 0°, 45°, and 90°, respectively.

It is obvious that an accurate accounting of the reflective losses and radiation attenuation with depth on a daily and seasonal basis can be quite involved. As a simple approximation for calculation purposes, Rabl and Nielsen (1975) suggested the use of a constant effective incidence angle i corresponding to the position of the sun at 2 p.m. at equinox, neglect of the distinction between direct and diffuse radiation, and the use of an effective (constant) extinction coefficient

$$\eta_j' = \eta_j/\cos r \tag{10.13}$$

They neglected the diurnal variations in the intensity of radiation incident on the pond surface, used a mean daily radiation, and assumed it to vary in a sinusoidal way on a yearly basis.

10.4 THEORETICAL ANALYSES OF THE TEMPERATURE DEVELOPMENT IN THE POND

Theoretical analyses of the thermal behavior of the solar pond can be used for predicting pond performance, and thus for optimizing the pond geometrical and operational parameters. A complete treatment would be exceedingly complex because of the large number of variables, and all past workers have idealized the problem to some degree. For practical size ponds, the lateral dimensions are large compared with the depths, the heat flow is primarily vertical, and it is reasonable to assume that the pond temperature is governed by the one-dimensional time-dependent heat conduction equation

$$\rho c(\partial T/\partial t) = (\partial/\partial x)(k\,\partial T/\partial x) + F(x, t) \tag{10.14}$$

where T is the temperature, t the time, x the vertical coordinate measured downward from the pond surface, and ρ, c, and k the density, specific heat, and thermal conductivity, respectively. Hirschmann and Gaete (1962) have presented the variation of the thermal properties with concentration for $MgCl_2$ solutions. In some analyses, ρ, c, and k have been approximated as being constant (see Rabl and Nielsen, 1975, and Weinberger, 1964) and in others, as linearly dependent upon concentration (Eliseev *et al.*, 1971, and Chepurniy and Savage, 1974). The source function

$$F(x,t) = -\partial H/\partial x \tag{10.15}$$

is due to the absorbed solar radiation $H(x, t)$ [see Eq. (10.11)]. As pointed out in the last section, the radiation incident upon the pond surface, the surface reflectivity, and the radiation path lengths to reach a given pond depth all depend upon incidence angle. Thus H, and hence F, depend not only upon the depth x, but on the time of day and year. The most sophisticated analysis, a numerical finite difference approach by Eliseev *et al.* (1971), considers in a complete way the daily variation in $F(x, t)$. Their calculations indicate relatively small daily temperature fluctuations; for an 80-cm deep pond the night time cooling at the pond bottom was about 6°C, a result that agrees with previous calculations by Weinberger (1964). In view of this, others have neglected the daily variations in $F(x, t)$ by using mean daily values, but have incorporated the seasonal variations for their calculations.

At the pond-free surface, solutions to Eq. (10.14) must satisfy the following boundary condition corresponding to a heat flux balance

$$k\,\partial T(0, t)/\partial x = q_c + q_e + q_r \tag{10.16}$$

where q_c is the heat loss by convection from the water surface to the atmosphere, q_e the heat loss by evaporation, and q_r the net long wave radiation from the water surface to the atmosphere. Details of the formulas, some of them empirical, to calculate the heat losses q_c, q_e, and q_r are given in Eliseev *et al.* (1971) and Anonymous (1972a). Weinberger (1964) noted that the mean daily air temperature is near (usually a few degrees centigrade above) the sea surface temperature in the Mediterranean and the Dead Sea. Thus, for the upper surface boundary conditions in his solar pond analysis, he used the simple approximation that the surface temperature was equal to the ambient air temperature, rather than determine it by the much more complicated Eq. (10.16).

A second boundary condition is required at the pond bottom, $x = h$. It should account for the radiation $H(h, t)$ reaching the bottom of the pond and assumed to be absorbed there, the heat flux q_s, lost to the soil beneath the pond, and the heat extracted at a rate q_u from the pond bottom; thus,

$$k\,\partial T(h, t)/\partial x = H(h, t) - q_s(t) - q_u(t) \tag{10.17}$$

Most workers have assumed the entire heat transfer problem including the soil beneath the pond to be one-dimensional. Instead of using Eq. (10.17), they account for the heat absorption $H(h, t)$ and heat extraction $q_u(t)$ at the bottom of the pond, assume that the soil, of thermal conductivity k_s, extends infinitely downward, and apply the boundary condition

$$\partial T(\infty, t)/\partial x = 0 \tag{10.18}$$

The boundary condition (10.18) implies that under steady state conditions there would be no heat losses to the ground. This result is a consequence of one-dimensionality and the real situations would involve heat losses, although these losses decrease with increasing pond size. A crude idea of the losses may be obtained from the work of Shelton (1975), who was concerned with heat losses from underground storage systems, and that of Chepurniy (1975). The steady state total heat flow from a hot constant temperature disk of radius a (corresponding to the pond bottom) to a cooler semi-infinite medium (the soil), having its surface insulated, divided by the disk area πa^2 is

$$Q/\pi a^2 = 4k_s\,\Delta T/\pi a \tag{10.19}$$

where ΔT is the temperature difference between the disk and the soil at large distances from the disk. The soil thermal conductivity varies widely, but assuming $k_s = 1$ kcal/h m °C for a moist soil (for dry sand $k_s \sim 0.15$ kcal/h m °C), $\Delta T = 70$°C, and $a = 10\,m$, we obtain an average heat loss per unit area of $Q/\pi a^2 = 21$ cal/cm^2 day, which is only a few percent of a typical daily mean solar radiation incident upon a horizontal surface of 500 cal/cm^2 day. Thus the assumption of one-dimensional heat transfer in the ground does not lead to large errors with regard to heat losses (except in very small ponds), whereas it provides a simple way to incorporate the effects of ground heat storage.

The solar ponds suggested by Rabl and Nielsen (1975) contain a deep convecting zone about 2 m deep separated from the overlying insulating layer by a partition. Their bottom boundary condition and ensuing analysis are somewhat more involved because of this deep convecting layer.

Solutions to the one-dimensional heat conduction Eq. (10.14) have been obtained by various authors both in terms of closed form analytical solutions (see Rabl and Nielsen, 1975; Weinberger, 1964; and Dake, 1973), as well as numerical finite difference schemes (see Stolzenbach *et al.*, 1968; Eliseev *et al.*, 1971; Chepurniy and Savage, 1974), the most sophisticated and complete approach being that of Eliseev *et al.* (1971). Figures 10.3 and 10.4 taken from Eliseev *et al.* (1971) illustrate the typical theoretical initial temperature buildup without heat extraction for a pond located in Tashkent, Uzbek and started on the first day of July. Figure 10.3 shows the temperature profiles in an 80-cm deep pond at four-day intervals. In 48 days the predicted bottom temperature is 145°C, and thus after 22 days heat must be extracted to prevent boiling of the solution in the bottom layer. Figure 10.4 shows the temperature buildup in the bottom layer for ponds 20, 40, and 80 cm deep. The shallower ponds heat up faster but reach a lower "steady state" temperature. The daily temperature fluctuations are smallest for the larger depths and about 6°C for the 80 cm pond. The times to reach steady state conditions are rather shorter than those determined by Weinberger, but his calculations were for ponds located in Israel and started in the spring and fall when the incident radiation was less.

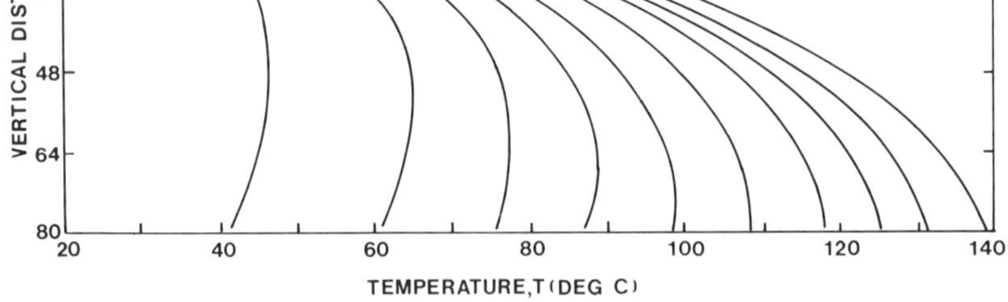

FIG. 10.3 Theoretical temperature build-up in an 80 cm deep pond located in Tashkent, Uzbek and started on the first day of July (after Eliseev *et al.*, 1971).

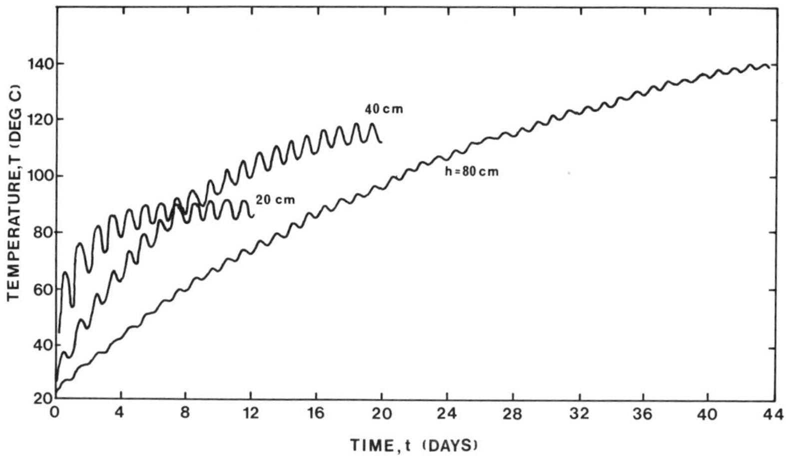

FIG. 10.4 Theoretical temperature build-up in bottom layers of 20, 40, and 80 cm deep ponds located in Tashkent, Uzbek and started on the first day of July (after Eliseev *et al.*, 1971).

10.5 EXTRACTION OF HEAT FROM THE POND

One way of collecting heat from solar ponds is to place heat exchangers (the simplest consisting merely of pipes) in the hot lower layers. However, this entails not only the initial installation cost but the continued pumping losses associated with the heat transfer fluid. A simpler approach is to take advantage of the stable density stratification and use the method of "selective withdrawal" (see Turner, 1973), a technique often applied for water quality control in large reservoirs. A sink placed in a stably stratified fluid withdraws fluid from a thin horizontal layer, much as one might draw a single card from a deck. This flow is quite different from that which would occur if the fluid was homogeneous. The phenomenon of selective withdrawal provides means by which, firstly, the stable density gradient required in the upper insulating layers can be corrected and maintained; and secondly, the hot lower layer can be removed, passed through a heat exchanger and returned at lower temperature to the bottom of the pond. Elata and Levin (1965) have studied the hydraulics of selective withdrawal of the bottom layer of a solar pond both in the laboratory and in field tests on a 55 × 25 m experimental pond. They found that, because of molecular diffusion and the presence of small mixing zones at the inlet and outlet, the interface above the bottom flowing layer rose slowly at an order of magnitude of 1 mm/day. This need not be a serious problem, since the rise can be balanced by utilizing the "falling pond" concept described in Section 10.2.

Nevertheless, in certain systems, such as the ponds with deep convecting lower layers as suggested by Rabl and Nielsen (1975), it may be desirable to separate physically the bottom layer from the upper insulating zone by a thin transparent partition. A thin plastic film of this kind was used with particular success in laboratory experiments of heat extraction from a model solar pond by Usmanov *et al.* (1973).

Weinberger (1964) has carried out studies of the heat extraction based upon his theoretical model of the solar pond. After the initial warm-up period, he finds that for a given extraction temperature T_u, there is a particular value of the pond depth h at which the rate of heat extraction q_u is a maximum. It transpires that this optimum rate of energy withdrawal is equal to the radiation reaching the pond bottom, and thus in this case the pond is heated only by the radiation absorbed throughout its depth. Weinberger has calculated the optimum extraction temperature and the collection efficiency (the ratio of the mean annual rate of energy withdrawal to the mean annual radiation incident upon a horizontal surface) as functions of pond depth h. Figure 10.5 shows his results for a pond in Israel having a mean ambient temperature of 26°C and a mean annual radiation of 505

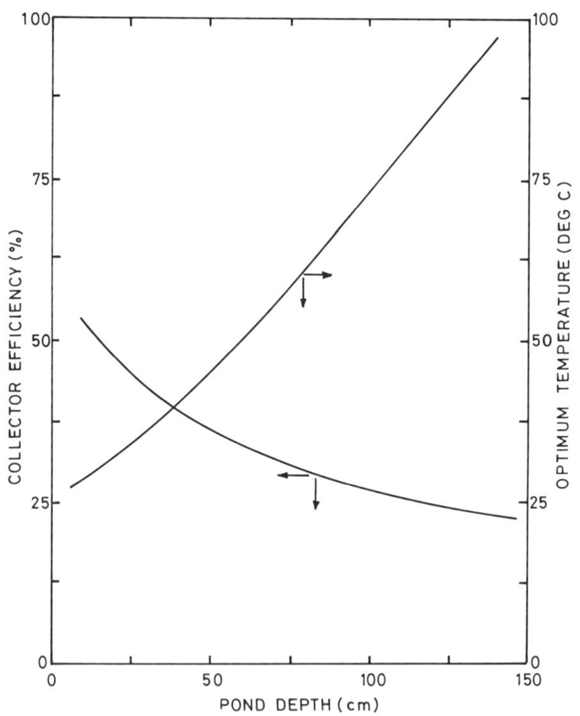

FIG. 10.5 Theoretical optimum extraction temperature and collection efficiency as functions of pond depth h (after Weinberger, 1964).

cal/cm^2 day on a horizontal surface. It may be economically advantageous to remove heat at temperatures higher than the theoretical optimum T_u. For example in a 100-cm deep pond the optimum extraction temperature is 73°C with a collection efficiency of 27%, but heat can be removed at 90°C with a slightly lower collection efficiency of 23%. The optimum depth for heat extracted at 90°C is 120 cm with an efficiency of 24.6%.

If the heat extracted from the pond bottom is converted to mechanical or electrical energy by means of a Carnot engine, then a quantity that should be optimized is the recoverable energy

$$J = q_u \eta_c \tag{10.20}$$

where

$$\eta_c = (T_u - T_0)/T_u \tag{10.21}$$

is the Carnot efficiency and T_u and T_0 are the absolute temperatures of the extracted fluid and the engine heat sink, respectively.

Weinberger (1964) found that the maximum value of J occurs for a depth of about 2 m, but the theoretical temperature of the lower layer then is greater than its boiling point. Since the theoretical energy recovery function J increases slowly for depths greater than about 1 m, an economically optimum depth is probably close to 1 m.

Although the feasibility of heat extraction has been confirmed in a few experiments [including laboratory tests as in the work of Usmanov *et al.* (1973), small scale field experiments as in the work of Saulnier *et al.* (1975), as well as tests in a large 55 × 22 m solar pond in India (see Jain, 1973)], none of these experiments have been either sufficiently extensive or reported in adequate detail to permit comparison with theoretical predictions. Carefully controlled field experiments should be performed over extended periods to determine whether the theoretically predicted heat extraction rates can actually be realized.

10.6 CONCLUSION

The basic idea of the solar pond has been verified—several ponds have been built, high temperatures close to boiling have been developed in the lower layers and heat has been extracted successfully from the pond. Generally, the potentialities of the solar pond appear promising. Nevertheless, there are still insufficient hard data available to make realistic economic comparisons with other types of collectors. A large scale pond should be constructed and operated continuously over a long period to gain operating experience and accurate information on collector efficiency as well as economics.

Tabor (1964; Tabor and Matz, 1965) has suggested that salt production is the most promising of the possible applications of low temperature heat supplied by the solar pond. His economic analyses show the solar pond to be quite competitive with other methods.

Rabl and Nielsen (1975) have made a preliminary (probably optimistic) economic study of solar ponds used for space heating. Taking the pond lifetime to be 20 years, they find the cost of heating is 1.1 US cents per kilowatt hour based upon a pond of 140 m^2 area to supply heat for a single house in Columbus, Ohio. By employing a larger pond ($20\times$ the area) to supply heat for 20 houses, the cost is reduced to 0.41 US cents per kilowatt hour, which is lower than the cost of heating by conventional fuels.

Another possible, but far more difficult, application is power generation. Some years ago, Tabor (1964; Tabor and Matz, 1965) made an economic analysis for the case of large ponds of a square kilometer in size and concluded that the solar pond was not competitive with large conventional power plants in Israel but that it could be promising for small units in developing countries. With the advent of new Rankine cycle engines (Girardier *et al.*, 1973, and Teagan and Sargent, 1973), which can operate with low temperature heat, and the recent increases in fuel prices, the case for power generation using solar ponds will have to be reexamined.

After a lull of a few years, following the curtailment of the important groundwork laid by Tabor and his co-workers, several groups are actively studying solar ponds again and many of the remaining open questions should be answered soon.

11

Solar Furnaces

TAKEMARO SAKURAI

SOLAR ENERGY LABORATORY
RESEARCH INSTITUTE FOR SCIENTIFIC MEASUREMENTS
TOHOKU UNIVERSITY
SENDAI, JAPAN

11.1 INTRODUCTION

A solar furnace is an instrument to get high temperatures by concentrating solar radiation onto a specimen. Children burn a black paper by focusing a sun image by a lens. The lens, in this case, is a solar furnace in miniature. It is historically known that in the eighteenth century, the French scientist Lavoisier used a lens system in order to melt various materials including platinum. More recently, a solar furnace provided with a concentrator composed of 19 lenses, 61 cm in aperture, was constructed at California Institute of Technology, (Loh *et al.*, 1957). Concentration by a lens, however, is not desirable when a high attainable temperature is wanted, because of the radiant loss caused by absorption in the lens material and by spherical and other aberrations. Therefore, in present-day furnaces, concentration by means of a lens is scarcely employed.

In 1921, a solar furnace composed of a paraboloidal concentrator and a lens was devised by the German scientist Straubel. In 1933, he built another furnace, in which the paraboloidal concentrator, 2 m in aperture and 86 cm in focal length, was fixed facing downward and the solar radiation was conveyed upward by turnable plane mirror or *heliostat* (Straubel, 1949). After World War II, Trombe *et al.* (1947) in France used a military searchlight mirror in a solar furnace. It was so successful that many furnaces fitted with searchlight mirrors were built in various places (Conn, 1954; Cohen and Hiester, 1957). Conn (1951) in the USA constructed a furnace using an aluminum concentrator, 3 m in aperture and 86 cm in focal length, in an equatorial mounting. The successive construction of solar furnaces at that time was stimulated by the requirement for the investigation of refractory materials necessary for jet, rocket, and other high-temperature processes. The furnaces constructed were found to give a high temperature of about 3000°C and many refractory materials were fused by using them. The high-temperature area obtained by these furnaces, however, was a circular portion, several millimeters in diameter, which was too small for some kinds of work. Thus, the construction of large furnaces began to be earnestly desired.

In 1952, a large furnace was constructed by Trombe (1952) in Mont-louis, France, composed of a heliostat and a concentrator with a horizontal optical axis, 10 m in aperture and 6 m in focal length. The solar energy incident on the furnace was some 70 kW. Thereafter, large furnaces, 60 to 100 kW in input solar energy, were built in Algeria (Juston-Coumat and Duburg, 1957), USA (Davies and Cotton, 1957), Japan (Sakurai *et al.*, 1964), and France (Meunier, 1973). These furnaces could give a high temperature area from 3 to 10 cm in diameter which was found to be sufficient for most scientific work. Using these furnaces, various investigations have been undertaken in the field of high-temperature science. In 1970, Trombe *et al.* (1973) established a huge furnace in France, with the intention of applying a solar furnace to high-temperature work for industry. The paraboloidal concentrator is 54×40 m in dimension and the input solar energy exceeds 1800 kW.

In this section, the development of the solar furnace has been surveyed historically. In the following section, arguments for solar furnaces will be given in some detail.

11.2 FUNDAMENTALS OF SOLAR FURNACES

A. Solar Radiation Available for the Furnace

The thermal radiation from the sun is propagated through space without any appreciable attenuation to arrive at the atmosphere of the earth. Roughly speaking, it shows a continuous spectra extending from the ultraviolet to infrared region as shown by curve (a) in Fig. 11.1 (see Handbook of Geophysics, Anonymous, 1960). Curve (b) is the spectral distribution of a blackbody at 6000°K, theoretically calculated, which will be seen to be in good agreement with (a). The solar constant denotes the intensity of solar radiation outside the atmosphere, incident normally upon unit area in unit time, and its measured value is 2.0 cal/cm^2/min which corresponds to 1.4 kW/m^2.

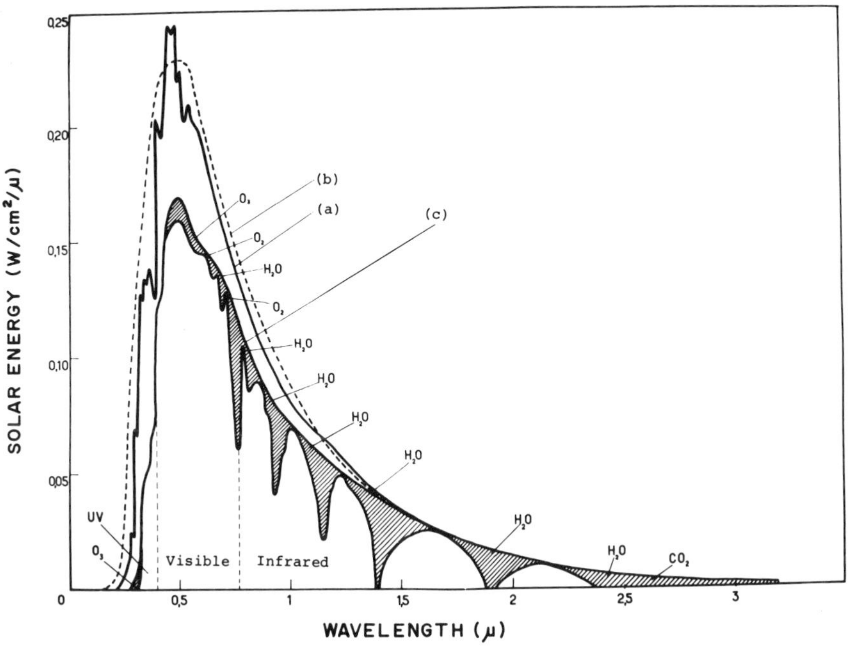

FIG. 11.1 Spectral distribution of solar radiation; (a)—outside the atmosphere, (b)—blackbody radiation at 6000°K, (c)—at sea level.

When the solar radiation penetrates the atmosphere, the ultraviolet region shorter than 0.3 μm is cut off by ozone, while many absorption lines appear in the infrared region due to water vapor and carbon dioxide in the atmosphere. Moreover, the radiation suffers a considerable attenuation by scattering in the atmosphere. Hence, the radiation that reaches the earth is only 70% or less of that outside the atmosphere. The spectral distribution of the radiation on the earth is represented by curve (c) in Fig. 11.1. It is well known that the intensity H_a of the direct solar radiation on the earth at noon takes the value of 1.3–1.4 cal/cm^2/min or 0.91–0.98 kW/m^2 without showing conspicuous change by latitude or season. More specifically, however, the direct radiation changes from place to place being affected by the content of atmospheric water vapor and turbidity especially arising from pollution. The diurnal change of the intensity is an exponential function of the optical path length in the atmosphere which is proportional to the cosecant of the altitude of the sun. It will be remarked also that the intensity increases with increasing height above sea level, which amounts to some 10% at 2000 m.

B. Concentrator

Since the angle α subtended by the sun is 32′ or 0.0093 rad as an annual average, the solar radiation is nearly parallel, and a paraboloidal mirror is adequate for a concentrator of the reflection type. The geometry of a paraboloid is defined by two parameters, say aperture D and focal length f, or aperture D and aperture ratio $n = D/f$. The total solar energy incident on a mirror is given by $\frac{1}{4}\pi D^2 \times H_a$, hence the aperture governs the input power of the solar furnace. On the other hand, the focal length and the aperture ratio determine the size and the intensity of the sun image. When D is kept constant, a paraboloidal mirror becomes deep with increasing value of n, and the circumference of the mirror coincides with the focal plane at $n = 4$. When a cylindrical or spherical specimen is heated by irradiation, a mirror with n larger than 4 might be effective. In practice, however, the target to be irradiated is flat in most cases; hence, furnaces with $n = 2$–3 are widely used.

The paraboloidal mirror is necessary to have an optically smooth surface with accurate geometry and high reflectivity. A portional defacement $\Delta\theta$ in angle effects a deflection of $2\,\Delta\theta$ to the reflected beam. Therefore, in order to restrict a confusion circle within, say $\frac{1}{5}$ of the sun image in diameter, defacements must be less than $\alpha/10$ or 1/1000 rad in inclination. In a heliostat-type furnace, the defacements of a concentrator and a heliostat mirror must be less than 1/2000 rad if they share the effect.

In some furnaces of the direct-incidence type, an aluminum mirror has been used to prevent an accident caused by the spill of melted material. However, it is very difficult to obtain an optically smooth surface of soft metals by grinding and polishing. The searchlight mirrors 1–2 m in D and 2–2.5 in n, made by polishing rhodium-deposited copper blank, are fairly good in image formation and available for a concentrator of a small furnace.

Glass is the most favorable material for reflecting mirrors because an optically smooth surface can easily be obtained. A paraboloidal mirror made by grinding and polishing a glass blank using optical methods and aluminized or silvered by vacuum evaporation to make a front-surface mirror provides an excellent concentrator, and a reflectivity of about 90% for solar radiation is expected. It is usual that the aluminized surface is coated with SiO film for protection, but a coating of thin plastic film is also effective. The method of fabricating paraboloidal mirror segments for a large concentrator appears in a later section. The substitution of paraboloidal mirror segments by toroidal mirrors will be very useful, because a sufficient approximation is obtained as long as the dimension is not large and a toroidal surface can be figured more easily by means of a tool that swings in two orthogonal directions with different radii. The substitution of spherical mirrors is not adequate when a large value of n is required because of the astigmatism of the image produced by marginal mirrors.

Front-surface mirrors, if they are coated, need considerable care for their maintenance; hence, back-surface mirrors are also used in spite of the radiant loss caused by the absorption of glass. The absorption of the solar radiation by commercial window glass with the thickness of 3, 6, and 12 mm amounts to 5, 9, and 17%, respectively. Therefore, when the solar radiation is incident normally on a back-silvered mirror 6 mm thick, the loss caused by the absorption amounts to 17% because the radiation passes the glass twice. Since the reflection loss at the glass-silver interface is about 8%, the total reflection loss amounts to some 25%. Back-silvered paraboloidal mirrors used in searchlights are also available for a concentrator of a small furnace, but at the expense of a reflection loss of a considerable amount.

In some furnaces, mechanically bent plane mirrors are used for paraboloidal mirror segments (Trombe, 1952; Trombe and Le Phat Vinh, 1973). When mirrors with imperfect reflecting surface are used as segments, it is important to divide a concentrator into a large number of small segments, by which the effect of imperfections can be minimized. If the dimension of the mirror is small compared with $\alpha\rho$, where ρ is the distance between the segment and the focus, even a plane-reflecting surface gives a correct sun image.

C. Heliostat

According to the procedures used to keep the solar radiation incident upon a concentrator always parallel to the optical axis, solar furnaces can be divided into two kinds. In a furnace of the direct-incidence type, the concentrator itself is directed towards the sun as seen in Fig. 11.2a, while in that of the heliostat type, the radiation is conveyed into a fixed concentrator by means of a turnable mirror or *heliostat* as shown in (b), (c), and (d). The direct-incidence type has the advantage in that the reflection loss by a heliostat is not introduced; however, there is a risk that melted material might spill during irradiation. A furnace of the heliostat type with vertical optical axis shown in (b) is most adequate for fusion, since the unmelted portion of material forms a crucible to hold the melted portion. With increasing aperture, however, the position of a concentrator becomes very high especially in low latitudes; hence this system is difficult to apply to large furnaces. The problem of height may be solved by adding another flat mirror as seen in (c), but such a system is handicapped by increasing reflection loss. A heliostat-type furnace with horizontal optical axis given in (d) is comparatively convenient and widely used in large furnaces.

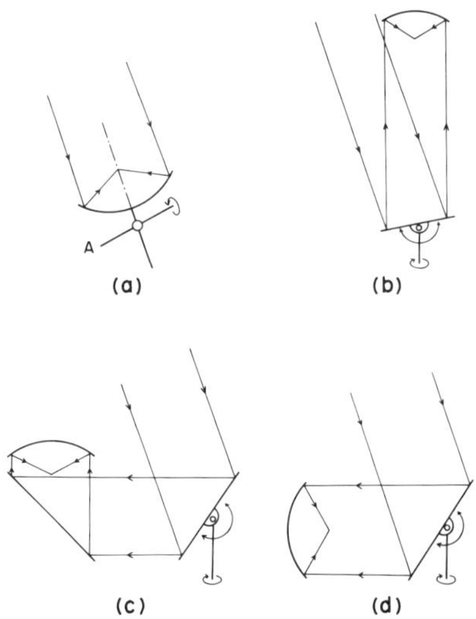

FIG. 11.2 Various types of solar furnace; (a) direct-incidence type; (b), (c), and (d) heliostat type. Axis A is set parallel to the earth axis.

A heliostat mirror must be large enough to convey the solar radiation over the whole aperture of a concentrator through the year. The necessary dimension of the mirror can be found by theoretical calculation (Jose, 1957), but roughly speaking, it is $1.4\,D \times 1.4\,D$ or less. The mirror is required to have a plane-reflecting surface, the quality of which is the same or better compared to that of a concentrator mirror. A mirror obtained by grinding and polishing a glass plate into an optical flat, aluminizing or silvering by vacuum evaporation, and coating with a suitable film, is most desirable. However, flat plates, less than 1/2000 rad in portional defacements, can be provided easily by selecting commercial window glass plates of the highest grade. A thickness of more than 1% of the dimension will be necessary to keep bending within 1/2000 rad. If a back-silvered mirror is preferred for durability, it is recommended that a silver film is sandwiched in between thin and thick plates; by this, the reflection loss can be minimized without sacrificing the rigidity. In most cases, a heliostat mirror is composed of a number of mirror segments; hence, great care must be taken in accurately mounting the segments parallel on a frame.

The change of elevation and that of azimuth can be obtained by the rotation of a frame about a horizontal axis and about a vertical axis, respectively. In order to rotate the frame, hydraulic or electric driving is used which is coupled with a servosystem or a time system for sun-following. The structure of a frame should be stable enough to keep the aberration of the sun image arising from the bending caused by the rotation within acceptable limits. This problem becomes very important in large solar furnaces. A method of minimizing the bending is to divide a heliostat mirror into several mirror bands that decline synchronously. Another method is to use many heliostats to convey the solar radiation into a concentrator.

D. Sun-Following

For the sun-following, three methods are used, namely, time system, servosystem, and manual operation. The time system for a direct-incidence type furnace is conveniently achieved by an equatorial mounting of a concentrator as illustrated in Fig. 11.2a. A diurnal change of direction is given simply by a constant speed rotation of the axis A set parallel to the earth axis. The seasonal change is effected by adjusting the angle between A and the optical axis of the concentrator, which becomes rectangular at the equinoxes. In an altazimuth mounting as used in a searchlight, the necessary elevation and azimuth of a concentrator are none other than the solar elevation h_s and azimuth A_s which can be given by

$$\sin h_s = \sin \delta \sin \Phi + \cos \delta \cos \Phi \cos t \tag{11.1}$$

$$\sin A_s = \cos \delta \sin t \sec h_s \tag{11.2}$$

where t is the hour angle (positive in the afternoon), δ the solar declination (positive in the summer for Northern hemisphere), and Φ the latitude of the location (positive in Northern hemisphere). Since the diurnal change of h_s and A_s is not simple, it is not easy to apply the time system to the concentrator in altazimuth mounting.

In the case of a heliostat, it is difficult to give the necessary change of direction by a single parameter. When the reflected solar radiation is required to be horizontal as seen in Fig. 11.2c and 11.2d, the necessary elevation h and azimuth A of a heliostat are given by

$$\sin h = \sin h_s / \sqrt{2(1 + \cos h_s \cos A_s)} \tag{11.3}$$

$$\sin A = \tan h \cot h_s \sin A_s \tag{11.4}$$

In Figs. 11.3 and 11.4, the diurnal change of h and A for solstices and equinoxes at $\Phi = 38°16'$ N (Sendai, Japan) is represented. In the Tohoku University furnace designed by the author, the diurnal change of elevation is achieved by a piston motion, obtained by crank system C in Fig. 11.5. The seasonal change is effected by the shift of the crank system, given by a driving system M_2. A constant speed rotation of a crank shaft by a driving system M_1 gives the change in elevation with the accuracy shown in Fig. 11.6. The necessary change of azimuth, on the other hand, is performed by a rotation

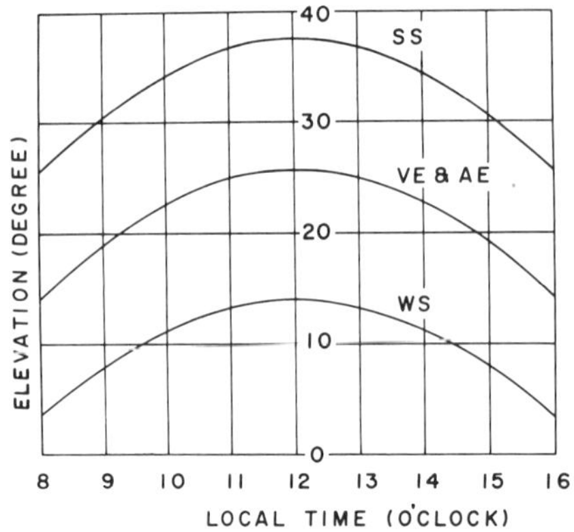

FIG. 11.3 Required elevation of heliostat as a function of local time at $\phi = 38°16'$N (Sendai, Japan); SS—summer solstice, WS—winter solstice, VE—vernal equinox, AE—autumnal equinox.

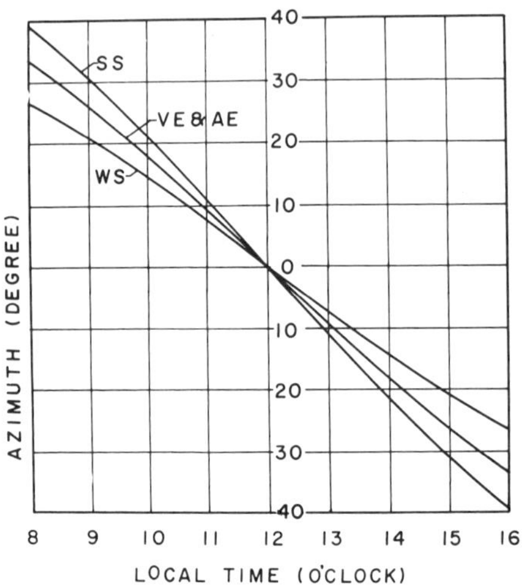

FIG. 11.4 Required azimuth of heliostat as a function of local time at $\phi = 38°16'$N (Sendai, Japan); SS—summer solstice, WS—winter solstice, VE—vernal equinox, AE—autumnal equinox.

of heliostat frame about a vertical axis at a constant speed (Sakurai *et al.*, 1964).

A servosystem is used to find a deviation of the radiation incident upon a concentrator by photoelectric methods and operates the elevation and azimuth driving systems to return the radiation parallel to the optical axis. One of the methods of detecting the deviation is to use an optical tube which

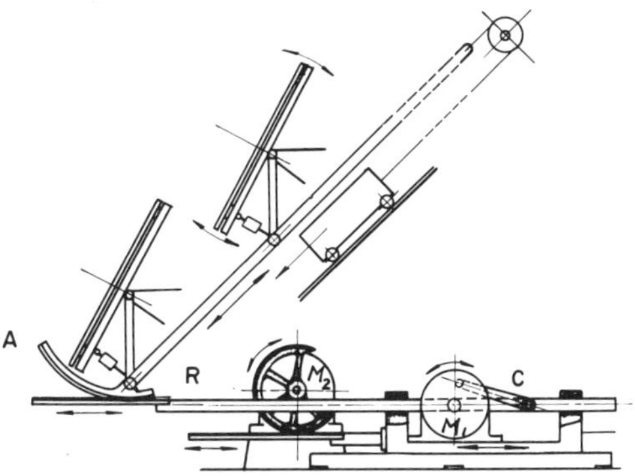

FIG. 11.5 Mechanism of changing the elevation of heliostat mirrors; C—crank, M_1—crank shaft driving system, M_2—driving system to shift the crank system. Rod R and arc A are connected by steel bands.

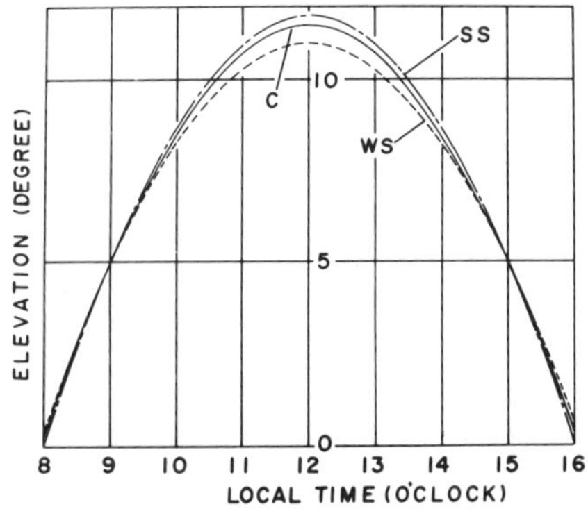

FIG. 11.6 Elevation change obtained by crank system (C) and that required at summer solstice (SS) and winter solstice (WS). The curve on equinoxes coincides with (C).

is fixed with its axis parallel to that of a concentrator. A beam of solar radiation introduced into the tube through a small opening at one end is incident on four photoelectric tubes at the other end. An imbalance of the voltage between two tubes up and down effects the change of elevation and that between two tubes right and left, the change of azimuth (Laszlo *et al.*, 1958). Another method of detecting the deviation is to place four phototransistors around a sun image obtained by focusing the solar radiation from a heliostat with a lens several meters in focal length (Sakurai *et al.*, 1964). The projection of the sun image is very convenient to the manual operation necessary for zone melting and other fine techniques in fusion.

11.3 THEORETICAL CONCENTRATIONS

The concentration of the solar energy by a paraboloidal mirror for a flat target has been theoretically treated by Hiester *et al.* (1957) and Hukuo and Mii (1957) on an assumption that the brightness in the solar disk is uniform over the whole area. When the axis of the paraboloidal mirror is directed towards the sun, the radiation reflected at a point of the mirror spreads as a cone having the same angle as that for the subtended sun, and the axis of the reflected cone always passes through the focal point F as illustrated in Fig. 11.7. The reflecting cone arising from the vertex A makes the sun image with the diameter αf on the focal plane, while that arising from an off-axial point $B(x, y)$, with the azimuthal angle θ and the distance ρ from F gives an elliptical image with a major and a minor axis $\alpha\rho/\cos\theta$ and $\alpha\rho$, respectively. From the geometry of the parabola, $\rho = f + x$; hence, the ellipse given by a reflected cone arising from every point of the mirror always includes the sun image. Therefore, the intensity of concentrated radiation on a focal plane becomes uniform and maximum within the sun image and decreases in the outside.

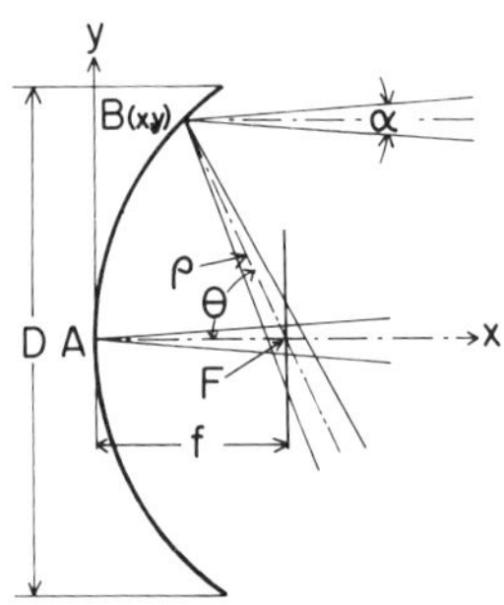

FIG. 11.7 Paraboloidal mirror with incident cone of solar radiation being reflected to focal plane.

In order to obtain the intensity H_m within the sun image, a differential ring between θ and $\theta + d\theta$ will be considered on the mirror. The reflected cone arising from every point of the ring makes an elliptical image with the same area S_θ that amounts to $\pi\alpha^2\rho^2/4\cos\theta$. The solar radiation H_θ incident on the ring is given by $2\gamma H_a \pi\rho^2 \sin\theta\, d\theta$, where γ is the reflectivity of the optical system; hence, the intensity I_θ within the sun image effected by the differential ring becomes

$$I_\theta = H_\theta/S_\theta = (8\gamma H_a/\alpha^2)\sin\theta\cos\theta\, d\theta \qquad (11.5)$$

Integrating I_θ from zero to the maximum azimuthal angle θ_m,

$$H_m = \frac{4\gamma H_a}{\alpha^2} \int_0^{\theta m} 2 \sin\theta \cos\theta \, d\theta = \frac{4\gamma H_a}{\alpha^2} \sin^2 \theta_m \quad (11.6)$$

The concentration ratio C is defined as the ratio of the intensity of the concentrated solar radiation to that of the incident radiation and is expressed as a function of a distance k from the focal point for a flat target. From the above equation, maximum value of the concentration ratio C_m becomes

$$C_m = (4\gamma/\alpha^2) \sin^2 \theta_m \quad (11.7)$$

By simple geometrical consideration, the following relation is obtained

$$n = D/f = \sqrt[4]{(1 - \cos\theta_m)/(1 + \cos\theta_m)} \quad (11.8)$$

hence, $\cos\theta_m = (16 - n^2)/(16 + n^2)$. Therefore, Eq. (11.7) becomes

$$C_m = (4\gamma/\alpha^2)\{1 - [(16 - n^2)/(16 + n^2)]^2\} \quad (11.9)$$

This relation shows that the maximum concentration ratio is governed by the aperture ratio but not by the aperture of a concentrator. The value of $C(k)$ in the outside of the sun image can be given also by calculation (Hukuo and Mii, 1957). In Fig. 11.8, the curves of $C(k)$ obtained by taking $\gamma = 1$ are drawn for various values of n.

It is known that the brightness in the solar disk is symmetrical about the center but considerably different along the radius. In Fig. 11.9, the observed values of brightness (Abbot *et al.*, 1922) are plotted against the solar radius q for various wavelengths by solid curves. In order to take the effect of limb

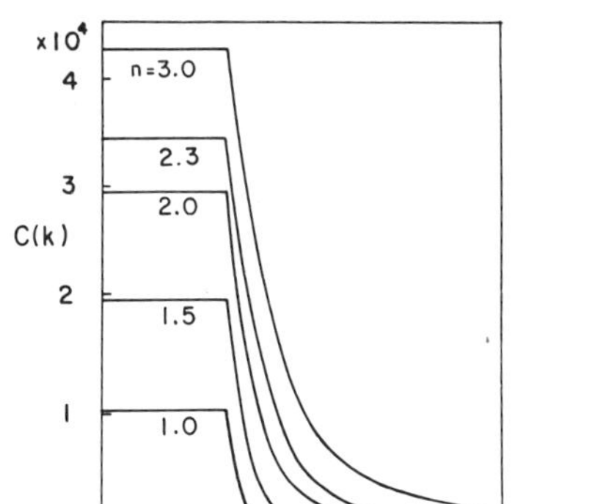

FIG. 11.8 Distribution of concentration ratio $C(k)$ for a flat target calculated by assuming the uniform brightness of the solar disk.

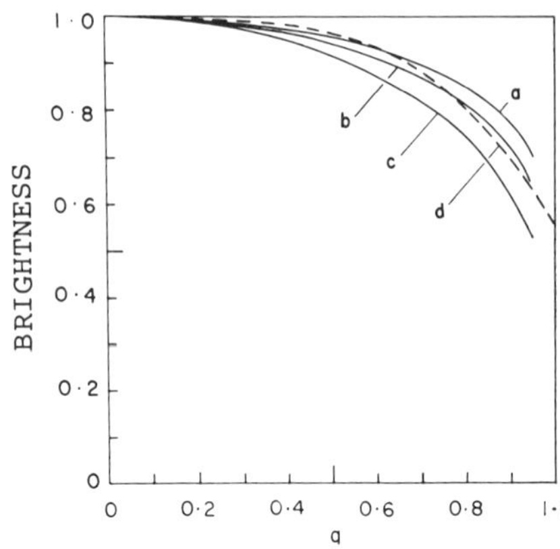

FIG. 11.9 Brightness distribution of the solar disk; a: $\lambda = 8500$ Å, b: $\lambda = 6702$ Å, c: $\lambda = 5062$ Å, d: distribution function $G(\xi)$.

darkening into consideration, Kamada (1965) introduced an empirical brightness distribution function.

$$G(\xi) = \cos(m\xi^2) \tag{11.10}$$

where $\xi = \frac{1}{2}\alpha q$, and m is given by the relation $\cos(m\xi^2) = 0.55$ at $\xi = \frac{1}{2}\alpha$. In the same figure, $G(\xi)$ is given by a dotted curve, which shows that the approximation is satisfactory. Aided by this function, the concentration ratio has been calculated taking the brightness distribution into account. The curves of $C(k)$ for a flat target obtained by assuming $\gamma = 1$ are represented in Fig. 11.10, which will be seen to be conspicuously different from those given in Fig. 11.8. The maximum values of the concentration ratio are 18% larger compared with those obtained assuming the uniform brightness of the solar disk. The concentration ratio practically obtained, however, is considerably smaller than that calculated because of the reflection loss and other imperfections in the optical system.

When the concentrated radiation is incident on a target surface, the radiant energy is absorbed to give rise to the increase of temperature, the amount of which is governed by the optical property of the target. On the other hand, the absorbed energy is lost by thermal radiation and conduction, the characteristic of which will be different case by case. Therefore, it is very difficult to give the exact temperature of the irradiated portion by theoretical treatment. However, by assuming

(1) the absorbed energy is lost only by thermal radiation,
(2) the emissivity is equal to the absorptivity, and
(3) the emission and absorption follow the Lambert's cosine law,

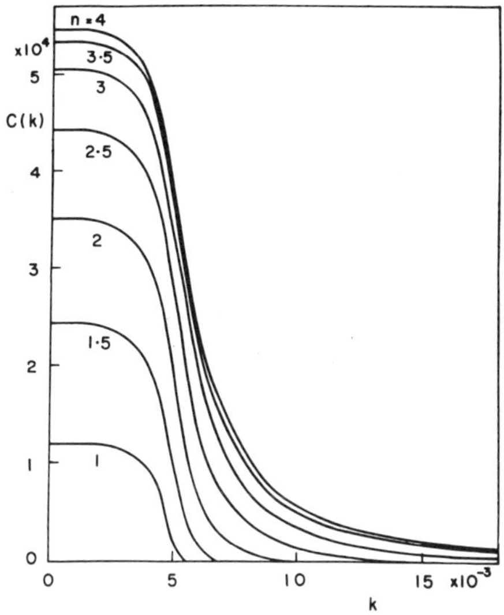

FIG. 11.10 Distribution of concentration ratio $C(k)$ for a flat target calculated by taking the brightness distribution of the solar disk into account.

it is possible to obtain an attainable temperature by theoretical calculation (Hiester *et al.*, 1957; Kamada, 1965). From assumption (3), the directional emissivity ε_θ and absorptivity κ_θ are given by $\varepsilon_n \cos\theta$ and $\kappa_n \cos\theta$, where ε_n and κ_n are the normal emissivity and absorptivity. When the uniform brightness of the solar disk is assumed, the intensity I_θ of the solar radiation at a focal point effected by the differential ring in Fig. 11.7 is given by Eq. (11.5). Hence, the radiant energy A_θ absorbed by the unit area of a flat target is

$$A_\theta = \kappa_n \cos\theta\, I_\theta = (8\gamma\kappa_n H_a/\alpha^2) \sin\theta \cos^2\theta\, d\theta \tag{11.11}$$

Integrating A_θ from zero to θ_m, the total energy A absorbed by the unit area becomes

$$A = \frac{8\gamma\kappa_n H_a}{\alpha^2} \int_0^{\theta_m} \sin\theta \cos^2\theta\, d\theta = \frac{8\gamma\kappa_n H_a}{3\alpha^2}(1 - \cos^3\theta_m) \tag{11.12}$$

On the other hand, when a flat target at the temperature of T_m follows $\varepsilon_\theta = \varepsilon_n \cos\theta$, the total radiant energy E emitted by the unit area is given by

$$E = \tfrac{2}{3}\varepsilon_n \sigma (T_m{}^4 - T_0{}^4) \tag{11.13}$$

where σ is the Stefan–Boltzmann constant and T_0 the ambient temperature which can be neglected in the present case. From assumption (1),

$$A = E,$$

while from assumption (2), $\varepsilon_n = \kappa_n$; hence,

$$T_m = \left[\frac{4\gamma H_a}{\sigma\alpha^2}(1 - \cos^3\theta_m)\right]^{1/4} = \left[\frac{4\gamma H_a}{\sigma\alpha^2}\left\{1 - \left(\frac{16 - n^2}{16 + n^2}\right)^3\right\}\right]^{1/4}$$

This relation shows that the target temperature T_m is governed by n but not by D or f.

When the brightness distribution of the solar disk is taken into consideration, the distribution of the attainable temperature for a flat target, taking $\gamma = 1$ and $H_a = 1.3$ cal/cm^2/min, is exemplified in Fig. 11.11. The maximum values in this figure are 4.3% larger compared with those given by the above equation. It will be remarked here that the temperature practically obtained is some 70% of the calculated values or less. The concentration ratio and the attainable temperature for cylindrical and spherical targets have been given also by Kamada (1965) taking the brightness distribution into account.

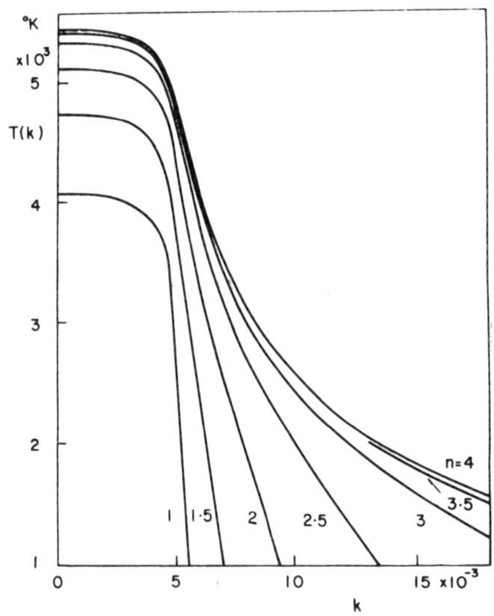

FIG. 11.11 Distribution of attainable temperature $T(k)$ for a flat target calculated by taking the brightness distribution of the solar disk into account.

11.4 DESIGN EXAMPLES

Some design examples of large solar furnaces in the world will be surveyed briefly. Figure 11.12 is the furnace of the direct-incidence type in Bouzareah, near Alger, Algeria (Juston-Coumat, 1957), provided with a paraboloidal concentrator in equatorial mounting. The concentrator, 8.4 m in

FIG. 11.12 Direct-incidence type furnace in Bouzareah, Algeria.

aperture and 2.5 in aperture ratio, is a mosaic of 108 aluminum mirror segments. An attenuator of the concentrated radiation for temperature control is composed of aluminum plates mounted radially about the optical axis, which can rotate to make a cylinder to interrupt the radiation.

The first large furnace constructed by Trombe (1952) in Montlouis, France, consists of a heliostat and a concentrator with a horizontal optical axis, 10 m in aperture and 1.7 in aperture ratio. The concentrator has been made economically by arranging 3500 back-silvered plane mirrors into a paraboloidal form as shown in Fig. 11.13, each of which is curved mechanically by means of a frame. The heliostat, 10.5 × 13 m in dimension, is composed of 540 back-silvered mirrors mounted to make a plane as seen in Fig. 11.14. This instrument has been used in various high-temperature investigations, but is not in use now.

FIG. 11.13 Concentrator of the furnace in Montlouis, France. (Reprinted with permission from F. Daniels and J. A. Duffie (eds.), "Solar Energy Research." Copyright 1955 by University of Wisconsin Press, Madison.)

The furnace of the US Army in Natick, Massachusetts shown in Fig. 11.15 has been designed by Davies and Cotton (1957). The concentrator, 8.5 m square in dimension, is a mosaic of 180 spherical mirrors, 10.9 m in focal length, which are arranged on a spherical surface with a center at the focus. The front surface of mirrors has been aluminized and protected by SiO coating. The aperture ratio is about 1. The heliostat is composed of 356 back-silvered mirrors arranged in a plane. An attenuator of the Venetian-blind type is placed in front of the concentrator. A concentrator of this kind is characteristic in that all segments give the sun image of the same size so far as the aperture ratio is small, but is not adequate when high attainable temperature is required because, with increasing aperture ratio, astigmatism appears in the images produced by marginal segments and the aperture ratio cannot exceed 2. This instrument is now in White Sands Missile Range, New Mexico.

A furnace in the Research Institute for Scientific Measurements, Tohoku University, Sendai, Japan (Sakurai *et al.*, 1964), was constructed in 1964 for

FIG. 11.14 Heliostat of the furnace in Montlouis, France. (Reprinted with permission from F. Daniels and J. A. Duffie (eds.), "Solar Energy Research." Copyright 1955 by University of Wisconsin Press, Madison.)

FIG. 11.15 Furnace in White Sands Missile Range, New Mexico, USA.

the purpose of carrying out basic and applied research in high-temperature physics. In order to obtain the highest attainable temperature for a flat target, a large aperture ratio, high reflectivity and correct geometry of the optical system were required.

The concentrator, 10 m in aperture and 3.2 m in focal length, is a mosaic composed of 181 mirror segments as illustrated in Fig. 11.16. Blanks for the mirror segments were cut from window glass plates 10 mm thick. They were curved by placing them on a mold and heated to yielding temperature in an electric furnace. To grind the blanks accurately into portions of a paraboloid, a paraboloidal surface grinder as shown in Fig. 11.17 was constructed. In this figure, T is an iron tube the axis of which forms an optical

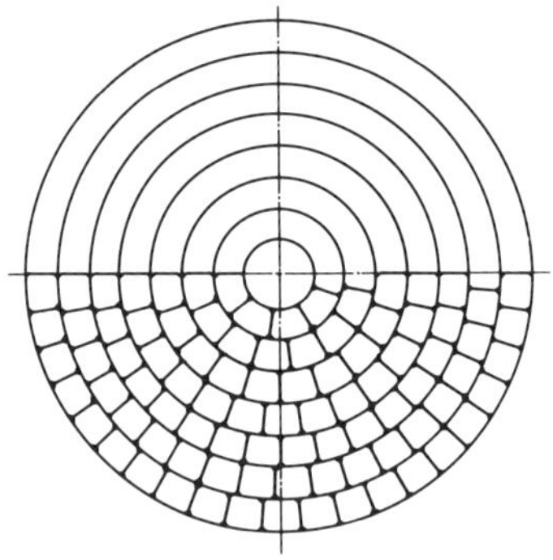

FIG. 11.16 Segmentation of the paraboloidal mirror.

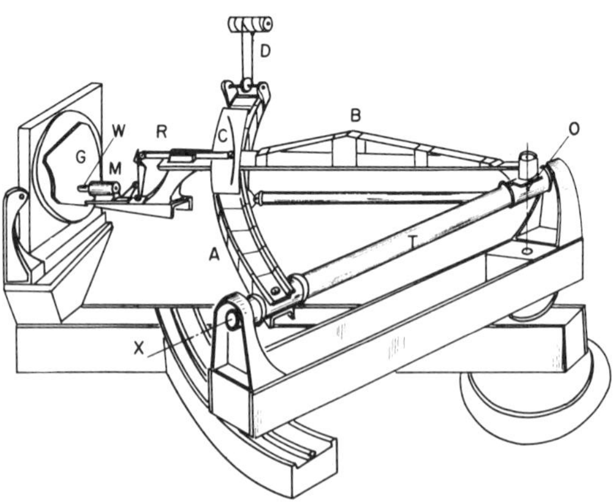

FIG. 11.17 Diagrammatic representation of the paraboloidal surface grinder.

axis X. The arc A is fixed to the tube and can be rotated about the optical axis by the winch D. The center of the arc is the point O at the end of the tube. The beam B moves on the arc by rack and pinion about the axis at the point O. The end of the beam is lowered to make a platform where a motor M is mounted to drive a diamond wheel W. G is a blank fixed on an adjustable table. As the beam moves on the arc, the diamond wheel will trace a circular track, but a cam C on the arc gives the rod R on the beam a displacement that is reduced in magnitude by the lever and transmitted to the motor; hence, the wheel traces a parabolic track. The blanks thus ground were smoothed and polished. By this procedure, surfaces with an accuracy of 1/2000 rad could be obtained. The finished surface was aluminized by vacuum evaporation to form a front-surface mirror and coated with a thin resin film, a few microns in thickness, by dipping in a lacquer-thinner solution of vinyl chloride and drying. The mirror segments are mounted on a concentrator support with a steel skeleton structure in a concrete building. Two V-shaped diaphragms, above and below, which can be brought together toward the optical axis, form an attenuator to control the temperature of the target. Figure 11.18 is the front view of the concentrator.

A heliostat has a steel skeleton frame that can rotate on a circular track to change the azimuth. Seven mirror bands 2 m wide and 15.5 m long,

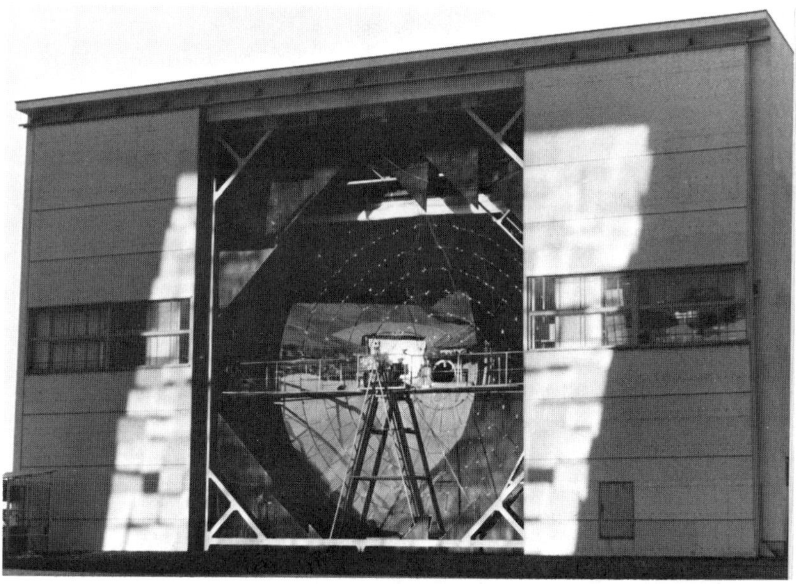

FIG. 11.18 Concentrator of the Tohoku University furnace, Sendai, Japan.

each with 34 mirror segments, are mounted on the frame and move synchronously to change the elevation. The blanks for the mirror segments, 100 × 90 cm and 10 mm thick, were selected from the window glass plates of the highest grade to meet the required accuracy of 1/2000 rad. They were also aluminized to form a front-surface mirror and coated with a thin resin film. Figure 11.19 is the front view of the heliostat. The sun-following is done by a time system, an image following servosystem, and a manual operation. The furnace can give a temperature as high as 4000°K in very fine weather.

FIG. 11.19 Heliostat of the Tohoku University furnace, Sendai, Japan.

The construction of the furnace was supported by a government grant of 30 million yens. Considering that the mirror segments were fabricated in the optical workshop of the Institute, the cost of the furnace would be 50 million yens ($170,000). The cost for the construction of a building was 15 million yens; hence, the total cost was 65 million yens ($220,000). Since the output power of the furnace is 50 kW, the cost per kilowatt amounts to $4400.

A huge furnace in the Solar Energy Laboratory in Odeillo was designed by Trombe and Le Phat Vinh (1973) with the intention of applying a solar furnace to high-temperature industry. Odeillo is in the French Pyrenees, 1800 m above sea level. In this furnace, the solar radiation is conveyed into

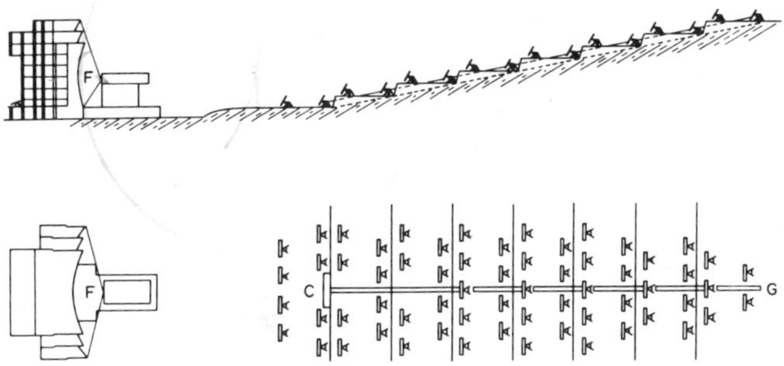

FIG. 11.20 Optical system of the furnace in Odeillo, France; F—focal point, C—oil station for heliostat driving, G—tunnel for oil piping.

a concentrator by 63 heliostats which are arranged on the slope of the mountain as shown in Fig. 11.20. A heliostat, 6 × 7.5 m, is composed of 180 back-silvered mirrors 50 × 50 cm in dimension. The direction of the heliostat is changed by hydraulic driving which is coupled with a sun-following servo-system using an optical tube.

A paraboloidal concentrator 40 × 54 m consists of 9500 back-silvered plane mirrors, 45 × 45 cm in dimension, which are curved mechanically by pulling the center from the backside as shown in Fig. 11.21. The effective area of the concentrator is some 1900 m^2 which provides the input solar energy of 1800 kW. The power of the concentrated radiation is estimated to be 1000

FIG. 11.21 Backside of a concentrator. Central portion of a mirror is fixed to a square plate by four screws and the plate is pulled by bolt nut.

FIG. 11.22 Whole view of the furnace in Odeillo, France (photo by Takebe, Asahi Press)

kW, taking the reflection loss into account. The focal length is 18 m; hence, the effective aperture ratio amounts to 2.8. The attainable temperature as high as 4000°K is obtained. Figure 11.22 is the whole view of the furnace.

There are two laboratories in the building which embraces the concentrator; one is the Solar Energy Laboratory and the other, the Ultra-refractory Laboratory. The building is also fitted with a pair of heliostat-type furnaces with vertical axis in nearly perfect mounting. Paraboloidal concentrators, 2 m in aperture and 2.4 in aperture ratio, are placed in a room on the sixth floor, which overhangs to the south and is provided with an opening in the floor, while a heliostat is mounted in the outside of the first floor. It is very interesting to see a huge furnace for industrial use and small furnaces for scientific use coexisting in the same building.

11.5 PROCEDURES IN HIGH-TEMPERATURE INVESTIGATIONS USING A SOLAR FURNACE

A. Methods of Fusion

Some procedures in high-temperature investigations using a solar furnace will be described based on the author's experience obtained by the operation of the Tohoku University furnace. The furnace is provided with a concentrator with a horizontal optical axis, 3.1 in aperture ratio and 3.2 m in focal length. When a flat target is irradiated, the temperature is highest within a circular portion about 3 cm in diameter and decreases toward the outside, as seen in Fig. 11.11. When a refractory brick SK35 is used as a target, the irradiated portion melts and flows down, and a crater as seen in Fig. 11.23 is formed within a few minutes. Since the melting point of SK35 is comparatively low, the diameter of the crater amounts to some 7 cm (see Fig. 11.11).

FIG. 11.23 Irradiated refractory brick (SK35).

Using the furnace, various oxides including CaO, Y_2O_3, ZrO_2, and ThO_2 have been fused in order to make crystals. In preparation, the powders are pressed into square bars 1 × 1 × 3 cm in dimension and fired for 100 h at about 1300°C in an electric furnace to prevent a crash by irradiation. The bar to be fused is erected at the focus of a concentrator, and the sun image is brought to it such that the center of the image coincides with the top of the bar. By irradiation, the bar begins to melt at the top and the melted portion flows down and solidifies at the middle of the bar to make a globule several millimeter in diameter, as seen in Fig. 11.24. By displacing

FIG. 11.24 Irradiated ThO_2 90%–ZrO_2 10% specimen (m.p. 3080°C). Pure ThO_2 specimen melts but also sublimes which makes it difficult to obtain a globule.

the sun image, rapid cooling takes place; the temperature decrease from 3000 to 2000°C, measured using ThO_2 specimen is about 8 s. More rapid cooling can be done by dropping the bar under irradiation into water.

When a large fused crystal is wanted, the powder is pressed into a stepped disk, 8 cm in diameter, which is mounted at the focus so that a flat surface might make an angle of 10–20° to the horizontal plane, as seen in Fig. 11.25.

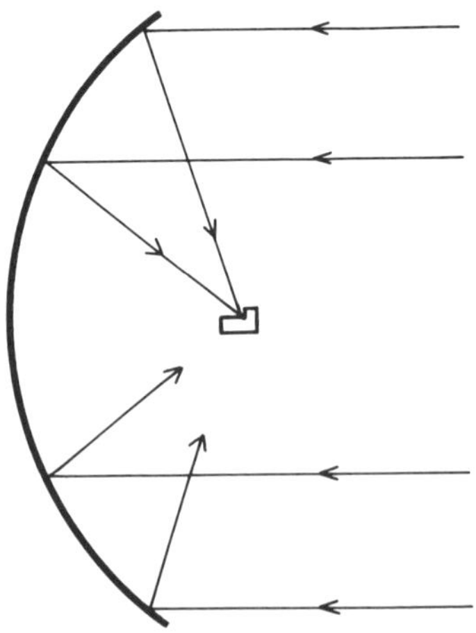

FIG. 11.25 Irradiation of a specimen shaped into stepped disk.

In this case, only the upper half of the concentrator is responsible for fusion; nevertheless, the concentrated radiation is sufficient to melt the surface from the elevated to the lower portion and to make a pool of molten material. Figure 11.26 is a stabilized ZrO_2 (m.p. 2677°C) specimen, thus fused. In solar-furnace fusion, the melted portion is supported by the unmelted portion of the specimen; hence, any contamination which might be introduced by reaction with a crucible does not occur. It is also quite favorable for oxides that a specimen is heated in air, because most oxides are not so stable at high temperatures and an oxidizing atmosphere is necessary to keep them stoichiometric.

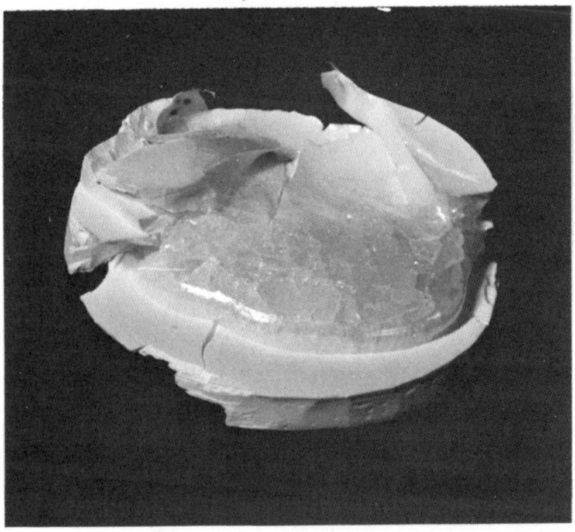

FIG. 11.26 Stabilized ZrO_2 specimen fused by irradiation.

Metals can be melted on a graphite column covered by a fused silica bulb as shown in Fig. 11.27, flushing with argon or helium to prevent oxidation. Figures 11.28a and 11.28b show a tantalum (m.p. 3030°C) and a tungsten (m.p. 3400°C) L-shaped plate, melted at their upper portion. Tungsten powder completely melted appears in Fig. 11.28c.

B. Temperature Measurements

Since the temperature attained with a solar furnace is extremely high, an optical method is now the only way available to measure the temperature of the target. The temperature decrease after the solar radiation is cut off

FIG. 11.27 Vessel for melting metals. Gas is introduced through a hole in the column.

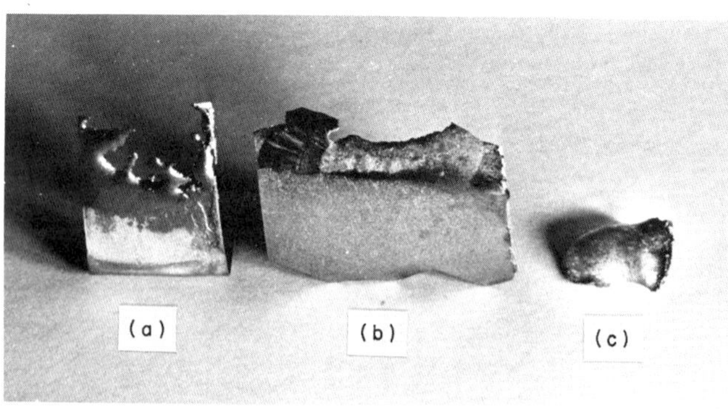

FIG. 11.28 Tantalum (a) and tungsten (b) plate irradiated at their upper portion. (c) A tungsten specimen melted completely.

can be measured by an optical pyrometer with rapid response (see Foex, 1960); but under irradiation, cannot be measured by an optical pyrometer as is, because, besides the thermal radiation of the target, a considerable amount of solar radiation enters the pyrometer by reflection or scattering from the target surface.

The effect of the reflected radiation can be removed by using a cylindrical radiation sector (Conn and Braught, 1954), which rotates about an axis parallel to the optical axis of a concentrator as shown in Fig. 11.29. When the cylinder comes in front of the target, it interrupts the solar radiation and the thermal radiation of the target is introduced into a pyrometer through the cylinder. This method, however, has the disadvantage that the attainable temperature is reduced by interruption of the solar radiation and by air cooling due to rotation of the cylinder.

In order to eliminate these difficulties, a pyrometric method, (Kamada, 1964), has been developed in the author's laboratory using the infrared radiation 1.38 μm in wavelength which is absent in the solar radiation owing to the absorption of atmospheric water vapor (see Fig. 11.1). When target material is opaque for the wavelength of 1.38 μm, this method gives the temperature of target surface with sufficient reliability. Many oxides, however, are highly transparent in visible and near-infrared regions and even a pressed powder specimen transmits the radiation to a considerable depth. In such a case, the temperature given by the pyrometry is not that of the target surface. Moreover, when the target has been partially melted, what a pyrometer indicates is not the temperature of the melted portion but that of the unmelted portion behind. These difficulties have been eliminated by using longer-wavelength radiation in the pyrometry.

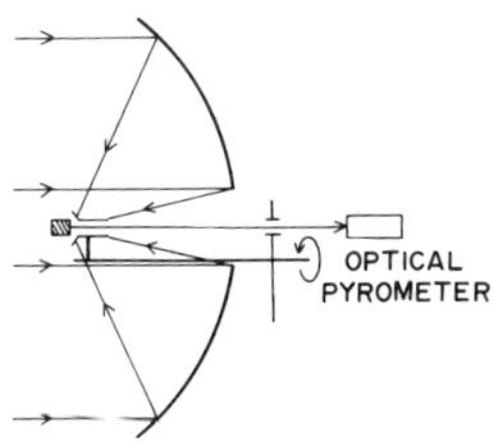

FIG. 11.29

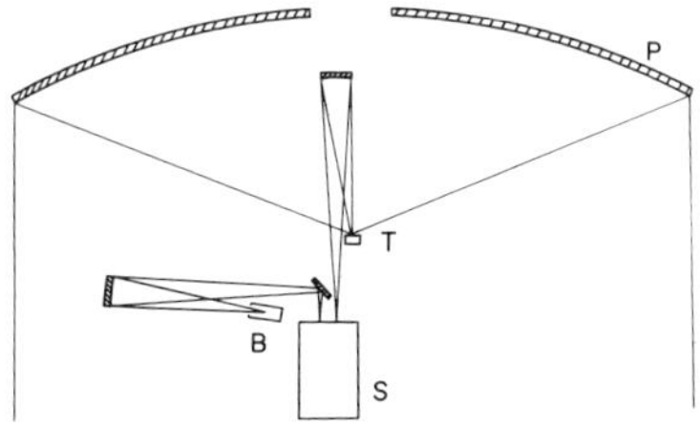

FIG. 11.30

FIG. 11.29 Cylindrical radiation sector. A tube in front of the specimen and a disk with a small opening in front of the pyrometer rotate about the same axis.

FIG. 11.30 Optical arrangement for infrared pyrometry; P—paraboloidal concentrator, T—target, B—reference source, S—double-beam spectrophotometer.

Because of a lattice vibration, dielectric crystals have a spectral region where reflectivity becomes practically null while the absorptivity is sufficiently high, which comes between 8 and 15 μm for polar crystals including metal oxides. The thermal radiation in this region is effected by a thin layer of the target surface, and the value of emissivity is very close to 1. Moreover, the solar radiation in these wavelengths is very faint and can safely be neglected for the pyrometry. Under these considerations a new infrared pyrometer (Arashi *et al.*, 1976) has been developed using a wavelength in this region. The optical arrangement used is schematically shown in Fig. 11.30. The thermal radiation from the target and that from a reference source heated in an electric furnace are introduced into a double-beam infrared spectrophotometer that records the ratio of the intensity of target radiation to that of reference radiation with time at a given wavelength by an optical null method.

C. Miscellaneous

A solar furnace is applicable to high-temperature x-ray diffractometry. Figure 11.31 shows the schematic diagram of a goniometer constructed for this purpose. A specimen is fixed at the focus with its surface normal to the optical axis of the concentrator. In order to rotate a source and a detector at the same angle in opposite direction about the goniometer axis, a link

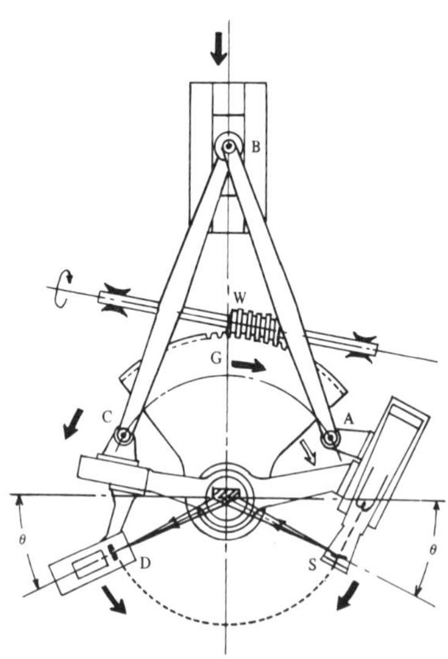

FIG. 11.31 Schematic diagram of the goniometer; S—x-ray source, D—detector, A–B–C—link system, W—worm, G—gear.

system has been introduced. The rotation of the source is driven by a worm gear. Using this apparatus, the thermal expansion of CaO (Kamada *et al.*, 1971) and SnO_2 (Sakurai and Takizawa, 1971) have been measured up to 2150°C and 1625°C, respectively. It is also useful to investigate the phase relationships in refractory materials at elevated temperatures.

The electrical conductivity of oxide crystals at high temperatures can be measured in an oxidizing atmosphere using a solar furnace and a microwave ellipsometry (Sakurai *et al.*, 1975). When a polarized microwave is reflected by a plane of conductive medium, amplitude and phase change take place which are different between the two components parallel and perpendicular to the plane of incidence; hence, the reflected microwave is elliptically polarized. The dielectric constant and conductivity of the medium can be deduced by analyzing the elliptically polarized light. Figure 11.32 is the schematic representation of this method. A 50-GHz microwave, polarized 45° to the plane of incidence is reflected by a flat target under irradiation and introduced into a receiver to be analyzed. Using this method, the conductivity of MgO crystals has been measured in the temperature range from 1800°C up to the melting point.

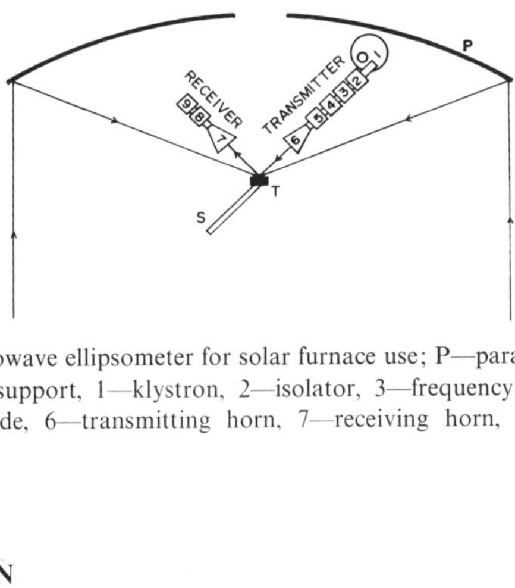

FIG. 11.32 Microwave ellipsometer for solar furnace use; P—paraboloidal concentrator, T—target, S—target support, 1—klystron, 2—isolator, 3—frequency meter, 4—attenuator, 5—45° twist waveguide, 6—transmitting horn, 7—receiving horn, 8—quarterwave plate, 9—analyzer.

11.6 CONCLUSION

In the preceding sections, a solar furnace as an instrument to get extremely high temperatures has been discussed in some detail. A solar furnace is a unique instrument provided with many valuable characteristics, which are difficult to achieve, as follows:

(1) an extremely high temperature can be obtained easily,
(2) contamination by reaction with a crucible does not occur in fusion,

(3) contamination by ions does not occur in fusion which might happen in the case of plasma or oxyhydrogen flame,

(4) an atmosphere desirable to the specimen can be provided,

(5) very rapid heating and cooling are possible, and

(6) an open specimen facilitates various property measurements.

It is true that a solar furnace is handicapped in that its use is limited to sunny days, and to 4 h from 10 a.m. to 2 p.m. if high attainable temperature is wanted. Speaking from the author's experience, however, sufficient scientific works can be done by the annual sunshine duration of about 2000 h (Sendai), because much time is needed for the preparation of specimens and for the property measurement of these specimens after irradiation.

Fundamental properties of various materials at high temperatures above 2000°C still remain to be investigated in the field of science. Using a solar furnace as a scientific tool, a number of fruitful results will be brought about in this field including valuable refractories indispensable for nuclear fusion and other high-temperature processes. By this, a solar furnace will be an industrial tool as well, the day of which is expected to be in the near future.

12

Photovoltaic Conversion

*CHARLES E. BACKUS**

MECHANICAL ENGINEERING FACULTY
ARIZONA STATE UNIVERSITY
TEMPE, ARIZONA

12.1 INTRODUCTION

One way of utilizing the energy of the sun is to generate electricity directly from sunlight by the photovoltaic process. The photovoltaic effect is defined as the generation of an electromotive force as a result of the absorption of ionizing radiation. Energy conversion devices which are used to convert sunlight to electricity by use of the photovoltaic effect are called solar cells.

Efficient solar cells were developed during the 1950s and have been primarily used as power sources for space craft. The more than 1000 satellites that are powered by solar cell arrays have provided tremendous changes in world-wide communications, weather forecasting, and natural resource evaluation. These arrays have typically converted 10–11% of the sunlight into electricity and have been used in sizes from a few watts to over 20 kW per satellite. Recent research has resulted in significant increases in efficiencies and decreases in costs. Presently this space power source is being studied in several countries for terrestrial applications. Starting in 1975 more terrestrial solar cells are being made than space solar cells.

The photovoltaic effect can be observed in nature in a variety of materials, but the materials that have shown the best performance in sunlight are the semiconductors. When photons from the sun are absorbed in a semiconductor, they create free electrons with higher energies than the electrons which provide the bonding in the base crystal. Once these free electrons are created, there must be an electric field to induce these higher energy electrons

* Formerly: Mechanical Engineering Department, Middle East Technical University, Ankara, Turkey.

to flow out of the semiconductor to do useful work. The electric field in most solar cells is provided by a junction of materials which have different electrical properties. In order to understand the characteristics of solar cells let us first review the properties of semiconductors.

12.2 SEMICONDUCTORS

The conductivities of solid materials vary over an enormous range. Representative values are from less than 2×10^{-17} mho/m for fused quartz, a good insulator, to about 6×10^{7} mho/m for silver, a good conductor. Materials with intermediate conductivity values are called semiconductors. The distinguishing features of conductors, insulators, and semiconductors are best explained in terms of what is called "band theory" or the energy band concept of materials.

12.3 THE BAND THEORY OF SOLIDS

Electrons in isolated atoms can exist only at discrete or quantized energy levels. Furthermore, the Pauli exclusion principle limits the number of electrons that can exist at any allowed energy level. When atoms are brought close together, as in a crystal, so that their potential functions overlap, the exclusion principle still holds and the energy level must split and form clusters of acceptable energy levels. These clusters, or bands, consist of a large number of closely packed discrete energy levels. There are as many levels in the bands as there are atoms in the crystal and as many bands as there are energy levels in an isolated atom of that material. Since there may be 10^{22} atoms in a crystal, the allowed energy levels within the bands can be considered as continuous. The energy of the electrons in the material can be represented on a one-dimensional energy diagram (see Fig. 12.1) showing various ranges of energies that electrons are allowed to have, and the ranges of energies in between the allowed bands where electrons are forbidden to exist. The energy widths of these allowed and forbidden bands depend on the particular material and its atomic spacing.

The number of electrons in a material is a small percentage of the allowed energy locations that are available. The electrons are constantly seeking the lower energy levels but are constantly being excited to higher states by interactions such as with phonons and photons. For most cases the electron distribution in the allowed levels can be described by the Fermi function. Applying the Fermi–Dirac statistics, the probability $f(E)$ that a state of energy E is occupied by an electron is given by

$$f(E) = \frac{1}{\exp[(E - E_{\mathrm{f}})/kT] + 1}$$

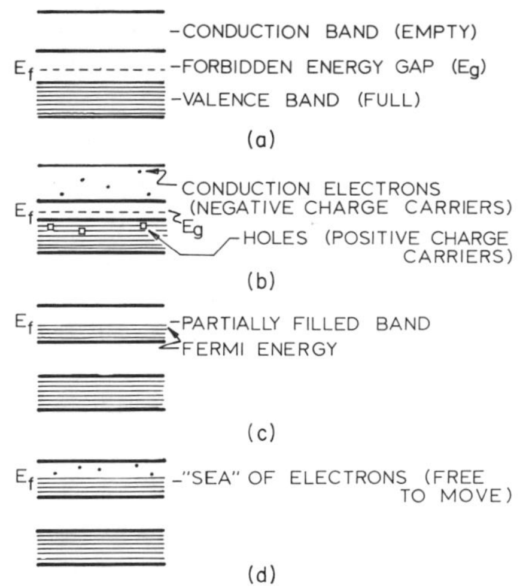

FIG. 12.1 One-dimensional energy band diagrams for different types of materials. (a) Intrinsic insulator or semiconductor at 0°K; (b) intrinsic semiconductor at $T > 0$°K; (c) metal or good conductor at 0°K; (d) conductor at $T > 0$°K.

where $f(E)$ is the Fermi function, E the energy of an allowed state, E_f the Fermi energy, k Boltzmann's constant, and T the absolute temperature. At room temperature the product of kT is equal to about 0.025 eV. The Fermi energy or Fermi level is by definition the energy at which the probability of a state being filled is exactly one-half. Another way of looking at it is the highest energy state an electron can have at 0°K. Perhaps the most important characteristic of the Fermi level is that, in thermodynamic equilibrium, it is always continuous across the contact between two materials.

The distribution of electrons in the outermost or highest energy bands determine most of the electrical and thermal properties of the material. This is similar to the outermost electrons in an atom, the valence electrons, that mostly determine the atom's chemical characteristics. If a crystal (for example, most metals) contains an outermost band which is partially filled, an externally applied electric field can shift the occupation of the energy levels and cause a current to flow. If a band of energy states is completely empty, there can, of course, be no contribution to an electric current by that band. Similarly, if a band is completely filled, there can be no contribution to an electric current by the band. These materials are then good electrical insulators (see Fig. 12.1).

The highest occupied band corresponds to the ground state of the outermost or *valence* electrons in the atom. For this reason the upper occupied band is called the *valence band*. In an insulator the valence band is full. In addition, the width of the forbidden energy gap between the top of the valence band and the next allowed band, called the *conduction band*, is so large that under ordinary circumstances a valence electron can accept no energy at all from an applied field, because there are no empty allowed states accessible to it. Semiconductors are similar to insulators, except that in semiconductors the forbidden gap is much narrower. For example, alumina (Al_2O_3) at room temperature has an energy gap (E_g) of 10 eV while the semiconductor germanium has an energy gap of only 0.7 eV. In a semiconductor at room temperature, though the valence band is full, some electrons have enough thermal energy or may receive enough energy from light so that they may jump the narrow forbidden energy gap into the empty conduction band. The higher the temperature or the higher the intensity of the light source, the larger the number of electrons that will be excited across the gap. (Recall the Fermi function.) The electrons that are so raised are then free to accept electrical energy from an applied field and to move through the crystal. In addition, the sites or "holes" left vacant in the valence band become charge carriers themselves. An electron near a hole can jump in and fill it, leaving a new hole in the place it had occupied, and this in turn can be filled by a neighbor, and so on. Current is actually carried by electrons moving in relays but it can equally well be pictured as a flow of positively charged holes moving in the opposite direction. Thus conduction is done by both electrons and holes. When the conduction of current is due only to those electrons excited up from the valence band to the conduction band, the material is called an *intrinsic semiconductor*.

Any disruption of the perfect crystal will disturb the periodicity of the system and result in additional energy levels within and between the allowed bands. Imperfections in crystals usually come from four sources:

(1) foreign atoms substituted into lattice sites;
(2) vacant lattice sites and interstitial atoms;
(3) dislocations of the crystal (or gain boundaries);
(4) the crystal surface.

The technology for producing low-defect crystals has been developed by using ultrahigh purity material and in slowly growing large, single grain crystals. However, with carefully controlled impurity levels it is possible to obtain desirable properties for semiconductors. By adding small amounts of impurities called *dopants* to semiconductor crystals, it is possible to choose the dominant type of conduction (either electrons or holes) in a material. When the conduction is due to impurities, the material is called an *extrinsic semiconductor* (see Fig. 12.2). Impurities can supply extra electrons, negative charge carriers, in which case they are called *n-type* materials (see Fig. 12.2a). If the impurities are deficient in valence electrons, they are called *p-type*, positive charge carriers (see Fig. 12.2b).

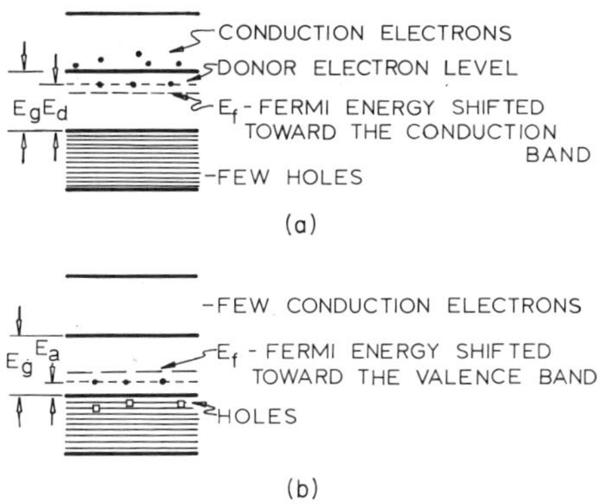

FIG. 12.2 Extrinsic semiconductor materials. *n*-type material results from impurities with excess electrons that can be donated to the conduction band. *p*-type material results from impurities with a deficient number of electrons which can accept electrons from the valence band. (a) *n*-type electron conductor; (b) *p*-type hole conductor.

Figure 12.3 shows the physical arrangement of added impurities to a two-dimensional silicon crystal. Figure 12.3a shows the normal bonds that exist by the shared electrons in a perfect silicon crystal. By substituting a phosphorus atom as in Fig. 12.3b, there is an extra electron available. Since it is only held in position by the coulomb attraction to the phosphorus nucleus, it can be removed and made a conduction electron with far less energy than that required to move a valence electron across the band gap (i.e., remove a shared electron used in normal bonding). In a similar way an impurity such as the aluminum in Fig. 12.3c has a deficiency of one electron that a silicon neighbor would like to share. A valence electron can jump into the location with far less energy than would be required to jump the forbidden energy gap, thus creating a hole in the valence band for positive charge conduction. These types of impurities thus provide allowable energy levels for an electron which are within the forbidden gap of the basic crystal. The extra electron case of Fig. 12.3b provides a "donor level" as in Fig. 12.2a that donates electrons as the majority charge carrier in this material called *n*-type. The extra hole case of Fig. 12.3c results in providing an acceptor energy level E_a as in Fig. 12.2b that accepts electrons from the valence band, thus making the holes the majority charge carrier in this *p*-type material.

In intrinsic semiconductors the Fermi energy is exactly in the middle of the forbidden energy gap ($E_f = E_g/2$) and there are the same number of conduction electrons as there are holes. As seen in Fig. 12.2, the Fermi energy in an extrinsic semiconductor is shifted toward the acceptor energy level (E_a) or the donor level (E_d) depending on the nature of the impurity. The exact location of the Fermi energy in these materials depends on the doping level (impurity atoms per cubic centimeter) and the absolute temperature. This

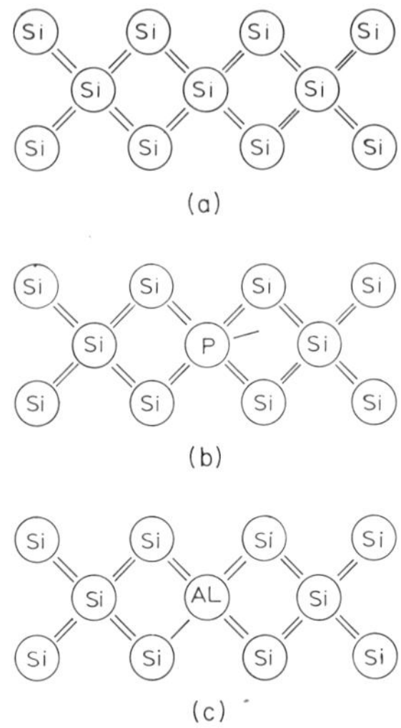

FIG. 12.3 Impurity atoms in a silicon lattice. (a) Normal silicon bonding. All four valence electrons in silicon form bonds with their silicon neighbors. (b) Excess electron impurity. Impurities such as phosphorus have five valence electrons and thus have an excess electron that can be easily removed to become a conduction electron. (c) Deficient electron impurity. An impurity such as aluminum has three valence electrons, providing a temporary site for one of the normal bonding electrons.

dependence is qualitatively shown in Fig. 12.4. Each impurity site provides one electron or acceptor location. Once these sites have been used, the material acts as an intrinsic semiconductor. This can be seen in Fig. 12.4. As the temperature increases, more impurity sites are used and eventually

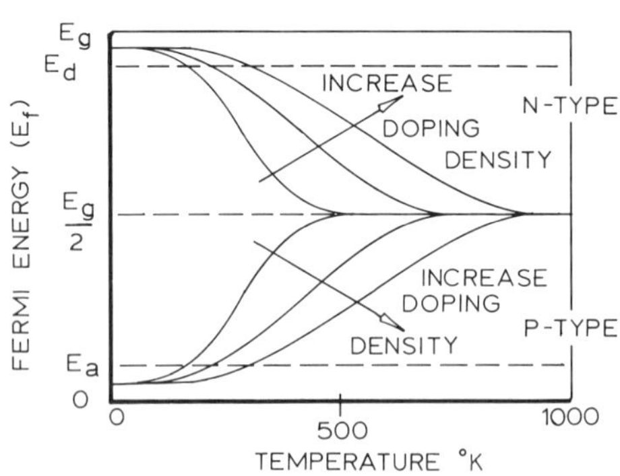

FIG. 12.4 Qualitative dependence of the Fermi energy on the temperature and doping level of the impurity. As all of the acceptor sites get filled with increasing temperature, or all the donor electrons get excited to the conduction band, the material approaches its intrinsic characteristics with the Fermi energy approaching the middle of the forbidden energy gap.

the number of electrons thermally excited across the forbidden ene is large compared with the impurity doping level and the Fermi energy approaches the intrinsic level of $E_g/2$. The effect increasing temperature would have on a p–n junction solar cell is to bring the Fermi energies of both sides closer, thus reducing the output voltage and efficiency.

12.4 OPTICAL CHARACTERISTICS

In order to obtain useful power from photon interactions in a semiconductor, three processes are required:

(1) The photon has to be absorbed and result in electrons being excited to a higher potential.

(2) The electron-hole charge carriers created by the absorption must be separated and moved to an edge to be calculated.

(3) The charge carriers must be removed to a useful load before they recombine with each other and lose their added potential energy.

The absorption of photons in a material is given as a function of distance into the material, x, as

$$I(x) = I(0)e^{-\alpha x}$$

where $I(x)$ is the intensity of photons at depth x, $I(0)$ the intensity incident on the material, and α the absorption coefficient. For example, if one takes a typical α of 10^4 cm^{-1}, then 90% of the photons would be absorbed in the first 2.3 μm of the material.

The incoming photons with energies greater than the forbidden energy gap can be completely absorbed by an electron and jump the gap. Any excess energy the photon has over the minimum required ($E_{Ph} - E_g$) is quickly given up to the lattice as thermal energy as the electron drops down to the bottom of the conduction band. For photons with energies less than the band gap, the material will appear transparent since there is no mechanism that would allow an electron to interact. Thus one would expect a step change in the absorption coefficient to take place as a function of incident photon energy at the energy of the band gap. This is indeed the case in many semiconductors and they are called direct absorbers. The steep absorption edge of several materials can be seen in Fig. 12.5. In order to explain the behavior of semiconductors that do not exhibit an abrupt change in α, one must go into band theory in more detail than in the simplified presentation above.

When one solves the quantum mechanical equations for a periodic potential in a crystal, the allowed energy bands are, in general, a function of a lattice parameter k. This parameter is a geometric factor related to the

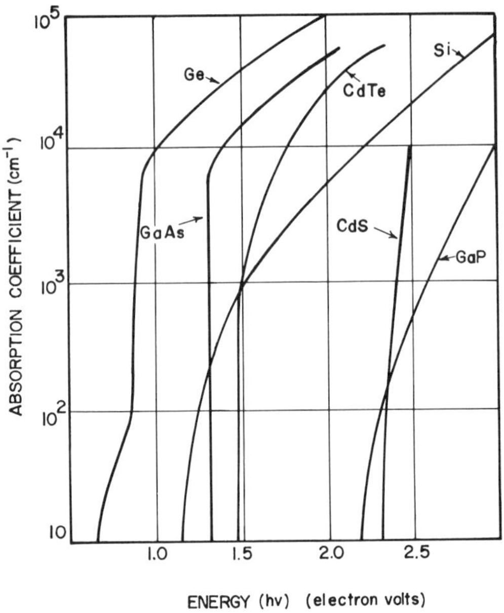

FIG. 12.5 The optical absorption coefficients as a function of photon energy for a number of materials of interest for solar cells. Materials having steep curves (sharp absorption edge) are indicative of direct absorbers and could be made into thin film solar cells.

periodicity of the crystal, but it can be shown to be related to the momentum. Figure 12.6 shows the allowed bands as a function of k for two different types of semiconductors. The forbidden gap is always defined as the minimum vertical distance on this type of plot between the bottom of the conduction band and the top of the valence band. For an indirect absorption of a photon to take place, it requires the interaction of a photon, an electron, and a phonon from the lattice to provide the momentum change. A photon having an energy greater than E_g may have a rather deep penetration before the interaction conditions are met. This results in an absorption coefficient that depends on the energy of the photon, and only starts to increase at the band gap energy and slowly increases to large values that would allow a direct transition to take place at the $k = 0$ distance between bands. Semiconductors that have their minimum in the conduction band and their maximum in the valence band at $k = 0$ are called *direct absorbers*. Semiconductors which have the maxima and minima at different k values are called *indirect absorbers*. Many of the materials being considered for solar cells are direct absorbers which will allow cells to be made from thin films. Silicon and germanium are indirect absorbers.

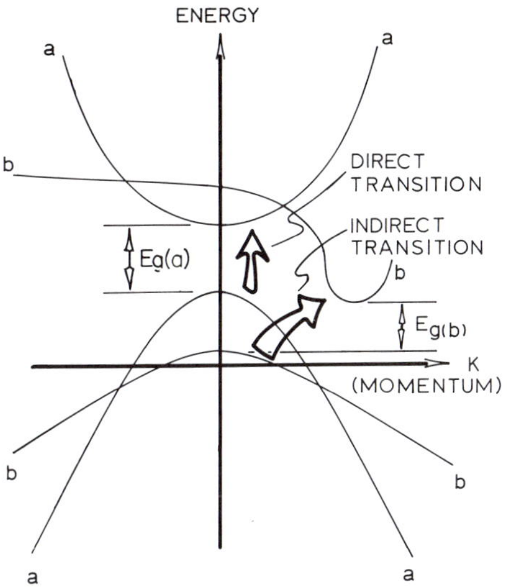

FIG. 12.6 The forbidden energy gap as a function of the lattice parameter k (related to the momentum) for two different semiconductor materials, a and b. Material a is a direct absorber. Also depicted are the electron transitions across the gap. Direct transition does not involve a change in momentum.

After the photons are absorbed and electron-hole pairs are created, the charges must be separated. In a typical solar cell this is done by use of p–n junction. Figure 12.7 demonstrates how this p–n junction provides an electric field that sweeps the electrons in one direction and the positive holes in the other. If the junction is in thermodynamic equilibrium, then the Fermi energy

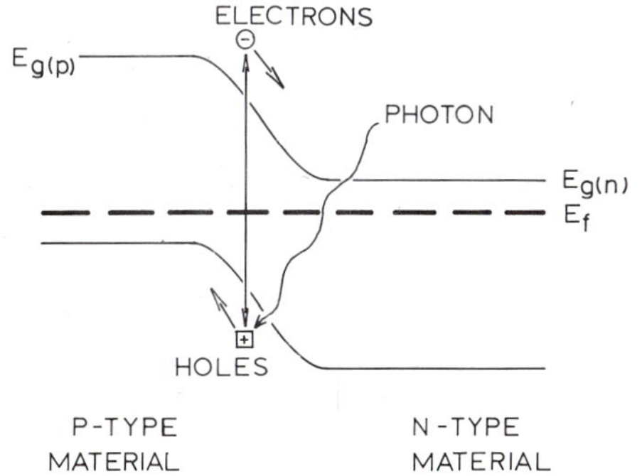

FIG. 12.7 The inherent electric field provided by a p–n junction which will separate the charge carriers after their creation by the absorption of a photon.

must be uniform throughout. Since the Fermi level is near the top of the gap of an *n*-doped material and near the bottom for the *p*-doped side, an electric field must exist at the junction providing the charge separation function for the cell.

12.5 TRANSPORT CHARACTERISTICS

Once an electron has been optically excited to the conduction band, it may either "recombine" with a hole or migrate to the edge of the semiconductor and be collected for useful work. Thus the properties of the material such as electrical resistance become important. Recall that resistivity $= \sigma = ne\mu$, where n is the number of electrons, e the electric charge per electron, and μ the mobility. The mobility of the charge carrier is an index to how rapidly the charge carrier will move on application of an electric field, or the average drift velocity per unit electric field. By comparison we might note the ranges of resistivities of:

Insulators	-10^{15} ohm-cm
Semiconductors	-10^{-2} ohm-cm
Metals	-10^{-6} ohm-cm

Other terms that are useful to know when considering transport effects are:

Lifetime The average time between electron-hole pair creation and recombination.

Mean free path The average distance traveled between collisions for a carrier moving through a material.

Diffusion length The average distance a carrier diffuses before recombination.

For specimens that have a high concentration of impurities, the scattering is frequent, the mean free paths are short, and the mobilities are low. Imperfections in the crystal due to dopants, vacancies, dislocations, and surfaces act as recombination centers and thus decrease the diffusion length and lifetimes. Undesirable impurities such as heavy metals provide energy levels in about the middle of the forbidden gap and are notorious for being electron lifetime killers. Indirect recombination occurs, which is an electron dropping down from the conduction band to an intermediate level and then on down to the valence band.

Table 12.1 shows many of the materials that have been considered for solar cells and some of their transport characteristics (Bylander, 1971).

TABLE 12.1

Solar Cell Material Candidates[a]

Major materials

Material	Energy gap (eV)	Mobility (300°K) (cm^2-V^{-1} sec^{-1}) μ_n	μ_p	Lifetime (sec) τ_n	τ_p	Experimental diffusion length (μm)
Si	1.1	1450	500	400		100
GaAs	1.4	8500	400	50	110^{-2}	5
CdS	2.6	340	615	100	10^{-2}	~1
B	1.5	6000	4000			
Se(Hex)	1.6	2	17			
SiC(β)	2.6	1000	10			
As_2Se_3	1.6	15	45			
Sb_2Se_3	1.2	15	45			
AlSb	1.6	400	400			
GaP	2.2	>200	150			
$CdAs_2$	1.0		100			
InP	1.2	4000	>600	2000	2	
ZnSe	2.6	500	16			
ZnTe	2.2	500	900			
CdSe	1.7	600				
CdTe	1.4	700	65			

Other materials

Material	Energy gap (eV)	Material	Energy gap (eV)
Ca_2Si	1.9	Ca_2Se_3	1.9
Cu_2O	2.0	In_2Te_3	1.2
Cu_2S	1.8	Sb_2S_3	1.5
Cu_2Se	1.3	InN	2.5
$CdAs_2$	1.0	BSb	2.6

[a] From Bylander (1971).

12.6 PHOTOVOLTAIC PRINCIPLES

A solar cell usually uses a *p–n* junction, as described earlier, in a physical configuration as shown in Fig. 12.8. The relationship between the current and voltage in an ideal *p–n* junction is given by

$$J_j = J_0[\exp(Ve/kT) - 1] \tag{12.1}$$

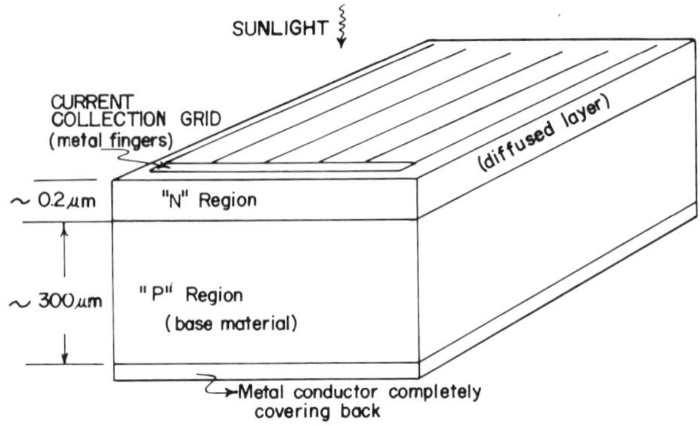

FIG. 12.8 Schematic of the physical configuration of a typical solar cell.

where V is the voltage imposed across the junction, e the electronic charge, k Boltzmann's constant, and T the absolute temperature. The saturation current J_0 (sometimes called the dark current), is obtained when a large negative voltage is applied across the diode. When light impinges on the junction, electron-hole pairs are created at a constant rate providing an electrical current flow across the junction. The net current is thus the difference between the normal diode current and the light generated current J_L. Figure 12.9 shows the simplified equivalent circuit for the cell. The internal resistance R_s is mostly due to the high sheet resistance of the diffused layer which is in series with the junction. The net current J is given by

$$J = J_L - J_j = J_L - J_0\{\exp[(V + JR_s)e/kT] - 1\} \tag{12.2}$$

The internal voltage drop in a cell can usually be minimized, and the assumption that is used for an ideal cell is that $R_s = 0$. The corresponding J–V curve and electron potential diagram for an ideal solar cell are given

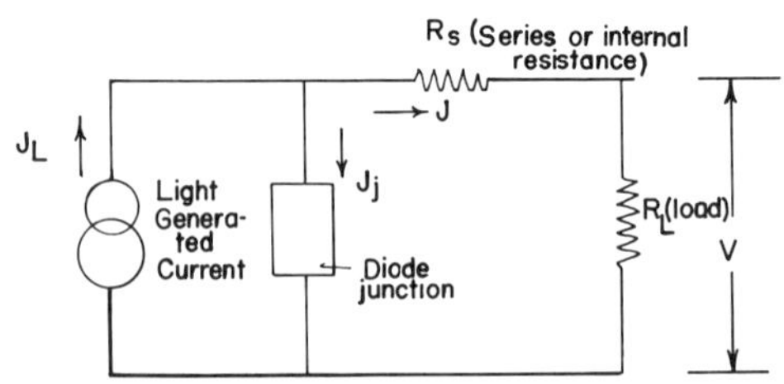

FIG. 12.9 The equivalent circuit for a solar cell showing the internal series resistance. The light-generated current acts as a constant current source supplying the current to either the junction or a useful load depending on the junction characteristics and the value of the external load resistance.

in Fig. 12.10. It can be shown that the open circuit voltage V_{oc} for the ideal cell is given by

$$V_{oc} = (kT/e)\ln[(J_L/J_0) + 1] \tag{12.3}$$

For normal operation, J_L is several orders of magnitude higher than J_0 and the 1 in the equation can be neglected. Care must be taken in the use of Eq. (12.3) for predicting the temperature dependence of V_{oc}, because J_0 is a strong function of temperature. In practice the open circuit voltage of a cell usually decreases linearly with increasing temperature.

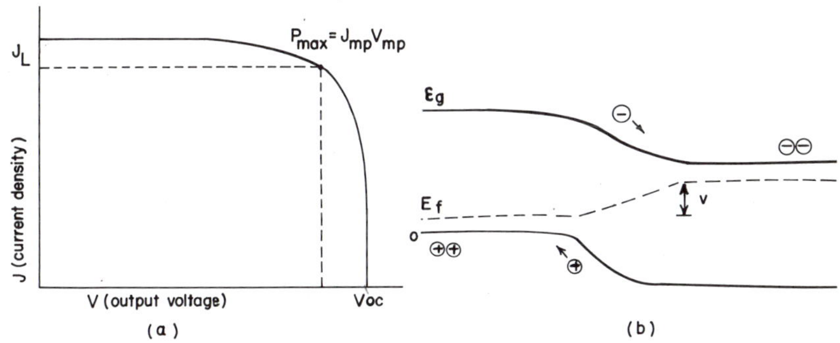

FIG. 12.10 (a) A typical current–voltage curve for an ideal solar cell with the short circuit current approaching the constant light-generated current J_L. The maximum power point is shown as P_{max}. (b) A typical electron energy diagram for a p–n junction under illumination of sunlight with the external resistance set such that the output voltage is at value V. This same voltage V is thus imposed on the junction which will reduce the current delivered to the external load.

The point P_{max} on the J–V curve in Fig. 12.10 represents the maximum power point. The corresponding current and voltage for this point cannot be explicitly solved, but it can be seen that the maximum efficiency for the cell is obtained by dividing $J_{mp}V_{mp}$ by the total power density of the sunlight P_{sun}. In the evaluation of solar cells it is often convenient to use the terms defined in the equation

$$\eta = J_{mp}V_{mp}/P_{sun} = (J_L E_g/eP_{sun})\underbrace{(J_{mp}V_{mp}/J_L V_{oc})}_{\substack{\text{Fill factor} \\ \text{(curve factor)}}}\underbrace{(eV_{oc}/E_g)}_{\substack{\text{Voltage} \\ \text{factor}}} \tag{12.4}$$

The fill factor is used as a measure of how well a junction was made in a cell and how low the series resistance has been made. A typical value of the fill factor for a good silicon cell is about 0.8. The voltage factor is determined by the basic properties of the materials in the cell and is typically about 0.5 for a silicon cell.

Inspection of the above equations in light of the properties of semiconductors, can give insight into characteristics of solar cells. First, the short circuit current is directly proportional to the light-generated current which in turn is linearly proportional to the intensity of sunlight on the cell. The open circuit voltage increases as the log of the light intensity if the temperature is constant. Thus if the fill factor remains the same, one should expect the efficiency to increase as the log of the intensity of sunlight. Also as the forbidden energy gap of the semiconductor is increased, the open circuit voltage of the cell and its efficiency should increase. However, the light-generated current is due to the electron-hole pairs created by the absorption of photons whose energies are greater than the band gap of the semiconductor. When sunlight is used as an energy source, the number of photons available is a function of photon energy. The light-generated current will monotonically decrease with increasing semiconductor band gap values. Thus as a function of band gap energies, the expected cell efficiencies should first increase due to the increase in voltage, pass through a maximum, and then decrease due to the decrease in the light-generated current available. These expected efficiency curves have been calculated by assuming typical *p–n* junction characteristics and the solar spectrum. Figure 12.11 gives the results of one of these calculations with various semiconductor materials identified at their appropriate band gaps. As can be seen from this figure, *p–n* junction solar cell efficiencies cannot be expected to exceed the 20–25% range. Most of the limitation on efficiency is due to the energy spectrum of sunlight. The excess of energy that photons have above the band gap of the semiconductor is lost to thermal heating as the excited electrons drop down

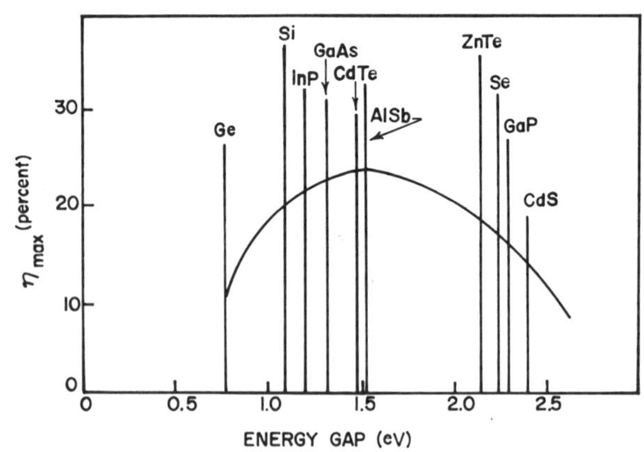

FIG. 12.11 A calculated curve of the maximum efficiency that can be obtained as a function of the energy gap of the semiconductor made into a *p–n* junction and illuminated with the solar spectrum outside the atmosphere, AMO (Loferski, 1963).

to the bottom of the conduction band. Photons with energies less than the band gap are all lost since they cannot excite any electrons. Because of this frequency mismatch, only about 44% of the energy in the solar spectrum is available to silicon for possible conversion to electricity (Wolf, 1971). Significant increases could be made in cell efficiencies if they are used with a monochromatic source or if the solar spectrum could be modified to different wavelengths (Backus, 1972).

12.7 CELL FABRICATION

This section briefly describes the processing of silicon cells for terrestrial applications, and a more complete discussion can be found in Chapter 13.

Silicon is the second most common element on earth and is usually obtained from sand (SiO_2). The steps with identifying characteristics are given in Table 12.2 for the 1972 type cell. The growing of the purified silicon into a single crystal is usually by Czochralski method. This process uses a single crystal *seed* which is lowered into a crucible of molten silicon containing a desired amount of doping material. The seed is slowly rotated and removed allowing the molten silicon to solidify and *grow* as it leaves the molten surface. As a result of this process, the finished silicon crystal is usually between 5 and 8 cm in diameter and about 75 cm long. This crystal is cut with diamond saws into wafers that are around 400 μm thick (see Fig. 12.12). The saws used are roughly the width of the wafers and leave the surfaces of the wafers with a great deal of damage to the crystal. Since the

TABLE 12.2

Economics of 1972 Solar Cell[a]

Form	Purity (%)	Use	\$/kg–silicon	\$/kW cap.*[b]
Sand	~90	building	<0.005	<0.01
Metallurgical silicon	>95	steel	0.60	1.40
$SiHCl_3$	>99.999	silicones	6.00	14.00
Polysilicon	>99.9999	semiconductor devices	60.00	140.00
Single crystal silicon	>99.999	semiconductor devices	250.00	600.00
Solar cell blanks	>99.999	solar cells	1300.00	3000.00

[a] From Currin *et al.* (1972).

[b] Assume 100 μm thick, 10% efficiency AM1.

FIG. 12.12 Single crystal silicon rods that have been crucible-grown and then sawed into thin wafers that can subsequently be made into solar cells.

damaged areas can act as recombination centers, the wafers are etched on both sides to a thickness of about 300 μm. This etched wafer is called the base region of the cell, and usually contains a p-type dopant such as boron. The p–n junction is obtained by putting the p-type base material into a diffusion furnace containing a gaseous n-type dopant such as phosphorous, and allowing the n-type dopant to diffuse into the surface about 0.2 μm. The junction is thus formed slightly below the planar surface of the wafer, the light impinges perpendicular to the junction, and a voltage is induced between the top and the bottom of the wafer. To facilitate the collection of the current, a conducting metal contact is evaporated over the entire back surface and collecting fingers are evaporated on the front. The surface coverage of the conducting collectors is the order of only 5% to allow as much light as possible to reach the junction area. An antireflecting coating is then applied to reduce the reflection coefficient of 0.3 for normal polished silicon. Each of these wafers will produce power of about 0.5 V with the current directly proportional to its area. The individual cells are connected in series–parallel combinations to meet the voltage, power, and reliability requirements of the particular application.

For terrestrial applications, the solar cell arrays must be encapsulated to protect them from atmospheric degradation. There is no known degradation process associated with the absorption and conversion of sunlight in a silicon solar cell. The degradation in the arrays would come from the oxidation of the metal contacts causing peeling and open circuits. As a result, all of the cells are assembled in panels that are weathertight with a clear front cover. This cover may be glass or materials such as acrylics, or silicon epoxy. The lifetime or degradation of the plastic-covered arrays will depend mostly on how much cracking and discoloration is caused in this cover material with long term exposure to sunlight and weather. A silicon solar cell array using a glass cover deployed in Chile in 1961 has not shown any degradation in electrical output (Dalibot 1974). Figure 12.13 shows a variety of terrestrial arrays from the main manufacturers. Shown are the different ways that the circular wafers can be cut for arrays. By cutting them into semicircles, quarters, or squares, a higher packing factor can be obtained.

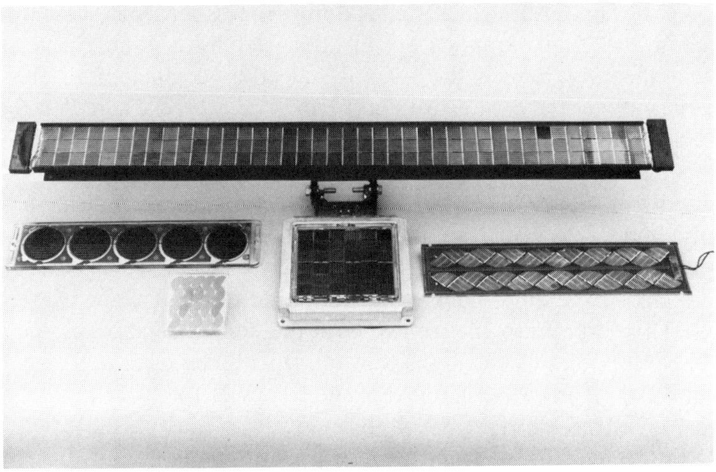

FIG. 12.13 A collection of commercially available terrestrial solar cell modules. Spectrolab, 3.6 W (top); Solar Power Corp., 1 W (left-center); Centralab, 1 W (bottom-center); Solarex Corp., 1 W (bottom-right); and SHARP (Japan), 0.36 W (bottom-left). Photograph courtesy of Jet Propulsion Laboratory, Pasadena, California.

A. Silicon Solar Cells

The silicon solar cell has been the most developed and used cell to date. In 1957 and 1958, the silicon cell demonstrated that it could convert about 14% of terrestrial sunlight into electricity. This prompted several companies, e.g., Hoffman Electronics and International Rectifier, into producing cells and anticipating large markets. Subsequently, some of the larger electronic companies such as RCA and Texas Instruments entered the field only to terminate their activities when the large markets did not develop. As a result, two relatively small companies, Centralab and Heliotek made most of the cells for the US satellites (Wolf, 1974).

It is rather peculiar that even though these silicon cells were used extensively in the space program since the late 1950s, very little research was performed on their basic nature. Comsat Laboratories in the US started a modest research program in the early 1970s that has resulted in significant advancements in cell efficiency. In 1972 they announced the *violet* cell (Lindmayer and Allison, 1972), which is a cell that has a shallower junction depth than most previous cells and a different antireflective coating. This cell has an increased efficiency because it responds to the shorter wavelength or *violet* part of the solar spectrum. The next major improvement in efficiency was announced in 1974 and was called the Comsat NonReflective (CNR) cell or sometimes called the *black* cell because of its appearance (Haynos *et al.*, 1974). This CNR cell uses an etching process that texturizes the surface in such a way that pyramids are left that require a double reflection and that significantly reduce the overall reflection coefficient of the silicon. Table 12.3 shows the comparison of the 1970 commercial silicon cell with the violet and CNR cells (Wolf, 1975). Also shown are goals that have been established as being practically obtainable with an outerspace solar spectrum (Wolf, 1975). Terrestrial operational characteristics are harder to identify because the solar spectrum is so dependent on atmospheric conditions. As a rough estimate, the terrestrial efficiency of a particular silicon cell can be expected to be 20% higher than the same cell in the space spectrum of AMO. This is because the atmosphere preferentially absorbs the infrared and the ultraviolet wavelengths that the silicon cell does not respond to at all. Thus the violet cell with an AMO efficiency of 14% would be expected to have a terrestrial efficiency of almost 17%.

It can be concluded from the previous discussions that the efficiencies of present silicon cells are approaching the expected limits and that major improvements in cell characteristics should not be expected. The major

TABLE 12.3

Performance Status of Current Silicon Solar Cells and Improvements Goals

	Efficiency contribution (Factor in the efficiency equation that is due to that attribute)			
Attribute	Commercial cell (1970)	"Violet" cell	"Black" cell	Practical goal
Voltage factor	0.522	0.535	0.531	0.61
Fill (curve) factor	0.82	0.825	0.822	0.86
Series resistance	0.96	0.985	0.984	0.97
Reflection	0.905	0.951	0.97	0.97
Grid finger coverage factor	0.96	0.95	0.95	0.96
Efficiency (AMO)	0.104	0.14	0.153	0.19

changes are expected in the reduction of the production costs of cells. The present costs of terrestrial arrays such as shown in Fig. 12.13 are 15–20 dollars per peak watt ($\$/W_p$). Typical applications of these cells are shown in Figs. 12.14 and 12.15. Several countries, including the US and Japan, have recently established major government programs to reduce these costs to the levels that these systems could have an impact on their national energy requirements. These programs are directed at cost reductions in each of the steps listed in Table 12.2 for silicon plus the use of concentration and other photovoltaic materials.

FIG. 12.14 A solar cell array being used to provide a voltage for cathodic protection of a natural gas line. Photograph courtesy of Spectrolab, Inc., Sylmar, California.

Silicon cells have traditionally been made from semiconductor grade silicon by the electronic industry. Solar cell production will not affect the silicon market for some time because the annual production of semiconductor grade silicon in the US alone is about 1500 metric tons per year. Investigations are being made to determine what purity of silicon is required for solar cells and what alternative processes may be used for purification from sand. Preliminary results are that a *solar* grade of silicon can have

FIG. 12.15 A solar cell array with accompanying battery storage under test by Southern Railway that operates railroad crossing bells and lights near Atlanta, Georgia. Photograph courtesy of Solar Power Corporation, North Billerica, Massachusetts.

three or four orders of magnitude more impurities than the semiconductor grade.

There are a large variety of options available for making purified silicon into acceptable solar cells. One method that appears very promising for the growth of single crystal material is a process called edge-defined, film-fed, growth (EFG). This is a process that uses a *die* made of silicon carbide located in the molten silicon crucible. By capillary action, the liquid silicon moves up in a narrow slit in the die. A single crystal *seed* of silicon is touched to the liquid silicon at the top of the slit and slowly removed. Thus a thin, continuous ribbon of single crystal silicon can be obtained at the rate of several centimeters per minute. The cost advantage of this process is due to three major characteristics: it is a continuous process rather than batch; the silicon is in a geometry that can be used in solar cells without waste and expense of sawing and etching; and multiple ribbon dies can be made from one melt, thus increasing the rate of production per machine. The cost attractiveness of the EFG process has prompted the Mobil Oil Company to set up a new company with the Tyco Corporation (Mobil–Tyco Corporation) whose sole purpose is to develop the EFG process in order to make silicon solar cells for terrestrial applications. Figure 12.16 shows a silicon ribbon grown from the EFG process that is about 300 μm thick, 3 cm wide, and several meters long.

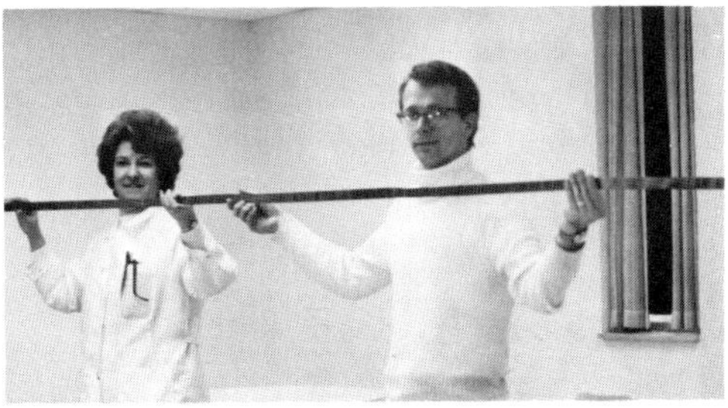

FIG. 12.16 A long ribbon of essentially single crystalline silicon grown with the edge-defined, film-fed growth (EFG) technique. Photograph courtesy of Mobil-Tyco Solar Energy Corp., Waltham, Massachusetts.

There are a variety of approaches being investigated that are aimed at developing cheaper silicon cells by using polycrystalline silicon. Past cells that have been made of polycrystaline material have had relatively low efficiencies, e.g., 2–3%. Grain boundaries inhibit electron flow and recombination results. However, if the grains can be made much larger than the thickness of the cell, it may approach the characteristics of a single crystal cell. Processes such as chemical vapor deposition (CVD) are being investigated that would deposit silicon in large sheets very inexpensively. Numerous techniques and substrates are being used to enhance large grain deposition. Recrystallization after deposition may be possible with the use of techniques such as electron beam scanning, stress grain growth, or even hot rolling.

B. Cadmium Sulfide/Copper Sulfide Cells

The CdS/Cu_2S is perhaps the type of solar cell that has the highest state of development next to silicon cells. It is not a simple p–n junction. It provides a charge separation field by the junction of two dissimilar materials that have different band structures (i.e., a heterojunction). This cell is commonly called the cadmium sulfide cell, but has an effective band gap similar to silicon ($\sim$1.1 eV) rather than that characteristic of CdS ($\sim$2.4 eV). The attractiveness of this cell is that it can be made thin ($\sim$20 μm) using chemical processing techniques and does not require single crystal material.

Cadmium sulfide/copper sulfide solar cells have been made for more than 15 years in the US. A typical efficiency for these cells is 5%, although 8% cells have been reported. Most of these CdS/Cu_2S cells have exhibited degradation in ambient air and at elevated temperatures. The degradation mechanism is currently not well understood. However, the French have indicated that retardation of the degradation can be achieved by controlling the composition of the copper sulfide and by proper doping of the CdS. So far only a laboratory pilot production of modest output has been attempted. Most of the US program is being directed at understanding how to control the processing of the cells to ensure stability and to develop mass production techniques that lend themselves to the production of cheap cells. The goals are to demonstrate cells costing less than 1 \$/W by 1980 and less than 20 ¢/W by 1985.

C. Gallium Arsenide Cells

As seen from Fig. 12.11, gallium arsenide with a band gap of 1.4 eV has the potential for higher conversion efficiencies than silicon. Gallium arsenide is also a direct absorber (see Fig. 12.5) and thus can be made in thin film thicknesses. Solar cells made of normal *p–n* junctions with GaAs demonstrated over 10% efficiency more than 15 years ago and presently only have about 12% efficiency in terrestrial sunlight. GaAs solar cells have been used on Russian satellite systems but have never economically competed with silicon for US satellites. These cells did not have higher efficiencies because of the lack of understanding of the material fabrication parameters that determine minority carrier lifetimes, thus resulting in GaAs crystals that have short diffusion lengths. The junctions were very close to the surface and there was an extremely high surface recombination rate. In 1972 a GaAs cell was developed that removed the junction from near the surface and resulted in efficiencies of about 19% (Woodall and Hovel, 1972). This new cell is really a heterojunction cell made of GaAs–$Ga_{1-x}Al_xAs$. The gallium–aluminum–arsenide layer is fairly transparent to the sunlight and is grown on top of a GaAs crystal by liquid epitaxy. The junction at the interface of the GaAs and the Ga–Al–As does not affect the crystal structure very much and thus does not introduce recombination centers near the junction (Woodall and Hovel, 1975).

The future development of GaAs is hampered by the present cost of materials and fabrication. However, its high demonstrated efficiency and its tolerance of higher temperatures make it very attractive for concentration systems.

12.8 PHOTOVOLTAIC SYSTEMS WITH SUNLIGHT CONCENTRATION

An alternate way to lower the cost per watt from solar cells is to concentrate sunlight on the cells. Recent experiments have shown that silicon cells can be made to have essentially the same efficiency at 100 suns as they have at one sun intensity if the temperature is maintained constant (Backus *et al.*, 1975). Also, GaAs cells have been experimentally operated at 19.1% efficiency under an illumination of 1735 suns (James and Moon, 1975). The present costs per watt can be considerably reduced because the area of the cells is reduced inversely proportional to the concentration ratio. The other components of the systems soon dominate the system costs. A variety of concentration systems are available (Backus, 1975), from ratios of one to over a thousand. The cells must be cooled for the higher concentration ratios but this does not appear to be a limiting factor (Florschuetz, 1975).

12.9 CONCLUSIONS

Photovoltaic devices were developed in the middle 1950s as the semiconductor industry was growing and proved themselves as useful power converters in the space programs. Their golden age is just now starting as their potential for contributing to the world's energy requirements is being recognized. The next few years should see drastic changes and rapid development of the understanding and usefulness of photovoltaic devices and systems. The reader may find additional general information on solar cells in references by Backus (1976), Rappaport (1959), Wolf (1974), Shirland (1966), Angrist (1971), Kettani (1970), and Fischer (1974).

13

Application of Solar Energy in Space

CHARLES E. BACKUS

MECHANICAL ENGINEERING FACULTY
ARIZONA STATE UNIVERSITY
TEMPE, ARIZONA

13.1 INTRODUCTION

The "space age" was formally started with the launching of the Sputnik by the USSR in 1957. That satellite and many more in the early years of the space programs did not use solar energy for power, but it was not long before solar energy became the dominant source of power for space craft. On March 17, 1958 the US launched Vanguard I which contained a small array of available silicon solar cells to power its 5 mW backup transmitter. Two months later the Russians launched a satellite that was totally powered by solar cells. Since the fall of 1959, essentially all space crafts with a mission duration of more than a couple of weeks used solar cells as their prime power supply. This includes more than 1000 satellites with power outputs up to 20 kW. The demand for solar cells from 1959 through 1974 was completely dominated by the demand for space craft power supplies. In the US the annual demand for space rated solar cells has held rather consistently at about 50 kW/a. In 1975 the demand for terrestrial solar cells exceeded the space demand and the future will probably show that the space market is a specialty within the general solar cell field. This could possibly turn around if some of the proposed systems of using huge satellites with thousands of megawatts of capacity are used for terrestrial demands.

Several other energy conversion systems have been investigated for coupling with the solar source in space, but thus far have not been able to compete with the solar cell arrays. These conversion systems include thermodynamic vapor cycles, thermoelectrics, and thermionics.

13.2 SOLAR CELL SYSTEMS

When one considers the selection of an electric power supply for a space craft application, it is considerably different than terrestrial applications. Although one still uses the usual engineering criteria of the lowest cost system to meet the constraints, the unusual requirements for reliability and not interfering with the mission are severe for space craft. The best power system for a particular mission depends primarily on the power level required and the length of the mission. For example, if the mission is less than one day with a modest power requirement of less than about 1 kW, then chemical storage batteries are probably the best. If the mission is between one day and one month in duration with power requirements of up to 100 kW, fuel cells would probably be the best choice. If the mission is more than a month, then the weight of the fuel for any system other than solar or nuclear

FIG. 13.1 The US Satellite Skylab. The solar cell arrays could provide more than 20 kW of power. Photograph courtesy of Jet Propulsion Laboratory, Pasadena, California.

makes it prohibitive. Radioactive-isotope thermoelectric generators (RTG) are used for some systems going away from the sun, such as a Jupiter flyby, or on missions that continuously require some power. Some of the solar cell powered satellites have a small RTG on board to supply power for the small amount of time the satellite may lose the sun, such as in an eclipse. For large power outputs, it would appear that a nuclear reactor would be more compact and cheaper than solar cell arrays. The anticipated crossover point in power size between a nuclear reactor and a solar array has been a function of time. In 1960, solar cell systems were only thought to be attractive up to power levels of 200 W. By 1963, this number had gradually been moved to about 500 W and indeed this power level of satellite was launched in August 1964 (Nimbus).

This was followed by the Orbiting Astronomical Observatory in 1966 with 1000 W of installed power, and eventually by Skylab (see Fig. 13.1) with more than 20 kW of power. Figure 13.2 shows one of the four panels that could supply a total of 100 kW to US satellites. Each of these quadrants is 30 ft wide and 100 ft long.

FIG. 13.2 One of four panels that will provide power to the US space station. Each of these panels is approximately 30 × 100 ft and will provide 25 kW. This array will "fold up" for launch and be deployed after the correct orbit is reached. Photograph courtesy of Lockheed Missiles and Space Co., Cupertino, California.

13.3 SOLAR CELL FABRICATION FOR SPACE

Solar cells used in space utilize the same principles of operation as described in Chapter 12 but are influenced by the following additional constraints:

Operation is with no maintenance.
Operation is in an intense radiation field of high energy particles.
Launch costs make light weight, high-efficiency cells attractive.
Lack of area and complexity of deployment makes high packing factors attractive.
The only cell cooling is by radiation.

These constraints are listed roughly in their order of priority and the first two have received almost all of the development effort. A high reliability requirement has imposed extremely tight specifications, inspections, and testings at each step of the processing. Each cell in an array must individually be tested in a solar simulator and display the identical voltage and current characteristics. A speck on the back of a cell that did not get covered with the evaporated contact or a chip off the corner of a cell, even if it has no effect on the electric output, would be grounds for rejection. A particularly troublesome area in reliability has been with the contact leads between cells. These interconnects must be reliably soldered on both the top and bottom contacts of the cells without shading a cell, increasing the space between the cells, or shorting out cells. They must withstand the vibrations of launch and the thermal cycles in space between no sun and full sun with no cooling other than radiation. They also must be transported and stored before launch in high moisture atmospheric conditions without corrosion or degradation in any way. Titanium–silver contacts, using flexible interconnects, have proven very reliable.

Solar arrays in space are bombarded by radiation of all types and energies. Meteorite penetration has not proven to be a problem as was anticipated before the start of the space program. The charged particles in the radiation belts around the earth have provided the major degradation of the cells. Now that these belts have been mapped, the exposure to a satellite in a particular orbit can be fairly well estimated. These belts contain electrons with energies over 1 MeV and protons with energies over 30 MeV. These high energy particles impinging on a solar cell physically dislodge atoms from their crystalline sites and create large numbers of vacancies and interstitial atoms. The crystal imperfections thus created act as recombination sites for the free electrons and holes lowering the overall efficiency of the cells. These types of defects could be fairly well annealed out, but this would be very difficult to do remotely in space. In order to minimize the effect of radiation, there has been rather extensive experimental investigations using particle accelerators to bombard solar cells (Curtin and Statler, 1975). Figure 13.3 shows typical results from these types of experiments (Allison *et al.*, 1975). The general conclusions shown in Table 13.1 from irradiation experiments have been substantiated by experience from flight data. These conclusions have resulted in most space cells being thin ($\sim$300 μm), *n* on *p* silicon cells made from 10 Ω-cm base silicon. In addition, they usually have a transparent cover glass over the cells to transmit the photons but which absorbs many of the protons and electrons. Lithium-doped cells have proven to be advantageous for some applications. Lithium has a much higher diffusion rate than most doping materials and seems to anneal some of the radiation damage.

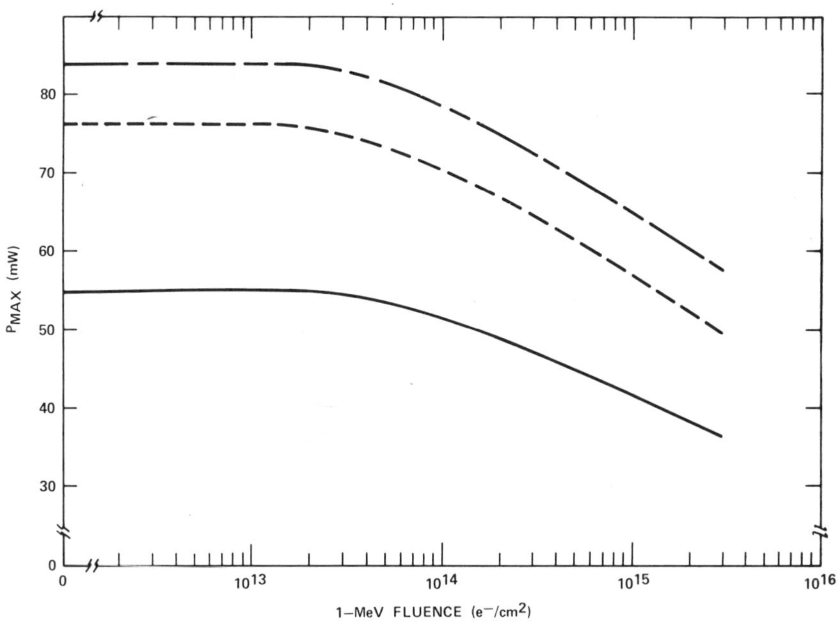

FIG. 13.3 Maximum power output of nonreflective (———), violet (– – – –), and conventional (—) cells as a function of 1-MeV electron irradiation fluence (Allison *et al.*, 1975).

The processing of silicon solar cells for terrestrial applications was described in Chapter 12 and is similar to the steps for space cells. In the US essentially all of the single crystal silicon is grown with the Czochralski crucible process from purified silicon. Until the last few years, all cells manufactured in Western Europe used float-zone refined silicon. In this

TABLE 13.1

General Conclusions Concerning Radiation Damage to Solar Cells[a]

High resistivity cells (10 Ω-cm) degrade less than lower resistivity cells (1 Ω-cm).
n on *p* base cells degrade less than *p* on *n* base cells.
Thin cells degrade less than thick cells.
Float-zone-refined silicon cells degrade less than crucible-grown silicon cells when exposed to protons and electrons.
Crucible-grown cells degrade less than float-zone silicon cells from photon exposure.
The radiation damage is manifested as a decreased infrared response of the cells.
The violet and CNR cells maintain their advantage over previous cells throughout irradiation.
Lithium-doped (*p* on *n*) cells recover better from damage than conventional *n* on *p* cells at temperatures >50°C.

[a] Curtin and Statler (1975).

method a R–F heater coil is moved slowly along a silicon bar maintaining a small molten zone of silicon within it. Silicon atoms preferentially solidify on the crystal behind the zone and the impurities build up in the molten zone as it moves. After several passes, the bar becomes a high-purity single crystal with all of the impurities concentrated at the ends which can be discarded. These single crystal rods are made to contain about 10^{15} atoms/cm^3 of a desired *p*-type dopant such as boron. This doping density will provide a base resistivity of about 10 Ω-cm.

Both crucible and float-zone silicon bars are frequently cut in their long direction in order to obtain a square cross-section bar with dimensions of about 4 × 4 cm. From this square bar, it is easy to cut individual cells as may be required for a particular satellite specification (e.g., 1 × 2, 2 × 2, or 2 × 4 cm). The rectangular cells are required to ensure a high packing factor for the cells on the array. The smaller cells are easier to attach to a curved surface. The surfaces of the rectangular wafers must be extremely smooth (lapped) to remove the saw marks and to ensure uniform thickness. The raw cells are subsequently cleaned and etched to remove any mechanical damage caused by previous processing steps. The blanks are then loaded into trays and put in a diffusion furnace to diffuse a *n*-type dopant, such as phosphorus, into the surfaces to provide the *p–n* junction. The junction depth is determined by the time and temperature in the furnace. The surface concentration of the impurity reaches about 10^{20} atoms/cm^3, which is approaching the silicon density, before removal from the furnace. Since only one side of the blank is desired to have a junction, it must again be etched on one side to remove the undesired junction.

The cells with junctions are mounted onto receptacles and covered with masks that have slits of the desirable front contact pattern. Figure 13.4

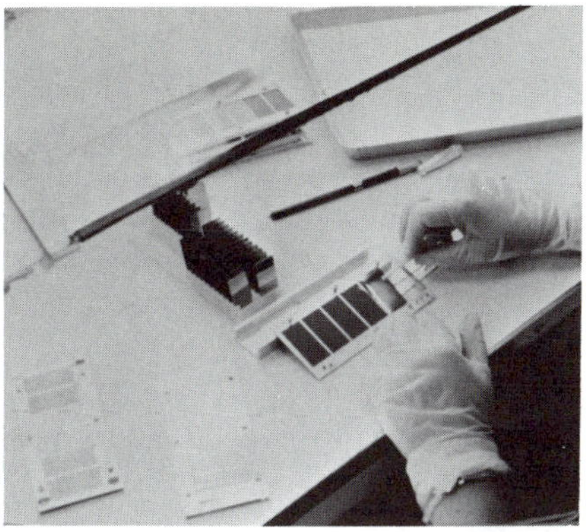

FIG. 13.4 Diffused silicon wafers are being loaded into fixtures for placing into the vacuum chambers for the evaporation of the metal contacts. Photograph courtesy of OCLI, Inc., City of Industry, California.

shows diffused wafers being loaded for application of front grids and back contacts. After being placed in a high vacuum chamber, titanium–silver contacts are evaporated through the front mask (on the junction side of the wafer) and evaporated completely over the back. The evaporation of the contacts does not make a good low resistivity bond to the silicon until it is heat treated (sintered) at about 600°C. The cells are then dipped into solder which will only adhere to the Ti–Ag deposition. The cells are again mounted onto fixtures and placed in a high vacuum for the evaporation of the antireflective coating.

Most space cells up through 1975 used SiO thin films as an antireflective coating. With the development of the violet cell that responds to wavelength down to 0.3 μm, the SiO was not transparent enough. Most cells using violet cell technology use Ta_2O_5 which is transparent to below 0.3 μm. Sometimes TiO_x is used which has its absorption edge between SiO and Ta_2O_5. The need for the higher transmission in the 0.3 to 0.4 μm range is seen in Fig. 13.5 that shows a comparison with the relative spectral responses of the conventional cell, the violet cell, and the CNR cell (Allison *et al.*, 1975).

After the antireflective coating, the cells are essentially complete and must undergo a series of tests to meet specifications. They are checked for their physical dimensions to meet given tolerances. They are each electrically tested to ensure identical and specified J–V characteristics. There is an adhesive tape peel test performed. The peel test is to see if the antireflective coating or the contacts can be pulled off. If the cell meets all of the specifications, then it is ready for assembly.

The cover glass is applied to the individual cells before they are soldered together. The added weight and shadowing effect these coverslips have on

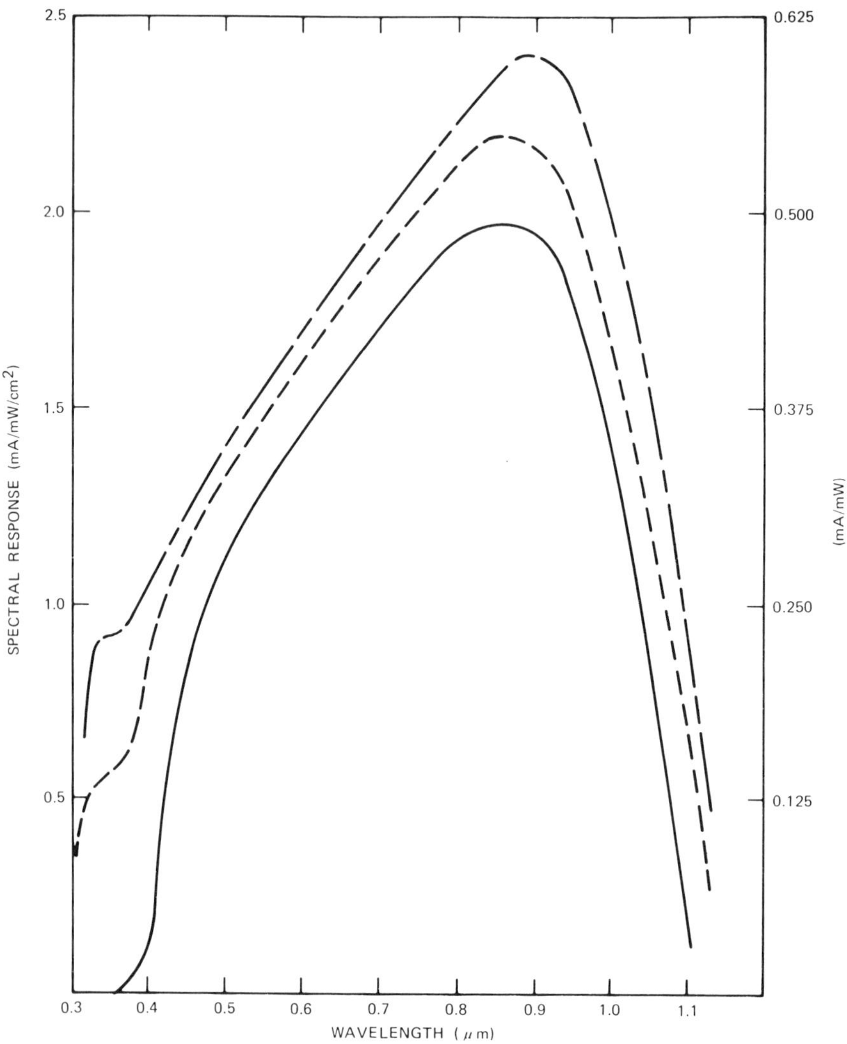

FIG. 13.5 Spectral response of nonreflective (———), violet (– – – –), and conventional (—) cells (Allison *et al.*, 1975).

the cells is justified by the more stable output of the cells since they are shielded from most of the radiation. The coverslips are usually about the same thickness as the cells, i.e., 300 μm, and are typically made of Corning 0211 microsheet or Corning fused silica. Sometimes the cover glasses have their own antireflecting coating on the outside surface and occasionally an optical filter on the underside to protect the coverslide adhesive from the solar UV irradiation. A clear adhesive must be between the coverslide and the cell to provide mechanical strength and optical coupling. Silicon adhesives are often used for bonding. Any degradation, such as discoloring, in either the coverslide or the adhesive due to the ultraviolet photons or the particle bombardment will, of course, result in a decrease of cell output. Figure 13.6 shows the adhesive being applied to some 2 × 2 cm silicon cells followed by the placement of the cover glass.

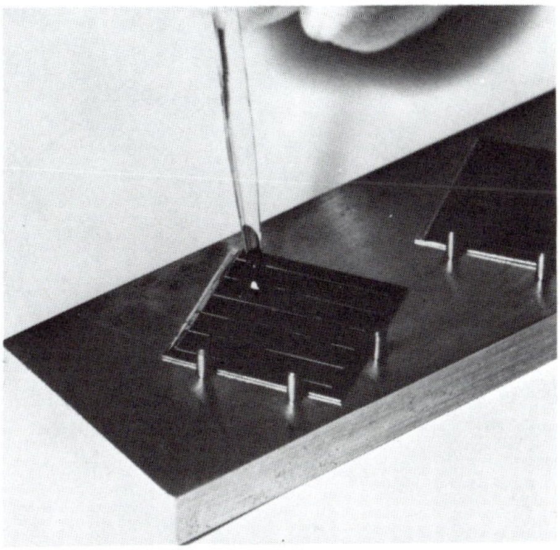

(a)

(b)

FIG. 13.6 (a) A clear adhesive is applied to bond the coverslip onto the cell and provide optical coupling. (b) The coverslip is applied to provide radiation shielding for space cells. Photograph courtesy of TRW Systems, Inc., Redondo Beach, California.

The individual cells are usually soldered together into small panels which can be later soldered into larger arrays. The interconnects that go from the top contact on one cell to the bottom contact of an adjacent cell can be seen in Fig. 13.7. Also seen is the application of the adhesive that bonds the cells onto the substrate material.

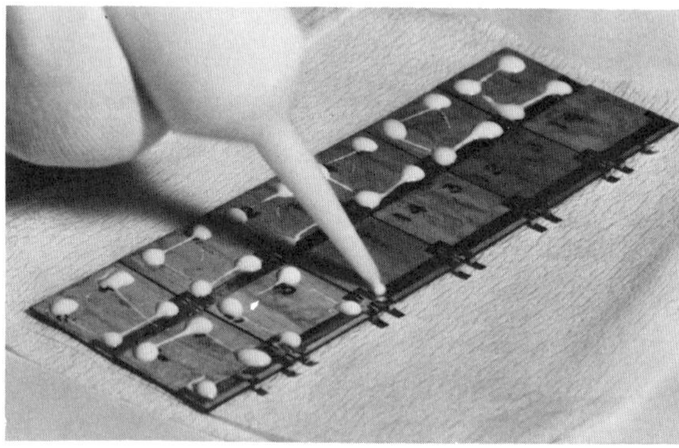

FIG. 13.7 Adhesive is applied to bond back of the cells onto the substrate. The interconnecting leads between the cells can be seen. Photograph courtesy of TRW Systems, Inc., Redondo Beach, California.

FIG. 13.8 The ESRO IV satellite using AEG–Telefunken cells that provide 60 W of power from the cells, which are mounted directly on the body of the satellite. Photograph courtesy of the European Space Research Organization, Noordwijk, The Netherlands.

The shapes of the final arrays vary a great deal depending on the power required for the mission, the physical constraints on the packing of the array for launch, and the deployment geometry in space so that it does not interfere with the mission function of the satellite. If the power requirement is not too high, the cells may be mounted directly onto the satellite's surface (a "body" mount) as seen in Figs. 13.8 and 13.9. This type of mounting is not an efficient use of cells since only about one third of the cells will be facing the sun at any given time, but it does not require any tracking or locking onto the sun's position. For higher power satellites one is forced to use extended surfaces such as "paddles" for the arrays as seen in Fig. 13.10. The French satellite shown in Fig. 13.11 uses a combination of body-mounted cells and paddles.

FIG. 13.9 A NASA satellite using body mounted cells from OCLI. Each of the flat section panels contains 770 cells and can produce 19.2 W. Each panel has two circuits of 7 parallel by 55 series connected cells. Photograph courtesy of TRW Systems, Inc., Redondo Beach, California.

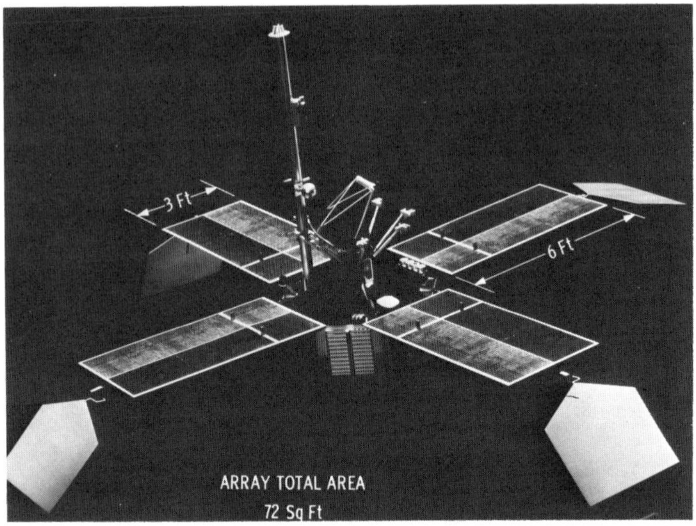

FIG. 13.10 The US Mariner 64 satellite using paddle mounted cells to provide more area for the cells than is available on the body of the satellite. Photograph courtesy of Jet Propulsion Laboratory, Pasadena, California.

FIG. 13.11 The French satellite EOEL which uses a combination of body mounted and paddle mounted cells. Photograph courtesy of Centre National d'Etudes Spatiales, Paris, France.

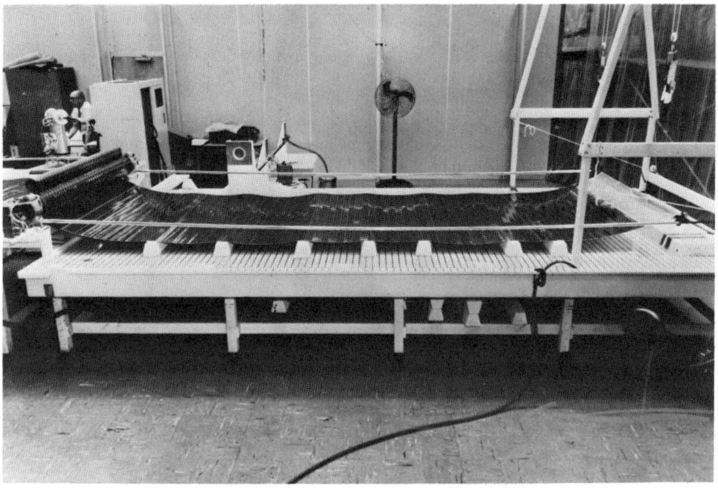

FIG. 13.12 This 1500-W flexible array can be rolled up like a window shade for launch. The flexible substrate is Kapton. Photograph courtesy of Air Force Aero Propulsion Laboratory, Wright-Patterson AFB, Ohio.

FIG. 13.13 USAF satellite using the roll up flexible arrays. Photograph courtesy of Air Force Aero Propulsion Laboratory, Wright-Patterson AFB, Ohio.

For large power requirements, the storage and deployment in space of the large arrays present a sizeable problem. There are two techniques that are commonly used for these panels. One method is shown in Fig. 13.12 which uses a mechanism like a window shade. This 1500-W flexible array is rolled up into its stored position on an 8-in. diameter magnesium drum. The two 16 × 5.5 ft panels of the flight model (see Fig. 13.13) contain 34,500 solar cells bonded to a Kapton fiberglass laminated substrate material. The other technique for deployment of large panels is to use an accordion-type packing as shown in Fig. 13.14.

FIG. 13.14 The ESRO fold-up array in which the panels fold in an accordion fashion for storage during launch. Photograph courtesy of the European Space Research Organization, Noordwijk, The Netherlands.

13.4 FUTURE SPACE CELLS AND SYSTEMS

The development of the violet cell and subsequently, the CNR cell has raised the efficiency of silicon cells to 15.3% when exposed to an air mass zero (AMO) solar spectrum. Future improvements could possibly raise this efficiency to about 19% (Wolf, 1975). It is obvious that the 19% cells will not reduce the area required for arrays very much when compared with 15.3% cells. The future advancements will most probably be in larger array technology, lighter weight cells, lower cost cells, and more radiation resistant cells.

One of the new types of cells developed in the last few years is called the vertical multijunction (VMJ) solar cell. The physical configuration of this cell can best be described by envisioning the cell that would result from stacking several *p–n* junction solar cells on top of each other, turning the stack on its side and then cutting slices out of the stack perpendicular to the junctions. Figure 13.15 shows a cell that was made by this method. The surface of the cell exposed to the sun contains several junctions that run in a direction parallel to the incoming photons (Rahilly, 1972). The junctions are usually connected in electrical series resulting in a cell with a very low current but with a high voltage output. For silicon, one would expect about 0.5 V per junction; if a 2 × 2 cell had 200 junctions, it would produce maximum power at about 100 V. It was initially thought that such a configuration would provide a higher efficiency cell since there would be a junction near photon absorption regardless of how deep it was (i.e., enhanced red response). In practice the cell has about the same or less efficiency when compared with a conventional cell.

The VMJ cells were proven to have desirable characteristics for two applications. They should perform well in terrestrial use for high sunlight

FIG. 13.15 A vertical multijunction (VMJ) solar cell made at NASA–Lewis by stacking several *p–n* junction cells together and then cutting perpendicular to the junctions. This 2 × 2 cm cell contains 94 vertical junctions in electrical series.

concentration systems since they would have very low currents for high illumination levels, and thus avoid the the concern for high currents that result in internal voltage drops in conventional cells. The other advantage that VMJ cells have exhibited is a high tolerance for irradiation experienced in space. Because of the latter characteristic, the VMJ cell is being developed in an effort to extend the operating life of space power systems or to extend the acceptable orbits into higher radiation regions. Research on VMJ cells has led to the development of open groove cells as shown in Fig. 13.16. These cells are made by using orientation dependent etching of the grooves into single crystal silicon wafers. These grooves are typically 5 μm wide and 100 μm deep (Lloyd, 1975). These devices have demonstrated the typical advantages of long-wave response while maintaining good blue response and have the high tolerance to radiation.

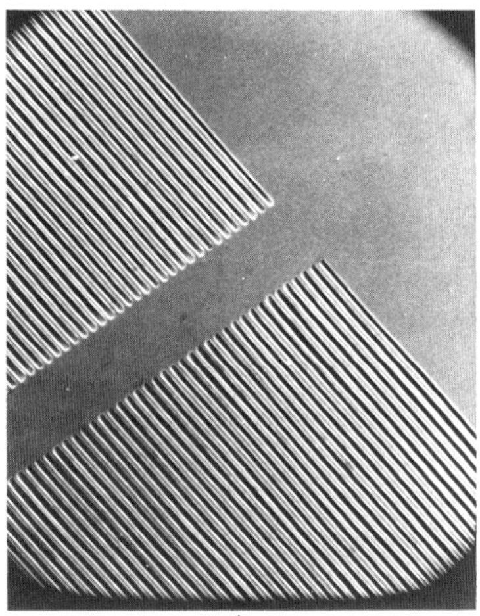

FIG. 13.16 The open groove VMJ cell fabricated by Texas Instruments. The grooves are etched into a single crystal silicon wafer and are about 5 μm wide and 100 μm deep. The diffusion is then done and results in vertical junctions (Lloyd, 1975). Photograph courtesy of Air Force Aero Propulsion Laboratory, Wright-Patterson AFB, Ohio.

The continuing success of larger solar cell systems in space has led to a proposal for a large station in space providing power to earth (Glaser, 1968). This concept is called the Satellite Solar Power Station (SSPS) and an artist drawing is shown in Fig. 13.17. It utilizes solar cell arrays that are 4 km^2 and a microwave transmitting antenna that is 1 km in diameter (Mockovciak, 1972). The satellite is in synchronous orbit and beams microwaves to a receiving antenna on the ground that is 7 km in diameter. The concept could provide 10,000 MW of ac electricity on a 24-h per day basis with no interruptions from clouds or other atmospheric conditions. The transmission efficiency from solar cell output to electricity on the ground with present technology is about 31% with anticipation that it could be raised to 69% with improved technology. This space system has many advantages for the utilization of the sun's energy on earth but is such a big step beyond present space systems that it is hard to predict when it will be built.

FIG. 13.17 The Satellite Solar Power Station (SSPS) concept of generating large scale power in synchronous orbit and transmitting it to earth by microwaves (Glaser, 1968). Photograph courtesy of Grumman Aerospace Corp., Bethpage, New York.

13.5 OTHER SOLAR SYSTEMS IN SPACE

During the early days of space exploration before the dominance of solar cells as the space power source, many other concepts were investigated that used the sun as its energy source. All of the systems investigated (none were ever launched) utilized large mirrors that concentrated the sunlight to obtain high temperatures to ensure higher efficiencies for the thermal energy converters. The energy conversion processes studied included thermoelectric, thermionic, and Rankine cycles. The most developed converter of these was the thermionic diode for coupling with a solar source in space.

The solar–thermionic system consisted of a mirror that concentrated solar radiation onto a cavity absorber. The emitters for the thermionic diodes were positioned around this cavity and were designed to operate at 1700°C. Diodes were built and tested for this system and demonstrated over 3000 h of lifetime at an efficiency of 7.7% (Fry, 1967). Neither thermionic nor thermoelectric converters ever demonstrated higher efficiencies than solar cells when combined with a solar source. The complexity and cost of making, deploying, and sun tracking with a lightweight, accurate mirror also made these systems uncompetitive.

14

Conversion of Solar Energy into Electricity

M. ALI KETTANI

ELECTRICAL ENGINEERING DEPARTMENT
UNIVERSITY OF PETROLEUM AND MINERALS
DHAHRAN, SAUDI ARABIA

14.1 INTRODUCTION

The sources of energy can be divided into two groups: energy capital and energy income. Energy capital arises from the storage of solar energy in chemical form by the action of living organisms millions of years ago, forming fossil fuels. These are exhaustible in due time. Energy income is continuously renewed and replenished, either instantaneously by the direct use of solar energy, or after a short period of storage in the form of winds, rivers, waves, etc.

All energy originates from matter; and, except for nuclear energy, all energy originates from the sun. Whether a specific energy is recognized as income or capital depends on its storage time scale (Fig. 14.1). For example, in the case of hydraulic energy, solar energy vaporizes sea water and carries it up into clouds; then the winds, which result from the temperature gradients in the atmosphere, carry the clouds to higher altitudes; precipitation results when the clouds cool; then water is collected in basins and falls down through rivers and eventually through hydraulic turbines to produce electricity (Kettani, 1970).

In this chapter we are concerned with the instantaneous conversion of solar energy into electricity. A multitude of devices and schemes have been conceived to transform solar energy into electricity. Figure 14.2 shows solar energy as the starting point and electrical energy as the end product. Three

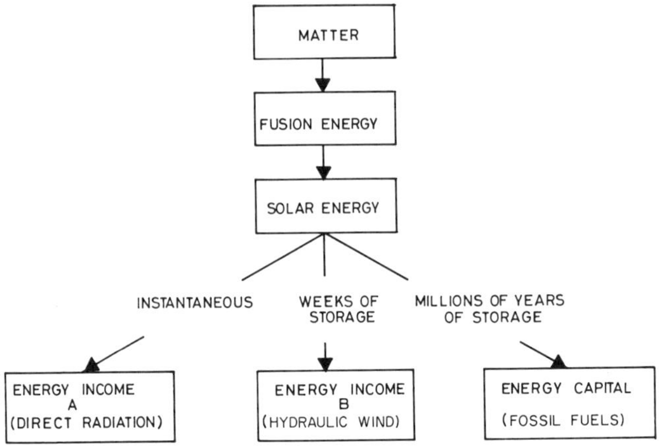

FIG. 14.1 Energy income and energy capital, all originating from matter. The difference is in the duration of the storage period.

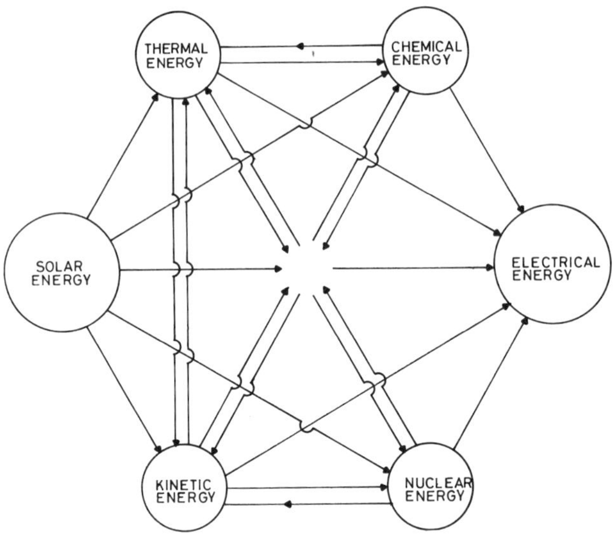

FIG. 14.2 Different means of converting solar energy into electricity.

intermediate stages are added: thermal, chemical, and kinetic. The arrows show the process of transformation from one form to another, starting from solar energy and ending in electricity. Any path can be taken, although it seems obvious that the most direct ones would be the most efficient. For instance, solar energy can be transformed into electricity by following a lengthy path. It can first be transformed into chemical energy by photosynthesis, then into thermal energy by combustion, then into kinetic energy by thermal expansion, and finally into electricity by the use of a conventional generator. On the other hand, solar energy could be converted directly through the photovoltaic effect into electricity (Daniels, 1974).

14.2 DIRECT ENERGY CONVERSION

To eliminate a lengthy series of stages in energy conversion, it seems attractive to follow the most direct process of conversion (Kettani, 1970). For solar energy, this implies the use of a physical process that converts directly the electromagnetic energy into electric form. Basically, such a process is not subject to Carnot's limitation and is potentially more efficient. Four main methods could be used for the conversion of solar energy directly into electricity:

(a) photovoltaic power conversion,
(b) photoemissive power conversion,
(c) photogalvanic power conversion, and
(d) photomagnetic power conversion.

(a) Photovoltaic power conversion—the photovoltaic effect can be described easily for a *p–n* junction in a semiconductor. In an intrinsic semiconductor such as silicon, each one of the four valence electrons of the material atom is tied in a chemical bond, and there are no free electrons at absolute zero. If a piece of such a material is doped on one side by a five-valence electron material, such as arsenic or phosphorus, there will be an excess of electrons in that side, becoming an *n*-type semiconductor. The excess electrons will be practically free to move in the semiconductor lattice. When the other side of the same piece is doped by a three-valence electron material, such as boron, there will be a deficiency of electrons leading to a *p*-type semiconductor. This deficiency is expressed in terms of excess of holes free to move in the lattice. Such a piece of semiconductor with one side of the *p*-type and the other of the *n*-type is called a *p–n* junction (Fig. 14.3). In this junction, the free electrons of the *n*-side will tend to flow to the *p*-side, and the holes of the *p*-side will tend to flow to the *n*-region to compensate for their respective deficiencies. This diffusion will create an electric field E from the *n*-region to the *p*-region. This field will increase until it reaches equilibrium for Ve, the sum of the diffusion potentials for holes and electrons. For more details see Chapter 12.

(b) Photoemissive power conversion—the photoemissive effect is the ejection of electrons from the surface of a solid or a liquid by incident electromagnetic radiation (Jensen and Limensky, 1960). Figure 14.4 shows

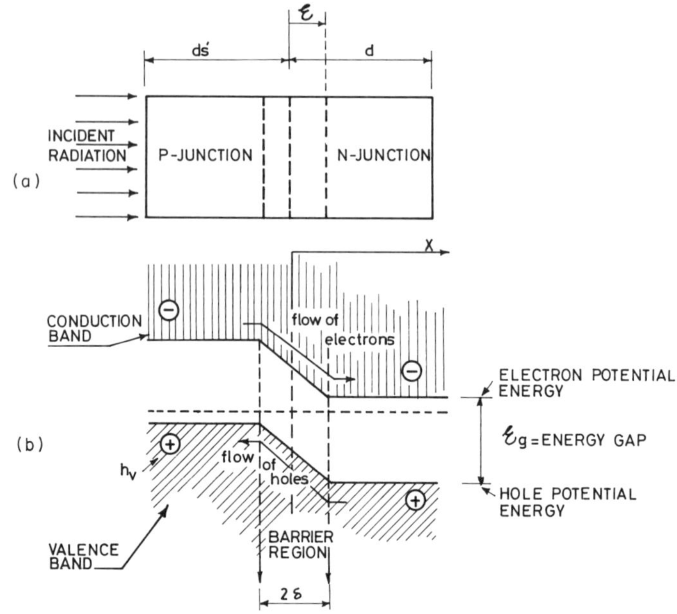

FIG. 14.3 The operation of a photovoltaic junction. (a) A sketch of *p–n* junction. (b) The energy level diagram.

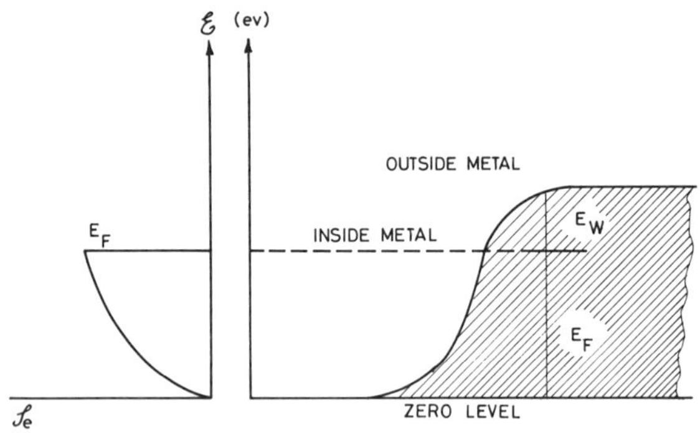

FIG. 14.4 The energy level diagram for the photoemission effect.

the energy-level diagram for free electrons within a metal at absolute zero temperature as well as the potential energy barrier at the surface of the metal. It indicates that electrons inside the metal exist in energy levels ranging from zero to the Fermi level ξ_F. There are no electrons beyond this highest level.

When a photon of energy $h\nu$ impinges on the surface of the metal, the most energetic electron would absorb the photon and could eventually escape from the surface with a kinetic energy equal to or smaller than $(h\nu - \xi_w)$, where ξ_w is the work function of the metal. Therefore the maximum energy of the escaping electron varies linearly with the frequency of the impinging radiation and is independent of the light intensity. An increase in light intensity implies an increase in the number of photons absorbed, which leads to a proportional increase in the number of ejected electrons per unit time.

The photoemissive effect is generally made of two subeffects: a surface subeffect in which the electron is coupled to the barrier at the surface during photon absorption, and a volume subeffect in which the electron is coupled to the internal periodic potential. Furthermore, photoemission could be enhanced when imperfections in the form of negative ion vacancies are introduced in the crystal. Thus, color centers are formed absorbing visible light. Photoemission in which a percentage of the ejected electrons are ejected by these centers is called extrinsic photoemission. The presence of color centers enhances the creation of excited states which in turn might transfer enough energy to color centers to eject electrons. Such a photoemission is called exciton-induced photoemission. These effects exist in some metals, semiconductors, and alkali halides.

The same remarks made for the photovoltaic effect could be made concerning the possibility of using materials of different work functions in a multilayer, or using a material with graded work function which would eventually match the solar spectrum.

A photoemissive cell would be similar to a thermionic cell. It would be constituted of an emitter made of a photoemissive material and a collector. This is, however, working at the same cold temperature. To avoid accumulation of space charge in the interelectrode space, this space should be evacuated and should be smaller. The necessity for the photoemissive surface to face the sun imposes special geometric configurations. One possibility is to have a rodlike collector surrounded by a shell-like emitter. Another possibility is to have a rodlike emitter surrounded by a shell-like collector mirror. In this way solar energy would be reflected on the emitter by the collector mirror which in turn would collect the emitted electrons. A third possibility is to have a gridlike collector covering the emitter, thus allowing passage of the sun's rays. A fourth possibility is to use magnetic fields to transport electrons from emitter to collector, the two being placed side by side or at an angle. All these possibilities are shown in Fig. 14.5.

(c) Photogalvanic power conversion—when aqueous solutions of some dyes with suitable reducing agents are illuminated, the reducing agent reduces the dye by transfering an electron to it. As soon as light is removed,

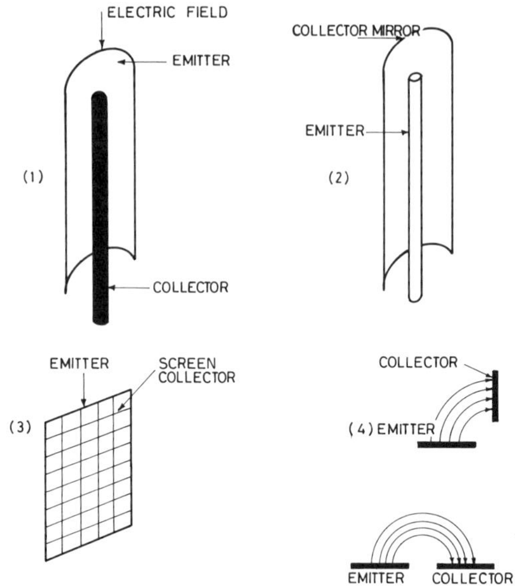

FIG. 14.5 Various configurations of photoemission cells.

the reaction is reversed. The reaction can be repeated again and again in an endless series of cycles (Saucer, 1958).

One such reaction uses thionine as dye and ferrous ions (Fe^{2+}) as reducing agents. Thionine is purple in color, but when illuminated the ferrous ions reduce it into a colorless leukothionine. Thus, each ferrous ion loses an electron and becomes a ferric ion. In the dark, the reaction reverses itself in several minutes, thus

$$2Fe^{2+} + \text{thionine} \underset{\text{dark}}{\overset{\text{light}}{\rightleftharpoons}} \text{leukothionine} + 2Fe^{3+} \tag{14.1}$$

The potential of the bleached solution is up to 0.4 V more negative than that of the purple solution.

A photogalvanic cell consists of two inert electrodes immersed in an aqueous solution of the dye and the reducing agent. One electrode is subjected to solar radiation, the other is left in the dark. Under the effect of solar radiation, electrons are generated at the illuminated electrode, and a difference of potential appears between the two electrodes. When an external load is placed between the dark and the illuminated electrode, a current flows in this external circuit due to the motion of electrons toward the dark electrode.

As a result of this transfer of electrons from the illuminated electrode to the dark electrode, there will be an accumulation of higher ions (such as

Fe^{3+}) in the illuminated part of the aqueous solution and an accumulation of lower degree ions (such as Fe^{2+}) in the dark part. This will create a potential that will tend to cancel the potential caused by the reduction of the dye. Several approaches have been proposed to eliminate this volume polarization. An obvious one would be to alternate the illumination of the electrodes, by changing their role after each cycle, thus producing ac power.

Another problem is caused by the competition between the electrode reaction and the solution reaction in the illuminated region of the solution. This has the effect of limiting the efficiency of the cell. This effect can be minimized by choosing a large electrode area to solution–volume ratio.

(d) Photomagnetic power conversion—the photomagnetic effect could convert light directly into electricity. When a plate of, say, cuprous oxide at cryogenic temperature is placed in a tangential magnetic field and illuminated axially, a difference of potential is created in a direction perpendicular to both light and the magnetic field (Kikoin and Moskov, 1934).

14.3 CONVERSION THROUGH A HEAT STAGE

When electromagnetic solar energy falls on a body, the body absorbs part and reflects the remainder. This results in an increase in temperature that will be stabilized only when losses by radiation equal the gain by absorption of solar energy. With special precautions, a material surface could absorb as much as 95% of the solar energy and convert it into heat. The difference between input heat and the rejected heat could be converted into electricity directly or indirectly. The entire process in this case is limited by Carnot's efficiency $\Delta T/T_h$, where ΔT is the difference between the source and the sink temperatures and T_h is the source temperature. Therefore, it is necessary to have T_h as large as possible and T_c, the temperature of the sink, as low as possible. This conversion through heat stages can be subdivided into three areas, which we now consider.

A. Conversion of Solar Energy into Heat

The function of the collector is to capture solar energy, to concentrate it, and to transform it into heat at a level high enough to allow an efficient operation (Duffie and Beckman, 1974). Two different methods are possible: the first is to use mirrors and lenses to concentrate solar energy before transforming it into heat; the second is to transform solar radiation into heat and then concentrate this heat to obtain the desired temperature. These methods were aptly described in previous chapters.

It was also proposed to heat large shallow ponds of water with a black bottom absorbing heat. If evaporation is prevented by one method or another, the pond would store more and more heat energy and its temperature would increase. A temperature of 90°C has been reached by placing at the bottom of the pond a permanent salt solution of high density. The solution sank in ordinary water and evaporation from it was thus avoided. All this was dealt with in detail in Chapter 10.

B. Conversion of Solar Heat Directly into Electricity

Once solar energy is converted into heat, the heat could be converted in turn to electricity. This process could be direct or indirect. There are many physical processes which are used for conversion of heat into electricity. These are thermoelectricity, thermionic effect, magnetohydrodynamics, electrogasdynamics, and ferroelectricity, among other methods.

Thermoelectric power generation is based on the Seebeck effect which states that a voltage is developed in a circuit constituted of two different metals or semiconductors when there is a difference in temperature between the two materials. Electrons at the hotter material have a higher kinetic energy and therefore higher speed than those in the colder material, which leads to both a transfer of energy and diffusion of electrons to the colder material. An emf will be established between the two ends as a result of this motion of electrons. This emf, the Seebeck voltage, will increase until the net flow of electrons ceases completely. If V is the open-circuit voltage thus developed and if T_h and T_c are the temperatures of the hot source and the cold source, respectively, then for small temperature differences $V = \Omega(T_h - T_c)$, where Ω is the Seebeck coefficient for the junction between the two materials. A typical solar thermoelectric system consists of an oriented mirror to concentrate solar radiation on a thermal storage material such as lithium hydride. The thermal energy stored in this material is then transferred to a radiator through the thermoelements.

Thermionic emission is the release of electrons from a hot metallic surface. Heat supplies energy to the electrons, allowing them to escape and migrate to a collector material. The hot material is called the emitter and the cold one is the collector. If an external load is placed between emitter and collector, an electric current is obtained. The free electrons of the emitter have an average energy related to the Fermi level. These electrons would be able to escape from the emitter if the amount of heat supplied to them becomes equal to the emitter work function. They will travel toward the collector with a certain velocity, and as they reach the anode, they are absorbed giving up energy to the work function of the collector in the form of heat, thus reaching the Fermi level of the anode. The difference between the emitter work function and the anode work function could then be transformed into electric energy if an external load is connected between emitter and collector. The type of thermionic geometry is dictated by the necessity of collecting solar energy. In general, a solar thermionic system would consist of a solar concentrator, a cavity-type thermionic engine, and a tracking device to follow the sun. One possible configuration for a cavity absorber and thermionic generator assembly is shown in Fig. 14.6. Solar radiation from the parabolic mirror crosses the glass dome into an evacuated blackbody cavity. Heat is rejected by a multitude of thin highly reflecting foils. The cavity could be cubic allowing the placing of five thermionic cells, one on each face of the cube, the sixth face being left open to admit the incoming light. Another configuration would be to place a thermionic cell in the center, surrounded by six others in a first ring and eight more in the second ring, thus putting together 15 cells in one single package. The cells would be connected in series inside the vacuum enclosure and only two leads would appear from outside the package. Another possibility is to integrate solar thermal conversion and thermionic thermoelectrical conversion. In this case the individual thermionic cells would have ceramic surfaces, exposed to light, in the form of cups. The cathode and its electrical lead are in the form of a plating made of a refractory metal outside the cup. It is completely surrounded by the anode and the radiator which acts as the heat sink.

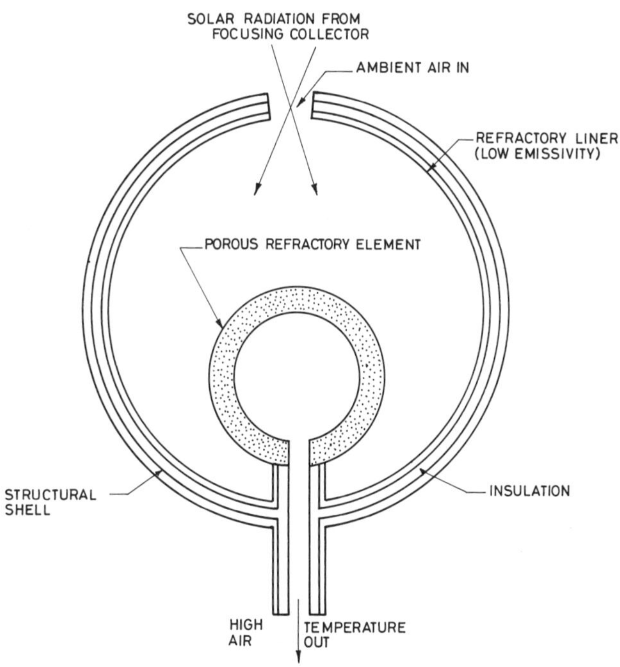

FIG. 14.6 An example of a cavity heater.

Heat used for the ionization of a gas could be converted to electricity by the magnetohydrodynamic (MHD) process. A MHD power generator converts the kinetic and thermal energies of a conducting material (ionized gas) flowing in the presence of a magnetic field directly into electricity. Solar energy can be used as the heat source of an electrogasdynamic closed-cycle system. This method is based on the fact that when positive ions are transported against an electric field by a flow of a hot neutral gas, they will do electrical work if they are allowed to flow through a load. The electromechanical coupling is provided by electric fields.

Ferroelectric power generation uses the change with temperature of the permittivity of a dielectric material to convert thermal energy into electricity. A temperature-induced change in permeability leads to ferromagnetic power conversion. Thermal energy could be converted to electricity through the Nernst effect in which an emf is generated between the two ends of some materials in a direction perpendicular to both the heat flow and the magnetic field present. Heat energy could also be converted to electricity by the magnetothermoelectric effect, by phase transition in a superconductor, by electron convection, etc.

C. Conversion of Solar Heat Indirectly into Electricity

The heat obtained from solar energy could readily be transferred to a fluid such as air or water. The fluid is heated by absorbing solar energy to high temperature, and contracts by liberating heat to low temperature. The difference between input heat and rejected heat goes into mechanical energy of expansion that would drive a turbine wheel or a reciprocating piston. These could be made to drive an electric generator to produce electricity. The entire process is limited by the Carnot efficiency.

When water is the working fluid, steam engines of the reaction type could be used (Farber and Prescott, 1973). The simplest case would be an engine in which the steam is ejected from at least one nozzle to rotate a wheel. For higher efficiency, the wheel should have a large diameter and should rotate at a very high speed. This in turn will create large friction losses limiting the efficiency. Higher efficiency could be obtained by condensing the rejected steam rather than discharging it in air. Another type of reactor steam engine uses a spherical shell of steel which is rotated by jets of steam. Both the boiler and the engine are combined in one single unit, for higher efficiency. However, such an engine would work until all water is vaporized, and then needs refilling.

When air is the working fluid, hot air engines of the Stirling type are most often used. A hot air Stirling engine is shown in Fig. 14.7. It works in a closed constant-volume cycle. It is made up of two cylinders connected at

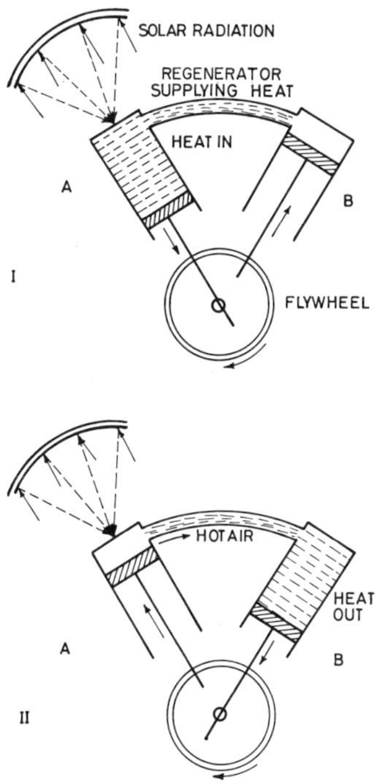

FIG. 14.7 An example of a Stirling solar hot air engine (from Daniels, 1964). I: First half cycle: piston A moves out, piston B moves in. II: Second half cycle: piston A moves in, piston B moves out.

their bottom for passage of air. In each cylinder there is a piston linked to a common flywheel. As air is heated in cylinder A, it expands and pushes out the piston that turns the flywheel clockwise. As the flywheel turns, it moves the other piston into cylinder B compressing air and forcing it through the link into cylinder A. Air is then further heated and expanded. Thus heat is absorbed at cylinder A and rejected at cylinder B, the walls of which should be cooled. When piston A reaches the end of its expansion stroke and piston B reaches the end of its compression stroke, the flywheel momentum reverses the direction of motion of the pistons. Piston B moves out and piston A moves in forcing the heated low-pressure air through the link into cylinder B; and the cycle goes on. Other variations of such an engine using single or multiple cylinders and different arrangements of the piston have been proposed. Engines working in a constant-pressure cycle could also be considered, while engines using an open thermodynamic cycle have also been proposed. Ambient air would be compressed to several atmospheres in a compressor. It is then expanded to subatmospheric pressures in the nozzle of an ejector, where it mixes with the high enthalpy air drawn in from the solar heater. The mixed air is compressed again in the diffuser section of the ejector and expanded through a turbine (Green, 1974).

The working fluid could be other than steam or air. Such a fluid should be chemically stable, noncorrosive, and cheap. It should also have high molecular weight and a small atomic number to allow for high efficiencies. Several fluids have been considered at one time or another; these are mercury, sulfur dioxide, freon and its compounds, monochlorobenzene (C_6H_5Cl), ammonia, and trichloromonofluoromethane, among others.

The working fluid should be capable of absorbing heat from the collectors and discharging it to the heat engine by one of three different methods:

(a) change in temperature,
(b) change in the physical phase, and
(c) change in chemical composition.

The first case is used in low temperature systems not involving any change in phase. An example is the Stirling cycle mentioned above, where the working fluid remains gaseous (air) during the entire cycle. The second case is the one used in the above mentioned steam systems. The change in physical state here is vaporization and condensation. The fluid is taken into the boiler as a liquid under pressure; it is vaporized by the addition of solar heat; it escapes as a high-pressure vapor to the heat engine, where it expands converting heat into mechanical work. The low-pressure vapor is discharged in the condenser to become liquid, and then pumped back to the boiler to close the cycle.

An example of the third case is given by the solar regenerative systems. Such a system would be made up of a dissociator to dissociate the working fluid and of fuel cell batteries for the fluid to recombine to produce electricity. Thermal dissociation can be a one-step process or a multistep process. An example of such a regenerative fuel cell is shown in Fig. 14.8. The working fluid is an MX_2 molten salt which is dissociated by electrolysis at high temperature in M(l) and X_2(g). Liquid metal is deposited on one electrode and gaseous X_2 at the other. These two materials are cooled by a heat exchanger and allowed to recombine in the fuel cell at lower temperature, producing an electric potential. A more pragmatic approach would be the dissociation of water at high temperature through a series of endothermic chemical reactions. In this case hydrogen could be produced, and become in itself an end-product fuel which would eventually be used for the production of electricity. A redox fuel cell uses an indirect regenerative process. It is an electrochemical generator on the electrodes of which react two oxydoreductor systems. Another type of regenerative fuel cell system is shown in Fig. 14.9 (Fuschillo *et al.*, 1966; Brosens, 1965; Gould, 1967).

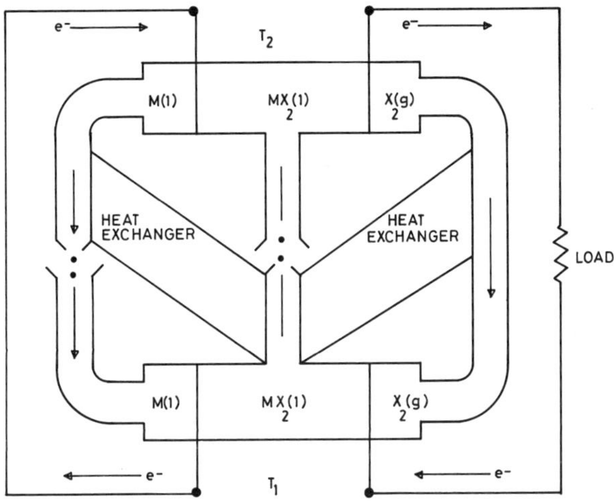

FIG. 14.8 Thermally regenerative system.

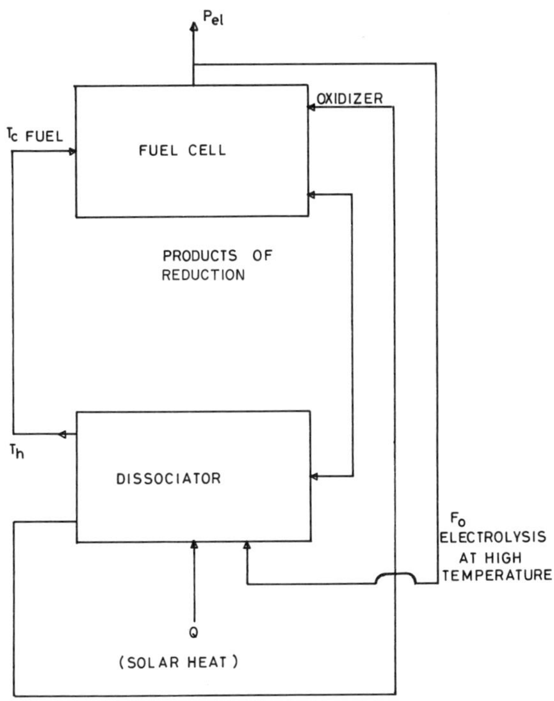

FIG. 14.9 A regenerative fuel cell system. The overall system is subjected to Carnot's limitation, since the dissociator is at a higher temperature than the fuel cell.

14.4 CONVERSION THROUGH OTHER STAGES

Solar energy could first be converted to mechanical or hydraulic energy before being converted into electricity. It could be converted to chemical or biological energy or any other form of energy.

A. Conversion of Solar Energy through a Mechanical Stage

Theoretically, no hydraulic energy can be tapped below sea level. However, there are many natural depressions around the world with an average depth below sea level. Admitting sea water into these depressions through hydroelectric turbines, and with a reasonable hydraulic head, electric energy may be generated. The difference between this plant and the one on a river is that in the latter water is disposed of by gravity to the sea; in the former, water is disposed of by evaporation resulting from the action of the sun. Basically, this concept which could be called heliohydroelectric (HHE) power generation utilizes the conversion of solar energy of evaporation to hydraulic energy, and then to electricity. A system of two reservoirs is required to establish the hydraulic head: one reservoir, an *infinite* constant level source of water; the other, a *sink* for evaporation (Kettani and Gonsalves, 1972).

There are two ways in which such a system may be developed and exploited. Natural depressions well below sea level exist in many parts of the world. Those in the proximity of open seas and situated in regions of abundant sunshine offer excellent prospects for one way to develop an HHE system. In this case, the *sink* is an "artificially created lake." Seas that are almost enclosed bodies of water offer a means of "artificially creating a depression" that would result in a situation similar to the above. This is done by building a dam to seal off a body of water from the open sea. The power produced by one of these two systems is given by

$$p_e = (g\eta)(\rho_i\lambda_s')(SH_e)$$

where g is the acceleration of gravity, η the efficiency, ρ_i the density of the incoming water, λ_s' the net evaporation rate, S the area of lake presented for evaporation, and H_e the hydraulic head. The area and the head are interdependent, when $H_e = 0$, S is maximum, and when H_e is maximum $S = 0$. Therefore, there will be a value of H_e for which the product SH_e is maximum. The head should be chosen at this maximum value.

One way of using solar energy, is by storing it in the form of hydraulic energy. There are many coastlines near the ocean in arid zones that could be used for hydraulic storage of solar energy. A dam could be built across a valley not too far away from the sea or a large body of water. Water could be pumped behind the dam by using solar energy. Electricity would then be produced by a conventional hydroelectric plant. This method has the advantage of avoiding the intermittence inherent in solar energy and of producing ac power directly. This system could be established in any valley, the water input should however be large enough to make up for losses by evaporation, which should be minimized (Seifert *et al.*, 1973).

The solar pump is not much different from a solar heat engine working in a low temperature cycle. The source of energy is the solar collector, and the sink is the water to be pumped. Figure 14.10 shows a schematic view of such an engine. Water flows within a solar collector; it is heated and due to the temperature gradient flows into a heat exchanger in the boiler and comes back to the collector. This water yields its heat to an intermediate fluid in the boiler. This fluid evaporates and expands in the engine before reaching the condenser where it condenses at low pressure. The condenser is cooled by the water to be pumped. The fluid is then reinjected in the boiler to close the cycle. The expansion engine is coupled to the pump and it could of course be coupled to an electric generator.

Solar energy could be converted into mechanical energy by using the thermal expansion of metals. Such an engine would consist of a metal tube mounted in a bearing at each end, and free to rotate about its axis. A flywheel would be fixed at the midpoint of the tube. When the upper surface of the tube is heated and the lower one is cooled, it will rotate. If solar energy is converted into mechanical energy, this energy could in turn be converted

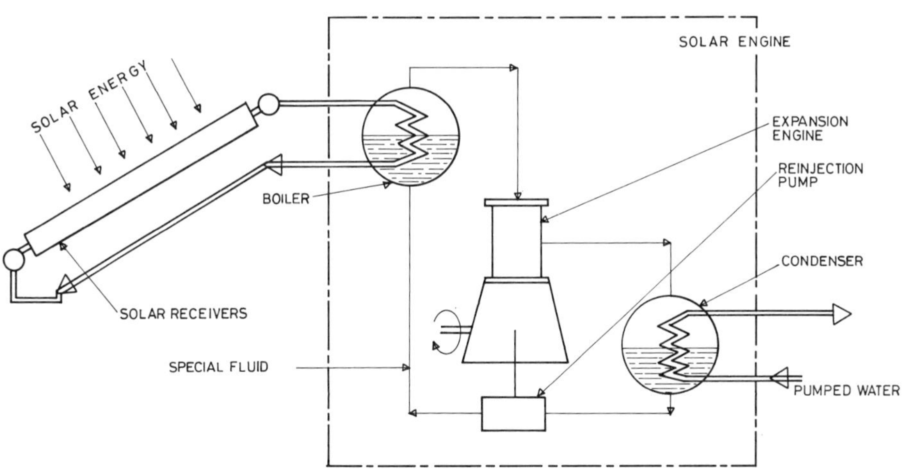

FIG. 14.10 A sketch of a solar pump.

into electricity through the piezoelectric effect. In asymmetrical crystals possessing a polar axis, a compression parallel to this axis polarizes the crystal. This effect is reversible. If first converted into mechanical energy, solar energy could be converted into electricity through the magnetostriction effect, or the electrokinetic effect, among other means.

B. Conversion of Solar Energy through a Chemical Stage

Basically, fossil fuels are the result of a photosynthetic process that took place millions of years ago. Fossil fuels are therefore no more than the storage of solar energy in a chemical form. The process could be reversed readily at high temperature yielding back the energy stored. Theoretically, it is possible to repeat such a process in the laboratory. Fuel could be synthesized on a large and much more efficient scale by photosynthesis. It could then be burned to produce electricity by the conventional methods. The chemical reaction of photosynthesis occurs in the presence of chlorophyll and can be summarized as follows:

$$CO_2 + H_2O + \text{light} \rightarrow H_2CO + O_2$$

The absorbed light is in the ultraviolet and infrared range. Visible light having a wavelength below 6800 Å is absorbed by the green chlorophyll which becomes activated and passes its energy on to the water molecules. A hydrogen atom is then released and reacts with the carbon dioxide molecule to produce H_2CO and oxygen. H_2CO is the basic molecule forming carbohydrates. Stable at low temperature, it breaks at high temperature releasing an amount of heat equal to 112,000 cal/mole.

$$H_2CO + O_2 \rightarrow CO_2 + H_2O + 112 \quad \text{kcal/mole}$$

The absorbed energy of the photons should be at least equal to this amount. There can be two photons of blue light or three photons of red light. But the chances of two blue photons or three red photons acting simultaneously on one single molecule are very slim. Therefore, the energy necessary for the release of the hydrogen atom from the water molecule is most often obtained from the excited chlorophyll molecules rather than directly from the sun.

In nature, photosynthesis is not optimized for energy conversion, natural optimization follows other criteria. In the laboratory, it is possible to produce one molecule of carbohydrate for every eight photons absorbed under optimum conditions. This corresponds to an efficiency of 30%. It is therefore possible to produce large amounts of carbohydrates by growing, say, algae under optimum conditions in plastic tubes or in ponds. The algae could be harvested, dried, and burned for production of heat that could be converted into electricity by conventional methods.

Photobiological effects could be used as a source of fuel. Methane has been produced by the anaerobic fermentation of algae. Algae could be grown in nutrient-loaded municipal sewage to produce methane as a result of their fermentation. Electricity could thus be produced on a large scale at a low cost. This method would in turn solve the problem of sewage treatment in large cities. The grown algae are first harvested, then treated to yield about 5% of solids, the rest being water. The solid material is subjected to anaerobic fermentation, and after about three weeks, half the organic material is converted into methane. This methane could be used as a regular fuel to produce thermal energy and eventually electricity (Golueke and Oswald, 1963a).

Electricity could be produced directly by using bioorganic matter as the source of fuel in a biochemical fuel cell. Air is injected at the cathode as an oxidizer and bioorganic matter at the anode as the fuel. Microorganisms are employed to catalyze the oxidation of the fuel. The electrolyte is usually an organic solution or an aqueous medium such as potassium hydroxide. The bacteria create new losses by consuming about half of the fuel. This consumption is necessary for their nutritional requirements. The electrical power output of a biochemical fuel cell is proportional to the bacterial metabolism rate. Theoretically, this rate may be extremely high, leading to biochemical performances similar to those of regular fuel cells.

C. Conversion of Solar Energy through Other Forms

Any solar-induced reversible reaction could be the basis of an energy conversion process. The energy involved should be equal to the energy necessary to excite a molecule, i.e., the energy should be in the 40 to 100 kcal/mol range. Therefore, only visible or ultraviolet light could thus be converted. The different types of primary photochemical processes could be included in the eight following classes:

1. intramolecular rearrangement of stable products;
2. photoisomerization;
3. photophysical processes;
4. photoionization;
5. decomposition of unexcited molecules;
6. photoconduction in solids;
7. photohomolysis or free radical formation;
8. electron transfer.

The first reaction could be ruled out since it is not usually reversible. The second is a low energy process, and could hardly be an efficient method of energy conversion. Photoionization is a high energy reaction, and could be applied to cesium for instance. Fluorescence is an example of a photophysical process that could be used for solar energy conversion. One of the two electrodes of a cell could be surrounded by a fluorescent material and subjected to solar light. This will result in a difference of potential between the electrodes.

Decomposition of unexcited molecules under the effect of light could be achieved by electron transfer or bond rupture resulting in the formation of free radicals. Electron transfer reactions occur most often under the effect of ultraviolet light. An example of such a reaction is the cerous–ceric electron transfer reaction with water. Hydrogen could be produced by this method, but since the reaction is very slow, the produced amount is negligible, even at very high solar energy concentrations.

Photovoltaic power generation is an example of photoconduction in solids. Photohomolysis is one of the most promising methods of solar energy storage, thus conversion. One such reaction is the photolysis of nitrosyl chloride which is both reversible and endothermic. Photogalvanic power conversion is a typical example of an electron transfer reaction (Swet, 1974; Szego *et al.*, 1972).

14.5 SOLAR ENERGY POWER SYSTEM

To obtain high power levels, it is necessary to connect individual cells in series and/or parallel. Connection in series gives a voltage equal to the sum of the individual voltages and the same current flows through the cells. Connection in parallel leads to a current equal to the sum of the individual currents, the overall output voltage remaining the same as for each cell. For both high current and high voltage levels, a combination of series and parallel connections in arrays becomes necessary.

Solar power generation systems should include concentrating mirrors and refrigerating arrangements, when necessary, to eliminate heat losses. This heat may eventually be used to generate more electricity.

Two basic problems are inherent to the conversion of solar energy: first, solar energy is dilute; second, it is available only during daytime. To solve the first problem, large areas of efficient photovoltaic cells should be deployed. As for the second problem, several methods are proposed in the case of photovoltaic power conversion. One possible solution is to store energy in one form or another during the day and then consume it during the night. A better solution is to make the solar power plant work in conjunction with conventional power plants. For instance, the sun could produce electricity during the day, whereas a hydroelectric power plant would function during the night only, thus leading to a large savings in the dam's water reserves.

To a large extent, HHE power plant would solve both problems and the energy produced would be dictated by demand and not by the energy available. The overall average of several years of consumption should equal the average energy capacity of the plant. HHE power generation affords therefore a unique way of storing solar energy and leveling out its diurnal and seasonal variations. Storage in the form of hydrogen, methane, carbohydrates, or water would solve all the problems altogether. It becomes necessary, however, to reach good storage capabilities and efficiencies.

One drastic solution to the above mentioned problems has been proposed by placing a photovoltaic converter system on a 50 km^2 satellite maintained in synchronous orbit, 36,000 km from the earth's equator. Thus, solar energy would be converted practically 24 h a day. The produced electricity would then be converted into microwave form and transmitted through a 1-km diameter antenna to a 7-km diameter antenna on earth. This antenna would convert the microwave energy back to electricity, thus generating up to 20,000 MW (Brosens, 1965). One of the basic advantages of solar energy is the fact that it is pollution-free, including global thermal pollution. In other words, converting solar energy does not upset the overall energy balance of the earth. However, a satellite solar power plant would eliminate this advantage, since energy that would never have reached the earth would be collected and transmitted in microwave form.

14.6 CONCLUSIONS

More efficient and cheaper solar converters should be produced if the sun is to become one day a potent source of electric energy. The efficiency of photovoltaic cells, one of the most developed solar energy converters, did not exceed 1% at the turn of the century, and remained a curiosity for a long time. The revival of interest in these cells in the 1950s brought this efficiency to 5%. Today the running efficiencies reach 18% for silicon cells, sometimes 20%, 14% for GaAs and high efficiencies for CdTe and CdS. The efficiency of solar heat engines barely reaches 20% and those of thermoelectric and thermionic solar cells rarely reach 16%. There is therefore room for increasing the efficiency dramatically.

The present cost of solar converters is still extremely high, or about \$20,000/kW for silicon photovoltaic cells, \$2000/kW for thermionic cells excluding the mirrors, \$2000/kW for a small HHE plant such as Dawhat Salwah. This cost could be brought down easily in many ways. For silicon photovoltaic power conversion, the use of much less material can be achieved by using thin films. New inexpensive techniques of cell manufacturing are being investigated. One of the most promising is dendritic crystal growing in which single silicon crystals are produced in a ribbon form. Use of concentrating mirror reflectors to minimize the number of cells would bring down further the cost of the installed kilowatt. For HHE power production larger schemes would be cheaper per installed kilowatt; for instance, the installed kilowatt of a Red Sea scheme is only $50. Such new techniques and automated production of the cells could bring the cost of the installed solar kilowatt down to $1000 in the near future or even less. If this is achieved, then the roofs of houses in sunny cities will produce all their energy needs, however industrialized they may be.

15

Storage of Solar Energy

M. ALI KETTANI

ELECTRICAL ENGINEERING DEPARTMENT
UNIVERSITY OF PETROLEUM AND MINERALS
DHAHRAN, SAUDI ARABIA

15.1 INTRODUCTION

Without storage, solar energy has to be used as soon as it is received. This is a serious handicap, because of the inherently intermittent nature of the energy received from the sun. Solar energy could be exploited along with other energy sources such as fossil, hydraulic, or nuclear. In such a hybrid system, the minimum needed power at any moment will often be larger than the maximum solar capacity available. The other sources would then be used whenever a deficiency in solar energy occurred.

For instance, steam turbines could be operated with water heated by both the sun and fossil fuels. A solar power plant could operate in conjunction with a hydroelectric plant. At every instant, the hydraulic turbines would admit enough water to produce the balance between the energy required and that produced by the solar plant. These arrangements would greatly increase the efficiency of the conventional plants.

Nevertheless, if solar energy is to be reliable, the need for storage cannot be avoided. Solar energy cannot be stored in its electromagnetic form. It must be first converted into another form. Assuming that the end product would eventually be electricity, solar energy could be stored after it is converted into electricity, or before.

If converted first into electricity, solar energy could be stored capacitively, i.e., in an electric field, or inductively, i.e., in a magnetic field. It could be more economical in some cases to convert electricity back to some other form, such as chemical, in which it would be stored. However, this extra electrical stage should generally be avoided.

Solar energy could be stored in the form of heat by heating a fluid or a large reservoir such as an aquifer, or by using the heat of fusion of a storing material.

Solar energy could be utilized to produce a fuel that could be used on demand. One such fuel, hydrogen, could replace gasoline in automobiles, airplanes, and other vehicles.

Solar energy could be used to pump water to a higher level and store it behind a dam to be used as hydraulic energy as required. Other ways of storage that might not now appear feasible could one day become the solution to the solar energy storage problem.

One important characteristic of a storage system is the length of time during which energy could be kept stored with acceptable losses. If solar energy is converted into a fuel such as hydrogen, there will be no time limit to the storage. Storage in the form of heat lasts for very short times because of losses by radiation, conduction, and convection. However, these losses could be minimized if thermal insulation is very good, such as in the case of an aquifer.

Another characteristic of a storage system is its volume per unit energy stored. The smaller the volume, the better is the storage system. Therefore, a good storage system should have a long storage time and a small volume per unit of stored energy (Clark, 1963; Inville, 1963).

15.2 ELECTRICAL STORAGE

Theoretically, capacitors could store large amounts of electrical energy for long periods. The total energy stored is

$$\xi_{\text{cap}} = \tfrac{1}{2}\mathscr{V}\varepsilon\mathbf{E}^2 \tag{15.1}$$

where $\mathscr{V}$ is the volume of the dielectric, ε its constant, and $\mathbf{E}$ the electric field strength. The latter is limited by the breakdown strength E_{br} of the dielectric. Therefore, the electric energy storable in a dielectric is limited. Table 15.1 shows the maximum energy storable per unit volume for several materials. At present, with the best dielectric materials available (mica), 3.4 m^3 of capacitor banks are necessary to store a mere kilowatt hour of energy. If energy were to be stored in impregnated paper, the volume would be 100 m^3.

Practically, because the conductivity of a dielectric is never nil, there will always be leakage losses. At present, capacitive storage is economical for times no longer than 12 h. To increase this time, much research is needed.

TABLE 15.1

Dielectric Materials

Material	ε_r	E_{br} (MV/m)	$\xi_{cap}/\mathscr{V}$ (kWh/m^3)
Mica	6.0	200	0.2945
Amber	2.7	90	0.0269
Quartz	5.0	40	0.00983
Teflon	2.1	60	0.00931
Paper (impregnated)	3.0	50	0.00922
Polyethylene	2.3	50	0.00706
Glass	6.0	30	0.00664
Titanium dioxide	100.0	6	0.00444
Bakelite	5.0	25	0.00383

Therefore, unless better and cheaper dielectric materials for storing electric energy are found, capacitive storage on a large scale will remain uneconomical.

Whereas capacitors store electrical energy at high voltage and low current, inductor storage is at low voltage and high current. The energy stored in an inductor is

$$\xi_{ind} = \tfrac{1}{2}\mathscr{V}\mu H_m^{\,2} \tag{15.2}$$

where μ is the permeability of the material, and H_m is the magnetic flux density.

For ξ_{ind} to be large, both μ and H_m should be large. Consequently, high magnetic fields are required. This will create large mechanical forces which should be supported by strong structures. The reverse operation of discharging the stored energy creates another problem, since it involves the opening of circuits carrying large currents.

Although energy densities in inductors can be 100 times higher than those in capacitors, large inductive storage is still impractical. Only storage on a small scale and for special applications is common. One main reason is that materials having high permeabilities also have large electrical resistivities, and since the energy stored is at a very high current, Joulean losses become prohibitively high. With advances in superconducting materials, this problem might be eliminated, thus allowing storage of large amounts of energy for long times. Some special magnetic alloys are listed in Table 15.2, along with the value of their maximum relative permeability and their resistivity.

Electrical energy could be stored in chemical form in the primary cell. This is the simplest storage device. It has no moving parts, its efficiency is high, and its output is in the form of electricity. It comprises two electrodes

TABLE 15.2

Magnetic Materials

Material	Maximum relative permeability	Resistivity (10^{-6} Ω-cm)
Supermalloy	800,000	60
Mumetal	110,000	42
78 Permalloy	100,000	16
1040 alloy	100,000	56
4-79 Permalloy	80,000	57
Hipernik	80,000	35
4750 alloy	32,000	45
45 Permalloy	25,000	50
Permendur	5000	7

(anode and cathode) and an electrolyte, which is an ionic conductor (Ewe, 1972; Wiederhold, 1963).

The energy stored by these cells is, however, very low for a given volume and the storing cost is expensive. Present research work is directed toward the improvement of the energy-to-weight ratio, compactness, and maintenance-free scaled batteries. The characteristics of some primary cells are shown in Table 15.3.

TABLE 15.3

Primary Cells

Cell type	Cell system	Open circuit C voltage (*V*)	Ampere hours	(Wh/kg)	(kWh/m^3)
Mercury	Zn–KOH–HgO	1.35	14	100	360
Silver–zinc	Zn–KOH–AgO	1.86	11	170	360
Air Cell alkaline	Zn–NaOH–O_2(air)	1.50	14	200	310
Air Cell neutral	Zn–NH_4Cl–O_2(air)	1.50	14	200	310
Lalande	Zn–NaOH–CuO	1.06	18	110	270
Manganese	Zn–KOH–MnO_2	1.50	8	76	210
Mercury–carbon	Zn–$ZnSO_4$–Hg dioxysulfate	1.36	8	76	200
Organic depolarizer	Mg–$MgBr_2$–*m*–dinitrobenzene	1.60	6.5	88	160
AgCl reserve	Mg–H_2O–AgCl	1.70	2.5	110	150
Mg dry cell	Mg–$MgBr_2$–MnO_2	2.05	4.5	66	130
CuCl reserve	Mg–H_2O–CuCl	1.60	2.5	110	120
Leclanché	Zn–NH_4Cl–$ZnCl_2$–MnO_2	1.65	5	57	120

15.3 STORAGE IN THE FORM OF HEAT

After being converted into heat, solar energy could be collected by any fluid such as air or water. Special organic fluids such as freon are used in closed systems. For this storage to be effective, losses of heat by radiation, conduction, or convection should be minimized.

Thermal storage often uses the heat of fusion of the storing material. The heat absorbed from solar radiation melts the material and it is stored as fusion heat. A system using this concept is shown in Fig. 15.1. This unit is devised for usage in space. It is a two-zone system, one at a lower temperature, the other at a higher temperature for the superheat and reheat storages. For low temperature (680°C), the heat-storing material is a mixture of lithium hydride (95%) and lithium metal (5%). For the high temperature stage (1024°C), the heat-storing material is sodium fluoride. Mixtures of sodium fluoride and sodium metal have also been considered for storage of heat at high temperature. Heat could be stored for hours in nonboiling fluids such NaK or mercury (Schroder, 1974; Telkes, 1955).

A good material for heat storage should have high values of heat of fusion and an acceptable fusion temperature. The material should be stable within the recycling range, i.e., it should not undergo supercooling, sedimentation, or phase change. Furthermore, the material should not be toxic or corrosive.

The phase change between ice and water (330 kJ/kg) has been proposed for storage, but its temperature level is low (0°C). The phase change of

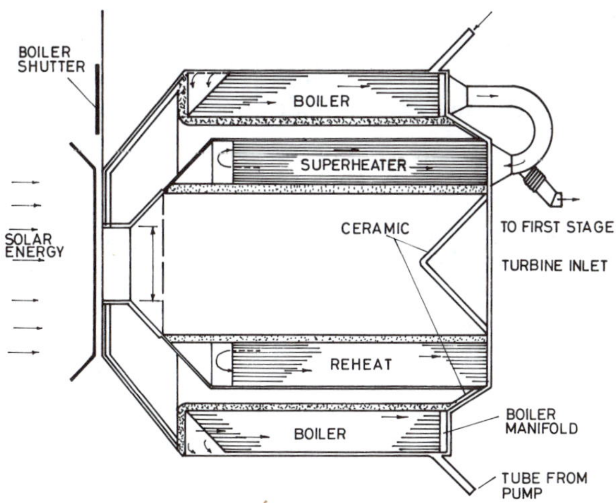

FIG. 15.1 Cross section of boiler, superheater, and heat storage.

$Na_2SO_2 \cdot 10H_2O$ has also been proposed (243 kJ/kg) mainly for storing domestic heat. The best materials proposed belong to a family of eutectics, which are stable fluorides of the alkaline earths. They have fusion heats ranging between 600 and 900 kJ/kg at temperature ranging between 450 and 850°C. Heat can be stored for a long time in these materials because of their low thermal conductivity (0.01–0.02 cal/cm s°C), and the cost of storage per kilowatt hour is relatively low. Table 15.4 gives the heat storage capacity of several materials to be compared with the values of Tables 15.1 and 15.3.

TABLE 15.4

Heat Storage

Material	Method of storage	Temperature level (°C)	Storage capacity (kWh/m^3)
Water	Phase change	0	92.6
Water	Sensible heat	20–90	81.3
Rock	Sensible heat	20–150	83.3
Eutectic Na/K/Mg/Li–F	Sensible heat and phase change	150–450 449	613.5
Eutectic Na/Ca/Mg–F	Sensible heat and phase change	150–850 745	1063.8

A system using such a solar energy storage method has been proposed by Taylor (1975) for the European Southern Observatory at La Silla, Chile. The system uses several sources of energy for charging the heat batteries: solar, wind, and fossil fuel. This system is shown in Fig. 15.2.

To store solar heat on a large scale, it has been proposed to inject hot fluids into deep permeable rock strata through wells. It has been calculated

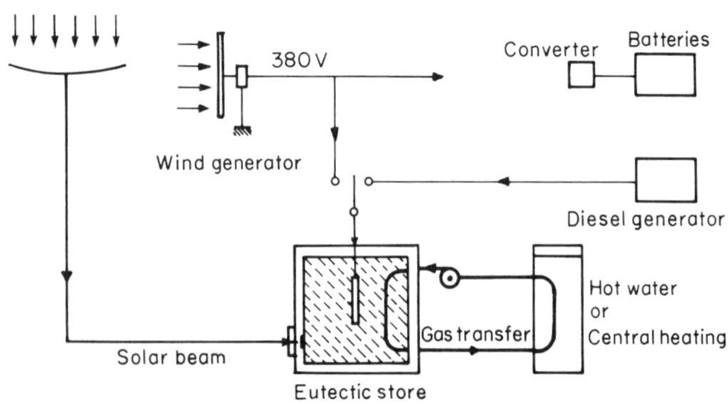

FIG. 15.2 Layout of the thermal storage system.

that, solar heat stored in such underground aquifers could be retrieved with losses barely reaching the one percent level (Collins and Davis, 1975). After steady state is reached, the temperature of the retrieved fluid would essentially be equal to that of the injected fluid.

Such a system, evidently using water as a working fluid, would have the advantages of storing solar heat at low cost in large amounts, and with long storage times. However, the choice of a site has to be dictated by compatibility of geological and insolation requirements.

15.4 STORAGE IN THE FORM OF FUEL

It is possible to devise a storage battery in which the reactant is regenerated by a photochemical reaction. In this case the converter itself acts as a storage battery. The battery is charged photochemically and discharged electrically whenever needed. However, there is competition between electrochemical transfer of electrons and their direct chemical transfer. This could be alleviated by choosing a small interelectrode space and a fast electrochemical reaction. Some of the reactions that could potentially be useful for the storage of solar energy are listed in Table 15.5 (Broda, 1975; Golucke and Oswald, 1963b).

Each one of these reactions has some shortcoming that should be overcome before it could be considered for energy storage: the rate of the reverse reaction is slowed down when one of its products is a precipitate; side

TABLE 15.5

Reactions for Solar Energy Storage

Overall reaction	Type of reaction	Maximum effective wavelength (Å)	ΔG° (kcal/mol at room temperature)	ΔH (kcal/mol at room temperature)
$NO_2 \rightarrow NO + \frac{1}{2}O_2$ (in CCl_4)	homolysis	4350	8.3	13.5
$NOCl \rightarrow NO + \frac{1}{2}Cl_2$ (in CCl_4)	homolysis	6370	4.8	9.0
$AgCl(s) \rightarrow Ag(s) + \frac{1}{2}Cl_2$	homolysis	4050	26.2	30.4
$AgBr(s) \rightarrow Ag(s) + \frac{1}{2}Br_2$	homolysis	4600	22.9	23.8
$AgI(s) \rightarrow Ag(s) + \frac{1}{2}I_2$	homolysis	4400	15.9	14.9
$2Fe^{++} + I_3^- \rightarrow 2Fe^{+++} + 3I^-$ (in H_2O)	electron transfer	5460	10.7	−8.6
$NO_2^- + H_2O + I_2 \rightarrow NO_3^- + 2HI$ (in H_2O)	electron transfer	5790	11.0	7.0
$H_2O + \frac{1}{2}O_2 \xrightarrow{ZnO} H_2O_2$	electron transfer	4000	23.1	12.1
2 anthracine → dianthracene	polymerization	3800	—	15 6

reactions might be present; the back reaction may be too fast; the compound may be insoluble; the energy storage might be small; or the exciting light might have too small a wavelength for the reaction to occur in the terrestrial atmosphere.

For a reaction to be seriously considered for solar energy storage, the following conditions should be met:

(1) The reaction should be reversible.

(2) The reactant, with or without addition of photosensitizer, should be able to use as much of the solar spectrum in the terrestrial atmosphere as possible.

(3) The energy stored in the reaction should be large enough, say of the order of 60 Wh/kg at least.

(4) The reactants used should be cheap.

There are many forms in which solar energy could be stored through these reactions. The products of the reaction could be separated and stored; an example is the formation of oxygen and hydrogen. In this case the reaction does not have to be reversible, since hydrogen recombines readily with oxygen in the air to produce thermal energy and water. Hydrogen is one of the most efficient and practical fuels.

If hydrogen is obtained from water, the economics of its production could become very attractive. This could be achieved in many ways: by photochemistry; by electrolysis, by dissociation at high temperature; or by a combination of two or more of the above methods (Abdel-Aal, 1975; Funk and Reinstrom, 1966; Veziroglu and Kakaç, 1975).

Water is dissociated at room temperature yielding hydrogen and oxygen, but the maximum effective wavelength, 1800 Å is too small for the reaction to occur on earth. Water could be dissociated through the following chain of reactions.

$$\begin{aligned} 2Ce^{4+} + H_2O &\rightarrow 2Ce^{3+} + 2H^+ + \tfrac{1}{2}O_2 \\ 2Ce^{3+} + 2H^+ &\rightarrow 2Ce^{4+} + H_2 \end{aligned} \tag{15.3}$$

The first reaction occurs at a maximum effective wavelength of 4000 Å and the second at 2537 Å, which is too short. The increase in enthalpy amounts to 68.4 kcal/mol at room temperature. Experiments have been run to produce hydrogen by this method in the apparatus shown in Fig. 15.3. However, the quantum yield did not exceed 0.13%.

Water can be cracked by adding work, or heat, or both. If solar energy is to be converted into heat for the cracking of water, extremely high temperatures would be needed. If thermal energy is used alone, a temperature of 5175°C would be required for complete cracking of water at atmospheric pressure. If solar energy is first converted into electricity, water could be

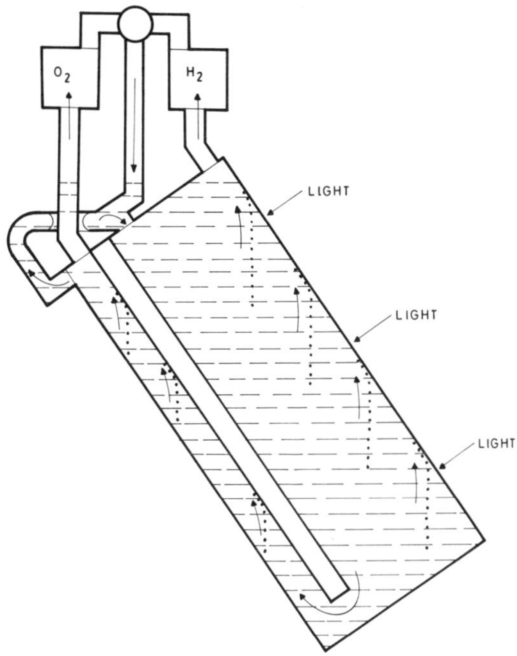

FIG. 15.3 A device for separately producing hydrogen and oxygen.

electrolyzed to produce hydrogen. A hybrid arrangement might be attained by electrolyzing water at an acceptable temperature, say 1500°C. At this temperature, 71% of the needed energy would be electrical and 29% thermal.

At present, electrolysis remains the most efficient method of water decomposition; potentially it is only about 80% efficient. A multistep process such as that of reaction (15.3) could be more efficient. Such multistep processes are divided into two categories: oxidation reactions

$$\begin{aligned} H_2O + X &\rightarrow XO + H_2 \\ XO &\rightarrow X + \tfrac{1}{2}O_2 \end{aligned} \tag{15.4}$$

and hybrid reactions

$$\begin{aligned} H_2O + X &\rightarrow XH_2 + \tfrac{1}{2}O_2 \\ XH_2 &\rightarrow X + H_2 \end{aligned} \tag{15.5}$$

Hydrogen can also be obtained by homolysis, such as in the reactions

$$2HBr \rightarrow Br_2 + H_2 \tag{15.6}$$

$$2NH_3 \rightarrow N_2 + 3H_2 \tag{15.7}$$

$$2HF \rightarrow F_2 + H_2 \tag{15.8}$$

Reaction (15.6) occurs at a maximum effective wavelength of 2537 Å, has $\Delta G^\circ = 12.7$ kcal/mol, and $\Delta H^\circ = 8.7$ kcal/mol at room temperature; it cannot be used on the earth's surface without a sensitizing agent, and its energy storage per unit volume is small. Reaction (15.7) occurs at an optimum 2537 Å as well, has $\Delta G^\circ = 4$ kcal/mol, and $\Delta H^\circ = 11.0$ kcal/mol; it needs a sensitizing agent and has a small storage capability. Reaction (15.8) occurs at 2800 Å, has $\Delta G^\circ = 64.7$ kcal/mol, and $\Delta H^\circ = 64.2$ kcal/mol; it also needs a sensitizing agent.

Multistage reactions where water is dissociated have been proposed. In one of these sequences the dissociation occurs at 730°C with a 50% efficiency. This sequence is

$$2H_2O + CaBr_2 \rightarrow Ca(OH)_2 + 2HBr \tag{15.9}$$

$$2HBR + Hg \rightarrow HgBr_2 + H_2 \tag{15.10}$$

$$HgBr_2 + Ca(OH)_2 \rightarrow CaBr_2 + HgO + H_2O \tag{15.11}$$

$$HgO \rightarrow Hg + \tfrac{1}{2}O_2 \tag{15.12}$$

This reaction is of the oxidation type, since it could be reduced to the two reactions

$$\begin{aligned} 2H_2O + CaBr_2 + Hg &\rightarrow Ca(OH)_2 + HgBr_2 + H_2 \\ HgBr_2 + Ca(OH)_2 &\rightarrow CaBr_2 + Hg + \tfrac{1}{2}O_2 + H_2O \end{aligned} \tag{15.13}$$

Solar energy could be used by the anaerobic fermentation of algae for the production of methane (CH_4). Methane is an excellent fuel, which is stable at room temperature and which reacts with oxygen to provide high temperatures, releasing the stored energy in thermal form (Pivic, 1958).

$$CH_4 + 2O_2 \rightarrow 2H_2O + CO_2 \tag{15.14}$$

Solar energy has been converted into chemical energy of methane with a 2% efficiency. Therefore, one square kilometer of algae fields could produce an amount of methane storing 4 MW of converted solar energy. Algae could be grown in nutrient-loaded municipal sewage to produce methane. The grown algae would first have to be harvested, then treated to yield about 5% of solids, the rest being water. The solid material would then be subjected to anaerobic fermentation and after three weeks, half the organic material could be converted into methane.

Some photochemical reactions have methane as an end product. One such reaction is the dissociation of H_3CHO under the effect of light, producing methane and carbon dioxide. This reaction is not reversible. However, since methane reacts with oxygen in air through reaction (15.14), this is not necessarily a major problem. Nevertheless, carbon monoxide is poisonous, which might be a shortcoming.

Other reactions might lead to other fuels (such as hydrazine, NH_2NH_2), unsaturated hydrocarbons (such as ethylene), or saturated hydrocarbons of higher order than methane (C_nH_{2n+2}).

Photosynthesis has been mentioned as a method of solar energy conversion. The products of the reaction are oxygen and H_2CO which is a fraction of carbohydrates (Neuwirth, 1958).

$$CO_2 + H_2O \rightarrow H_2CO + O_2 \tag{15.15}$$

in the presence of light and chlorophyll. The carbohydrates are stable at room temperature; but at high temperature the reaction is reversed, releasing the stored energy in thermal form.

The quantum yield is defined as the ratio of the number of molecules reacting over the number of photons absorbed. In nature, this reaction is not optimized for maximum storage of energy, and photosynthesis in agriculture has usually a productivity of about 800 tons of dry plant material per square kilometer per year. One could calculate that with a quantum yield equal to unity, it is potentially possible to produce plentiful quantities. It should be noted that at a maximum effective wavelength of 1470 Å, the following reaction is possible

$$CO + H_2 \rightarrow H_2CO \tag{15.16}$$

with $\Delta G^\circ = 6.5$ kcal/mol and $\Delta H^\circ = -1.3$ kcal/mol at room temperature. This wavelength is however too small for the reaction to occur without a sensitizing agent. Furthermore, the amount of energy stored is relatively small. On the other hand, it is far more economical to let hydrogen react with oxygen in air to produce energy. Consequently, the photosynthetic reaction seems to be the only practical way of storing solar energy in carbohydrates.

15.5 STORAGE IN THE FORM OF POTENTIAL HYDRAULIC ENERGY

One way of storing solar energy in the form of hydraulic energy is through heliohydroelectric (HHE) power conversion. To a large extent, the energy produced by an HHE power plant would be dictated by the demand and not by the energy available. The average over several years of consumption should equal the average energy capacity of the plant. More specifically, consider Fig. 15.4 representing the Red Sea basin capacity versus the hydraulic head. From this figure, the optimum production rate of the HHE plant is 5.7×10^{11} kWh a year. Energy could be produced at, say, ten times this nominal rate, reaching 5.7×10^{12} kWh a year. HHE power generation therefore affords a unique way of storing solar energy in hydraulic form.

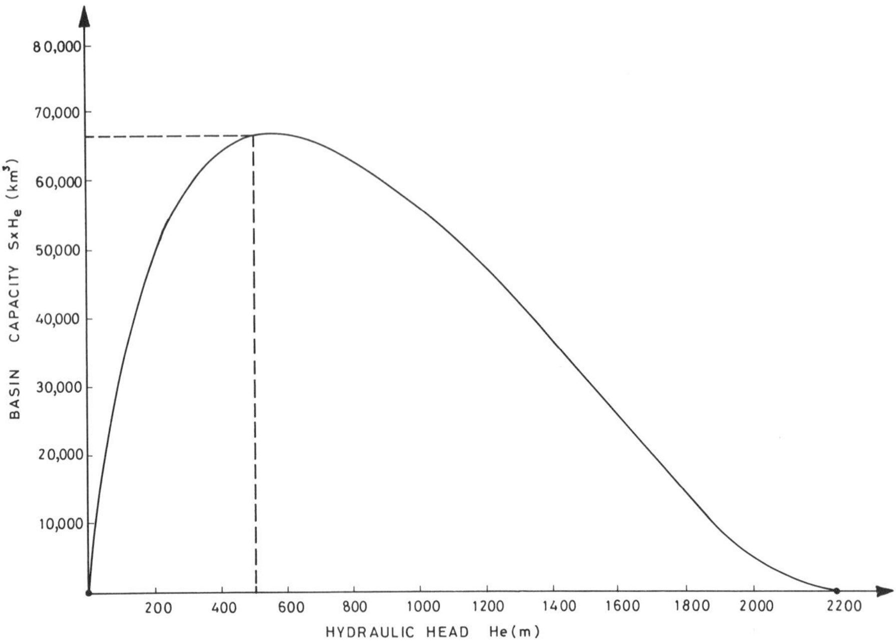

FIG. 15.4 Optimum head for maximum power in the Red Sea reservoir. Its value is 500 m.

However, this is not the only way. In many countries around the world, there are high mountains near shorelines of open seas. Depending on the topography of the region, solar energy could be stored in hydraulic form in many ways. First, consider the case of a natural depression near a shoreline. One means of bringing water from the open sea to the depression is by digging an open channel across rocks and hills. However, this method would often be extremely expensive. Consider the Qattara Depression in the Western Desert of Egypt. The equivalent surface area of this depression at sea level is about 19,500 km^2, and the average depression depth is 60 m, with the lowest point at 134 m. Admitting seawater from the Mediterranean to the depression through hydraulic turbines would produce electric energy. Many proposals have been put forward for bringing this water across an 80 km distance, reaching at its highest point an altitude of 230 m. It was first proposed to dig a 10-km-long open channel followed by two 16-m-diameter linked tunnels, each 70 km long, the water flowing practically at sea level (from 0 to 6 m below sea level). Another proposal required the nuclear excavation of a sea level open channel 70 km long. A better solution seems to be to pump seawater to, say an altitude of 200 m. The energy used for pumping would be retrieved (minus the losses) as the water falls in the depression. The capital cost of the pumping station would certainly be much below the cost of digging open channels or tunnels (average flow 650 m^3/s).

The same method could be used for the exploration of such depressions as the Dead Sea (bringing seawater from the Red Sea or the Mediterranean), the Caspian Sea, Death Valley (California), Salinas Grandes (Argentina), Djerid-Melrhir (Algeria), or Lake Eyre (Australia). In this case there is no need for direct solar pumping. At a first stage, electric energy from another source could be used to start pumping water from the sea. This source could be a very small one; as water starts falling in the depression, the energy produced could be used entirely for more pumping until the steady state is reached. The original exciting source of power could then be disconnected, the energy feedback would be maintained for the pumping of the needed 650 m^3/s average flow. A schematic arrangement to this approach is shown in Fig. 15.5 (Ball, 1933; Garbell, 1963; Kettani and Gonsalves, 1972).

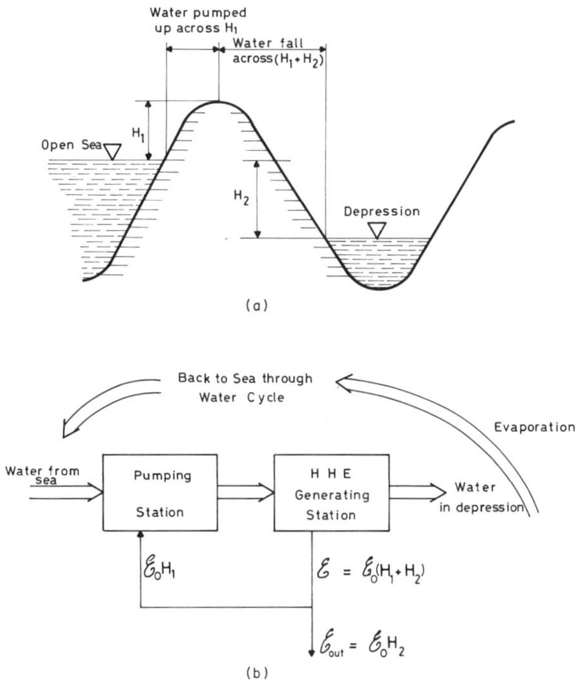

FIG. 15.5 Solar energy storage in an HHE system. (a) Cross section of the system. (b) Block diagram of the operation representing the energy feedback for the pumping station. To start operation, this feedback is made positive.

There are many coastlines near oceans in arid zones that could be used for hydraulic storage of solar energy. A dam could be built across a valley not too far away from the sea (or a large body of water). Seawater could be pumped behind the dam by using solar energy and electric energy would then be produced by a conventional hydroelectric plant. This method of solar energy storage has the advantage of avoiding the intermittence inherent to solar energy and of producing ac power directly. One could spot hundreds of locations around the world for such a scheme: along the shores of the Mediterranean, the Red Sea, the Pacific, etc. This concept is schematized in Fig. 15.6 (Kettani and Lam, 1974; Kettani and Scott, 1974; Kettani, 1975).

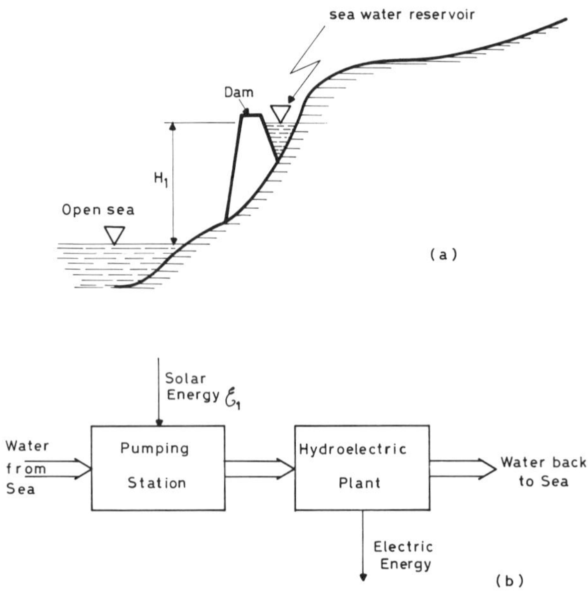

FIG. 15.6 Storage of solar energy by pumping water behind a dam. (a) Cross section of the system. (b) Block diagram of the operation.

In this scheme, water acts only as a working fluid. Once the initial volume of water is secured, the only flow needed would be that necessary to make up for the loss of water by evaporation (and seepage). Therefore, the scheme does not have to be near large bodies of water.

The technique of solar water pumping is still in its infancy. But recent developments show that there is no inherent technological limitation to solar pumping on a large scale. The solar pump is no more than a solar heat engine working in a low-temperature cycle. The source of energy is the solar collector and the sink is the water to be pumped. Water flows within a solar heat collector, it is heated and due to temperature gradients, it flows into a heat exchanger in the boiler before coming back to the collector. This water yields its heat to an intermediate fluid in the boiler. The fluid evaporates and expands in the engine before reaching the condenser where it condenses at low pressure. The condenser is cooled by the water to be pumped and the fluid is then reinjected in the boiler to close the cycle. The expansion engine is coupled to a conventional pump.

15.6 CONCLUSION

One could think of many other ways in which energy could be stored, at least temporarily. Solar energy could be converted into mechanical energy and stored in the inertia of a gyrating body. Heat energy from the sun could be stored in any one of the physical processes reviewed for possible power conversion, such as ferroelectricity, pyroelectricity, and ferromagnetism. It could be stored in the heat of evaporation of a material, or in its thermal expansion. Mechanical energy could be stored in a piezoelectric material as well as a flywheel in the form of shaft power.

Nature could be the best model in this respect. Indeed, plants photolyze water, and use hydrogen to cover all their energy requirements. If the reducing and the oxidizing products are separated by a membrane as in photosynthetic cells, hydrogen could be produced as a fuel by the use of solar energy.

Proteins are also a form of solar energy storage. This brings us to a concept that should remain clear to the scientist. Is energy stored in any form good for burning as a cheap fuel? It is obvious that proteins locked in the meat of sheep and fish are better used as food, i.e., fuel for our own life-sustaining natural energy conversion scheme, than burned as charcoal for the running of a steam engine! Is it not as obvious that oil, this rich mixture of hydrocarbons that took millions of years to be formed, could well find a better usage than being burned as a fuel? Should we wait for an energy crisis to stop wasting such a wealth while such a free source of energy as the sun is just waiting to be tapped?

16

Refrigeration and Air Conditioning

B. J. BRINKWORTH

SOLAR ENERGY UNIT
DEPARTMENT OF MECHANICAL ENGINEERING AND ENERGY STUDIES
UNIVERSITY COLLEGE
CARDIFF, UNITED KINGDOM

16.1 INTRODUCTION

Of all the possible applications of solar energy, its use for cooling seems the most fitting. This is one of the few uses in which the supply of energy and the demand for it are closely matched. In the distillation of water for drinking and the pumping of water for irrigation, the greatest need also occurs during periods of greatest insolation, but with cooling requirements, the matching extends in many situations even to the variation within the day.

In developed countries, a society has grown up that is largely urban and is sustained by an industry having high productivity, based on the intensive use of energy. The refrigeration of foodstuffs and many other commodities is a normal part of its commercial and domestic life. Air conditioning of homes, factories, and offices has become sufficiently extensive to have disturbed the usual pattern of demand for electricity, causing the peaks to occur on summer afternoons and evenings, and even at weekends. Cooling is thus beginning to appear as a significant element in the energy economy. With the realization of the vulnerability of fuel supplies and an awakening to the need to conserve fuels in the face of their rapid depletion has come a new interest in the utilization of solar energy, with the prospect of solar cooling a tempting prize.

Even in the era of cheap fuel, solar cooling was considered a worthy objective. A British NPL report (Anonymous, 1952) on the utilization of solar energy included the development of air conditioning equipment among the few topics felt to be worth pursuing at that time. It appears that the authors had in mind its application in the developing countries in particular. Today, the needs in those countries spread over a wide spectrum (Anonymous, 1974b). In urban areas the need may not be very different from those of the developed countries, though the opportunities for meeting those needs will often be affected by lower availability of electrical power and water supplies. The bulk of the population, however, is widely dispersed so that neither the generation of power at central sites and reticulation to users, nor the transport of fuels for local use can be economic. The very limited capital available for the extension of communications and services into rural areas has been further reduced by the burden of increased fuel prices. Many of the developing countries are without indigenous fuel supplies or lack the resources to exploit them, yet these are the countries which could benefit most from refrigeration and air conditioning. The high ambient temperatures, often combined with humidity, lead to the spoiling of much food by microbiological actions which could be slowed down by quite modest cooling. Air conditioning could lead to a significant increase in industrial and commercial productivity in urban areas and to greater effectiveness in social advancement through education and medical care. The problems of establishing viable rural industries and community life would be somewhat eased, and areas in which the climate is at times totally enervating would be made more endurable.

Although the need for cooling is due everywhere to excessive insolation, the problems are only superficially similar. Quite different constraints will apply from place to place; for example, in the availability of electricity and water which determines the kind of equipment that may be used. Moreover, the economic scale varies very widely. A householder in a developed country might be willing to pay $4000 for a solar air conditioner, but in another place such a sum would be an impossible burden for an entire community. The opportunities for local manufacture, or even the assembly of prefabricated kits, will vary widely with the availability of materials, technical skills, equipment and tools. These constraints will have a major influence on the engineering of solar cooling. Though the rest of this chapter deals only with technical factors, it is not inappropriate to begin with a reminder that social factors are often equally important.

16.2 REQUIREMENTS

Typical requirements form the background against which the practicability of solar cooling has to be judged, and these are now reviewed briefly.

A. Refrigeration

Many foodstuffs, including fruits, vegetables, meats, fish, and dairy produce, can be maintained in a fresh state for periods of the order of weeks at temperature of 0 to 3°C. For longer-term storage, deep freezing at −18 to −25°C is necessary. In most instances, these conditions are obtained by the circulation of air or chilled brine, cooled by vapor-compression refrigerators (Stoecker, 1958).

Food storage and transit take place in chambers covering a wide range of sizes, from central cold rooms through freight cars to retail storage and display cabinets and domestic refrigerators. Evidently there is an economy of scale in the amount of cooling required, since the ratio of surface area to volume decreases as the size of the chamber increases. The continuous rate of extraction of heat accordingly varies from as little as 5 W/m^3 for a large commercial store to over 500 W/m^3 for a small domestic deep-freeze cabinet. In the latter case, the operation is usually intermittent, and the peak capacity of the unit may be an order of magnitude greater.

A unit commonly employed in the rating of cooling equipment is the refrigeration ton. This is defined as the rate of extraction of energy which would cause water at 0°C to be converted into ice at 0°C at a rate of one ton per day. It is equivalent to about 3.52 kW (12,000 Btu/h).

B. Air Conditioning

Air conditioning is the provision of a comfortable indoor environment by replenishing, heating, cooling, humidifying, or dehumidifying the atmosphere. In all real situations this atmosphere is a mixture of noncondensible gases and a vapor: air and steam. The state of this mixture thus requires three properties for its definition, one of which is usually the moisture content. This is defined numerically as the mass of steam per unit mass of air, but it may conveniently be represented in terms of the relative humidity (RH), the ratio of the partial pressure of steam to the corresponding value for saturated air at the same temperature. Because of the effect of barometric pressure, the RH is not exactly equal to the ratio of the moisture content to its value for saturated air, though the difference is slight. The properties of moist air are embodied in psychrometric charts and tables of various kinds (Anonymous, 1971a) such as that outlined in Fig. 16.1.

An alternative method of expressing the moisture content is in terms of the wet bulb temperature, WBT. This is the temperature indicated by a thermometer whose bulb is kept covered with a film of water. When the air is caused to move relative to the bulb, heat and mass transfer produce an equilibrium temperature which is determined by the air temperature and

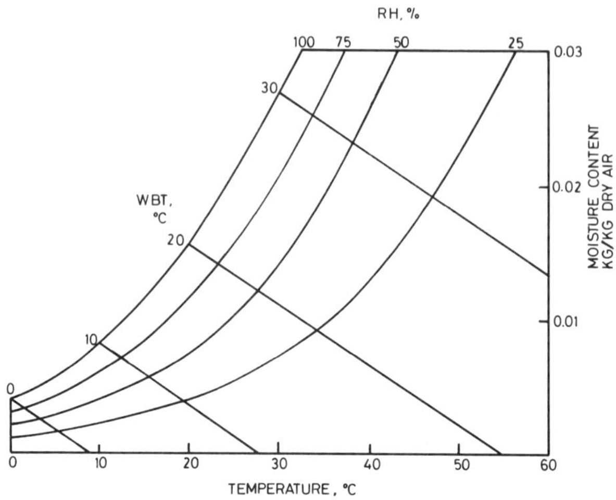

FIG. 16.1 Psychrometric chart.

the RH. Thus the difference between the temperature and the WBT is related to the RH. In American practice, use is more often made of the adiabatic saturation temperature, which is the equilibrium value that would be reached if the air were maintained in contact with free water at the same temperature. It so happens with the air–steam system that the WBT and the adiabatic saturation temperature are nearly equal (Jones, 1973).

Comfort is perceived subjectively, and the relationship between comfort and the thermodynamic state of the environment cannot be fully defined. The human being maintains a surface temperature of around 33°C by heat and mass transfer with the surroundings, and comfort is evidently related to the ease with which this can be done. As a result only a narrow range of environmental temperature, say 18 to 27°C, is considered to be comfortable. The tolerance for humidity is much wider (about 25–60%), with a tendency for lower RH to be preferred at higher temperatures, presumably to assist heat loss by evaporation. Boundaries of temperature and RH, within which most people would feel comfortable, are drawn differently by different authorities. A consensus of the requirements for continuous occupation and light activity would be roughly as shown in Fig. 16.2.

Conditions well outside these boundaries can, and often must, be tolerated. Outdoor air temperatures in excess of 50°C occur and 100% RH is common in coastal regions (Anonymous, 1972b). Sheridan (1961), writing of Australia, draws attention to the widely differing climates of coastal and inland areas. Near the coast the average maximum temperature varied during the year between 25 and 32°C, with RH between 60 and 70%; although

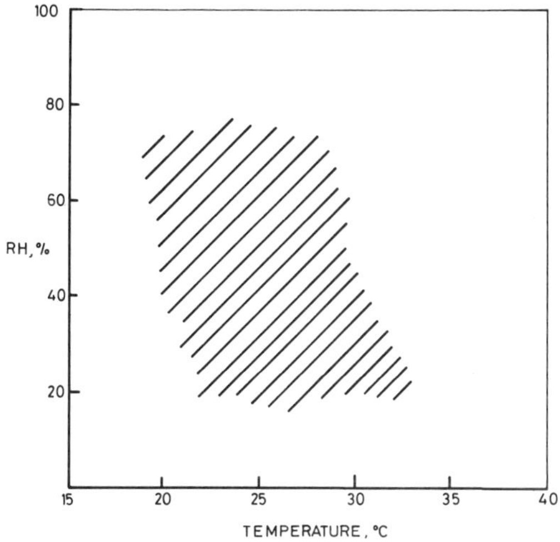

FIG. 16.2 Region of human comfort.

the corresponding range inland was from 24 to 43°C, the RH was only 20 to 30%. For indoor conditions, the recommended design values of the IHVE Guide (Anonymous, 1971a) for tropical summer climates extend up to 25.5°C at 70% RH and 30°C at 40% RH, for transient exposure. However, there is evidence of a loss of efficiency and increased proneness to accidents if conditions such as these are prolonged (Billington, 1967–1968). A drop in industrial output, typically of the order of a few per cent per °C, occurs above approximately 20°C, reaching about 50% at 43°C and 50% RH (Kapur, 1960). There are, however, no universal rules about acceptable conditions which can be held to apply in every case. Jones (1973) points out that acclimatization increases tolerance in the direction of the outdoor conditions and, as a rule-of-thumb, suggests that indoor temperatures should be between 4 and 11°C lower than the outdoor temperature, with RH of about 50%. Even if this cannot be achieved, it is certain that any reduction in temperature will give some feeling of benefit on entering the conditioned space.

It is difficult to state in general terms the likely rate at which energy must be extracted from a cooled space to maintain the required interior conditions. A common size of conditioner for American dwellings has a 3 ton (10.56 kW) capacity. Kapur (1960) considers that a 5 ton (17.6 kW) capacity would be needed for a house of 100 m^2 floor area in India. These represent about 50 W/m^3 extraction at peak load. Halmos (1975) quotes 78 W/m^3. Since the main part of the energy input comes from the solar radiation incident on the external surfaces, the shape, disposition, and radiative characteristics of these surfaces and of the fenestration might be arranged in new buildings so as to minimize this input (Hoffman, 1973). Conduction through external walls and infiltration of external air should be minimized and the thermal capacity of the structure chosen so as to limit the excursions of interior temperature. Matthael (1975) finds that measures to minimize solar heat gains can reduce the cooling load to less than 10 W/m^3. Regenerative cooling may further limit the plant capacity required for a particular case (Robeson, 1970).

C. Psychrometry

The desired conditions in a room may be represented by a point on a psychrometric chart, such as *R* in Fig. 16.3. It would be inappropriate here to consider in detail how this condition may be maintained, but some aspects must be examined so that the state required of the air supply may be visualized.

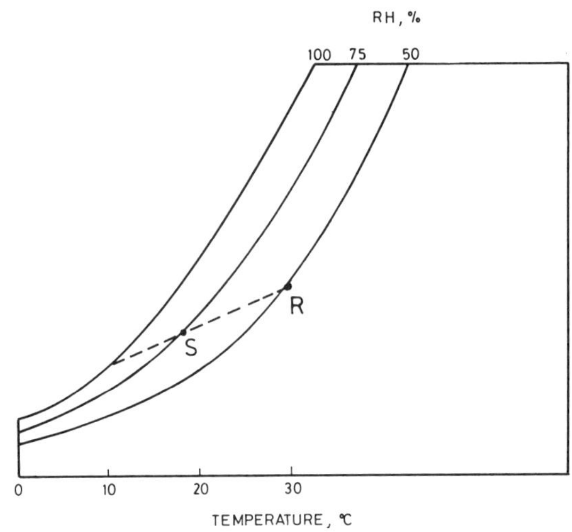

FIG. 16.3 The room ratio line.

It will be possible to maintain the desired conditions only if the supplied air is at some state, represented by *S*, from which the change in its temperature and its moisture content, on diffusing into the room, will take it to the room condition *R*. The supplied air is thus absorbing the sensible heat and the moisture (or latent heat) released in the room. The line containing *S* and *R* is the room ratio line; it will be evident that the slope of this line is determined only by the ratio of the rates of release of sensible heat and latent heat in the room. Thus, the state of the supplied air must always lie on the room ratio line passing through the desired condition *R* if this condition is to be maintained. The required mass flowrate of the supplied air will depend upon how far the entry state-point *S* is from *R*. If air is supplied from outside, whether or not it is partially mixed with recirculated air, it will normally need to be further conditioned in some degree so as to lie on the room ratio line. Certain processes, of which a few basic ones are outlined below, enable this to be done.

Suppose that the state of the available air is represented by the point *A*. This may be moved to a suitable point on the room ratio line by methods such as the following.

(a) Sensible cooling (Fig. 16.4a)—passage over a cooling coil carrying a chilled liquid reduces the temperature at constant moisture content. The coil temperature is required to be somewhat lower than T_s.

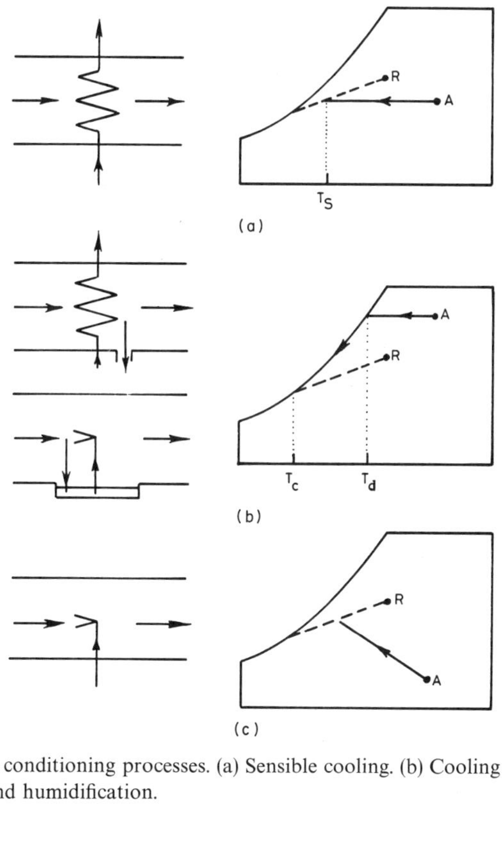

FIG. 16.4 Air conditioning processes. (a) Sensible cooling. (b) Cooling and dehumidification. (c) Cooling and humidification.

(b) Cooling and dehumidification (Fig. 16.4b)—if the coil temperature is below T_d, the dew point corresponding with the state A, the air becomes saturated and liquid water condenses on the coil. Cooling may continue down to T_c, with dehumidification; the air-exit temperature T_s is somewhat higher. A similar effect can be obtained with a recirculated spray of chilled water in place of the coil.

(c) Cooling and humidification (Fig. 16.4c)—if the conditions are such that the sprayed water is all evaporated into the air, cooling is obtained by the reduction of sensible heat at the expense of an increase in the latent heat. In the limit, the process approaches adiabatic saturation. Whatever the initial water temperature, it takes place approximately along a line of constant WBT (Jones, 1973). If regenerative cooling is employed, the temperature may be reduced a few degrees below the WBT (Hogg, 1971).

One of these processes, or a combination of them, can be made to bring the air supply condition onto a suitable point on the room ratio line. If the WBT of the available air is sufficiently below that of the room condition, evaporative cooling alone may be adequate. In other cases, chilled water will be required to operate a coil or a spray. In many practical cases, the water temperature required is in the range 5–10°C. A brief consideration is now given to the ways in which this is normally obtained.

16.3 REFRIGERATION SYSTEMS

A. Ideal Refrigerator

The ideal refrigerator is a reversed heat engine with reversible heat and work transfer, as in Fig. 16.5. For the expenditure of work W, heat Q_1 is withdrawn from the cold space at temperature T_1, and Q_2 rejected into the

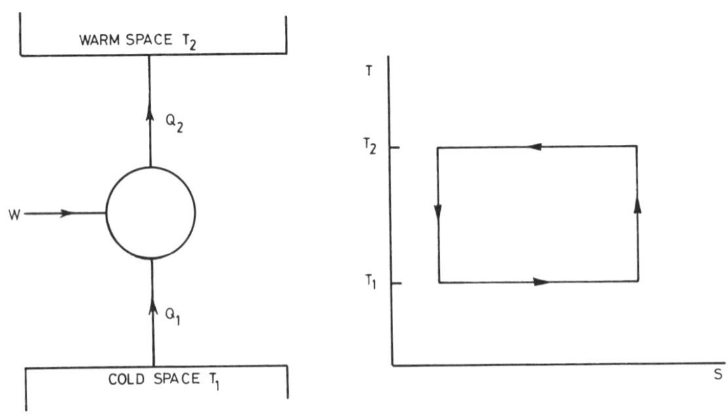

FIG. 16.5 Ideal refrigerator.

hot space at temperature T_2. The effectiveness of the device is defined by the coefficient of performance COP, where

$$\mathrm{COP} = Q_1/W \tag{16.1}$$

As there is no net change of energy,

$$W + Q_1 - Q_2 = 0 \tag{16.2}$$

and as there is no net change of entropy and the heat transfers are reversible,

$$(Q_1/T_1) - (Q_2/T_2) = 0 \tag{16.3}$$

It follows that

$$\mathrm{COP} = T_1/(T_2 - T_1) \tag{16.4}$$

To give chilled water in the range of temperature required, the cold space temperature might have to be 3°C, say, so that the heat exchange area would not be too large. The warm space temperature will be determined by that of the heat sink available. For a pond or cooling tower sink, the temperature might be 30°C, and for an ambient air sink, perhaps 50°C. For these conditions, the COP is found to be 10.2 and 5.87, respectively.

If the machine had been driven by an ideal heat engine operating from a source at temperature T_h and rejecting heat at the same temperature T_2 as the refrigerator, the work W would be obtained by the supply of heat Q_h given by

$$W/Q_h = (T_h - T_2)/T_h \tag{16.5}$$

Then the thermal ratio (TR) for the combined engine/refrigerator is

$$\mathrm{TR} = Q_1/Q_h = (T_1/T_h)[(T_h - T_2)/(T_2 - T_1)] \tag{16.6}$$

Table 16.1 shows the resulting values of TR for several values of source temperature.

TABLE 16.1

Thermal Ratio of Ideal Engine/Refrigerator Combination

Source temperature (T_h, °C)	Heat rejection at temperature (T_2, °C)	
	30	50
70	1.19	0.34
80	1.45	0.50
90	1.69	0.65
100	1.92	0.79
150	2.90	1.39
250	4.30	2.25

Although these figures are for ideal systems, they illustrate clearly the central problem in the effective use of solar energy for cooling. A refrigerator requires little power to drive it, but if this power is provided by a solar engine, the overall thermal ratio is poor except when the input can be provided at high temperature.

B. Vapor-Compression Refrigerator

Many practical refrigeration machines operate on a simple vapor-compression cycle, illustrated in principle in Fig. 16.6. Heat is extracted from the cold space to evaporate the refrigerant fluid at low pressure, and is rejected in condensing it at higher pressure. A pump provides the pressure rise between evaporator and condenser, but the drop is obtained in a throttle valve or capillary tube. Departure of the T–s diagram from the rectangular form of the ideal cycle indicates that the COP will be somewhat lowered. It may usefully be expressed in terms of the enthalpy at the extremities of the cycle:

$$\mathrm{COP} = (h_1 - h_4)/(h_2 - h_1) \tag{16.7}$$

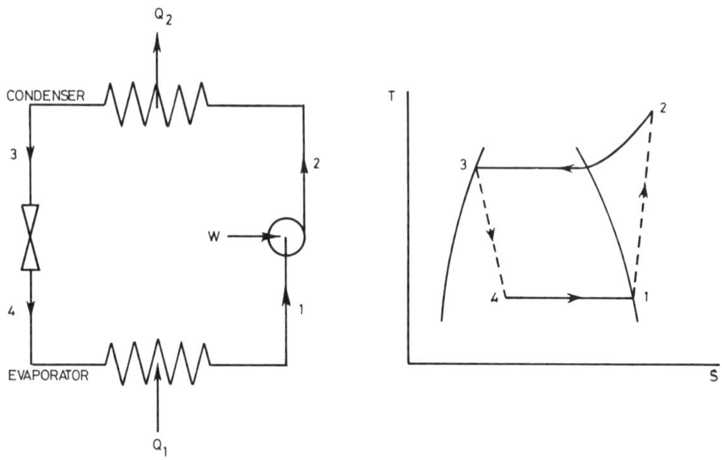

FIG. 16.6 Vapor-compression refrigerator.

In practice, the value obtained from Eq. (16.7) will usually be between 0.75 and 0.85 of the Carnot value, given by Eq. (16.4), the factor depending largely upon the properties of the refrigerant fluid used. In many respects, water is a good refrigerant, but it has a very low vapor pressure (0.0076 bar, 0.76 kN/m^2 at 3°C) so that the system has to be large and physically robust and is difficult to keep free from leaks. Moreover, temperatures below 0°C cannot be obtained. Many machines use one of the fluorinated hydrocarbons, often called by the tradename Freons-trichloromonofluoromethane (Freon 11) or dichlorodifluoromethane (Freon 12). Whatever fluid is employed, the overall coefficient of performance based on the work supplied to the machine is further reduced by the mechanical efficiency of the pump and unavoidable heat transfers to and from the surroundings. For planning purposes, a value of about 50% of the Carnot COP may usually be assumed.

C. Absorption Machines

In many situations, the need for a work input is a great disadvantage, and the absorption cycle provides a means of refrigeration requiring a thermal input only. This is of particular interest from the point of view of solar cooling. The arrangement of an absorption machine is shown in Fig. 16.7, but the cycle cannot readily be illustrated on a *T–s* diagram, since two working fluids are involved, the refrigerant and the carrier fluid or absorbent.

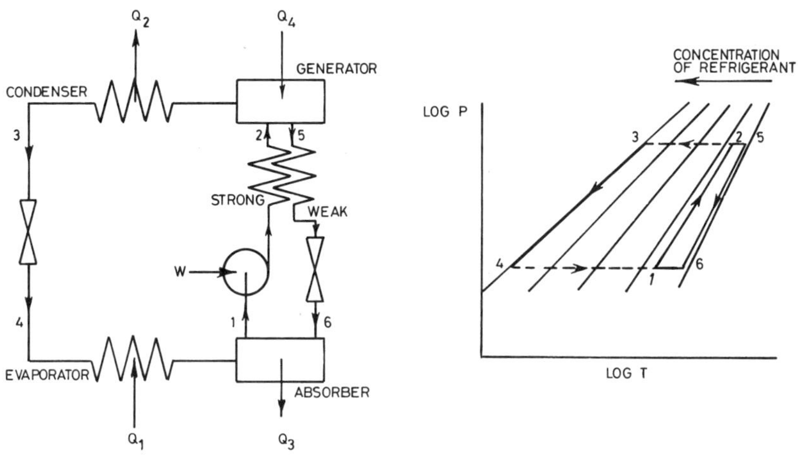

FIG. 16.7 Absorption refrigerator

Refrigeration is again obtained by the evaporation of the refrigerant at low pressure and rejection of heat during condensation at higher pressure. The pressure rise now occurs between two new elements, the absorber and the generator. The absorbent has a high affinity for the refrigerant, so that the refrigerant vapor has a lower pressure over a solution in the absorbent than over its own liquid. The solution is pumped to the generator, where the temperature is maintained by heating at a level at which the vapor pressure equals the saturation pressure in the condenser. Separated refrigerant then passes into the condenser while the depleted solution returns to the absorber via a heat exchanger and a throttle valve. Since the pump is transferring liquid only, very little work is required—pump ratings are typically very few watts per kilowatt of refrigeration.

A convenient representation of the change of state of the fluids can be made in terms of the thermodynamics of mixtures. The logarithm of the vapor pressure of many pure liquids is found, to a good approximation, to vary linearly with the reciprocal of the absolute temperature, as indicated by kinetic theory. When it is dissolved in a nonvolatile absorbent, in many cases it tends to follow Raoult's law; that is, its vapor pressure is proportional to its mole fraction in the solution. A diagram showing the vapor pressure, temperature, and concentration can thus be cast in the form of a series of straight lines, upon which the cycle can conveniently be traced, as in Fig. 16.7b.

Originally, the refrigerant used in absorption machines was ammonia, with water as the absorbent. With this system, there is appreciable transfer of absorbent with the refrigerant, and the combination most widely used now in commercial machines is water as refrigerant and lithium bromide as absorbent. Lithium bromide is a solid at ordinary temperatures but it is highly hygroscopic, and the mixture becomes liquid at a water concentration of about 30%. The higher vapor pressures which make ammonia a preferable refrigerant to water can be retained if the absorbent is sodium thiocyanate, and good performance has been obtained with this combination (Swartman *et al.*, 1973a).

It is apparent in Fig. 16.7 that the absorber/generator system replaces the compressor of the vapor-compression machine, acting like a heat engine working between a source at generation temperature T_g and sink at absorber temperature T_{ab}. The overall TR of an ideal machine is thus given by an expression similar to Eq. (16.6), if the heat-rejection processes in the absorber and condenser take place at the same temperature T_2. Absorption machines are usually employed when a waste heat source is available. In remote areas where there is an insufficient electricity supply to work compression machines, they are sometimes operated from a gas or kerosene burner, providing a simpler and cheaper system than a compression machine driven by a heat engine. In a solar cooling system, the generator would be heated directly by the sun or by a heat-transfer fluid from a collector. Some idea of the highest TR possible is given by Table 16.1; this is expanded at the lower generator temperatures specifically to illustrate the low values obtained when the generator temperature is of the order to be expected from flat plate collectors.

The behavior of real absorption refrigerators is, however, profoundly different from that of the ideal machine, and the TR achievable is even less than might be expected from the tabulated values. This is attributable to the thermodynamic characteristics of real refrigerant/absorbent mixtures. To begin with, there is a threshold value of generator temperature which must be exceeded if the machine is to function at all. If the desired evaporator temperature is specified, this determines the evaporator pressure and hence the pressure in the absorber. This in turn determines the refrigerant concentration in the absorber, at the specified heat-rejection temperature. Similarly, the heat-rejection temperature determines the pressure in the condenser and hence in the generator. At this pressure, there is a unique relationship between temperature and refrigerant concentration in the generator. However, the system cannot function unless the concentration of refrigerant is lower in the generator than in the absorber. This fixes a lower limit to the generator temperature at which operation becomes possible. For example, in the instances already cited, with the evaporator temperature of 3°C, it is found that the generator temperature must exceed about 60°C for operation to be possible with a heat-rejection temperature of 30°C, using the ammonia/water combination. When the heat-rejection temperature is 50°C, the threshold for the generator temperature rises to about 110°C. Wilbur and Mitchell (1975) find similar values also for the water/lithium bromide system.

In most machines, the sensible heat addition to the absorber and generator is recycled via the heat exchanger in the return line. An approximation to the behavior is that the cooling effect is equal to the enthalpy of evaporation at evaporator temperature and the heat supplied equal to the enthalpy of evaporation at generator temperature. For an ideal fluid the enthalpy of separation of the refrigerant from the absorbent would be zero, and the enthalpy of evaporation independent of temperature. Thus the TR would be expected to have the value unity at all generator temperatures above the threshold for operation. Actual values depend on the thermodynamic characteristics of the fluids employed (Badylkes, 1966) and are often substantially less than unity, though the nearly constant value expected above the threshold generator temperature, which is in marked contrast with the rising values of the ideal refrigerator, is a well-known characteristic of real absorption machines. Wilbur and Mitchell (1975) find a value of TR of about 0.7 for an ammonia/water system just above the threshold, falling to about 0.6 at a generator temperature of 130°C. This level of TR is confirmed by Chinnappa (1974). A value of about 0.8 is obtained with water/LiBr, though there is an upper limit of the usable generator temperature, when the water concentration falls to the value at which solid LiBr begins to crystallize; this is placed at around 85°C in a typical case.

One variety of absorption refrigerator, the *Electrolux*, is designed to operate with no work input at all. A third, inert fluid is present in the low-pressure side of the circuit, maintaining the total pressure at approximately the same value as in the high-pressure side. Liquid traps, in the form of U-bends, keep this fluid (usually gaseous hydrogen) in place. The work required to pump the refrigerant is then very low and is accomplished by a "bubble lift" in which a stream of vapor bubbles carries slugs of fluid along a narrow tube into the generator. The available TR is somewhat lower than that obtained with the pumped cycle.

D. Resorption Cycle

In Fig. 16.8 is shown a development of the absorption machine, in which the pressure drop takes place in a second pair of devices similar to the absorber and generator, called the desorber and resorber. The refrigerant is now not completely separated from the absorbent, and the system operates at lower pressures. At the expense of greater complexity, this results in a TR which approaches the ideal value a little more closely. Still more complicated cycles can give yet higher values (Chinnappa, 1974).

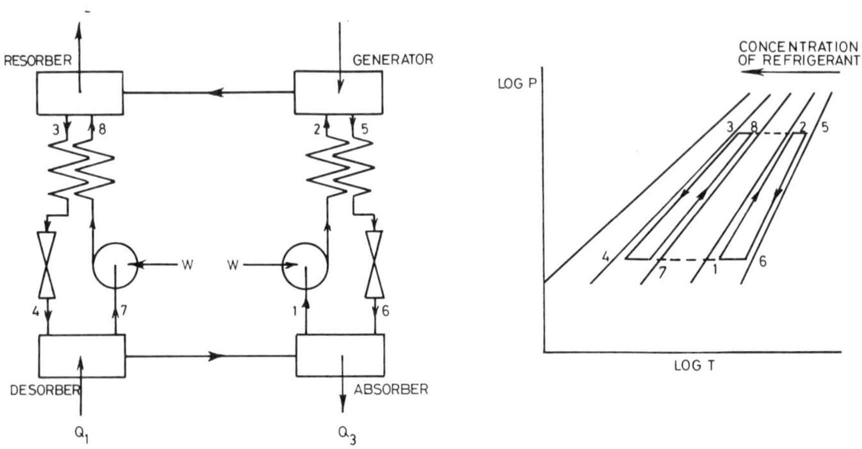

FIG. 16.8 Resorption refrigerator.

E. Dehumidification

For the operation of air conditioning systems, the moist air itself can be made to take part in a cycle, as shown in Fig. 16.9. Moisture is removed from the air by a solid adsorbent or desiccant, such as silica gel, or a liquid

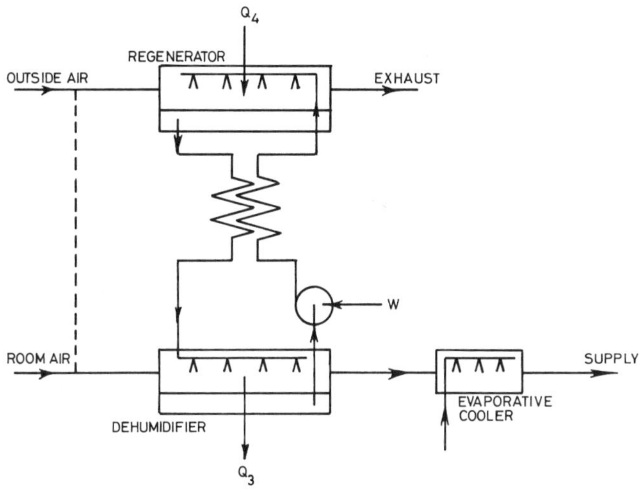

FIG. 16.9 Solution dehumidification cooler.

absorbent, such as ethylene glycol or lithium chloride. The heat of adsorption or solution is rejected and the air further cooled by an evaporative cooler. Heat is then applied to reconstitute the absorbent by driving off the moisture into a stream of outside air. Other arrangements in which water vapor is driven off into the atmosphere are also sometimes referred to as dehumidification.

F. Intermittent Operation

Simple absorption and adsorption refrigerators have been made, which operate discontinuously rather than in a cycle (Keith 1930). Two connected vessels, as in Fig. 16.10, act alternately as a generator/condenser combination and as an absorber/evaporator combination. Heating is first applied to vessel 1, raising the vapor pressure of the refrigerant, which passes at constant pressure into vessel 2 where it is condensed. This continues until the concentration of refrigerant in 1 falls to a value at which the vapor pressure equals the saturation value corresponding with the condenser temperature. If the valve is now closed, the system can be kept until required in a charged state and may be moved to another place. When the valve is reopened, the refrigerant is caused to return to 1 by its solution in the absorbent, and evaporation occurs in 2 at a low temperature, determined by the lowered pressure.

During this intermittent operation, the thermodynamic processes are far from reversible, and are dependent upon the properties of the refrigerant–absorbent combination used and the proportion of refrigerant present at the commencement of generation. Williams *et al.* (1957) discuss some of the

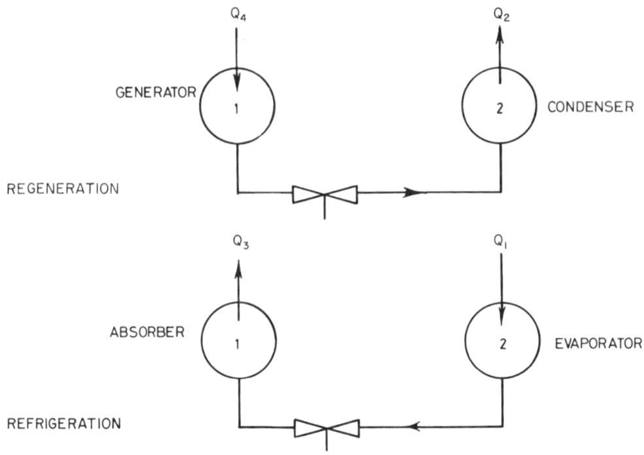

FIG. 16.10 Intermittent absorption cooler.

processes in detail and show that a major source of irreversibility is the addition of sensible heat to the refrigerant–absorbent solution during generation. As the solution becomes depleted, the temperature must rise so that the vapor pressure of the refrigerant over it is maintained at the value determined by saturation at condenser temperature. The sensible heat absorbed, which must eventually be rejected to the sink, becomes sufficiently great as the maximum generator temperature is increased for the TR to cease rising with it and even to pass through a maximum. Values of maximum TR for various refrigerant–absorbent combinations, obtained by assuming an idealized cycle, are in the range 0.2–0.6 only.

16.4 SOLAR COOLING

A. Collector/Refrigerator Combination

It is apparent from the data in Table 16.1 that respectable values of TR can be obtained only with relatively high source temperatures. Moreover, it was seen later that the performance to be expected from ideal continuous absorption machines is less than that of ideal engine-driven compression machines, given by Eq. (16.6). When the energy for the system is provided by a solar collector, further losses occur, which are likely to be high at the source temperatures known to be required. It will be useful to obtain an impression of the maximum possible performance that could be obtained from a collector/refrigerator combination. This might be expressed in terms of the system thermal ratio, STR, defined as the ratio of the energy withdrawn from the cooled space to that falling upon the collector, i.e.,

$$\mathrm{STR} = Q_1/H \tag{16.8}$$

If the collector delivers the source energy Q_h, then

$$\mathrm{STR} = (Q_h/H)(Q_1/Q_h) = (Q_h/H)\,\mathrm{TR} \tag{16.9}$$

An ideal collector would deliver to the refrigerator all the energy falling upon it from the sun. Then the STR would be the same as the refrigerator TR, given by Eq. (16.6). This case will be an upper limit, approached to some degree by systems driven by concentrating collectors, in which the losses are made small by ensuring that the area from which they occur is much less than the collection area. On the other hand, the characteristics of a flat plate collector are determined largely by the losses. It would be more reasonable to define an ideal flat plate collector as one in which the only performance decrement is due to thermal losses.

It is well known (Hottel and Whillier, 1955) that the output of a flat plate collector may be written in the form

$$Q_h = D[\tau\alpha H - U(T_h - T_a)] \tag{16.10}$$

where $\tau\alpha$ is the product of the transmissivity of the cover and absorptivity of the collector surface. U is an overall loss coefficient. The constant D is primarily determined by the effectiveness of the collector itself as a heat exchanger, but a value can be assigned to it when the collector is feeding a secondary system (Brinkworth, 1975). It is apparent that for an ideal collector defined as above, the characteristics would take the form

$$Q_h = H - U(T_h - T_a) \tag{16.11}$$

so that the STR of an ideal compression machine is given by

$$\begin{aligned}\mathrm{STR} &= Q_1/H\\ &= [1-(U/H)(T_h - T_a)](T_1/T_h)[(T_h - T_2)/(T_2 - T_1)]\end{aligned} \tag{16.12}$$

Since the TR of the refrigerator increases with increasing T_h while the output of the collector falls, it is apparent that the STR has a maximum. From Eq. (16.12), it is found that this occurs when

$$(T_h/T_2)^2 = (T_a/T_2) + (H/UT_2) \tag{16.13}$$

An ideal collector/refrigerator would be one operated at the optimum source temperature at all times. The STR of such a system is given for a range of values of insolation H in Table 16.2. As before, two values of heat-rejection

TABLE 16.2

System Thermal Ratio of Ideal Collector/Refrigerator

Insolation (W/m^2)	STR ($T_2 = 30°C$)	STR ($T_2 = 50°C$)
200	0.390	0.048
400	0.728	0.247
600	1.024	0.422
800	1.288	0.578
1000	1.526	0.719

temperature T_2 have been used (30 and 50°C); the evaporator temperature is 3°C and the ambient temperature T_a has been taken to be 30°C. A value of 4 W/m^2K is used for the overall loss coefficient U, representing approximately a collector with two cover plates and neutral absorption characteristics (Whillier, 1967).

It is assumed for this ideal system that the mean collector operating temperature is equal to the refrigerator source temperature.

From calculations like these, it is possible to establish the variation of STR when a directly coupled ideal collector/compression refrigerator combination is operated throughout an ideal tropical day. For present purposes it will be sufficient to assume a sinusoidal variation of insolation (Hirschmann, 1974) over a period of 12 h, with a total incident energy of 7.5 kWh/m^2 (Brinkworth, 1972). The daily average STR for the two examples used is then found to be 1.23 and 0.55, and the corresponding collector areas required to give unit cooling are 0.11 and 0.24 m^2/kWh day, respectively. These might be regarded as the upper limits of expectation for systems driven by flat plate collectors, and values half as good might be considered very satisfactory in practice.

In the case of the idealized absorption machine, it might be assumed for the purpose of comparison that a constant value of TR is obtained whenever the collector temperature can reach the generator threshold temperature, being otherwise zero. If the TR assumed is 0.7, it is found for the same conditions as above that the daily average STR is 0.57 and 0.38, with required collector areas of 0.23 and 0.35 m^2/kWh day, respectively.

Having regard to the typical cooling loads quoted earlier, these STR figures imply that for the air conditioning of a dwelling the collector area required would be some tens of square meters. There is much scope for ingenuity in the arrangement of the system so as to minimize the area of collectors, which represent a significant part of the cost. A critical element in optimization is the arrangement (if any) for energy storage. This can take place at various points in the system. Storage of hot fluid decouples the collector from the refrigerator, while storage of cold fluid decouples the refrigerator from the cooled space. Some degree of storage may be possible within the refrigeration system itself. For example, in an absorption system concentrated refrigerant might be stored after separation; in this case, storage is possible at low volume and ambient temperature so that there is no heat exchange with the surroundings.

Experimental systems in great variety have been described in the literature over the last century, and experience with solar refrigeration is now accumulating rapidly. It is difficult to find among these reports a common basis on which to compare the systems described. Performance is sometimes given in terms of the instantaneous TR of the refrigerator or the STR of the system; sometimes the daily average of these values is given, or some other measure of the average performance, such as the quantity of ice produced per day. Only rarely are insolation data quoted for the time of the tests. Nevertheless, it is possible to obtain some idea of the degree to which the performance approaches the ideal values for various classes of system.

B. Compression Machines

The vapor-compression refrigerator driven by a solar heat engine has been one of the least studied of the devices considered. Daniels (1974) reports that Kirpichev and Baum operated a system of this type, employing a steam boiler at the focus of a parabolic mirror, but it appears that the refrigerator was actually an absorption machine (see Kirpichev and Baum, 1954).

Farber and Prescott (1973) showed that it was technically feasible to develop a heat engine driven by flat plate collectors if the working fluid is suitably chosen. The coupling of such an engine to a compression refrigerator is discussed by Teagan and Sargent (1973). Prigmore and Barber (1975) describe a combined system in which the engine is a gas turbine. This 3-ton capacity system has a TR of 0.5 with a source temperature of about 102°C, decreasing to 0.3 at 79°C. As these are for an evaporator temperature of 7°C and heat-rejection at about 35°C, they represent about 30–33% of the values for an ideal system given by Eq. (16.6).

C. Absorption Machines

According to Heywood (1953), Mouchot and Pifre are reported to have been the first to operate a solar absorption refrigerator, in 1878, successfully producing a quantity of ice. The system described by Kirpichev and Baum (1954) produced 130 kg of ice in a period of 7 h.

Swartman *et al.* (1973b) reviewed many of the systems of this type which have been subsequently described in the literature. A commercial 3-ton water/LiBr cooler, linked to a flat plate heater, was tested at the University of Wisconsin and is described by Löf (1963). A TR value of 0.6 was obtained with a generator temperature of about 79°C. Augmentation of the solar heating by another source makes it difficult to evaluate the STR, but Löf predicts a value for the daily average of 0.3, and later suggests a collector area requirement of about 0.4 m^2/kWh day. A very compact system of a similar kind, designed at the University of Florida, is described by Farber (1970). This produced ice at a rate of 12.5 kg/day per m^2 of collector from water at 24°C, representing an average STR of about 0.1 and specific area of 0.66 m^2/kWh day. Later, this is reportedly reduced to about 0.39 m^2/kWh day (Farber, 1973).

Asher and Reti (1961) however, advancing a continuous system of some complexity, predict a specific area of about 0.54 m^2/kWh day with heat rejection at 29°C, and about 0.60 with rejection at 35°C.

A comprehensive study of factors affecting the design of solar-heated absorption coolers is given by Duffie and Sheridan (1965), covering particularly the effects of heat exchangers in the system. For a system based on a commercial LiBr refrigerator, they obtain specific collector areas between 0.37 and 0.67 m^2/kWh day.

At least one manufacturer is now marketing a continuous absorption cooler adapted for operation from solar collectors. A 3-ton model is said to have a TR value of 0.65 when the generator is heated by water entering at 99°C. In this case the condenser and absorber are cooled by water entering at 29°C. Operation at reduced rating (1.71 ton) is possible at a hot water inlet temperature of 82°C and cooling water at 24°C.

D. Intermittent Absorption

Williams *et al.* (1957) and Chinnappa (1967) discuss at some length the reasons for expecting lower performance from intermittent systems than those obtained with continuous ones; the latter shows very satisfactory agreement between experimental values of TR and those given by analysis of an assumed cycle. Williams *et al.*, working with a paraboloidal concentrating collector of 0.77 m^2 area, obtained average STR values over periods of 1.5–4.5 h of about 0.15 with ammonia and water and about 0.06 with Freon 21 and ethylene glycol. (It should be pointed out that these rather low values are due partly to poor collection efficiency, the average TR being about 0.39 and 0.22, respectively.) A similar system, described by Trombe and Foex (1957), utilized a reflector in the form of a parabolic cylinder with presented area of 1.5 m^2. This produced ice at a rate of 4 kg/m^2 day, equivalent to a specific area of approximately 2 m^2/kWh day.

Chung and Duffie (1961) put forward a design for an advanced ammonia/water system with a parabolic cylinder reflector, claimed to be capable of producing about 9.5 kg of ice per m^2 of collector, with an average STR of 0.18 and specific area about 0.87 m^2/kWh day.

Several tests with intermittent systems driven by flat plate collectors have been reported. Chinnappa (1962), using a collector of area 1.4 m^2, obtained average TR values of about 0.3, with average STR 0.057; the specific area was about 4.9 m^2/kWh day. This was a very simple device, and it was estimated that with the addition of a heat exchanger, the STR could be raised to 0.09. These tests were made with ammonia and water. Sargent and Beckman (1968) suggested that the ammonia/sodium thiocyanate combination would be superior, and tests with this are reported by Swartman *et al.* (1973a). With a collector area of 1.5 m^2, values of STR between 0.05 and 0.14 were obtained with ammonia/water and 0.11 to 0.27 with ammonia/NaSCN. A simple flat plate heated system, employing ammonia/sodium chloride, produced about 1 kg of ice/day per m^2 of collector, equivalent to a specific area of about 8 m^2/kWh day (Muradov and Shadiev, 1971).

E. Dehumidification

An ethylene glycol absorption dehumidifier, operated in conjunction with a solar air heater, is described by Löf (1963). Water, absorbed in the glycol, is driven from solution by air heated to 60–70°C. It is concluded that the average STR of the system is in the range 0.19–0.26, with specific area of 0.44–0.82 m^2/kWh day. An open-cycle system, originally due to Altenkirch, is described by Dannies (1959). In this system, the absorbent—typically a solid such as silica gel—is contained in the east and west walls of the building to be cooled. Each wall acts as absorber and desorber in turn, as it is successively in shadow and irradiated. An experimental building constructed on these lines (Adler *et al.*, 1961) produced negligible cooling, though it was concluded that a TR of about 0.1 was possible, with specific collector area of 7.1–9.5 m^2/kWh day. This might be reduced by a factor of 4 if desorbtion took place in the roof rather than the walls.

The flexibility of systems of this kind was considered to make them the preferred ones in the study of Kapur (1960) and numerous variants have been investigated. Hollands (1963b) shows that desorption can be carried out in the conventional solar still, and Mullick and Gupta (1974) describe a simple desorber in which the dilute solution trickles down a tray beneath a glass cover while the drying air passes by natural convection in the opposite direction. A specific collector area of about 0.5 m^2/kWh day is indicated for cooling systems using this type of collector.

Solar desorption has been much studied by workers in the USSR. A system closely investigated is one in which chilling of water is obtained in a conventional evaporator/absorber system, similar to that of an absorption refrigerator, but desorption of the absorbent takes place by evaporation directly into the atmosphere, assisted by solar heating.

Kakabaev *et al.* (1972), for example, give a detailed analysis of the performance of systems linked to glazed desorbers. This shows how the specific collector area varies with the required evaporator temperature and ambient air humidity, represented by the temperature of chilled water produced and the vapor pressure p_0 of atmospheric steam. For insolation of about 810 W/m^2, for example, the instantaneous STR obtained under optimum conditions is found to be 0.40, 0.34, and 0.24 for $p_0 = 6$, 13, and 18 mm Hg, respectively, when the chilled water temperature is 6°C. Baum *et al.* (1972) show that good performance may be obtained even with desorbers which are open to atmosphere. The maximum STR values obtained are quoted only for production of water at 10°C, being about 0.52, 0.41, 0.30, and 0.19 for $p_0 = 5$, 10, 15, and 20 mm Hg, respectively [corresponding values given for a glazed desorber (Mullick and Gupta, 1974) at 6, 13, and 18 mm Hg are approximately the same: 0.48, 0.40, and 0.30, respectively]. Baum *et al.* (1972) also give data from which the specific collector area may be determined, giving, for chilled water production at 5°C, areas of 3.0, 4.7, and 10.5 m^2/kWh day when the atmospheric water vapor pressure is 5, 10, and 15 mm Hg, respectively. Kakabaev and Golaev (1971) discuss other variants, including desorption by spraying into a solar-heated air stream.

16.5 A PLACE FOR SOLAR COOLING

Typical performance data for many different systems have been cited above though it is to be noted that these were largely experimental and that no long-term experience has so far been obtained with any solar-driven cooling system. The practical implementation of solar cooling in areas of need will depend upon social acceptability, economic competitiveness, and technical suitability under local conditions. Among the factors affecting this are the availability of electric power and cooling water and tolerance of water loss in open systems.

It seems unlikely that solar cooling could compete with vapor-compression refrigeration where electric power is plentiful. The energy input requirements for these machines are quite modest, their reliability is well established and they are already manufactured on a large scale. Moreover, this type of system is probably the only one readily available for obtaining the low temperatures needed for deep freezing and long-term storage of foodstuffs. There is a stronger case for considering solar cooling with absorption systems for air conditioning and space cooling of domestic dwellings, factories, public buildings, and food stores, except where ample quantities of waste heat are available. This case has been made repeatedly during recent decades, but is strengthened now by the increase in the price of fuels and a greater awareness of the need to economize in the consumption of capital resources.

The main controlling factors in absorption refrigeration are very well summarized by Tabor (1962) in his review of papers on solar cooling presented at the 1961 UN Conference on New Sources of Energy. He points out that with absorption systems, the temperature difference $(T_g - T_2)$ is always greater than the difference $(T_2 - T_1)$. Accordingly, as shown herein, air conditioning temperatures may be obtained with generator temperatures appropriate to efficient operation of flat plate collectors only if the heat-rejection temperature T_2 is fairly low. This implies that where cooling water is not available, higher generator temperatures are needed, requiring the use of concentrating collectors. The rapid technical advances in collector design during the last few years has not yet been consolidated by economic appraisal, so that the relative costs of various systems are far from clear. Tabor (1962) made a rough analysis of the economics of solar cooling, on the assumption that the system, apart from the collector, would have the same cost and the same performance as a fuel-operated system. He found that the solar system would then be competitive if the cost per square meter of the collectors were less than that of about 0.7 ton of fuel. Even allowing for the crudeness of this assessment, it seems certain that the economics have now moved substantially beyond this in favor of solar-driven systems. Moreover, it is likely that even the simplest intermittent systems, capable of being fabricated with very limited resources, would now be economic in places in spite of their large specific collector area. There is also a renewed interest in the possibility of economic competitiveness even in areas well supplied with cheap electric power. This is a consequence of the realization that combining summer cooling with winter heating may be beneficial to the economics of both. Löf and Tybout (1974), for example, found that the cost per kilowatt hour of combined heating and cooling ranges from 17 to 95% of that of cooling alone, for domestic uses in a number of American cities. They concluded also that a further doubling of gas prices, beyond the 1970 level, would make solar heating and cooling competitive with the use of fuel in almost all locations in the US. Butz *et al.* (1974) identify the conditions under which a combined system would be competitive with a fuel-operated one for Albuquerque, New Mexico. It is found that the required collector areas lie generally within the range 20–80 m^2. Ward and Löf (1975) give full details and a cost breakdown of a combined system with a collector area of 71.3 m^2, installed in a residential-type building at Colorado State University.

The climatic indications for the introduction of solar cooling are reviewed by Baum and Kakabaev (1972). They first point out the desirability of designing buildings so as to minimize cooling requirements or to fit existing buildings with window shades, night-time ventilation and so on, which have the effect of reducing the period over which cooling is necessary. In many areas, cooling requirements can then be met by evaporative coolers, the indications being that these will be satisfactory wherever the atmospheric water vapor pressure is less than about 10 mm Hg, corresponding to an RH of about 22% at ambient temperature of 35°C. Bahadori (1973) describes a very simple system, working under these conditions. Up to about 12 mm Hg (RH of 29% at 35°C), it is claimed by Baum and Kakabaev (1972) that flooding or spraying of roofs with water is an effective method of cooling. In the range 12–15 mm Hg (RH up to 35% at 35°C), they recommend the use of absorption systems with open solar desorbers and state that these are economically superior to systems requiring thermal and electrical power inputs in those conditions. For hot moist climates, systems with closed desorbers or fully solar-heated generators are indicated and are again claimed to be more economical than conventional systems. It is claimed that the collector area requirements do not exceed the aggregate roof area, even for three- and four-story buildings. These conclusions are largely in accord with those of Kapur (1960).

A review of the work of recent decades, much of which is reported here, thus shows that sufficient basic investigation has been made to establish the technical feasibility of solar refrigeration and air conditioning and to delineate the climatic indications for its use. It appears that it is also economically competitive. The way seems open for the design of fully engineered systems and their introduction into service.

17

Solar Heating and Cooling of Homes

JOHN I. YELLOTT

COLLEGE OF ARCHITECTURE
ARIZONA STATE UNIVERSITY
TEMPE, ARIZONA

17.1 INTRODUCTION

The heating of residences by the use of solar radiation is a very ancient concept, and it has been used since man first began to build habitations. Aristotle advised that houses should have very little fenestration on the north side, but generous openings on the south side, provided with horizontal extensions to shade these openings during the summer. The principle enunciated by "the master of them that know" is still appropriate and widely used today, although the introduction of mechanical refrigeration to provide air conditioning in summer and the plentiful supply of fossil fuels which had existed until recent times caused building designers to forget the admonitions of Aristotle and many who followed him. One of the outstanding examples of natural solar heating is the Indian cliff dwelling known as "Montezuma's Castle," located some 80 miles north of Phoenix, Arizona. This remarkable structure, begun in A.D. 700 and completed in A.D. 1300, is a masonry building, tucked into the side of a south-facing white limestone cliff, and protected against both the summer sun and the depredations of marauders by an overhanging ledge. During the winter, the sun shines directly upon the south-facing surface of this remarkable "apartment house," heating the masonry, and, because of the time lag imposed by the thick walls, providing heat to the structure at night. During the day, the residual heat in the walls and the additional heat that entered through the relatively small windows was apparently enough to produce comfortable conditions.

During the summer, natural ventilation was adequate, since the elevation at the location of Montezuma's Castle is some 4000 ft above sea level, and summer temperatures are moderate by Arizona desert standards.

The plan of the ancient and now vanished city of Mohendro-Daro, in Pakistan, indicates that the same principles were used by those who planned this remarkable collection of dwellings over 3000 years ago. Many primitive peoples, including other American Indians, oriented their simple dwellings to take advantage of the warming rays of the sun in winter and the cooling breezes in summer.

There was very little apparent activity in the field of solar heating until the beginning of interest in the utilization of solar energy for many different purposes which occurred about a century ago. One outstanding example was the use by Professor E. S. Morse, of Salem, Massachusetts, in 1881, of "a surface of blackened slate under glass, fixed to the sunny side of a house with vents in the walls, so arranged that the cold air of a room is let out at the bottom of the slate and forced in again at the top by the ascending heated column between the slate and the glass." This concept is used widely in France today.

The first scientific work on solar heating began in 1940, at Massachusetts Institute of Technology, Cambridge, Massachusetts, when a team of engineers, under the direction of Dr. Hoyt Hottel, with financial support from the Godfrey L. Cabot Foundation, began to study the performance of solar collectors and fluid heating systems. The first definitive paper on flat plate collectors for this service was published by Dr. Hottel and his graduate student B. B. Woertz in 1942, and other papers followed to lay a solid foundation for the analytical study of the type of collector which is most widely used in solar heating and cooling of residences (Whillier and Hottel, 1955; Whillier, 1967). Also of interest was the house built near Cambridge in 1945–1946 by Dr. Maria Telkes, solar engineer, and Miss Eleanor Raymond, architect, for Miss Amelia Peabody. This house used warm air heaters, mounted on the vertical south-facing wall behind glass; heat storage was accomplished by the use of heat-of-fusion materials stored in small barrels in a special section of the house (Telkes and Raymond, 1946).

During the two decades following the end of World War II, interest in solar-heated residences grew slowly throughout the United States and in other parts of the world. Mention should be made of the pioneering work of Dr. Felix Trombe and his colleagues at the Solar Laboratory in Montlouis, and the solar houses built in Japan by Mr. M. Yanagimachi. More recently, solar houses have been built in Australia, in England, and because of the serious shortage and rapidly rising prices of petroleum products throughout most of the industrialized world, solar heating has become a matter of great interest in most of the developed countries of the world.

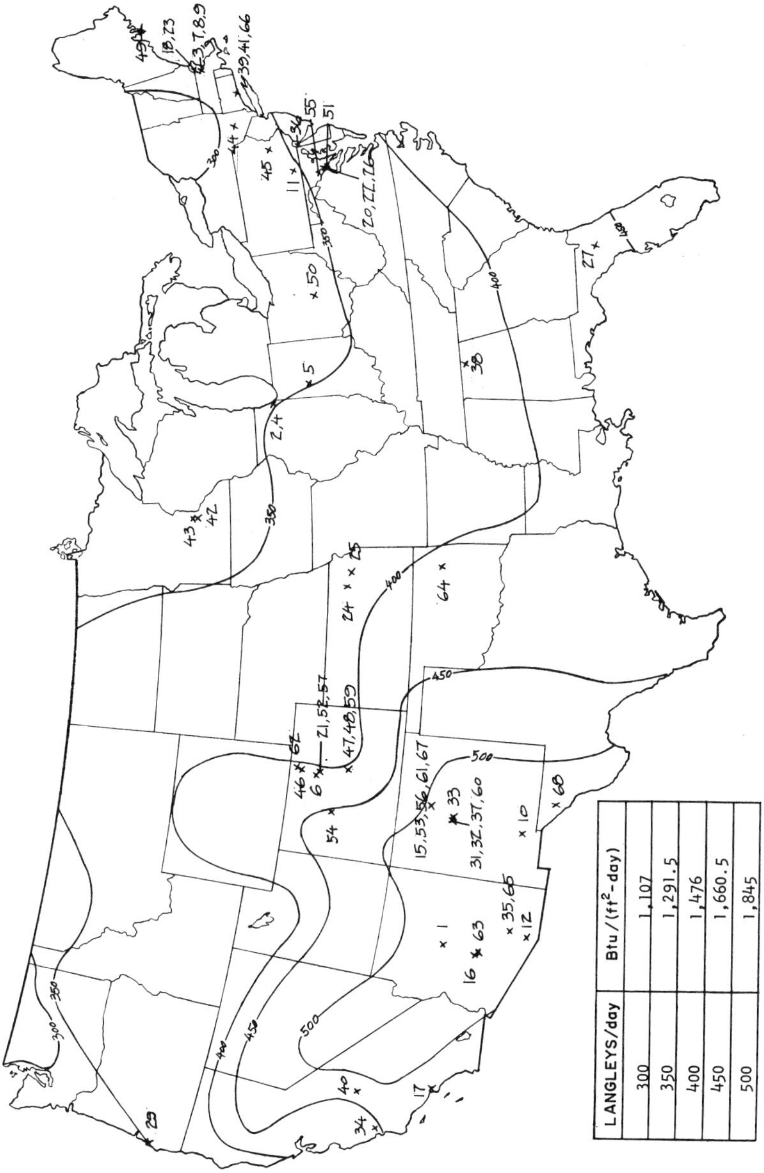

LANGLEYS/day	Btu/(ft^2-day)
300	1,107
350	1,291.5
400	1,476
450	1,660.5
500	1,845

FIG. 17.1 Location of solar buildings numbers refer to the identification table in Yellott *et al.* (1975). Isopleths show annual average daily horizontal insolation in Langleys (1 ly/day = 3.69 Btu/(ft^2 · day)).

The distribution of solar buildings in the United States is shown by the map presented in Fig. 17.1. The map also shows the average insolation on horizontal surfaces throughout the US, presented in the meteorological units known as Langleys/day, and converted into Btu/ft^2/day. To convert these further into watts per square meter, multiply the US unit by 3.15.

A tremendous amount of activity has been initiated in the US as a result of the fuel crisis which arose shortly after the oil embargo of October 1973, and a gigantic program, by solar standards, is now under way to create a viable industry in the area of solar heating and cooling of buildings of all sorts. The demonstration of the feasibility of solar house heating takes high importance in this program, and it is anticipated that some hundreds of solar-heated and, eventually, solar-cooled buildings will be constructed in the US during the three years following 1977. The sections presented below will explain the systems which are currently in use for solar heating and cooling, and will look ahead to the future when the generation of electricity on each individual residence will become a reality.

17.2 SOLAR SYSTEM TYPES

There are two primary categories into which virtually all solar heating systems may be divided. The first is passive systems, in which solar radiation is collected by some element of the structure itself, or admitted directly into the building through large, south-facing windows. The second is the active systems, which generally consist of (a) separate solar collectors, which may heat either water or air, (b) storage devices which can accumulate the collected energy for use at night and during inclement days, and (c) a backup system to provide heat for protracted periods of bad weather. In some cases, the system will be designed to provide somewhat less than the full heating and cooling requirements of the buildings at design conditions, in which case an auxiliary heating or cooling device will be called into play. Heat is transferred from the collectors or from the storage means by conventional equipment, such as fan-coil units, when hot or cold water is provided; fans, ducts, and air outlets when the heat transfer medium is air; and radiant means when heating is the only task which must be accomplished.

17.3 SYSTEM COMPONENTS

A. Collectors

In active solar heating systems, a separate collector is generally used, and this may be built into the structure, as a part of the roof or a south-facing wall, or it may be a separate device, which, in some cases, is mounted in some convenient location that is not actually attached to the residence.

A wide variety of solar collectors has been developed over the years, starting with the simplest version, which is simply a zigzag pattern of copper tubes soldered to a flat copper plate, and proceeding through many other variations to the most recent development, which uses the "Roll-Bond" process, in which the tubes are integral with the plate. This can only be done with a malleable metal, such as aluminum or copper. Many versions of galvanized steel collectors have been made, and air heaters have been developed which use glass, appropriately blackened, as the collector material. Yellott (1974) shows 14 different varieties of solar water and air heaters, and others have been developed since that time. Collection efficiency is dependent upon the transmittance of the cover glass, the absorptance for solar radiation of the absorber plate, and the longwave emittance of the plate. Collection efficiency varies with the time of day, and is dependent upon the number of hours away from solar noon, for south-facing collectors. This subject is treated in detail by Hottel and Woertz (1942), Whillier and Hottel (1955), Yellott (1974), and Chapters 6–8 herein.

B. Storage Devices

The storage of heat or cold is an essential part of a complete solar heating and cooling system, since there is no solar energy available at night, and, in virtually every location throughout the world, there will come a time when there will be three or more almost sunless days in succession. The types of heat storage systems are few in number, consisting primarily of tanks full of water, some times with an antifreeze liquid additive, or containers filled with small rocks. These operate by the specific heat process, with water having a specific heat of 4.187 kJ/kg°C (1.0 Btu/lb°F) and rocks having approximately 0.837 kJ/kg°C (0.2 Btu/lb°F). The rock bed system is used with air as the heat transfer medium, and this combination has the advantage that it cannot freeze and leaks are generally not serious. The water system has the advantage of being able to store far more heat in a given volume and weight, but precautions must be taken to prevent freezing, and leaks can be a very serious matter.

A system which is still under development uses the heat of fusion of various materials, particularly the eutectic salts represented by *Glauber's salt*, sodium sulfate decahydrate. This substance has a latent heat of somewhat more than 232.6 kJ/kg (100 Btu/lb) and it melts and solidifies at approximately 40°C (104°F). This was used by Dr. Telkes in the house mentioned, and it was here that some of the difficulties of employing this well-known and relatively inexpensive chemical were discovered. Water is another heat-of-fusion material which, when it freezes at 0°C (32°F), gives up its latent heat of fusion, 335 kJ/kg (144 Btu/lb). When it melts, the same amount of heat must be added to it. This procedure has been used for many years in the dairy industry to enable a relatively small refrigeration system to cool a large amount of milk in a short period of time by the melting of ice which has been stored up during off-peak periods. Water is one of the few liquids which, when it freezes, takes on a larger specific volume in the solid phase than it had when it was a liquid. Most of the heat-of-fusion storage materials have the opposite characteristic in that, when they freeze, the solid component is considerably heavier than the liquid, and so the solid falls to the bottom of the container, leaving the liquid at the top. When melting takes place, it is frequently difficult to reestablish a homogeneous fluid, similar to the water–ice relationship. Dr. Telkes developed a number of additives to nucleate the material, enabling it to solidify more readily, and adding another component which minimizes the tendency to stratify when solidification (freezing) takes place. The containers used by Dr. Telkes in her latest solar residence *Solar One*, to be described later, are made of plastic, with a relatively small thickness and a large surface area to improve the heat transfer characteristics of the storage device.

C. Distribution Systems

When water is used as the heat transport medium, the heat is carried into the structure by means of supply and return pipes, and the air within the structure is heated by blowing it over a finned coil, using a fan–coil combination of the type widely used in the heating and air conditioning industry. Radiant heating may also be used, employing either thin tubes along the upper portions of a wall, just below the ceiling, or using tubes embedded within ceilings or floors. The latter is not widely used in the US because most solar heating systems of the active type are being designed so that they can eventually be used as cooling systems as well, and this requires the use of a fan–coil unit. When air is the heat transport medium, the fan which draws the air through the collector and blows it into the storage may be used also to cause the air to flow to the various rooms which are being heated. Conventional air outlets will then distribute the warmed air properly. When cooling is attempted, the fan–coil system is the most widely used type, since it can both heat and cool, depending upon the temperature of the water flowing through its tubes. Radiant heating is not widely used, and it is not successful when cooling must also be done, since reduction of the air temperature below its dew point would cause droplets of moisture to form on the radiating surface, and this is not a permissible situation.

D. Auxiliary Heating and Cooling

In the terminology currently used in the US, an auxiliary heater or cooler is one which is called upon to carry part of the load during periods of highest heating or cooling demand. It may be a fuel-burning heater, an electric heater, or a compression refrigeration system operating as a heat pump in the heating mode. For cooling, the heat pump offers a very good solution to the problem, since it can both cool in the summer and heat in the winter. Electrical resistance heating can be used, but it is relatively expensive, since each kilowatt hour produces only 860 kcal (3413 Btu), whereas the heat pump, with a coefficient of performance of 4 or 5, will yield from 13 to 17 Btu/kWh and it can also provide cooling in summer.

E. Standby Heating and Cooling

Once again, to use current American terminology, a standby system is one which can carry the entire heating or cooling load and could be used when the solar system is not functioning at all. In past years, a fuel-burning system would have been used, preferably one burning natural gas. In future years, natural gas is not likely to be available in the US for this purpose, nor will light oil of acceptable sulfur content be available except at extremely high prices. Therefore, most users will turn towards electricity, which can provide as much heat as may be needed, although, again, the cost will be very high. Once again, a heat pump offers a good solution, except in areas where extremely low outdoor temperatures are experienced. A combination of heat pump and sun-warmed water forms a very economical way to solve this problem, as has been demonstrated by Yanagimachi in Tokyo (Yellott, 1960).

F. Cooling Systems

The one practical means of providing refrigeration with solar radiation as the source of energy is the absorption system. This may use water as the absorbant and ammonia as the refrigerant, or the situation may be reversed, and lithium bromide may be used as the absorbant, with water as the refrigerant. The latter system is by far the most widely used in solar cooling systems, but it has the drawback of requiring a working temperature close to 93.3°C (200°F), and this is difficult for a flat plate collector to achieve for more than 6 h/day.

17.4 SURVEY OF SIGNIFICANT SOLAR-HEATED AND COOLED RESIDENCES

A. Passive Systems

An increasing number of residences using various passive systems have been built and are under construction in the United States. Some of the more important of these will be described in alphabetical sequence. Figure 17.2 shows the very unusual house constructed near Albuquerque,

FIG. 17.2 The Baer house near Albuquerque, using water barrels mounted horizontally behind insulating panels which can be lowered during the day time to admit solar radiation, and raised again at night to conserve the collected solar radiation.

New Mexico, by Steven Baer as his family residence. This consists of four interconnected dome-type segments, called *zomes*. The south-facing wall of each of these elements is glazed with single glass, and behind the glazing is a *water-wall*, consisting of 20 to 25 steel drums, with their exposed ends painted black to absorb the incoming solar radiation. The indoor surfaces of the barrels are painted in various colors, since black is not needed for longwave radiation, and the drums provide heat to the interior of the structure primarily by radiant transfer. In front of these south walls are flat panels of insulating material, hinged at their lower edges, and provided with a simple means by which they can be lowered when the sun begins to strike them, and raised again at night or during bad weather. No other source of heat, other than a fireplace, is needed to keep the residence warm during the very cold and windy winters which Albuquerque experiences. Domestic hot water is supplied by a battery of four flat plate collectors on the east side of the house, and a windmill supplies a modest amount of electricity, so that the house can become at least partially self-sustaining. Mr. Baer has also devised a very ingenious window–wall, which can be turned into an insulating element of the house at night by filling the space between the two glasses with beads of styrofoam. During the daylight hours, when it is desired to admit solar radiation for the purpose of heating, the beads can be withdrawn from the wall cavity and stored in containers. This system, called, appropriately, the *bead wall* is one of the first approaches toward making a wall which is transparent during the daytime and still possesses good insulating qualities at night and during periods of bad weather.

Figure 17.3 shows another residence near Albuquerque, adjacent to the Baer home. This house, built by Dr. and Mrs. Paul Davis, uses a conventional air heater and a rock bed storage bin located beneath the patio on the side of the house. Solar heating is accomplished partially by allowing direct

FIG. 17.3 Paul Davis residence in Albuquerque, with air heaters providing hot air to charge, by thermosyphon action, rock bed storage enclosed beneath patio. The house is warmed at night by heat stored in the rock bed, and a fireplace provides the necessary auxiliary heat. Cooling in summer is adequately provided by ventilation through open windows.

radiation to enter through the windows, and these are provided with insulating shutters on the inside so that, somewhat like the bead wall, they can be blocked off at night to prevent excessive heat loss. The unique feature of this house is the fact that it uses thermosyphon action to cause circulation of warm air upward to the collector during the sunlit hours of the day. The air then passes into the top of the rock bed, and moves slowly downward through the rocks, giving up the absorbed heat to the storage system. The cooled air reenters the collector, and this action continues throughout the day. At night, the warm air from the rock bed is admitted to the house, and recirculating ducts bring the cooled air back from the house to the bottom of the rock bed. The system has worked well throughout two winters and, since it requires no power whatsoever in its operation, it has a very low annual cost. The house is well insulated, and the collector plus a wood-burning fireplace provide all the heat that is needed to maintain comfort throughout the winter.

One of the most original developments in solar heating and natural cooling is the *Skytherm* system, invented by Harold Hay of Los Angeles and tested very extensively in a prototype structure at Phoenix, Arizona, for 18 months in 1967–1968. The system proved to be able to keep the house within the "comfort zone" for all but about 2% of the hours during this protracted period. This is fully explained in references by Hay and Yellott (1969) and Yellott and Hay (1969). The structure features a steel panel ceiling–roof which supports ponds of water enclosed within plastic bags. The water is in thermal contact with the roof–ceiling, and the house is heated by radiation from the roof to the floor and walls, and to the occupants. It is cooled in summer by radiation from the floor and walls and the occupants to the ceiling which transmits the absorbed heat to the water. The water in turn loses heat by radiation to the sky and by evaporation, if additional water is allowed to flood over the bags, and is thus exposed to the sky. The system operates because it has a thermal valve in the form of insulating panels that can roll horizontally to cover the water ponds during winter nights and summer days. The solar radiation is absorbed by a black lining beneath the water ponds, and this in turn heats both the roof-ceiling and the water. During the summer, the panels cover the water ponds during the daytime and their aluminum coating reflects the solar radiation back to the sky. On summer nights, the insulation is rolled away and cooling takes place. In the original prototype, the panels were moved manually, but in the full-scale residence built at Atascadero, California (Fig. 17.4), the panels are moved by a motor that is controlled by an ingenious thermostat. This system is primarily applicable to areas that are free of snow in the winter and favored with relatively low dew point temperatures during the hot summer months.

FIG. 17.4 *Skytherm* residence built at Atascadero, California, for its inventor, Harold Hay of Los Angeles. Architecture by Professor Kenneth Haggard, California State Polytechnic University at San Luis Obispo.

The house at Atascadero (Fig. 17.4) has been in operation since the fall of 1974, and it has remained within the comfort zone throughout that entire period, with no need for auxiliary heating or cooling, despite the fact that the climate of Atascadero is considerably colder than that of Phoenix. The steel roof decks are each 4 ft wide and 51 ft long, and they tie the walls of the house together in such a secure manner that it has already gone safely through at least one moderately severe earthquake. No sign of the solar heating and natural cooling system can be seen from the adjacent street, and this is a valuable asset in areas that forbid the mounting of any portion of the heating or cooling system so that it is visible above the roof of the dwelling.

FIG. 17.5 Trombe–Michel house, developed for use in the south of France, with a massive concrete south-facing wall used as the heat absorber, insulated by double glazing, with convection space provided between the concrete and the glazing. The vertical south wall provides both heat storage and heating of the building at night.

Figure 17.5 shows the house originated by Dr. Felix Trombe, Director of the Solar Energy Laboratory at Odeillo, and designed by architect M. Michel. The Trombe system has been studied since 1956, and patents have been granted to the National Center for Solar Research, in the name of Dr. Felix Trombe, in 1956 and again in 1971 and 1972. The system is passive, because the solar radiation is absorbed by a heavy concrete south-facing wall, which is covered with a single glazing. The concrete wall extends below the floor of the living space, and entry ports are provided which enables cool air from the floor level to flow downward, and then, as the concrete is heated by the sun, the air rises in the space between the glass and the wall, and returns to the house through ports located near the roof. An overhang protects the wall from excessive insolation during the summer. Electric strip heaters are provided for standby purposes during excessively cold weather and during days with little sunshine. The location, adjacent to the resort town of Font Remeu, enjoys more sunlight than any other portion of France, and the summer weather is pleasantly cool, so that, by simply opening up dampers at the top of the windows, air can be circulated through the house from openings on the north side into the solar-heated space between the glazing and the concrete, to be discharged through the dampers to the atmosphere at night. This system is extremely simple, but it has proven its effectiveness, and much more elaborate structures than that shown in Fig. 17.5 have been constructed in the same region.

Another completely passive heating system is that used by Mr. and Mrs. David Wright in the house which they have built at Santa Fe, New Mexico (Fig. 17.6). This house is constructed with heavy double adobe walls, with

FIG. 17.6 The David Wright house, Santa Fe, New Mexico, employing large south-facing glass, internally shaded at night to minimize heat loss, and using the floor of the building as the primary heat storage means.

insulation between two 12-in. thicknesses of adobe block. The floor is also made of 24 in. of adobe, with insulation below it. The roof is heavily insulated, and the entire south wall consists of double-glazed windows. An overhang shades the windows from the summer sun, but permits the low-angle winter sun to enter freely. In order to seal the house against heat loss at night, a horizontal insulating curtain is provided, which can be retracted into space at the top of the window during the daytime, and then lowered to any desired extent to control the admission of solar energy in season when the house would tend to be overheated without this provision. At night, the insulating curtain is lowered completely, so that all of the glass is covered, and the combination of glass and curtain achieves a very high insulating value.

Domestic hot water is provided by a double-glazed thermosyphon system located on a hillside adjacent to the structure, and summer ventilation is accomplished simply by opening the windows in the east and west sides. The interior of the house is quite open, and air circulates freely to enable the heat, which is stored in the floor and walls of the building, to flow upward to the sleeping quarters on the second level. A wood-burning fireplace is the only auxiliary heat that is needed, and cooling is not needed during the summer, because, at the 6000-ft altitude of Santa Fe, high temperatures are rarely encountered, and night temperatures are always relatively cool.

The residences described in the foregoing section are treated in much greater detail in "Solar Oriented Architecture," which was prepared by the writer and a team of graduate students in the College of Architecture, Arizona State University, Tempe, Arizona, for the American Institute of Architects Research Foundation, under a contract with the National Science Foundation.

B. Active Systems

Most of the hundreds of solar-heated residences built throughout the world use the active system, in which separate collectors are used to gather the solar radiation, transfer it to air or water, and store it in tanks of water or rock piles, or both. The air and water are circulated by fans or pumps, and conventional means are used to distribute the heat to the interior of the residences. The oldest of these in terms of operating history is the structure shown in Fig. 17.7, which was built by Dr. George Löf, one of the pioneers in the solar energy movement in the United States, in a southern suburb of Denver, Colorado. The altitude at this point is approximately 5000 ft above sea level, and the winters are quite severe. The summers are characterized by brilliant sunshine and relatively cool nights, so air conditioning is not a necessity. The Löf house (Fig. 17.7) uses two long air heaters, mounted within a parapet on the roof. The heaters use the overlapping glass plate principle, invented by Dr. Löf during World War II and first tested by him in a much smaller cottage in Boulder, Colorado.

FIG. 17.7 The Löf house, erected in Denver in 1958. Two banks of overlapping glass collectors are used to warm air, which in turn heats gravel in two storage cylinders. Heating at night and during cloudy weather is accomplished by the stored sun-collected heat in the rock-filled cylinders.

The collectors utilize glass plates, with the upper half transparent and the lower half blackened. The plates are set at an angle to the direction of air flow, so that the solar radiation, entering through a cover glass, also passes through several thicknesses of clear glass and is then absorbed on the blackened lower segment of each glass plate. Temperatures as high as 76.7°C (170°F) can be attained, and the heated air is blown through two large cylinders which are filled with rocks approximately 5 cm (2 in.) in diameter. This constitutes the storage system, which is charged up during each day. Usually the house is kept sufficiently warm by solar radiation entering through south-facing windows and through sun-heated walls. At night, the house is heated to approximately 40% of its requirements on the coldest winter nights by heat that is stored in the rock-filled cylinders.

The entire structure was subjected to a very careful testing procedure during the winter of 1974–1975, and it was found that there was very little deterioration in performance of the system despite the fact that it had been in operation approximately 17 years. The backup system is a gas-fired air furnace, and domestic hot water is also provided by a gas-burning heater. It is probable that, when natural gas becomes unavailable in the Denver area, electricity will be used in resistance heaters, or in a heat pump, to provide the backup energy source.

One of the most successful and simplest solar heating systems is that devised by Dr. Harry Thomason of Washington, D.C. His *Solaris* system uses an open-flow collector made of corrugated aluminum sheets that are painted with a waterproof black. A single cover glass is used, and the collector is mounted directly on an insulated and water-proofed roof (Fig. 17.8). The primary storage is a 1600-gal water tank, located in the basement of the house, and this is surrounded by some 50 tons of rocks, which provide both additional storage and a very large amount of heat transfer surface. The rocks are warmed by the outer surface of the tank, which in turn is heated by the sun-warmed water. A differential thermostat with one sensor fastened on the collector surface and the other in the water tank directs the pump to start whenever the sun has warmed the collector until its temperature is above the temperature of the water in the tank.

The water is pumped to a distribution pipe at the top of the roof and runs down by gravity through the troughs of the solar collector. The system

FIG. 17.8 The Thomason residence in Washington, D.C., which uses the system shown in Fig. 17.9, to provide heating by relatively low-temperature sun-warmed water. This house now has more than eighteen years of experience to demonstrate its feasibility.

is "fail-safe," because the water drains from the collector whenever the pump stops for any reason. Warm air is used as the heat transport medium, and this is circulated by a small blower. For summer operation, a small compression refrigeration system is used to cool the water in the tank. A compressor operates only during off-peak periods, and cool air is circulated through the house during the daylight hours, when air conditioning is required.

The first of the Thomason houses was built in 1959, and no less than 20 were in operation by the end of 1975, in locations ranging from California and Colorado to Minneapolis, located in one of the coldest sections of the United States. Figure 17.9a, b shows a cross section of the *Solaris* system and an open-flow collector.

Representative of the high-technology systems which are now proliferating throughout the United States is *Phoenix of Colorado Springs* (Fig. 17.10), which was built during the winter of 1973 to show the citizens of Colorado Springs, Colorado, that the sun could provide most of the heating requirements of a large, well-insulated residence, despite the fact that this city is located at an elevation of more than 5000 ft and encounters very severe winters with large amounts of snow. The collectors are located on the south wall of the second floor of the residence and on a free-standing extension above the roof. The roof has a white topping applied above the waterproofed surface to reflect additional solar radiation onto this collector when the sun is low in the sky, as it is during the winter months. A 16,000 gal tank, buried on the north side of the house, provides the storage, and the heat transport fluid from the collector to the storage tank consists of a mixture of water and ethylene glycol. A heat exchanger transfers the heat from the sun-warmed transport fluid to the contents of the storage tank, which are then circulated through the air handlers within the basement of the house. Standby heating and summer cooling are provided by an air-to-air heat pump, also located in the basement. Because of the very large size of the storage tank, *Phoenix of Colorado Springs* can successfully go through a week of inclement weather without having to call upon the auxiliary heat source.

A number of active solar-heated and -cooled residences are now in operation in the United States. These use double-glazed high-performance flat plate collectors to heat water, freeze-proofed with ethylene glycol or a similar substance. In winter, the water is used directly in the heating systems of these buildings, and in summer, the hot water is used to activate a lithium bromide absorption refrigeration system. Difficulty has been experienced in obtaining a sufficiently high collection efficiency at the 93.3°C (200°F) temperature level to make such systems economically desirable at the present time, but with the introduction of a practical sun-following concentrating collector using a linear Fresnel lens, these systems offer far more promise of

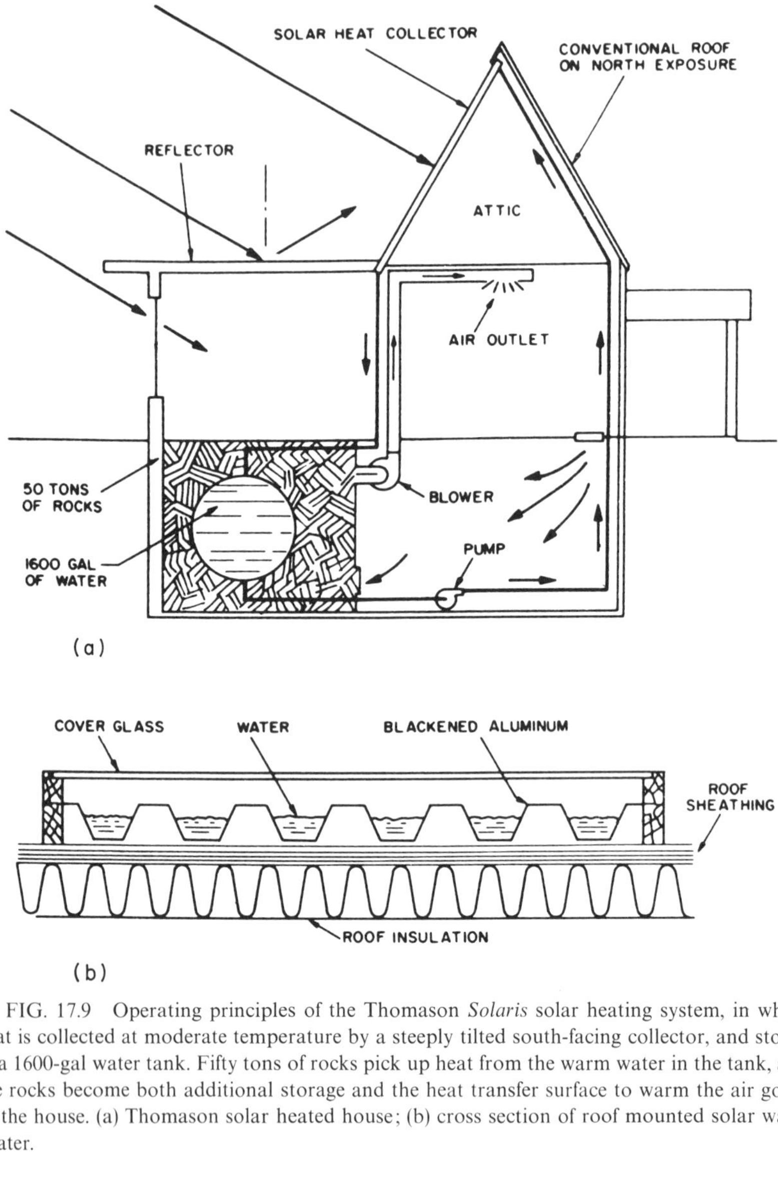

FIG. 17.9 Operating principles of the Thomason *Solaris* solar heating system, in which heat is collected at moderate temperature by a steeply tilted south-facing collector, and stored in a 1600-gal water tank. Fifty tons of rocks pick up heat from the warm water in the tank, and the rocks become both additional storage and the heat transfer surface to warm the air going to the house. (a) Thomason solar heated house; (b) cross section of roof mounted solar water heater.

being successful both technically and financially. Flat plate, south-facing collectors can generally operate in this temperature range for only about 6 h/day at the maximum, since, during the summer, the sun rises in the northeastern quadrant of the sky, and a south-facing collector does not

FIG. 17.10 *Phoenix of Colorado Springs*, a solar-heated and heat-pump-cooled house erected in Colorado Springs, Colorado in 1973. Flat plate collectors provide the heat required in winter, and air circulation is provided by the indoor fan of the heat pump, which serves as the auxiliary heat source and as the summer cooling means.

receive a substantial amount of direct radiation until close to 0900 h on a summer day. The sun moves again to the northwest by 1500 h, giving only approximately 6 h of collection time. The concentrating collector, on the other hand, can turn towards the east and pick up the early morning solar radiation, and follow the sun across the sky until it is about to set in the northwest. Such collectors are now undergoing extensive tests at a manufacturing plant near Dallas, Texas, and many are already in commercial use.

Representative of the highest technology yet applied to a solar house is *Solar One*, built by the Institute of Energy Conversion, University of Delaware, Newark, Delaware. Shown in Fig. 17.11, this house uses vertical south-facing air heaters to provide most of the winter heat requirements. In addition, two groups of collectors are mounted on the roof, which is tilted at an angle of 60° to the horizontal. The photovoltaic cells are intended to provide the electrical power required by the residence, and surplus power

FIG. 17.11 *Solar One*, the solar-electric house erected at Newark, Delaware, by the Institute of Energy Conversion, University of Delaware. Solar heat is provided during the winter by vertical south-facing panels in bay windows, and by solar-electric panels on the roof, tilted at 60° to the horizontal. Cadmium sulfide cells produce both electricity and heat when irradiated, and since they must be kept relatively cool, the heat is stored in heat-of-fusion devices, located in the basement. The electricity generated during the day will be stored in 120-V d.c. storage batteries, for use at night to run a heat pump which will cool the building and have its excess cooling power stored in another heat-of-fusion storage bin. This is the only Type IV house in existence at the present time, but it is certainly the shape of things to come.

generated during the day is stored in a bank of storage batteries. Cadmium sulfide cells cannot sustain temperatures much above 82.2°C (180°F), and so during summer days, the cells must be cooled, even though the heat which they collect cannot be used. Vents are provided to release the heated air to the atmosphere.

Another novel feature of *Solar One* is the use of two types of heat-of-fusion materials for energy storage. The first, similar in composition to the well-known Glauber's salts, stores heat at about 40.6°C (105°F), for use when the building requires heating. The second stores heat at 12.8°C (55°F), and this is used during the summer to store coolness, produced by a heat pump during the day when the solar cells are producing electricity, for use at night. The heat pump also serves as the backup heating and cooling system to keep the house comfortable when the solar apparatus is not functioning. Although it is recognized that *Solar One* is still very much in the experimental stage, it affords a preview of what the future may bring in the way of houses that are completely self-sustaining, producing both heat and electricity from the sun, with only minimal need for assistance from the local electrical utility. Cadmium sulfide cells are recognized as being less efficient by a factor of 3 to 1 than the silicon cells which have proved their capability in the space programs of the US and USSR. However, cadmium sulfide is extremely cheap to produce, in contrast with the ultraexpensive silicon cells. There is hope and even expectation that the cost of silicon cells will be reduced by a factor first of 10 to 1 and then 100 to 1 within the coming decade, and when this takes place, it is not too optimistic to expect that systems similar to that used in *Solar One* will be in much wider use in the sunlit portions of the world.

17.5 CONCLUSION

Solar radiation can produce most of the heat required by dwellings in the temperate and hot regions of the world. Perhaps a dozen houses throughout the world are currently being provided with most of their thermal requirements by means of passive systems, in which solar radiation heats some element of the structure itself, and this in turn both stores heat and transfers it to the interior of the building; or by active systems, in which the essential components, collector, storage, standby, and energy distribution, are separated, with air or water used as a transport fluid to carry the collected heat from the collector to the storage and from the storage to the interior of the building. Most of the solar-heated houses which have been built thus far use the second system, because it is more versatile and can be applied in virtually any part of the world that has a reasonable amount of sunshine. Costs are generally lowest for the passive systems, since they avoid the use of pumps, fans, and expensive controls. For the active systems, the least expensive are the *Solaris*, advocated by Dr. Harry Thomason, of Washington, D.C., and the *Skytherm* system, invented by Mr. Harold Hay of Los Angeles. The latter can provide both heating and cooling without the necessity for any electrical power, while the former is restricted to heating, since natural cooling cannot be produced at the present time when the solar heat collector is glazed. Vigorous efforts are being made to produce collectors at a cost lower than \$8 to \$10 (US currency, 1977 value), which represents the selling price of the least expensive flat plate collectors. Although glass is preferred for the glazing material because of its ability to withstand both high temperatures and intense ultraviolet radiation, transparent shatter-resistant fiberglass is now coming into use, which has slightly higher solar transmittance than glass and is apparently able to offer a life of at least 20 years. It is anticipated that the decade between 1975 and 1985 will see new types of collectors with higher efficiency than those available today, while other developments will bring about lower life cycle costs for collectors and thus make the use of solar energy more attractive from the economic point of view. The technology is available today, and the major remaining problem is to bring down the cost of these systems until they offer financial incentives much greater than those offered by electricity, which is likely to be the primary energy source of the future in most parts of the world.

18

Solar Production of Hydrogen

T. N. VEZIROGLU

UNIVERSITY OF MIAMI
CORAL GABLES, FLORIDA

S. KAKAC

MIDDLE EAST TECHNICAL UNIVERSITY
ANKARA, TURKEY

18.1 INTRODUCTION

Energy is an essential ingredient for increasing productivity in both agriculture and industry. Increase in energy consumption of a country gives a positive impact to the economic as well as social development of that country. For all countries, a cheap, abundant, and clean supply of energy is very important, since it plays a vital role in the development of industry and in increasing the agricultural output.

Today, there are three important world problems. They are shortage of energy, depletion of fossil fuel sources, and increasing pollution. These are related problems. The peoples of the world are aspiring to improve their standards of living. The standard of living is directly proportional to the energy consumption. Consequently, the peoples of the world are trying to increase their energy consumption faster than the population increase. The world has a finite size. Its resources, including the fossil fuels (coal, petroleum, and natural gas), are finite in quantity and are being depleted at an ever-growing rate. The fossil fuels have adverse effects on environment. An important type of pollution, air pollution, is caused mainly by fossil fuels used to obtain energy for transportation, power stations, etc.

Solar energy is being seriously considered to satisfy part of the demand. If a convenient way of utilizing the solar energy can be found, it could be an answer to the aforementioned problems, since it is almost undepletable and is clean. However, there are some difficulties for large-scale utilization of solar energy. It is available only during the daytime when the sky is clear. Even then, its intensity changes with time. Therefore, it becomes important to find a means of storing the solar energy when it is available, so that the stored energy can meet the demand when the solar energy is unavailable or is of low intensity. Producing hydrogen for storing and transmitting solar energy looks like an attractive solution to these difficulties. Hydrogen can be used as a fuel, can be used to produce electricity, can be transmitted over long distances economically, and is a clean form of secondary energy.

This chapter considers various methods of producing hydrogen using solar energy, viz., thermochemical, electrolytical, thermal, and photosynthetical. These methods are compared, and the advantages of each are stated. The advantages of using hydrogen for storing solar energy over the other synthetic fuels are discussed.

Indirect forms of solar energy such as ocean thermal, ocean current, wind, and hydraulic are also considered as primary energy sources for producing hydrogen.

18.2 SOLAR-HYDROGEN ENERGY SYSTEM

Figure 18.1 illustrates the solar-hydrogen energy concept advocated by many energy researchers (Escher and Hanson, 1975; Swet, 1974; Paleocrassas, 1975; Hildebrandt, 1975; Ohta *et al.*, 1975). As can be seen from the figure, some of the solar energy is converted to electricity (e.g., by means of a solar boiler–steam turbine system) to meet the local baseline load. A greater part of solar energy is used to produce hydrogen from water by means of one of the hydrogen production methods such as direct thermal cracking, thermochemical production, electrolysis, etc. Some of the hydrogen is used in the local power plant to meet the peak load demand and also the demand when the sun is unavailable. The rest of the hydrogen is transmitted to the centers of energy consumption by means of pipelines (gaseous H_2) or by means of supertankers (liquid H_2). Hydrogen is then used in homes (for cooking and heating), to power transport vehicles (cars, buses, trucks, trains, ships, airplanes, etc.), to produce electric power, and as a feedstock in industry. If oxygen is produced as a by-product while hydrogen is produced, it could be used in sewage treatment facilities to accelerate treatment, in oxygenating the *dead* rivers and lakes to rejuvenate them, and in industry. Hydrogen after use (after combining with oxygen) becomes

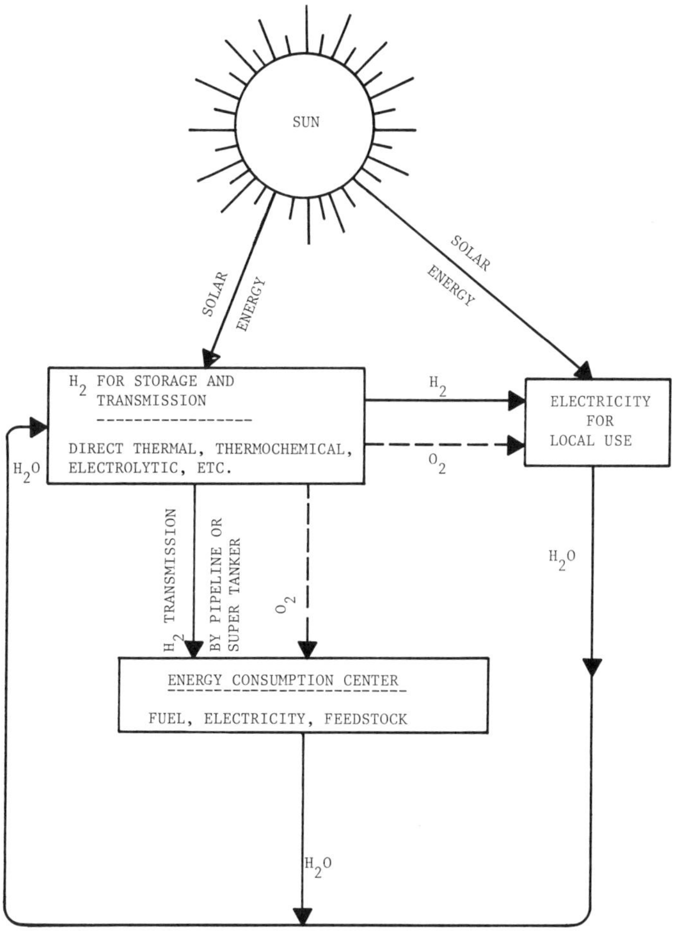

FIG. 18.1 Schematic of solar hydrogen energy system.

water vapor or water, and completes the cycle by condensation (rain) or directly. This described energy system is nonpolluting and depends on a source that is almost undepletable.

18.3 SOLAR HYDROGEN PRODUCTION

Today, the world hydrogen production is 15 billion SCF per year. It is equivalent to 1.26×10^7 cal/year in energy units. Some 90% of the hydrogen produced is used as a chemical intermediary in chemical processes to produce hydrogen containing products such as ammonia, methanol, and refined petroleum fuels and chemicals. When hydrogen assumes its complex energy as well as nonenergy roles, the hydrogen production will have to be increased by many orders of magnitude. If three quarters of the energy demand is to be met by synthetic fuels, such as hydrogen, then the hydrogen production rate would reach 5.9×10^{20} cal/year by the year 2050 when the world equilibrium conditions are expected to be reached (Veziroglu *et al.*, 1975). This rate of hydrogen production could be obtained using 0.5 to 1% of the earth's surface area for solar collectors, depending on what hydrogen production method is used.

There are four main hydrogen production methods in which solar energy can be used, viz., direct thermal, thermochemical, electrolytic, and photolytic. These methods of hydrogen production are considered here.

A. Direct Thermal

If water (steam) is heated up to about 3000°K or above, hydrogen and oxygen will start dissociating. This water decomposition reaction can be expressed as

$$H_2O + (\text{heat}) \rightarrow \eta_1 H_2O + \eta_2 H_2 + \eta_3 O_2 \tag{18.1}$$

where the substances are in gaseous form, and η_1, η_2, and η_3 are the mole fractions. The degree of dissociation of H_2 versus process temperature has been evaluated at four different pressures (Bilgen, 1975) and the results are presented in Fig. 18.2. It can be seen that extremely high temperatures are required for equilibrium dissociation of water to hydrogen and oxygen and that mole fraction or degree of dissociation increases with decreasing equilibrium pressure. Therefore water should be decomposed at a fairly high temperature combined with a reduced pressure. The total thermal energy required, heat rejected and the thermal efficiency versus temperature of dissociation of water are shown in Fig. 18.3.

The energy for dissociation of hydrogen can be obtained from the solar radiation. For this purpose, an optical system which collects solar radiation and concentrates it into a small area could be utilized. The concentration ratio C which is defined as the ratio of the heat flux within the small area to the actual flux received depends basically on the temperature requirements in the small area or image. For a parabolic mirror, for instance, the theoretical concentration ratio is given as a function of rim angle α (Duwez, 1963) as

$$C = 46{,}200 \sin^2\alpha \tag{18.2}$$

The direct solar energy flux or the solar constant is about 0.14 W/cm^2 in the outer atmosphere and it can be taken as 0.1 W/cm^2 on the earth, although it is somewhat variable. If the direct solar energy flux at normal incidence on the earth and concentration ratio are known, the heat flux

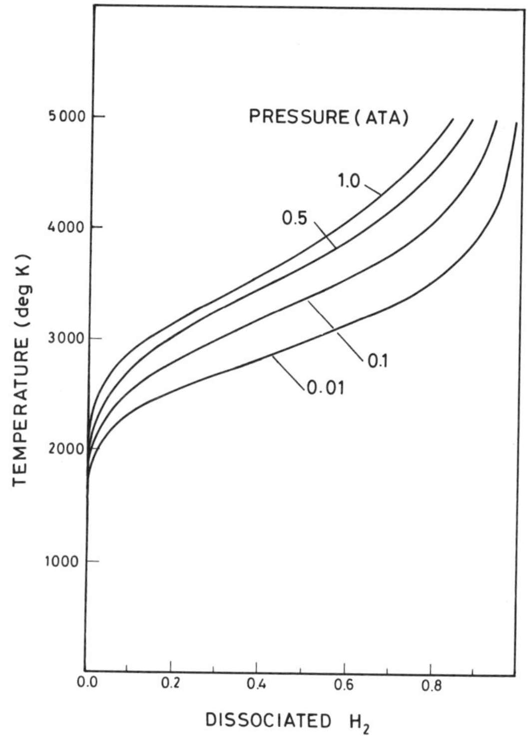

FIG. 18.2 Water decomposition temperature versus mole fraction of dissociated hydrogen. $H_2O(g) \rightarrow \eta_1, H_2O(g) + \eta_2 H_2 + \eta_3 O_2$

in the image can be calculated and the maximum attainable temperature can be estimated from the Stefan–Boltzmann law of radiation.

In order to account for the reflection losses, the reflection coefficient η_r can be introduced (Trombe *et al.*, 1973). The attainable temperatures as a function of concentration ratio are shown in Fig. 18.4. Several optical concentration systems are also indicated in the same figure, viz., Mont Louis, Odeillo, Natick, Algiers, and the proposed tower top point focus solar energy concentrators (Trombe, 1961; Davies and Cotton, 1957; Hildebrandt, 1974). For a reasonable yield in hydrogen production by direct decomposition of water, temperatures around 3000°K or above are required, and as a general rule, in order to operate at these temperatures, higher temperatures should be attainable at the image. High temperatures can be obtained in optical systems such as in large solar furnaces. Minimum requirements would be a concentration ratio around or in excess of 10,000. Above 10,000, the construction of the optical system necessitates a careful engineering design (Trombe and Le Phat Vinh, 1971; also see Chapters 8 and 10).

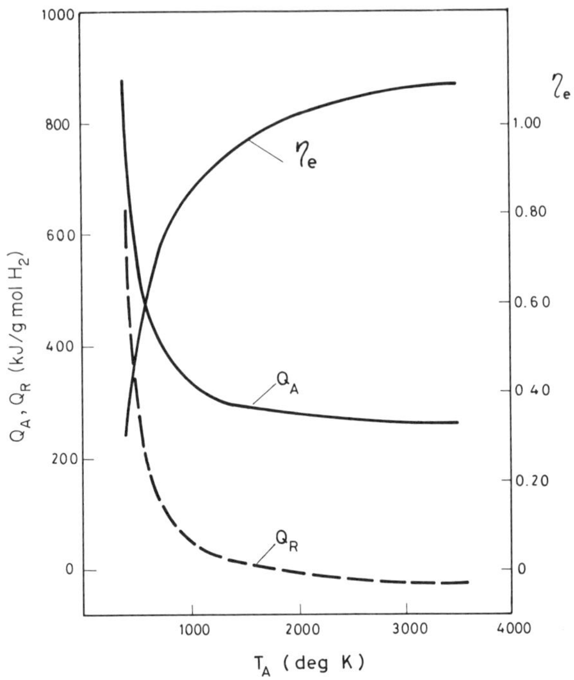

FIG. 18.3 Water decomposition temperature versus heat absorbed and heat rejected per mole of hydrogen and thermal efficiency.

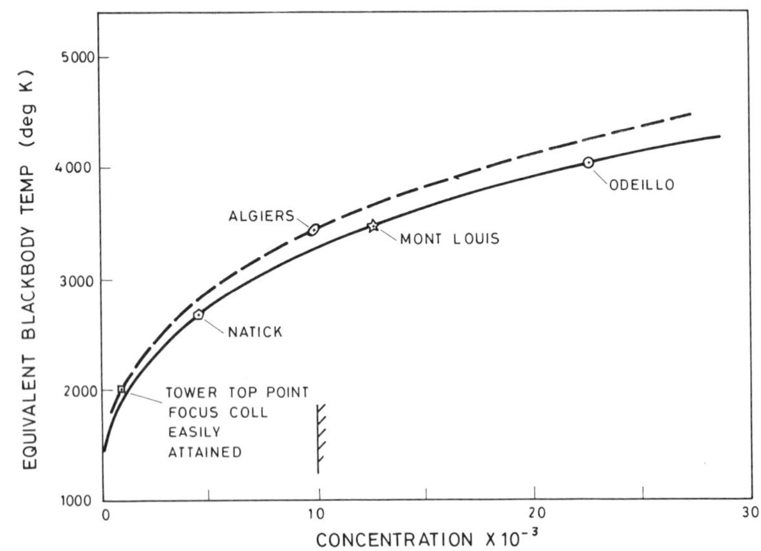

FIG. 18.4 Blackbody temperature at image versus concentration ratio. Incident direct sunshine $H = 0.1$ W/cm^2. Double reflection (——): $\eta_r = 0.65$; single reflection (– – –): $\eta_r = 0.80$.

Direct thermal production of hydrogen offers the following advantages: (1) high thermal efficiency, (2) little or no environmental impact, and (3) not using intermediary chemicals. On the other hand, for commercial realization of this method extensive research is needed because of the high temperature requirements.

B. Thermochemical

One disadvantage of the direct thermal method of hydrogen production is that it requires very high temperatures. This difficulty has led many scientists (Swet, 1975; Funk, 1975; De Beni, 1975; Hickman *et al.*, 1975; Chao and Cox, 1975) to search for thermochemical reactions to decompose water. In this method, first water and one or more chemical elements or compounds react while heat is being added, resulting in a combination of hydrogen and/or oxygen of water with compound(s) and either hydrogen or oxygen being released as a gas. Then, in one or more chemical reactions the new chemical compound(s) of the first reaction are reduced to their original composition with the help of other intermediary chemicals and/or heat, releasing oxygen or hydrogen. Some work is needed to separate the resulting chemicals. Consequently, the only inputs are heat, water, and work; and the only outputs are hydrogen, oxygen, and lower-grade heat. The intermediary chemical elements and/or compounds are regenerated and recycled.

As can be seen from the above description of this method, two or more chemical reaction steps are needed for a complete thermochemical hydrogen production cycle. The following are examples for such two- and three-step cycles:

Two-step cycle:

$$H_2O + CH_4 \rightarrow CO + 3H_2 \tag{18.3}$$

$$\begin{array}{c} CO + 2H_2 \rightarrow CH_4 + \frac{1}{2}O_2 \\ \hline H_2O \rightarrow H_2 + \frac{1}{2}O_2 \end{array} \tag{18.4}$$

Three-step cycle:

$$H_2O + Cl_2 \rightarrow 2HCl + \frac{1}{2}O_2 \tag{18.5}$$

$$2HCl + 2CrCl_2 \rightarrow 2CrCl_3 + H_2 \tag{18.6}$$

$$\begin{array}{c} 2CrCl_3 \rightarrow 2CrCl_2 + Cl_2 \\ \hline H_2O \rightarrow H_2 + \frac{1}{2}O_2 \end{array} \tag{18.7}$$

Other examples of thermochemical cycles for hydrogen production can be found in the literature cited.

Table 18.1 lists some thermochemical cycles, their steps, maximum temperatures needed for the reactions, and the thermal efficiencies based on

TABLE 18.1

Evaluation of Some Hydrogen-Producing Thermochemical Cycles

Reference	Cycle title	Steps	Max temp (°K)	Thermal efficiency (%)
Funk (1975)	Vanadium chloride	4	1000	71.8
	Iron chloride	3	1200	26.2
	Copper chloride	3	973	29.4
	Iron chloride–iron oxide	5	1073	32.5
	Carbon–steam–iron	3	1673	75.5
	Steam–iron–carbon dioxide	5	923	24.6
	Arsenic iodide–arsenic oxide	4	600	53.3
Pangborn and Sharer (1975)	IGT cycle C-5	4	1200	61.3
	IGT cycle A-2	5	1200	33.9
	Euratom cycle MARK-9	3	925	45.2
	General Motors	4	1000	17.5
Chao and Cox (1975)	Cesium oxide	4	1323	48.0
	Calcium bromide	4	1000	59.0
	Fe chloride oxide	5	923	53.0

higher heating value of hydrogen. As can be seen from the table, for the 14 cycles listed, the reaction steps range from 3 to 5, the maximum temperatures required from 600 to 1673°K, and the thermal efficiencies from 17.5 to 75.5%.

An important factor to be considered in thermochemical processes is the recovery of the reagents or the intermediary chemicals. It is estimated that in each cycle, recovery yields of 99.9 or even 99.99% (Joly, 1975) have to be achieved if the thermochemical processes are to be viable. Otherwise, the hydrogen production cost would rise because of the cost of the reagent replacements and important environmental problems would rise because of the pollution caused by the spoilage of the reagents.

C. Electrolytic

A method of hydrogen production, for which the technology exists today, is to electrolyze water using dc electricity. In this method an electrolysis cell is used to produce hydrogen and oxygen from water. The cell consists of two electrodes immersed in an electrolyte (water plus some conducting chemical) and connected to a dc power supply. When sufficient potential is applied between the electrodes to cause a current to flow, oxygen is liberated at the anode and hydrogen at the cathode. The net, overall effect is to decompose water into its elemental components of hydrogen and oxygen. The electrochemical reaction taking place can be described as follows:

$$\text{Cathode:} \quad 4H_2O + 4e^- \rightarrow 2H_2 + 4OH^- \tag{18.8}$$

$$\text{Anode:} \quad 4OH^- \rightarrow 2H_2O + O_2 + 4e^- \tag{18.9}$$

$$\text{Sum:} \quad 2H_2O + (\text{electrical energy}) \rightarrow 2H_2 + O_2 \ (\text{heat}) \tag{18.10}$$

Since this is a straightforward way of obtaining hydrogen (and oxygen) from water and since the technology is already established, several researchers (Bockris, 1975; Hildebrandt, 1975; Anderson, 1975) considered using it for storage and transmission of solar energy. In this method, the solar energy must first be converted to dc electric power, and then to hydrogen through electrolysis. Hence, it is especially suited for coupling with ocean–thermal, wind, hydro, and photovoltaic forms of solar energy since in these cases the solar energy is usually (or always) converted to electricity.

Presently, the conversion efficiencies of the electrolyzers are about 60–70%, resulting in an overall thermal (heat to hydrogen) efficiencies of 25–30%. It is expected that with advances in technology (e.g., electrolysis at higher temperatures and pressures), the efficiencies in the foreseeable future could go up to 40–42%.

D. Photolytic

The sun's photons, under certain circumstances, can be absorbed by water molecules, and when the energy absorbed reaches a certain level (68.3 kcal/mole of water), hydrogen can be released. This phenomenon, photolysis, could be used to produce hydrogen from the sun's rays or photons. Photons in the ultraviolet region of the radiation spectrum possess the energies needed for the direct photolysis of water. Most of the ultraviolet radiation is absorbed in the upper atmosphere and mainly by water vapor. Consequently, the intensity of the solar radiation reaching the earth is very low in the ultraviolet region of the spectrum, but high in the visible region. On the other hand, water is almost transparent to the visible light. Hence, photocatalysts must be used for the photolysis of water on earth. The function of such a catalyst is to absorb the solar radiation and then transmit the energy to water in order to decompose it. Using a photocatalyst "X" the photolysis reaction can be expressed as

$$H_2O + 2X + \text{photon energy} \rightarrow 2\,(\text{reduced X}) + 2H^+ + \tfrac{1}{2}O_2 \tag{18.11}$$

$$2\,(\text{reduced X}) + 2H^+ \rightarrow 2X + H_2 \tag{18.12}$$

$$\text{Sum:}\quad H_2O \rightarrow H_2 + \tfrac{1}{2}O_2$$

It is important to note that the photocatalyst X is not consumed, but is regenerated and available for reuse. Three different groups of photocatalysts have been considered (Paleocrassas, 1975). Table 18.2 shows their actual and theoretical conversion efficiencies. It can be seen that the efficiencies are low, and those of the semiconductor catalysts look a little better.

TABLE 18.2

Photocatalysts and Energy Conversion Efficiencies

Photocatalyst	Actual efficiency (%)	Theoretical efficiency (%)
Compound salts	0.0024	3
Compound semiconductors	0.82	28
Photosynthetic dyes	?	7

Biochemists (Mitsui, 1975) have found out that there exist biological photocatalysts too. These are intact cells of some species of blue-green, green, and red algae and some photosynthetic bacteria. Basically, solar radiation is absorbed by a pigment in the cell, e.g., the light harvesting chlorophyll protein.

This energy is transmitted to an electron through the reaction center chlorophyll protein. The source of this electron is a hydrogen atom donated by some electron donor compound such as water. Under certain physiochemical conditions, such energized electrons combine with $2H^+$ to form H_2. The energy conversion efficiencies of such processes are very low since the excited electrons are used in many biological processes in addition to producing hydrogen. Research efforts are now concentrated for producing hydrogen, using cell-force photochemical systems in order to increase the hydrogen yield.

18.4 CONCLUSION

It is evident that the hydrogen energy concept fits very well with solar energy for storing, transmitting, and utilizing it. Among the four basic methods for producing hydrogen from solar energy, the direct thermal method has the potential of highest thermal efficiency, followed by thermochemical, electrolytic, and photolytic methods. It is expected that in the near future solar hydrogen could be produced using electrolysis, and later, thermochemical processes; and in the years 2000s, by direct thermal water splitting. All the methods will benefit from further research and development work. However, more of the efforts should be devoted to the direct thermal and thermochemical processes because of their potentials for greater energy conversion efficiences.

19

Solar Energy Measuring Equipment

BYARD D. WOOD

MECHANICAL ENGINEERING FACULTY
ARIZONA STATE UNIVERSITY
TEMPE, ARIZONA

19.1 HISTORICAL BACKGROUND OF SOLAR RADIOMETRY

The various radiation fluxes to and from the earth's surface are among the most important terms in the energy economy of the earth as a whole. In comparison to all other known proven sources of energy, it is consoling to note that the life expectancy of this constant source of energy is at least another 50 million years (Backus, 1976). The measurement of these fluxes is of essential value for the following purposes (Guide to Meteorological Instrumentation and Observing Practices, (Anonymous, 1971b).

"The study of instantaneous variations and timewise constancy of the radiation emitted by the sun.

The study of the variation in time and space of the transformation of energy within the earth–atmosphere system.

The study of the effect of local atmospheric conditions (e.g., emission particulates, dust, and water vapor) on the direct and diffuse insolation in order to calculate the distribution and variations of incoming, outgoing, and net radiation.

A compilation of direct and diffuse insolation data (time and location) for the satisfaction of biological, medical, architectural, agricultural, and industrial activities with respect to solar radiation (Anonymous, 1971b)."

In order to optimize the design of any solar application, the daily and seasonal variations in each radiation constituent must be well understood along with their relationships with other meteorological elements.

The definitions used in this chapter are primarily based on the "Terminology of Radiation Quantities and Measuring Instruments" adopted by the World Meteorological Organization (WMO) Commission for Instruments and Methods of Observation (Anonymous, 1971b).

A. Classification of Radiation Measuring Instruments

Pyrheliometer A pyrheliometer is an instrument for measuring the intensity of direct solar radiation at normal incidence; it can either be a primary standard instrument or a secondary instrument scaled by reference to a primary instrument. The latter have sometimes been called actinometers.

Pyranometer A pyranometer is an instrument for the measurement of the solar radiation received from the whole hemisphere. It is suitable for the measurement of the global or sky radiation. Sometimes the term solarimeter is used instead of pyranometer.

Pyrgeometer A pyrgeometer is an instrument for the measurement of terrestrial radiation only.

Pyrradiometer A pyrradiometer is an instrument for the measurement of both solar and terrestrial radiation, i.e., net atmospheric radiation on a horizontal upward facing black surface at the ambient air temperature.

B. International Pyrheliometric Standard 1956

For many years prior to 1958 two standard absolute pyrheliometers (Ångström, 1919), which were developed independently, were used as the primary reference for solar radiation measurements. The radiation scales of these two instruments (the Ångström and the Abbot or Smithsonian standard pyrheliometers) did not agree because of small systematic errors. In 1956 an international Radiation Conference at Davos, Switzerland, recommended the adoption of a new scale, the International Pyrheliometric Scale (IPS), 1956. It was concluded that measurements made according to the Smithsonian Scale, 1913 should be reduced by 2.0% while measurements made with the Ångström Scale, 1899 should be increased by 1.5%. For a detailed discussion of the discrepancy between these two standards, see IGY Instruments Manual, 1958. All pyrheliometers manufactured since 1956 have been calibrated in accordance with IPS 1956.

In 1970 at the Third International Comparisons of Pyrheliometers organized by the WMO and the Davos Observatory, Switzerland, two new absolute radiometers were introduced by Kendall and Berdahl (1970) and

Geist (1971). Both of these instruments use cavity-type receivers and are calibrated by substitution of electrical power. In comparing these two instruments with the participating 28 Ångström compensation pyrheliometers and three Smithsonian pyrheliometers, there was a significant difference of 1.8 ± 0.4% (Fröhlick *et al.*, 1973). These data illustrate that the problem of an absolute pyrheliometric scale still needs further attention. The development of the instrument technology and measurement systems for these observations has not yet reached the point where it is possible to be absolutely confident that the calibration of the instruments used throughout the world is correct to within the accuracy desired.

C. International Pyranometer Standard

The WMO has sponsored several intercomparisons of international and national standard pyrheliometers but has not sponsored similar intercomparisons of standard pyranometers. It was only as recently as 1971 that this was accomplished through the leadership of the International Solar Energy Society (ISES). The results of this comparison of 21 working standard pyranometers from meteorological institutions throughout the world are detailed by Thekaekara *et al.* (1972). The major issue to be resolved was the selection of the standard that should be adopted and against which the pyranometers should be compared.

"It was on the basis of past knowledge of the stability of the relationship between the pyranometer standards of the Eppley Laboratory, the USA National Weather Service and the Canadian Atmospheric Environmental Service... that it was agreed to recommend the adoption of a North American Mean" (Thekaekara *et al.*, 1972). This value was the average of three specific pyranometers that were manufactured by the Eppley Laboratory and were of a similar type. Table 19.1 is a summary of the results for selected clear sky time periods.

A close examination of the data revealed small systematic changes in the mean ratio for the different time periods, which may be connected with changes in the solar elevation (cosine response), radiation intensity, or ambient temperature. The Chile pyranometer data strongly suggest that errors have crept into the calibration systems adopted by that country. There were ten Eppley Model 2 or 15 pyranometers, the mean ratios of which varied from 0.979 to 1.011. The calibrations used for these radiometers were those assigned by the manufacturer during a three-year span. This is encouraging because it shows that it should be possible to limit scatter due to calibration techniques to within ±2%. As with pyrheliometers, there is still a great deal of research required to develop an absolute pyranometer measurement.

TABLE 19.1

Comparison of International Working Pyranometer Standards for Selected Clear Periods[a]

Country/Instrument	Weighted mean	Standard deviation
USA–Eppley 7577A1 (reference)	0.994	0.003
Argentina–Kipp 490398	1.056	0.008
Belgium–Eppley 6217A	1.000	0.005
Chile–Eppley 4308	0.839	0.015
United Kingdom–Kipp 2508	0.989	0.011
USA–Eppley 5363	0.965	0.012
India–Kipp 1154	1.051	0.019
Switzerland–Davos 71st01	1.011	0.004
USA–Eppley 4381A (reference)	1.004	0.005
USA–Kahl 703	1.040	0.017
Argentina–Eppley 10566F4	1.008	0.005
Australia–Groiss 11	1.019	0.028
Canada–Eppley 7122D1 (reference)	1.002	0.005
Australia–Eppley 10148F4	0.971	0.010
Israel–Eppley 10567F4	1.011	0.007
USA–Eppley 10154F4	1.010	0.003
USA–Eppley 9880F4	0.984	0.003
USA–Eppley 2445	1.072	0.010
USA–Eppley 5426	1.052	0.011
USA–Eppley 903001	0.979	0.004
USA–Eppley/Harrison 7834	1.078	0.002

[a] This comparison is the ratio of the instrument reading to the North American mean (average of the three instruments identified as reference). The weighted mean of five selected clear periods and the standard deviation is given. See Thekaekara *et. al.* (1972) for more details.

D. World-Wide Solar Radiation Measuring Stations

In order to ensure the reliability and comparability of radiation observations on a national and international scale, the World Meteorological Organization has recommended the establishment of a system of world, regional, and national radiation centers based on the International Pyrheliometric Scale, 1956. Each national radiation center should satisfy the following requirements (Anonymous, 1971b):

"It should possess at least one working standard pyrheliometer of the Ångström compensation or silver disk type for use as a national reference standard for the calibration of radiation instruments in the national network of radiation stations. The national working standard pyrheliometer should be compared with a regional working standard pyrheliometer at least once in five years. The accuracy of calibration of the auxiliary measuring equipment should be of the same order as that of the pyrheliometer.

It should have the necessary facilities and equipment for checking and for the study of the operation and performance of instruments used in the network.

The staff of the center should provide for continuity and should include a qualified scientist with wide experience in radiation."

Additional requirements for world and regional centers are given in the previous reference. Table 19.2 is a listing of the active world and regional radiation centers.

TABLE 19.2

Location of the World Meteorological Organization World and Regional Radiation Centers

World radiation centers

Davos (Switzerland)
Vocjkov (USSR)

Regional radiation centers

Region I	Region IV
Khartoum, Sudan Kinshasa, Zaire Pretoria, South Africa Tunis, Tunisia Tamanrasset, Algeria	Toronto, Canada Washington, USA

Region II	Region V
Poona, India Tokyo, Japan	Aspendale, Australia

Region III	Region VI
Buenos Aires, Argentina	Davos, Switzerland Kew, UK and North Ireland Leningrad, USSR Stockholm, Sweden Uccle, Belgium

19.2 BASIC PRINCIPLES OF SOLAR RADIOMETRY

A. Classification of Radiation Fluxes

The term *radiation* is applied generally to all kinds of processes that transmit energy by means of electromagnetic waves, whether visible or invisible. Although all portions of the electromagnetic radiation spectrum are of general engineering interest, the principal interests of those concerned with solar radiometry are in the wavelength region of about 0.2 to 100 μm. The fundamental physical processes involved in radiation transfer entails a knowledge of both quantum mechanics and electromagnetic theory. For example, electromagnetic theory is used to predict the radiation properties of surfaces, whereas the properties of absorbing and emitting media are described using quantum theory. Albeit this, from the viewpoint of an engineer, it is often sufficient to make use of specific results from these disciplines without pursuing their derivation. We shall proceed with this in mind.

The net radiation to a detector from a particular source is a function of the spectral radiative properties of the source, detector, intervening medium, and other surfaces which the detector may see. According to the laws of thermodynamics, any detector (or measuring device) will come into equilibrium with its environment by using all possible energy paths, for example, conduction, convection, and radiation. For the purposes of this discussion we shall concern ourselves only with the latter and assume that conduction and convection losses can be minimized in the design or they can be accounted for in the calibration.

The total solar radiation reaching a terrestrial surface H is the sum of the direct, diffuse or scattered, and reflected components, i.e.,

$$H = H_{\mathrm{DN}} \cos\theta + H_{\mathrm{d}} + H_{\mathrm{r}} \tag{19.1}$$

where H_{DN} is the direct radiation incident on an area normal to the sun's rays with an intensity peak at 0.48 μm, θ the incident angle between the solar rays and a line normal to the surface, H_{d} the scattered solar radiation which reaches the surface as diffuse radiation with an intensity peak at 0.45 μm, and H_{r} the direct solar radiation which is reflected from an adjacent surface. The radiation fluxes can be conveniently divided into two spectral regions, i.e., solar or shortwave radiation originating from the sun at an equivalent blackbody temperature of about 6000°K, and terrestrial or longwave radiation originating from sources (ground, atmosphere, sky, and other surfaces) at temperatures of 200–400°K.

As the sun's shortwave radiation traverses the earth's atmosphere, the direct beam is attenuated by absorption by oxygen, ozone, water vapor, and carbon dioxide, by reflection by dust, emission particulates, and water droplets, and by scattering by air molecules. The oxygen and ozone at high levels of the atmosphere absorb the ultraviolet radiation to the extent that the lowest wavelength at which any detectable radiation has been received at the earth's surface is 0.286 μm. The longwave radiation lower limit is about 4 μm. The spectral region for longwave radiation is defined as 3–100 μm (Anonymous, 1971b).

Measurements of solar radiation reaching a terrestrial position are conveniently divided into the following seven main classes:

(i) Direct radiation from the sun measured at normal incidence.

(ii) Global radiation received on a horizontal surface. This includes both radiation received directly from the sun and also radiation that has been scattered or diffusely reflected in traversing the atmosphere.

(iii) Sky radiation. This is the second component mentioned of the global radiation.

(iv) Reflected solar radiation (from the ground and other surfaces).

(v) Direct, global, or sky radiation in restricted portions of the spectrum. Included in this category as a special case is the measurement of daylight illumination.

(vi) Measurements are also made of the solar radiation falling on a spherical receiving surface. These attempt to simulate the solar radiation environment of an isolated object such as a flying object.

(vii) Also occasionally made for special purposes are measurements of the global radiation received on a fixed plane surface other than the horizontal; for example, on a vertical surface facing south.

B. Types of Solar Radiometers

The objective of solar radiometry is to measure the instantaneous values of each of the terms in Eq. (19.1) as they relate to the seven classes of solar radiation for a surface which may be tilted at any angle to the horizontal. One must always keep in mind that the radiometer indicates the net radiation to the detector, i.e., the incident radiation minus the emitted radiation from the detector. In the case of shortwave radiation, where the detector temperature is greatly different from that of the sun, the emission from the detector is negligible. On the other hand, for terrestrially originated longwave radiation, detector emission is very significant and must be evaluated carefully.

There are basically two classifications of infrared radiation detectors which have proven their suitability as solar radiometers—thermal detectors and photon detectors.

1. Thermal Detectors

Thermal detectors are distinguished as a class by the observation that the heating effect of the incident radiation causes a change in some physical property of the detector. Unlike semiconductors, they respond equally to a wide band of wavelengths. The constructional materials for thermal detectors are not in themselves good radiation absorbers and must usually be blackened by applying an absorbing coating in order to increase their responsivity, but only with increased time constant.

As a general rule for most thermal detectors, high responsivity R requires a long time constant τ, and a low heat capacity C in the junction and receiver assembly (Hudson, 1975), i.e.,

$$R \propto \tau/C$$

Fortunately, a time constant of several seconds can often be tolerated in solar radiometry and this gives the manufacturer considerable latitude in choosing his detector materials and coatings.

The thermal effect can be evaluated absolutely by calorimetric or electric compensation methods as will be discussed later. Also, with an appropriate calibration, the radiation intensity can be interpreted by thermoelectric or thermal expansion techniques.

2. Photon Detectors

Photon detectors convert some of the incident radiation directly into electricity which is proportional to the intensity of radiation. They have a detectivity that is one or two orders of magnitude greater than that of thermal detectors. This higher detectivity (sensitivity) does not come for free, however. These detectors have a nonuniform spectral response and usually respond in a relatively narrow wavelength region. Moreover, many photon detectors will not function unless they are cooled to cryogenic temperatures. Because of the direct interaction between the incident photons and the electrons of the detector material, the response time is very short, of the order of microseconds.

Almost all semiconductors are photoconductive, i.e., their electrical resistance decreases when they are exposed to light. This effect is particularly strong in germanium, silicon, selenium, metal sulfides, oxides, and halides (Wolfe, 1965). The basic transducer circuit is shown in Fig. 19.1. The photoconductive effect can be monitored electrically and calibrated in terms of the incident radiation. A review of the physical processes in photoconduction devices is given by Rose (1956). A disadvantage of the photoconductive detector is the fact that it is non-self-generating, i.e., a current from an external power source must pass through it to indicate a resistance change.

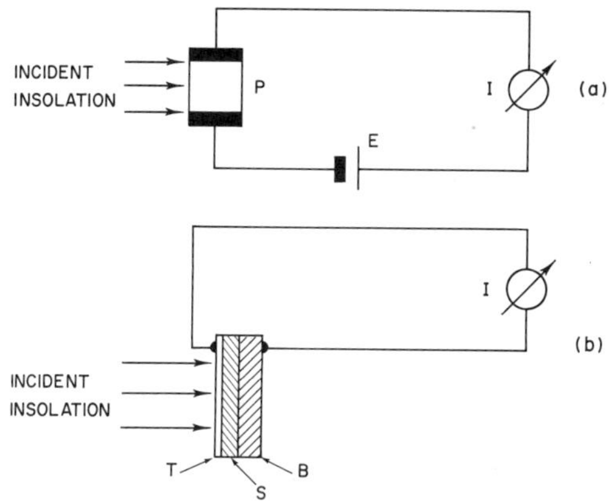

FIG. 19.1 Basic transducer circuits for (a) photoconductive detector and (b) photovoltaic photon detector.

Figure 19.1 also shows a photovoltaic device which is self-generating and therefore needs no external power supply. The photovoltaic cell consists basically of three solid layers: a base plate, a semiconducting layer S, and a transparent metal electrode T. Proper processing produces a thin insulating barrier layer between the semiconductor and the transparent metal layer. Light striking the barrier causes a voltage to be generated between the output terminals—the base plate being positive, the transparent electrode, negative. Cells with silicon, germanium, and selenium as the semiconducting material are in common use.

C. Broadband Spectral Measurements

The simplest way of measuring the incident insolation within a broad wavelength band is to use filters in front of the detector. Three specific filters have been adopted for use internationally. These Schott filters have the following characteristics:

OGI opaque up to 525 nm, transparent 525 to 2800 nm
RG2 opaque up to 630 nm, transparent 630 to 2800 nm
RG1 opaque up to 700 nm, transparent 700 to 2800 nm

The amount of solar radiation in the various bands can be deduced by the comparison of measurements with these filters.

D. Minimizing Errors Due to Convective Heat Exchange

Whenever it is necessary to measure longwave radiation, the emission of the detector is important and some material other than quartz must be used which will allow the detector to exchange the radiation with all appropriate parts of the hemisphere above the detector.

Four main solutions have been proposed:

(i) by providing a cover or window which is transparent to the terrestrial radiation;
(ii) by providing an artificial convective heat loss which is so great that the effect of a variable natural wind can be neglected (this is usually done by blowing a jet of air over a freely exposed plate);
(iii) by a compensation method, i.e., by introducing a second surface, which is not exposed to identical convective heat exchange conditions as the fully exposed surface, measuring the energy required to maintain the two surfaces at the same temperature; and
(iv) by making the response time of the radiometer very long, so that short-period variations of the convective heat loss are smoothed out, and correcting the instrument readings for the effect of longer term variations in wind speed.

19.3 PYRHELIOMETERS

Current practice in solar radiometry relies primarily on thermoelectric transducers. However, relatively low-cost photovoltaic transducers are becoming increasingly popular. To measure the direct solar radiation, the receiving surfaces must be normal to direct solar rays, i.e., a line joining the sun and receiver. Quite frequently an equatorial mounting is used to provide continuous recording. The adjustments in both azimuth and elevation should be correct to within $\frac{1}{4}^{\circ}$.

The descriptions of the various types of pyrheliometers which follow have been taken primarily from (Anonymous, 1971b) and Yellott (1967). Any exceptions are noted.

1. Ångström Compensation Pyrheliometer

In the Ångström compensation pyrheliometer which was invented in 1899 by A. K. Ångström, a thin blackened shaded manganin strip ($20 \times 2 \times 0.1$ mm) is heated electrically until it is at the same temperature as a similar strip which is exposed to solar radiation. A schematic of this technique is given in Fig. 19.2. Under steady state condition (both strips at identical temperature) the energy used for heating is equal to the absorbed solar energy. Thermocouples on the back of each strip, connected in opposition through a sensitive galvanometer or other null detector, are used to test for the equality of temperature. The energy H of the direct radiation is calculated

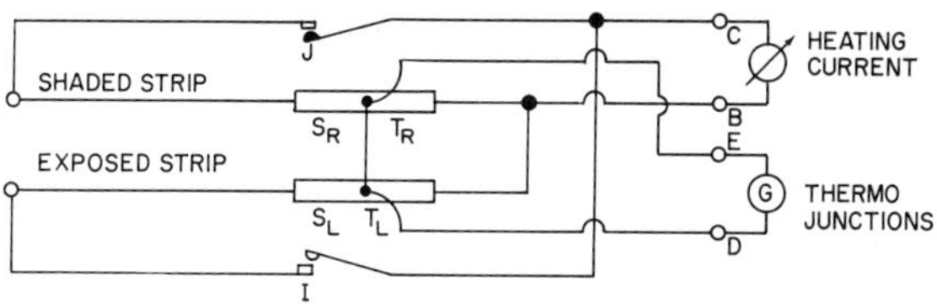

FIG. 19.2 Electrical circuit schematic of the Ångström standard absolute Pyrheliometer. The S_R strip is shaded and the S_L strip exposed to solar radiation (Drummund, 1967).

by means of the formula $H_{DN} = ki^2$, where i is the heating current in amperes and k a dimensional and instrumental constant. The rectangular apertures in use are $24° \times 6°$ for the short-tube instrument and $6° \times 3°$ for the long-tube instrument.

Although such an instrument can be made absolute by the calculation of k, in practice most Ångström instruments are calibrated by comparison with one of a few standard instruments (usually another Ångström pyrheliometer, the constant of which has been determined absolutely).

When the pyrheliometer is used absolutely, k is derived as follows:

$$k = R/W\alpha \qquad (H \text{ in W/cm}^2)$$

where R is the resistance per unit length of absorbing strip (Ω/cm), W the mean width of absorbing strip, and α the absorption coefficient of the absorbing strip. The auxiliary electrical equipment is important. The dc milliammeter should be of specific accuracy ($\pm 0.25\%$ or better) and individually calibrated so that corrections can be applied. If less precise meters are used, it is essential that they be checked each time they are used. It is also preferable that multirange instruments be used so that the measurements can always be made at more than half of the full scale deflection. The temperature-sensing device needs to be of adequate sensitivity; in general it should be capable of detecting a potential of at least 1 μV, but for the highest accuracy it should respond to $\frac{1}{4}$ μV. The distance between the milliammeter and the zero galvanometer should be sufficient to avoid mutual interference.

2. *Abbot Silver-Disk Pyrheliometer*

The silver-disk pyrheliometer (Aldrich, 1949), which was designed at the Smithsonian Institution in Washington, D.C. by C.G. Abbot, won immediate acceptance when it was announced in 1913. It consists essentially of a blackened silver disk positioned at the lower end of a tube with diaphragms to limit the whole aperture angle to 5.7°. [In an earlier design (1909) using an aperture angle of 10.6°, it was found that for reading on a very hazy day the results were as much as 2.5% too high due to the relative increase in sky radiation. This problem was eliminated in 1926 using tubes 32 cm long in place of the original 15 cm.]

A mercury-in-glass thermometer is used to measure the temperature of the disk. A shutter made of three polished metal leaves is provided at the upper end of the tube to allow solar radiation to fall on the disk at regular intervals, and the corresponding changes in temperature of the disk are measured. The thermometer stem was bent through 90° so that it lies along the tube to minimize its exposure to the sun. A precise routine must be followed and the temperature readings at exactly 100 s intervals require corrections for air, stem, and bulb temperatures. The timing of the operation of the shutter in particular must be very accurate; a consistent error of one second may, for instance, result in errors of about 1% in the final result.

The instruments must of course be calibrated against a primary standard, but their stability has been found to be very good and they are widely used for calibrating pyranometers.

3. *Michelson Bimetallic Pyrheliometer (Actinometer)*

In the Michelson bimetallic pyrheliometer, the deflection of a very fine bimetallic strip, when heated by solar radiation, is observed through a low-power microscope. The rectangular aperture of the instrument has angles of approximately 10° and 25°. It is a portable self-contained instrument and is especially suitable for daily *spot* measurements and as a traveling substandard, provided great care is exercised in its transportation.

The instrument constant is obtained by calibration against a primary standard instrument. The time required for essentially full response is 20–30 s. With proper calibration this instrument, which was first introduced in 1908, is said to give reproducible results within $\pm 1.0\%$.

4. *Linke–Feussner Pyrheliometer (Actinometer)*

The Linke–Feussner pyrheliometer uses a Moll thermopile protected by a thick shell of copper. This heavy mass acts as a shield from air currents and ensures that the thermopile is in a constant temperature enclosure apart from the conical aperture of about 11° in angle. In the newer instruments the thermopile is made of two equal sections connected in opposition, one of which is shaded from the radiation to be measured and one exposed, but both of which are equally exposed to temperature fluctuations caused by quasiadiabatic air pressure changes which are particularly troublesome in gusty winds. In the newer model the time for 99% response to a sudden change is about 10 s.

5. Eppley (Normal Incidence) Pyrheliometer

The sensitive element in an Eppley pyrheliometer is a temperature-compensated 15-junction bismuth–silver thermopile mounted at the base of a brass tube, the limiting diaphrams of which subtend an angle of 5.7°. During manufacture the tube is filled with dry air and is sealed with a crystal quartz window which is removable. A filter wheel (for the three standard meteorological filters) is standard. The response time for over 99% signal is about 20 s. The older version of this instrument has a copper–constantan uncompensated thermopile (response time about 6 s).

It is a stable instrument and provided proper care is taken can be used as a substandard. This instrument has found wide acceptance within the USA and many parts of the world.

6. Moll–Gorczynski Pyrheliometer (Actinometer)

The Moll–Gorczynski pyrheliometer incorporates a Moll-type thermopile. Two main models exist. The first consists of a thermopile in a short tube and is used for spot reading. In the second model, used for recording, a wire frame carries three circular diaphrams so spaced from the thermopile that the angular aperture is about 8°. The thermopile is permanently protected by a glass cover. This instrument is widely used in Europe, Asia, and Africa.

7. Yanishevsky Thermoelectric Pyrheliometer

The Yanishevsky thermoelectric pyrheliometer incorporates a symmetrical star-shaped thermal battery, and has an angular aperture of 10°, which is twice the diameter of the sensor. Battery sensitivity is about 7–7.5 mV/cal cm^{-2} min^{-1}. The 99% response time is about 25–30 s.

8. Photovoltaic Silicon Pyrheliometer

Silicon solar cells have the property that the short-circuit current is proportional to the radiation intensity at normal incidence (see for example, Chapter 12). Yellott (1964; Selçuk and Yellott, 1962; Yellott *et al.*, 1964) has successfully designed a silicon cell pyrheliometer and pyranometer. Its advantages are relatively lower cost and faster response time for instantaneous measurements. The major disadvantage is the highly specular dependence of the cell output. The spectral response is such that it is most sensitive to the red and near-infrared portion of the solar spectrum. Also, the the calibration varies with the angle of incidence of the radiation.

9. Active Cavity Radiometer

A series of cavity radiometers has been developed at the Jet Propulsion Laboratory, Pasadena, California, for the accurate measurement of irradiance in absolute units (Willson, 1971). This so-called active cavity radiometer (ACR) is the newest innovation in reference standard pyrheliometers. At one solar constant uncertainties are claimed to be less than $\pm 0.5\%$.

The essential physical features of the ACR are shown in Fig. 19.3. The ACR's conical cavity detector is connected by a low thermal impedance to a heat sink. The heat sink is insulated from the external environment to minimize the rate of change of its temperature when subjected to fluctuating thermal environment.

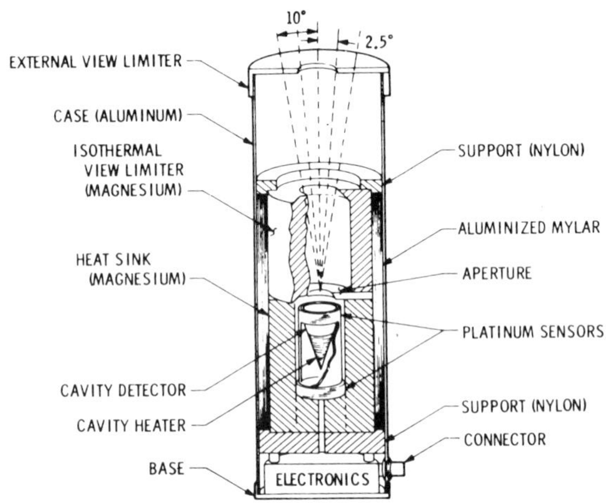

FIG. 19.3 JPL active cavity radiometer type III (Willson, 1971).

The interior surface of the conical cavity detector is coated with a surface material whose effective absorptance for solar radiant flux is high ($\alpha_S = 0.95 \pm 0.02$). The cavity's 2-cm diameter base is partially shaded by a round aperture that both defines the detector area for the radiometer and enhances the effective cavity absorptance. The aperture can be shaped precisely and its size determined with great accuracy. The cavity absorptance has been analytically shown to be 0.996 ± 0.002, effecting a tenfold decrease in the uncertainty of the cavity surface absorptancy. The field of view for the cavity is isothermally confined to 20° by the heat sink or to 5° with the addition of an external view limiter.

The dissipation of a fixed amount of power in the cavity will produce a constant temperature drop across the thermal impedance. This drop, transduced by platinum windings, can be calibrated as a direct measure of the irradiant power.

Two intercomparison tests were made of the JPL ACR and the IPS 1956 (Fröhlich *et al.*, 1973; and Willson, 1971). Based on the absolute uncertainty of ± 0.5 for the ACR, both test results suggested that the IPS 1956 contains a systematic absolute error of $-2.2 \pm 0.5\%$.

19.4 PYRANOMETERS

1. Eppley (180° Pyrheliometer) Pyranometer

Kimball and Hobbs (1923) of the US Weather Bureau designed a pyranometer which produced a thermoelectric voltage proportional to the intensity of incident radiation. Since its introduction in 1923, it has become the most widely used working pyranometer in the USA and is known as the "Eppley 180° pyrheliometer."

The receiving surface of the Eppley pyranometer consists of two concentric silver rings; the inner one is coated black (Parson's optical black lacquer) and the outer one is coated with white (magnesium oxide). The temperature difference between the two rings is measured with a thermopile whose junctions are in good thermal contact with the lower surfaces of the ring, but are, of course, electrically insulated from them. The whole assembly is hermetically sealed inside a specially blown spherical lamp bulb filled with dry air. The magnesium oxide is a good reflector of solar radiation but is a strong absorber of terrestrial radiation. This construction minimizes the effect of longwave radiation from the glass envelope. The time required for about 98% response to a sudden change is about 30 s.

In 1951, MacDonald observed and reported some systematic errors in this pyranometer. He showed that the response decreased with increasing ambient temperature by approximately 0.1%/°C. When the detector is oriented vertically, the convective pattern within the glass enclosure changes resulting in as much as a 5% reduced output.

2. Moll–Gorczynski Pyranometer (Kipp Solarimeter)

The Moll–Gorczynski pyranometer is a thermopile instrument, in which the receiving surface is covered by two concentric ground and polished glass hemispherical domes, 2 mm in thickness. The outer dome is removable and can be exchanged with filter domes. In the later models the necessary adjustments for cleaning and maintenance have been much facilitated. As the thermopile surface is rectangular, particular attention must be paid to orientation. A complete response of 98% to a sudden change takes about 30 s. The instrument is subject to zero-depression errors. The temperature coefficient is approximately $-0.2\%/°C$.

3. New Eppley Pyranometer

MacDonald's work and other tests provided the basis for a new Eppley pyranometer (Model PSP) (Manufacturer's 1975). A thermister-compensated electrical circuit reduced the temperature dependence to $\pm 1.0\%$ constancy from -20 to 40°C. The cosine response is $\pm 1\%$ for 0–70° angle of incidence and $\pm 3\%$ for 70–80° angle of incidence. The output of thermopile is unaffected by detector orientation. The outer precision-ground hemisphere can be either optically-clear glass or one of several filter glasses. Other innovations are the replacement of the two concentric-ring detectors by one of radial wirewound-plated construction and the use of more stable 3M (trade name) black coating and of barium sulfate as the whitening agent.

4. Volochine Pyranometer

The Volochine pyranometer has a thermopile of different construction to the Moll–Gorczynski instrument, and has, as the new Eppley pyranometer, a circular black flat receiving surface. The 98% response time to a sudden change is about 20 s. The temperature coefficient is approximately $-0.1\%/°C$.

5. Dirmhirn–Sauberer Pyranometer (Star Pyranometer)

The Dirmhirn–Sauberer pyranometer uses black and white segments alternatively mounted in the form of a star. The construction and operation of this instrument is discussed by Dirmhirn (1958). The measurement is accomplished similarly to that of the new Eppley pyrheliometer. The 98% response time to a sudden change is about 30 s. The temperature coefficient is negligible.

6. Yanishevsky Thermoelectric Pyranometer

The Yanishevsky thermoelectric pyranometer uses black and white thermojunctions painted with soot and magnesium, the odd junctions being black and the even junctions white. The set of thermojunctions (overall dimensions 3 × 3 cm) is covered with a glass hemisphere and kept dry with silica gel. The lower part of the glass is protected with a black shield to eliminate the effect of radiation from the earth's surface. The sensitivity of the instrument is 7–7.5 $mV/cal\ cm^{-2}\ min^{-1}$. The 98% response time is about 25–30 s.

7. Bimetallic Actinographs of the Robitzsch Type

Bimetallic actinographs are widely used as simple self-contained recorders. A mechanical linkage is used to record the temperature difference between a black-coated bimetallic strip exposed to solar radiation and two similar bimetallic strips either painted white or shielded from solar radiation.

Because of the relatively large mass of the bimetallic strips, the response time of the instrument is large (10–15 min for 98% response) and the instrument is only suitable for the purpose of obtaining daily totals. Even for this purpose, because of the difficulties in reducing to reasonable amounts the errors due to variation in azimuth, cosine response, and temperature, the accuracy is much less than that of an electrical pyranometer.

There are several improved models available but the errors are still still much greater than those of electrical pyranometer types.

19.5 SUNSHINE RECORDS

Daily and hourly records of the amount of sunshine (or cloud cover) are necessary to optimize the design of a particular solar collector. For example, if a location consistently reports an increased frequency for afternoon cloud cover thereby providing less hours of sunshine than in the morning, nontracking collectors can compensate for this phenomenon by facing southeast to collect more of the early morning radiation. This section provides a description of the following four most widely used sunshine recorders:

Campbell–Stokes recorder: focused solar radiation to burn a trace in a chart.

Jordan recorder: focuses sunlight onto photographic paper.

Marvin recorder: a thermoelectric switch-actuated chronograph pen to trace the sunshine hours.

Foster recorder: a photoelectric switch.

The first two act as sundials with no external power supply required.

1. Campbell–Stokes Sunshine Recorder

The Campbell–Stokes sunshine recorder consists essentially of a glass sphere about 10 cm in diameter mounted concentrically in a section of a spherical bowl, the diameter of which is such that the sun's rays are focused sharply on a card held in grooves in the bowl. The method of supporting the sphere differs according to whether the instrument is required for operation in polar, temperate, or tropical latitudes. Three overlapping pairs of grooves are provided in the spherical segment to take cards suitable for different seasons of the year. If comparable results are to be obtained, it is necessary that both the spherical segment and the sphere should be made with great precision and that the mounting be so designed that the sphere can be easily and accurately centered in it.

The chief requirements of the sphere are that it should be of uniform and well-annealed colorless or very pale glass. For a standard pattern instrument in which the radius of the spherical segment is approximately 73 mm, the principal focal length of the sphere for sodium *D* light should be approximately 75 mm. These figures correspond to an index of refraction of 1.52 ± 0.02. This special equipment is used for testing sunshine spheres (Bilham, 1929).

2. *Jordan Sunshine Recorder*

The Jordan sunshine recorder consists of two semicylindrical cameras each having, in its flat side, an aperture through which the sun's rays are admitted. The cameras are mounted side by side, one with its aperture towards the east and the other towards the west, on a leveling base plate which can be adjusted for latitude by means of a graduated arc. The record is obtained on photographic paper, such as the ferro-prussic type, placed around the cylindrical wall of each camera. The instrument requires adjustment for level and meridian as well as latitude. It has been found that measurements of the records are open to considerably more uncertainty than are the measurement of Campbell–Stokes records. One of the difficulties associated with the instrument is that of ensuring constant sensitivity of the photographic paper.

3. *Marvin Sunshine Recorder*

The Marvin sunshine recorder is essentially a differential air thermometer with one clear and one black bulb in an evacuated glass jacket. The bulbs are separated by a column of mercury and alcohol that closes an electric circuit when sufficient radiation falls on the instrument. The circuit can be made to operate a chronograph.

The Marvin instrument should be set in the meridian plane at such an angle that the mercury column just closes the electric circuit during the time when the sun's disk can be just faintly seen through the clouds. This adjustment should be made in spring or winter, but it does not provide an exact standard for the comparison of different instruments of the same type. Like the Campbell–Stokes instrument, the Marvin is subject to errors due to faulty installation and adjustment and does not record weak sunlight when the sun is within 5° of the horizon. It is affected by diffuse radiation as well as by direct solar radiation.

4. The Foster Sunshine Switch

The Foster sunshine switch is essentially a difference instrument sensing radiation by means of a shaded and an unshaded photovoltaic cell, both mounted inside a translucent tube supported by a simple equatorial mount that allows seasonal adjustment. A shading band concentric to the translucent tube protects one cell from direct sunlight. The differential output increases during periods of sunshine sufficiently to activate a relay which in turn is used to mark on a recorder or totalizer the duration of sunshine in 1 min units. The lag of the instrument is negligible and its sensitivity allows reliable measurements at dawn and dusk.

19.6 STANDARDIZATION OF SOLAR RADIOMETERS

A. Standard Classification and Accuracy Specifications

From the early measurements of insolation by the US Weather Bureau, virtually all calibrations were traceable directly to the Abbot waterflow pyrheliometer (Abbot and Aldrich, 1932). The Abbot waterflow pyrheliometer was essentially a calorimeter that measured the insolation absolutely by measuring the temperature rise of a precisely controlled flow of distilled water which carried away the solar heat absorbed by a blackened metal cone located at the bottom of a long tube. Apparently, the use of this standard diminished with increased reliance on the Abbot or Smithsonian silver disk and the Ångström pyrheliometers (Hill, 1966).

The WMO has established accuracy limits for four classes of radiometers: reference standard pyrheliometers, first class, second class, and third class instruments. These accuracy limits are given in Table 19.3. Additionally, each of the pyrheliometer and pyranometer types in Sections 19.3 and 19.4, respectively, have been rated as follows:

Reference standard pyrheliometer
Ångstrom compensation pyrheliometer
Silver-disk pyrheliometer
Active cavity radiometer

First class pyrheliometer
New Eppley pyrheliometer
Michelson bimetallic pyrheliometer
Linke–Feussner iron-clad pyrheliometer
Yanishevsky thermoelectric pyrheliometer

Second class pyrheliometer
Moll–Gorezynski pyrheliometer
Old Eppley pyrheliometer

TABLE 19.3

Classification of Accuracy of Solar Radiometers[a]

	Sensitivity ($mW\ cm^{-2}$)	Stability (%)	Temperature (%)	Selectivity (%)	Linearity (%)	Time constant (max)	Cosine response (%)	Azimuth response (%)	Errors in auxiliary equipment		
									Galvonometer	Milliameter (%)	Chronometer
Reference standard pyrheliometer	±0.2	±0.2	±0.2	±1	±0.5	25 s	—	—	0.1 unit	0.1	0.1 s
Secondary instruments											
1st class pyrheliometer	±0.4	±1	±1	±1	±1	25 s	—	—	0.1 unit	0.2	0.3 s
2nd class pyrheliometer	±0.5	±2	±2	±2	±2	1 min	—	—	0.1 unit	±1.0	—
									Errors in recording apparatus		
1st class pyranometer	±0.1	±1	±1	±1	±1	25 s	±3	±3		±0.3	
2nd class pyranometer	±0.5	±2	±2	±2	±2	1 min	±5–7	±5–7		±1	
3rd class pyranometer	±1.0	±5	±5	±5	±3	4 min	±10	±10		±3	
									Errors due to wind		
										%	%
1st class net pyrradiometer	±0.1	±1	±1	±3	±1	½ min	±5	±5		±0.3	±3
2nd class net pyrradiometer	±0.3	±2	±2	±5	±2	1 min	±10	±10		±0.5	±5
3rd class net pyrradiometer	±0.5	±5	±5	±10	±3	2 min	±10	±10		+1	±10

[a] Established by World Meteorological Organization in 1971. Reprinted with permission from "Guide to Meteorological Instrument and Observing Practices," WMO Publ. No. 8 TP.3, 4th ed., Geneva, 1971.

First class pyranometer
Selected thermopile pyranometers

Second class pyranometer
Moll–Gorezynski pyranometer
Eppley pyranometer
Volochine thermopile pyranometer
Dirmhirn–Sauberer pyranometer
Yanishevsky thermoelectric pyranometer
Spherical Bellani pyranometer

Third class pyranometer
Robitzsch bimetallic pyranometer

B. Methods of Standardization

Standardization of pyrheliometers is straight forward provided a day with reasonably steady solar radiation is chosen and sufficient number of observations are made. Traditionally the observed method of standardizing a pyranometer is by comparison with a working standard pyrheliometer using the occulting or shading technique (IGY, 1958). On a clear day the global solar radiation is alternately recorded with the diffuse component which is obtained by shading the pyranometer with a disk mounted at the end of a slender arm held some distance away. For example, if the shading disk is 10 cm in diameter, it should be held 1 m from the receiver. The glass bulb should be completely shaded to compensate for the circumsolar sky radiation falling on the pyrheliometer detector surface. The required calibration constant for the pyranometer is then given by

$$k = H \sin \beta/(T - D)$$

where H is the observed direct solar radiation, β the mean solar elevation, T the instrument output in shaded condition, diffuse sky radiation, and D the instrument output in shaded condition, diffuse sky radiation. Calibration problems arise due to lengthy instrument exposure period (up to several days), varying ambient temperature, and significant departures from the Lambert cosine law (especially in turbid atmosphere).

At the Solar Energy Conference in Boston, in March 1966, three companion papers (Hill, 1966; Latimer, 1966; Drummond and Greer 1966) were presented which describe an indoor procedure for the calibration of pyranometers using an integrating sphere or hemisphere. This technique creates an isotropic radiation field in a constant ambient temperature. It is now possible for two men to calibrate a dozen pyranometers per day reproducible to better than $\pm 1.0\%$ accuracy. The standard or reference pyranometer in the sphere or hemisphere is necessarily traceable to the traditional calibration technique.

There are two other but less acceptable methods for standardizing a pyranometer. The first is the comparison of the pyranometer with a similar previously calibrated pyranometer with the sun as the source. This method requires several days of exposure to get a proper comparison. The second is to bench mount the two (calibrated and uncalibrated) in the laboratory and use a collimated artificial light source at various altitudes and azimuth angles to measure the hemispherical response.

C. Sources of Error

There are four basic error types (GMIOP, 1971; IGY, 1958; Riches, 1975) associated with solar radiometry: data processing, sensor exposure, sensor design, and recording instrumentation.

The first two can be minimized with an adequate routine maintenance program. The importance of the US National Weather Service (NWS) solar radiation measuring network was not appreciated until the oil embargo to the US by the Arab oil-producing nations. Solar energy suddenly emerged as a viable source of energy. The past insolation records were necessarily scrutinized and found to be lacking. Apparently, due to insufficient resources there was for a number of years (particularly 1940–1970) inadequate routine maintenance and station inspection. The results were frequent equipment malfunction and poor quality control. All these have resulted in poor data for many stations, so poor in fact, that in September 1972 NWS requested the National Climatic Center to cease publishing data. Errors as high as $\pm 30\%$ have been found (Jessup, 1974).

NWS management has initiated an intensive program to upgrade its entire radiation measuring network. The Eppley pyranometers in at least 36 of the 60 NWS stations are being phased out in favor of the Spectrolab, Inc. (Sylmar, California) SR-75 pyranometer which is a first class instrument.

The last two error types can be grouped into systematic errors and sensor degradation. Systematic errors can be corrected by making the appropriate adjustment in the calibration constant. These errors are primarily associated with the calibration techniques and the calibration level of the reference pyrheliometer or pyranometer as the case may be.

In 1956 the Eppley Laboratory, Inc., began using Parson's optical black lacquer instead of lampblack on all its "180° pyrheliometers." This resulted in the so-called Parson's black problem. The problem was that the black lacquer on those instruments with extended sunshine exposure gradually turned to gray then to green. Consequently, the sensitivity (or calibration) decreased with the color changes. Unfortunately, this problem did not surface until about 1966. It took several years for the color to change. Moreover, the color change did not apparently concern the stations where it was occurring This was consistent with the previously mentioned poor quality control on the equipment maintenance. At any rate, in 1966 these pyranometers with discolored detectors indicated irradiance values 5–14% lower than a traveling standard (Flowers, 1974). Corrections on the data have been done by assuming that the change in sensitivity with time was linear.

Hanson (1974) has prepared a table of error estimates on the NWS field pyranometers 1954–1972. This is presented as Table 19.4 and does not include the Parson's black problem. Note that the largest systematic error is associated with the comparison of different calibration techniques. Lampblack detectors show no difference between the integrating sphere and sun calibrations. However, the Parson's black detectors showed a difference of

TABLE 19.4

Solar Irradiance Error Estimate for NWS Pyranometer Network, 1954–1972[a]

	Error (%)	
	Systematic	Random
Instrument		
(a) Standardization to absolute scale	+1	
(b) Reproducibility of standardization		1
(c) Crossmatching on standardization[b]	+7 and 0	
(d) Temperature response	+1	2.5
(e) Cosine and azimuth response		3
(f) Linearity		1
(g) Degradation with time		3
Instrument exposure		
(a) Incomplete field of view		1
(b) Frost, dew, etc.		1
Recording system		
(a) Standardization		0.5
(b) Stability		1
Data processing		
(a) Integration (visual)		2
	+2 to +9	5.8

[a] From Hanson (1974).

[b] Can be compensated for at individual locations.

almost 7.5%. Hanson concluded that about half of the NWS calibrated instruments had a systematic error of +2% and +9% for the other half.

Random errors are associated with many factors: orientation, quality control in the individual components used in manufacture, human variations, electronic instrumentation, fluctuations in ambient temperature, and instrument time constant. A random error of 5% is equivalent to "good and careful work" (Latimer, 1971; Nicolet, 1948).

TABLE 19.5

Commercially Available Pyrheliometers

Manufacturer	Type	Detector type	Comments
The Eppley Laboratory, Inc., 12 Sheffield Avenue Newport, Rhode Island	Normal incidence pyrheliometer (NIP)	Thermopile	
The Eppley Laboratory, Inc.	Ångström compensation pyrheliometer	Thermocouples	Instrument used to maintain IPS-56.
The Eppley Laboratory, Inc.	Eppley–Kendall	Thermocouples	Absolute radiometer
Spectrolab, Inc. Sylmar, California	SR-70	Thermopile	
Hy-Cal Engineering 12105 Los Nietas Road Santa Fe Springs, California	Hy-Therm Pyrheliometer	Thermopile	Water cooled
Matrix, Inc. 537 S. 31 Street Mesa, Arizona	MK-3 Sol-A-Meter	Silicon cell	
Radiometrics Corporation P.O. Box 251 Altadena, California	Active cavity radiometer	Thermocouples	Absolute radiometer
Kipp and Zonen, Ltd. Delft The Netherlands	Linke–Feussner	Thermopile	
Mashpriborintorg Smolenskaya-Sennaya Square 32/34 Moscow G-200, USSR	AT-50	Thermopile	
The Swedish Meteorological and Hydrological Institute P.O. Box 12108 S-102 23 Stockholm, Sweden	Ångström compensation pyrheliometer SMHI Design 1973	Thermocouples	Instrument used to maintain IPS-56

19.7 COMMERCIAL SOLAR RADIOMETERS

The commercially available pyrheliometers and pyranometers are all designed similar to one of the types described in Sections 19.3 and 19.4. The detector is a thermocouple, thermopile, silicon cell, or bimetallic strip. As discussed before, the thermocouples and thermopiles are calorimetric sensors, silicon cells are quantum detectors, and the bimetallic strips are calorimetric sensors which depend upon the thermal expansion of metals.

Table 19.5 which is taken directly from Hoyt (1974) lists ten commercially available pyrheliometers. Hoyt also compiled the list of 25 commercially available pyranometers given in Table 19.6. Either listing may not be exhaustive but does represent the major suppliers of solar radiometers.

TABLE 19.6

Commercially Available Pyranometers

Manufacturer	Model	Detector Type	Response time	Comments
The Eppley Laboratory, Inc. 12-Sheffield Avenue Newport, Rhode Island	Precision spectral pyranometer	Thermopile	1 s	Double dome
The Eppley Laboratory	Black and white	Thermopile	3–4 s	Star type, double dome.
The Eppley Laboratory	Model-15	Thermopile	—	Bulb type, black and white rings. Single dome. No longer manufactured.
Spectrolab, Inc. Sylmar, California	SR 75	Thermopile	1 s (63% point)	Early models have poor cosine response. Double dome.
Kahl Scientific Instrum. Corp. P.O. Box 1166 El Cajon, California	28AM100	Thermopile (Dirmhirn)	30 s	Star type, single dome.
Yellow Springs Instrum. Co. Yellow Springs, Ohio	—	Silicon cell	10 μs	Primarily for agricultural use.
Matrix, Inc. 537 S. 31 Street Mesa, Arizona	Mark 14	Silicon cell	1 ms	Primarily for agricultural use.

TABLE 19.6 (*Cont.*)

Manufacturer	Model	Detector Type	Response time	Comments
Matrix, Inc.	Mark 1-G	Silicon cell	1 ms	Primarily for agricultural use.
Herbert A. Groise and Co. Rear 1 Gordon Grove Malvern, Australia	—	Thermopile (Trickett)	5.6 s	Primarily for agricultural use.
Middleton & Co. Pty. Ltd. 8 Eastern Road South Melbourne, Australia	—	Thermopile	—	Double dome.
Rauchfuss Instruments & Staff Pt. Ltd. 11 Florence Street Burwood, Victoria, Australia	Integrating pyranometer	Silicon cell	—	Primarily for agricultural use.
Solar Radiation Instruments 21 Rose St. Altona, Victoria, Australia	Solarimeter S.R.I.3	Thermopile	99% in 90 s	Single dome.
Swissteco Pty. Ltd. Instrument Division 26 Miami St. Hawthrone East, Victoria, Australia	Tube pyranometer	Thermopile	98% in 25 s	Primarily for agricultural and botanical use.
EKO Instruments Trading Co. Ltd. 2-1 Ohtemachi 2-Chome Chiyoda-Ku Tokyo 100, Japan.	—	Thermopile (Moll)	99% in 7–8 s	
EKO Instruments	—	Thermopile	99% in 45 s	Bulb type, black and white rings. Single dome, similar to Eppley Model 15.
EKO Instruments	—	Thermopile	99% in 20 s	Star type, single dome.

TABLE 19.6 (*Cont.*)

Manufacturer	Model	Detector Type	Response time	Comments
Ishikawa Sangyo Co. Ltd., 4-6-13 Shinkawa Mitake City Tokyo, Japan	—	Thermopile	98% in 30 s	Bulb type, black and white rings. Single dome similar to Eppley Model 15.
Nakano-Seisakusho Ltd., Hiyashimito 1525 Ueno City, Mic Prefecture, Japan	—	Thermopile (Noshi–Denski)	—	Single dome.
C.F. Casella Co. Ltd. Regent House Brittania Walk London, N.I., UK	Bimetallic actinograph	Expansion of blackened bimetallic element.	Slow	
Litronic Ltd. 54–58 Bartholomew Close London, E.C.1, UK	—	Thermopile	30 s	
R. Feuss D-1 Berlin 41 Dunther Str. 8 Postfach 350, Germany	Bimetallic actinograph No. 58 dc	Expansion of blackened bimetallic element.	Slow	
Kipp & Zonen Ltd. Delft The Netherlands	—	Thermopile (Moll)	90% in 2.8 s	Double dome.
Mashpriborintorg Smolenskaya-Sennaya Square 32/34 Moscow, G-200, USSR	—	Thermopile	Less than 40 s	Black and white squares. Single dome. Also known as Yanishevsky pyranometer.
Phillip Schenk A-1212 Wien Postfach 3, Australia	Star pyranometer	Thermopile (Dirmhirn)	—	Star type, single dome.
Phillip Schenk	—	Thermopile	—	Star type, single dome. Also measures radiation on four vertical surfaces.

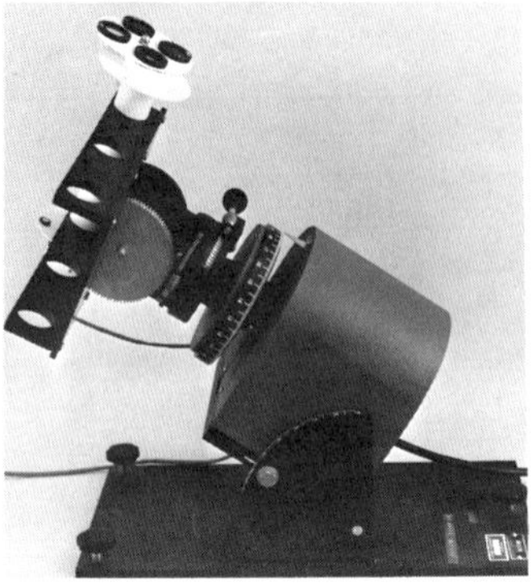

(a)

(b)

(c)

(d)

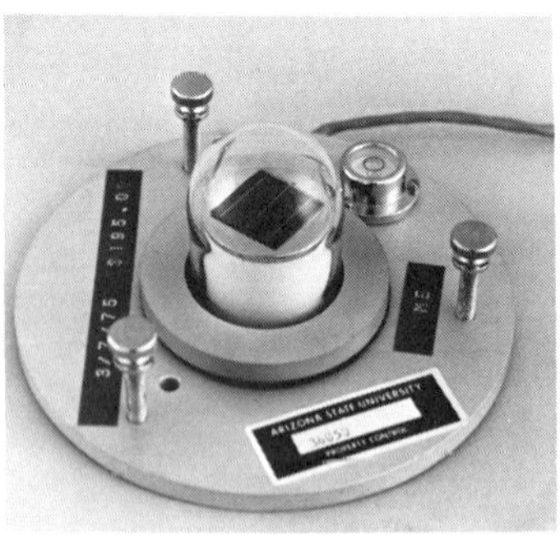

(e)

(f)

FIG. 19.4 Examples of commercially available solar radiometers. (Courtesy of Mechanical Engineering Laboratory, Arizona State University, Tempe, Arizona.) (a) Eppley NIP with Schott filters and equatorial mount; (b) Spectrolab SR-75 Pyranometer used by US National Weather Service; (c) Eppley ultraviolet detector; (d) another type of pyrheliometer; (e) Yellott solarimeter; (f) old Eppley 180° pyrheliometer (no longer manufactured).

19.8 FUTURE INSTRUMENTATION

According to Hickey (1974) of the Eppley Laboratory, Inc., the basic instrumentation of the future will probably look very much like the instrumentation of the past and present. The newer instruments are retaining the original or similar geometries with the major emphasis on detector development. The pyrheliometers and pyranometers will continue to use thermopile detectors but with reduced time response and increased ability to withstand degradation.

Perhaps the most significant changes to be implemented will be an improved instrument maintenance and calibration surveillance, particularly by the US National Weather Service (Riches, 1975; Hickey, 1974; Ångström, 1919). It appears that with proper care and maintenance, present instrumentation is sufficient for most of the desired measurements related to terrestrial solar energy applications.

We shall see increased applications of spectral measurement equipment, particularly in urban areas. Aerosols and other emission particulates do affect the spectral distribution of the insolation on a terrestrial surface, the long-term extent of which has not been documented.

An effort to solve the question of an absolute pyrheliometric measurement will continue. Currently those of the active cavity radiometer type are becoming recognized as the radiometers with the smallest experimental uncertainties. Figure 19.4 contains pictures of several solar radiometers.

19.9 SUMMARY OF MILESTONES IN SOLAR RADIOMETRY

Coulson (1975) has identified the following list of significant events in the development of solar radiation instrumentation from the early nineteenth century until 1971. This author has added to the list starting with 1966:

1825 Herschel's pyrheliometer (actinometer) developed: first instrument in which the rate of cooling was introduced into solar radiation measurements.

1837 Invention of Pouillet's pyrheliometer: first use of the term "pyrheliometer," solar constant measurement by Pouillet yielded 1.76 cal cm^{-2} min^{-1}.

1838– Invention of first photographic sunshine recorders by J. B. Jordan.

1840 Sir John Herschel obtained the first photograph on a glass plate (a picture of a telescope).

1845 First daguerrotype of the sun was taken by Foucault and Fizeau.

1859 Development of Kirchhoff's law.

1879 Experimental development of Stefan (later called Stefan–Boltzmann) law. Device introduced by Stokes for use of cards with the Campbell–Stokes sunshine recorder since that time.

1881 Observations from top of Mt. Whitney extended known solar spectrum to 1.8 μm. Value of solar constant estimated to be greater then 3 cal cm^2 min^{-1}.

1883 Becquerel obtained map of solar spectrum out to 1.4 μm.

1884 Boltzmann's derivation of Stefan–Boltzmann law on basis of thermodynamic theory.

1885 Invention of photographic sunshine recorder by J. B. Jordan. Invention of new form of polarimeter and of pole-star recorder by E. C. Pickering.

1886 Abney published table of 429 lines of solar spectrum at wavelengths below 1 μm. Use of first thermoelectric pyrheliograph by Crova. Invention of new type of radiometer by A. K. Ångström.

1891 Concept of Maring–Marvin sunshine recorder by D. T. Maring.

1893 Invention of electric compensation pyrheliometer by A. K. Ångström. Improved versions described in 1896 and 1899.

1894 Development of Wein displacement law.

1897–1898 Langley's determination of location and approximate strength of over 700 lines of solar spectrum between 0.4 and 6 μm.

1898 Invention of Callendar pyranometer; improved version brought out in 1905.

1900 Development of Planck's law. Measurements of total solar radiation begun in Washington, D.C. by Smithsonian Institution.

1902 Monochromatic observations begun by Langley in Washington, D.C. Estimates of solar constant varied between 1.75 to 4 cal cm^{-2} min^{-1}. Work started on a mercury pyrheliometer, which was later developed into the secondary standard silver-disk pyrheliometer of Abbot.

1903 Smithsonian Institution started construction of waterflow pyrheliometer, which was perfected into the primary standard by about 1910.

1904 First attempt, by Langley, to correlate weather with solar radiation.

1905 Einstein developed quantum theory of radiation. Ångström electrical compensation pyrheliometer adopted as a standard at Meteorological Conference, Innsbruck, and by Solar Physics Union, Oxford. Smithsonian solar observation expedition to Mt. Wilson. Additional expeditions in 1906, 1907, and 1908.

1906 Best estimate of solar constant: 2.1 cal cm^{-2} min^{-1}. Measurements of reflectance of stratus cloud near Mt. Wilson gave value of 65%.

1907 Weather Bureau solar radiation observatory established in Mt. Weather, Virginia.

1908 Development of Mie theory of scattering. Invention of Michelson pyrheliometer. Start of study of transmission of water vapor by F. E. Fowle.

1909 Abbot's first expedition to Mt. Whitney. First construction of

Abbot silver-disk pyrheliometer (modified to long tube in 1927). Summary of observations from 1905 to 1909 gave value 1.924 cal cm^{-2} min^{-1} for solar constant.

1910 Invention of Marvin pyrheliometer. Value of solar constant estimated at 1.922 cal cm^{-2} min^{-1}. Pyrheliometric observations started on Mt. Fuji, Japan, and at Madison, Wisconsin.

1911 Solar observation station established at Bassour, Algeria; solar radiation station established at Lincoln, Nebraska.

1912 Radiation measurements in Algeria showed 20% reduction of sunlight due to volcanic ash from Mt. Katmai eruption in Alaska. Experiments started with "water-stir" pyrheliometer. Observation station established at Harvard College Observatory, Arequipa, Peru.

1913 Mean value of solar constant given as 1.933 cal cm^{-2} min^{-1}. First revision of Smithsonian pyrheliometric scale, based on the water-flow pyrheliometer. Abbot constructed the hemispherical pyrheliometer, and a recording balloon pyrheliometer which was carried to 45,000 ft altitude. First daylight illumination measurements made at Mt. Weather, Virginia, with Sharp–Millar photometer.

1914 Balloon flights yielded solar constant of 1.88 cal cm^{-2} min^{-1}.

1915 First vacuum bolometer constructed.

1916 Pyranometer for measuring global radiation devised. Abbot's *solar cooker*, using direct solar radiation for cooking food, installed.

1917 Ultrasensitive vacuum bolometer constructed. Observation station installed on Hump Mountain, North California.

1918 Solar observation station installed near Calama, Chile. New instrument constructed for comparing brightness of sun and sky. Balloon measurement, by Aldrich, of cloud albedo at Arcadia, California, gave value of 78%. "The short method" of solar constant determination devised and put into use.

1919 Honeycomb (melikeron) pyranometer constructed. Marvin pyrheliometer and A. K. Ångström's electrical compensation pyrgeometer invented.

1920 Calama, Chile, station moved to Mt. Montezuma, Chile, where it was operated continuously until 1955. Mt. Wilson station moved to Mt. Harquahala, Arizona; abandoned in 1925; mean of 1244 observations at various locations from 1912 to 1920 yielded a solar constant of 1.946 cal cm^{-2} min^{-1}.

1922 Dorno's pyrheliograph invented.

1923 Invention of Kimball–Hobbs pyranometer (forerunner of first Eppley pyranometers). Invention of Moll thermopile. First weather forecasts based on (indicated) variations of solar constant made by H. H. Clayton.

1924 Moll thermopile used by Gorczynski for first Moll–Gorczynski pyranometer, later known as Kipp solarimeter.

1925 Solar observation station on Table Mountain, California, established (discontinued 1958). Strong questioning of Smithsonian Institution's determinations of solar variability begun.

1926 Solar observation station established at Mt. Brukkaros, South West Africa; abondoned in 1931.

1927 Double waterflow pyrheliometer developed by Shulgin.

1929 Method introduced by Kalitin for measuring radiative flux from various zones of the sky. Richardson developed method of measuring reflectivity of natural surfaces from aircraft.

1931 Method developed by Kalitin for measuring albedo of natural surfaces from ground.

1932 Double waterflow pyrheliometer was adopted as standard by Smithsonian Institution, and first revision of 1913 pyrheliometric scale. (Although the revised scale was reconfirmed in 1934, 1947, and 1952, and showed the 1913 scale to be approximately 2.3% too high, it was never generally adopted.) First standard design of Robitzsch pyranometer developed.

1933 Solar observation station established on Mt. St. Katherine on Sinai Peninsula; abandoned in 1937.

1939 Solar observation station established on Burro Mountain, New Mexico; abandoned in 1946.

1945 Solar observation station established at Camp Lee, Virginia; abandoned in 1947.

1948 Menzel developed sensitive sky photometer.

1950 Observation begun at Table Mountain with Menzel sky photometer.

1952 The silver-disk pyrheliometer recommended as an instrument for measuring direct solar radiation by sub-commission on Actimetry, World Meteorological Organization, Brussels. Best estimate of solar constant: 1.94 cal cm^{-2} min^{-1}.

1953 Invention of sunshine switch by Foster and Foskett of US Weather Bureau.

1954 Weather Bureau method of calibrating Eppley pyranometers changed. The old method utilized natural solar radiation; the new method uses artificial radiation in an integrating sphere (similar methods later introduced by Canadian Meteorological Services and Eppley Laboratory). Reevaluation of solar constant, by Johnson, through use of new rocket observations in ultraviolet and revised estimates in infrared. Best estimate given as 2.00 cal cm^{-2} min^{-1} with a probable error of 2%.

1956 Statistical study of 30 years of solar radiation data of Mt. Montezuma and Table Mountain by Sterne and Dieter showed root mean square of real changes of the solar constant no greater than 0.0032 cal cm^{-2} min^{-1}. "International Pyrheliometric Scale 1956" recommended by International Radiation Conference, Davos, Switzerland.

1957 "International Pyrheliometric Scale 1956" put into effect; original Ångström scale increased by 1.5% and 1913 Smithsonian scale decreased by 2.0%.

1961–1968 Balloon program for solar constant measurements conducted in Soviet Union.

1962 Adoption of the Campbell–Stokes sunshine recorder as an "interim reference sunshine recorder," by the commission on Instruments and Methods of Observation, WMO. Evaluation of solar radiation pressure on motion of satellite, by Jacchia and Slowey, provides value of solar constant as 2.00 cal cm^{-2} min^{-1}.

1965 Introduction of Eppley precision pyranometer. Development of automatic control of Ångström pyrheliometer by Marsh.

1966–1969 Program using aircraft, balloons, and spacecraft for direct measurement of solar constant and its spectral distribution conducted in the US. Parson's black problem with old style Eppley pyranometers documented.

1968 Value of solar constant is 1.952 cal cm^{-2} min^{-1} obtained by Laue and Drummond from rocket aircraft flight to 83 km altitude on 17 October 1967.

1969 Introduction of Eppley black and white (star-type) pyranometer. Previous model, based on Kimball–Hobbs design, discontinued. Development of the JPL active cavity radiometer.

1971 Announcement by Thekaekara and Drummond that 1.940 cal cm^{-2} min^{-1} adopted as an engineering design value of the solar constant by the US National Aeronautics and Space Administration.

1972 US National Weather Service requested the National Climatic Center to stop publishing insolation data due to 5 to 30% errors in data acquisition.

1973–1974 US Government initiates major commitment to solar energy research and development. International comparison of pyrheliometers at Davos Laboratory showed the difference between the IPS 1956 and active cavity radiometer to be 1.8%. US National Weather Service reprogrammed resources to upgrade entire solar radiation measuring program.

20

Fundamentals of Water Desalination

E. D. HOWE AND B. W. TLEIMAT

SEA WATER CONVERSION LABORATORY
UNIVERSITY OF CALIFORNIA
RICHMOND, CALIFORNIA

LIST OF SYMBOLS

C_{gs}	heat capacity of transparent cover and supports per unit of basin area
C_{wb}	heat capacity of water basin and contents per unit of basin area
C_p	specific heat at constant pressure
F_s	shape factor
g	acceleration of gravity
Gr	Grashof Number
H	solar radiation flux on a horizontal surface
h_b	equivalent heat transfer coefficient, basin to ground and surrounds
h_c	heat transfer coefficient by convection, water to transparent cover
h_c'	heat transfer by convection, transparent cover to ambient air
K	constant, Eq. (20.7)
k	thermal conductivity
L	latent heat of evaporation
M_A	molecular weight of water
M_B	molecular weight of dry air
M_f	molecular weight of humid air far away from water surface
M_w	molecular weight of humid air at water surface
M	initial amount of water in still, Eq. (20.20)
m	product, Eq. (20.20)
$\dot{m}$	productivity of still, Eq. (20.19)
n	constant, Eq. (20.7)
p_T	atmospheric pressure
p_w	vapor pressure of water at temperature of water surface
p_{wg}	vapor pressure of water at temperature of transparent cover
q_b	heat flux, still basin to grounds and surrounds
q_c	heat flux by convection, water surface to transparent cover
q_e	heat flux by evaporation/condensation, water surface to transparent cover
q_{ga}	heat flux, transparent cover to air
q_r	heat flux by radiation, water to cover
Re	Reynold's Number
t	time
T	final temperature of water, Eq. (20.20)
T_a	ambient temperature
T_f	temperature of humid air far away from water surface
T_g	temperature of transparent cover

LIST OF SYMBOLS (cont.)

T_{in}	inlet water temperature, Eq. (20.19)
T_0	initial water temperature, Eq. (20.20)
T_w	water temperature
v	ambient air velocity
w_d	product flux in still
W_w	feedwater rate, Eq. (20.19)
x	glass to water spacing, Eq. (20.7)
x	width of still, Eq. (20.18)
α_g	absorptivity of transparent cover for radiation
α_w	absorptivity of water to radiation
β	expansion coefficient of fluid
Δ	increment, e.g., ΔT is increment in T
ε	emissivity of transparent cover
λ	constant in Eq. (20.20)
μ	viscosity of fluid
ρ	density of fluid
σ	Stefan–Boltzmann coefficient

20.1 INTRODUCTION

Two basic approaches have been proposed for using solar energy to distill saline water. One approach reproduces on a small scale the pure water separation portion of the *hydrologic cycle* while the other uses the solar energy to energize distillers which are also capable of operating on other types of energy. Although both of these approaches will be discussed, it is only the first approach that has been used to any great extent up to the present time.

The hydrologic cycle, by which all freshwater is produced, is large scale, natural, solar distillation. Its four essential phases are:

(1) the production of vapor from oceans, lakes, and other large bodies of water;

(2) the transport of this vapor as humidity of the air above the land to cooler regions by convective winds;

(3) the condensation of this vapor and its precipitation as rain and snow; and

(4) the return of the water and melted snow by means of rivers to the oceans, lakes, and other large bodies of water.

Solar energy is the motivating energy for this hydrologic cycle. It penetrates the surface of the water, warms it and causes its evaporation. It also produces the winds which transport the vapor to the cooler regions where the vapor condenses as liquid. Man's use of this water occurs between the third and fourth phases of the above cycle.

Reproduced on a very small scale, the first three phases of the hydrologic cycle produce pure water in the simplest solar still, generally designated as the "basin-type solar still." Such a still is shown schematically in Fig. 20.1. It consists of a blackened basin containing saline water at a shallow depth, over which is a transparent, airtight cover that completely encloses the space above the basin. The cover, which is usually glass but sometimes

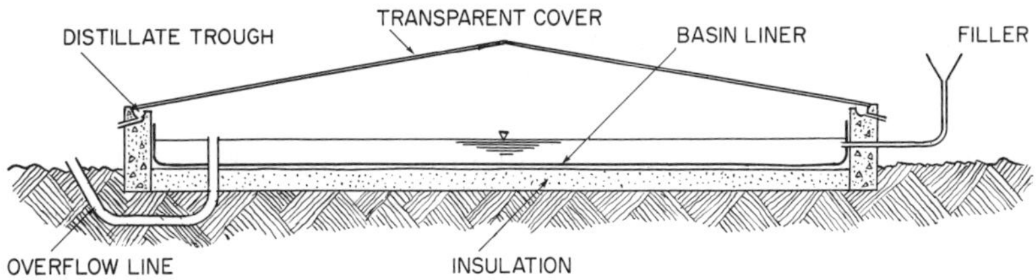

FIG. 20.1 Schematic diagram of a basin-type solar still.

plastic, is sloped toward a collection trough. In operation, solar energy, readily passing through the transparent cover, is absorbed by the water and the basin liner, and thereby warms the water causing evaporation. Because the transparent cover material does not absorb much solar radiation, it will be cooler than the water surface, producing convection currents in the air trapped inside the enclosure. These currents bring the humid air into contact with the relatively cool cover and result in condensation of some of the humidity on the surface of the cover. This condensate slides down the slope, flows along the trough and drains out of the enclosure through appropriate piping. Replacement of saline water in the operation of this process can be either *continuous* or by *batches*. Although basin-type solar stills have been constructed in many different configurations, their basic theory is identical. Such stills represent the most developed process for applying solar energy to fresh water production.

There have been occasional attempts to construct solar stills that would utilize the solar energy more efficiently than even the most effective basin-type solar still. These attempts have been motivated by the observation that even the best basin-type solar stills require the interception of solar radiation amounting to several times the latent heat of the water to be produced. In contrast, distillation plants, using heat energy from other sources, operate with supplies of heat energy amounting to only a small fraction of the latent heat of the water produced. Because of this condition, there are various proposals to construct solar devices which simulate in one way or another the reuse of heat exemplified by conventional distillers using nonsolar heat. The details of some of the proposals will be treated later in this chapter.

20.2 HISTORICAL SUMMARY

It would appear that one of the most desirable applications of solar energy should be the distillation of pure water from saline sources. The reason for this conclusion is the fact that water production need not be coincident with water usage, as is the case with power development or house heating. In house heating, solar-heated water must be stored for later use when solar energy is unavailable or is at too low an intensity, thus requiring expensive insulated tanks or other storage facilities. In spite of this desirable feature, however, solar stills have been found to be practical in relatively few instances. The reasons for the lack of solar distillation plants can be inferred from a review of the past history of research and development efforts in solar distillation.

Historically, instances of the application of solar energy to fresh-water production have been recorded throughout the world and at various times during the past century. An excellent summary of these efforts appears in the treatise by Talbert *et al.* (1970), a few highlights of which indicate the course of development.

The modern period in solar distillation began in 1872 with the installation of a large basin-type solar still near Las Salinas in northern Chile. Designed by a Swedish engineer, Carlos Wilson, this 4700 m^2 glass-covered still served the needs of a nitrate mining community for many years. The original still basin described by Harding (1883) was constructed of wood with bays 1.14 m wide by 61 m long, the wood being blackened with logwood dye and alum to absorb the sunlight. The salinity of 140,000 ppm (parts per million) in the feedwater resulted in very rapid saturation of the unevaporated brine, thus making flushing arrangements very important. Figure 20.2 is taken from a 1908 photograph of a solar still at this location and shows the still to be

FIG. 20.2 Solar still at Las Salinas, Chile (from Telkes, 1956a; used with permission).

quite different from that described by Harding. From this photo, the individual bays seem to be about 10 m long, with eight of these being grouped together. By counting the groups of bays in the photo, it is estimated that the area of the still in 1908 had been reduced to less than 3000 m^2. Due to other means of purifying water, the only remains of this old still in 1965 were reported to be "some of the foundation and numerous pieces of glass."

The next surge of interest in solar distillation began during World War II with the development of inflated plastic stills for use in connection with the emergency life rafts of the US Navy and Air Force. These units consisted of an inflatable plastic envelope containing a felt pad and a distillate collector bottle connected to the bottom of the plastic envelope. In use, the felt pad was to be saturated with seawater, and the whole assembly floated in the sea alongside the raft after inflation. Solar energy, striking the felt pad, would produce vapor which would condense on the inside of the plastic envelope and drip into the bottle at the bottom of the assembly. The developer of this device, Dr. Maria Telkes, reported that over 200,000 of these units were produced during the war.

During World War II and the decade immediately following, sustained drought conditions in many parts of the world brought water supply problems into sharp focus. The use of desalination methods seemed to promise relief from the drought. In the USA, the State of California and the national congress instituted research programs directed toward the development of various methods for desalting saline water. In Europe, the Organization for European Economic Cooperation set up a working party to accomplish similar objectives. Solar distillation was one of the many methods considered by these programs. However, it received relatively little of the available financial support because research on desalting methods involving electrical and fossil–fuel energy seemed likely to reach a commercial phase sooner than solar distillation. Research and development involving mechanical equipment was more compatible with the capabilities of existing manufacturing companies, so that most of the experimental effort was placed on such studies.

When the University of California commenced its investigation of desalination methods in 1951–1952, it seemed that processes using nonfuel sources of energy should be given important priority. This conclusion was based on the observation that the then existing desalination methods required large amounts of energy. For example, if seawater were distilled for supplying all the needs of a city as large as Los Angeles, the multiple-effect type of distiller then in use would have required the energy from more than all of the oil fuel produced in Southern California. Noting that Dr. Maria Telkes had continued her wartime interest in solar distillation and had constructed in the late 1940s a small solar still at Cohasset, Massachusetts, she was engaged to institute work on solar distillation at the University of California's

Engineering Field Station in Richmond, California. Following her short-term assignment in this work, solar distillation investigations at the University continued for the next few years at a vigorous rate. Various configurations for simple solar stills were built and tested in trying to reduce capital costs and improve efficiency. The use of warm cooling water from engines and industrial processes was studied as a means of augmenting the output of solar stills. However, it was eventually concluded that the basin-type solar still in any of the forms considered could not compete economically with large plants using other desalination processes. Since about 1960, the work on solar distillation has proceeded at a diminishing rate with emphasis on the development and improvement of small plants suitable for supplying drinking water to isolated residences and small communities. A summary of these University of California investigations has been presented by Howe and Tleimat (1974).

Other research centers, both in the USA and abroad, have reached similar conclusions. Active investigations have been centered in many universities and institutes. The USA program, under the Office of Saline Water, carried on a great deal of work by contract with these universities and industrial groups. One special experiment station for large-scale development work in solar distillation was set up on government property in Daytona Beach, Florida. Solar distillation experiments have been carried on in many other countries, Greece, Algeria, Australia, etc., in their searches to solve their pressing water problems. All of these programs were directed toward the development of large-scale basin-type solar installations, the largest one being the still at Patmos, Greece, with a maximum capacity of about 40 m^3/day. However, it was finally clear that other methods of desalination could produce desalted water more cheaply than the basin-type solar stills for the large capacities, 4000 m^3/day, and larger, which seemed to be the minimum capacity of interest.

As a result, most of the USA efforts and much of that in other countries in constructing large-scale solar stills were discontinued about 1970. In contrast, medium-scale solar distillation development has continued to take place in Australia, under the auspices of the Commonwealth Scientific and Industrial Research Organization (CSIRO). This organization carries on a continuous program attempting to reduce the cost of solar-distilled water in many of its arid locations. One of their recent ventures has been the use of cast plastic basins with single-sloped covers, adapted from the stills constructed by Hensley and Young, University of California (1972), which seem to offer the possibility of high efficiency and low cost because of factory fabrication. Another of their ventures is the use of waste heat from diesel engines in the solar still enclosure, which makes it possible to have more constant production independent of the weather.

The tangible remains of the solar still development of the 1950s and 1960s are a few large solar stills in different countries scattered over the world and an unknown number of small stills similarly dispersed. The large basin-type solar stills thought to be presently operating are listed in Table 20.1, much of the data being obtained from Talbert *et al.* (1970).

TABLE 20.1

Basin Type Solar Stills

Country	Location	Date built	Basin area (m^2)	Productivity (m^3/day)
Australia	Muresk II	Nov. 1966	372	0.895
	Coober Pedy	1969	465	1.1[a]
	Caiguna	Dec. 1966	372	0.834
	Hamelin Pool	Dec. 1966	557	1.30
	Griffith	Aug. 1967	413	0.977
	Townsville	1969	232	0.56[a]
Cape Verde Islands	Santa Maria do Sol	1965	743	2.28[a]
Chile	Quillagua	1968	107	0.38[a]
	Quillagua	1969	103	0.36[a]
	Pisagua	1971	124	0.44[a]
Greece	Salamis	Oct. 1965	388	1.18[a]
	Patmos	July 1967	8640	28.08
	Kimolos	1968	2508	8.14
	Nisiros	1969	2043	6.51
India	Bhavnagar	Oct. 1965	377	0.89
Mexico	Baja California Natividad Island	Mar. 1969	95	0.41[a]
	Baja California Sur			
	Puerta Chale	1974	300[a]	1.0
	Punta Eugenia	1974	470[a]	1.5
	Sonora- Punta Chueca	1974	470[a]	1.5
Spain	Las Marinas	Mar. 1966	869	2.77
Tunisia	Chakmou	1967	439	0.57
	Mahdia	1968	1300	4.48
U.S.S.R.	Turkmenia Bakharden	Apr. 1969	599	1.75[a]
Caribbean Area	Petit St. Vincent	Feb. 1967	1709	5.29
	La Gonave, Haiti	June 1969	223	0.81[a]

[a] Estimated values.

20.3 THE COST OF SOLAR-DISTILLED WATER

The cost of water produced by any desalination process depends on the capital cost of equipment, the cost of the energy and the costs of maintenance and operation other than energy. In the case of solar stills, the cost of energy is a very small fraction of the total, since the energy other than solar is only that required to operate pumps and controls. Thus, the major share of the water cost in solar distillation is that of amortization of the capital cost. The production rate is proportional to the area of the solar still, so that the cost per unit of water produced is nearly the same regardless of the size of the installation. This is in contrast with conditions for freshwater supplies as well as for most other desalination methods, where the capital costs of equipment per unit of capacity decrease as the capacity increases–the 6/10 power rule applying to many of the components. According to this rule, the capital costs of similar devices of different capacities would vary as the 6/10's power of the capacity. Because of this decreasing unit cost of water from desalination methods other than solar, it can be inferred that such methods are superior to solar distillation when large sizes are to be considered. This rule implies also that solar distillation may be more attractive than other methods for small sizes. Comparisons made between several methods tend to verify this fact, as is shown in Fig. 20.3 in which the cost of water versus capacity is plotted for solar distillation, vapor-compression

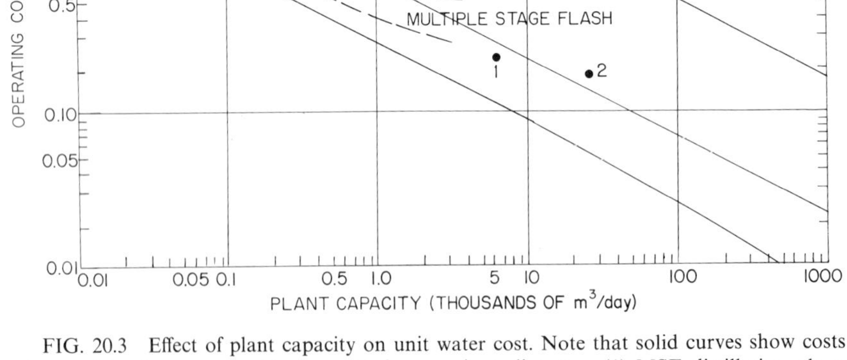

FIG. 20.3 Effect of plant capacity on unit water cost. Note that solid curves show costs of normal fresh water supplies brought from various distances. (1) MSF distillation plant, Key West, Florida; (2) MSF distillation plant, Tijuana, Mexico.

distillation and multiple-stage flash distillation. From this figure, one may conclude that solar distillation has an apparent advantage in sizes less than 200 m^3/day. However, economic engineering of plants of this size has not been demonstrated as yet since the largest solar plant, so far, located at Patmos, Greece, has a capacity of only about 40 m^3/day.

Cooper and Read (1974) published the most recent cost information in which they showed that the cost of large solar stills, installed in remote areas, is US\$33.4/m^2. Assuming a useful life of 10 years, and an average productivity of 1.22 m^3/year/m^2, gives a water cost of US\$2.73/m^3 for capital amortization alone. To this must be added the cost of maintenance and operation. For very small units, Cooper and Read gave only slightly higher per unit costs, thus substantiating the lack of variation of unit cost with size. The cost figures appearing in Fig. 20.3 were obtained before the present period of inflation, so that they give only comparison figures rather than absolute figures. It is noted that prices reported for bids on large distillation plants in various parts of the world have risen from a low of about US\$1 per daily gallon of capacity before inflation to over US\$6 per daily gallon in the period since 1970, most of the increase being due to inflation in material and labor prices. It is likely that the capital cost of solar distillers will be proportional to that of other types of distillers, so that the comparison seen in Fig. 20.3 may remain valid.

FIG. 20.4 Some types of solar stills tested at the University of California, Richmond, California.

20.4 CONFIGURATIONS OF BASIN-TYPE SOLAR STILLS

Efforts to decrease the cost of water produced by basin-type solar stills have been directed toward reducing the initial capital cost without unduly shortening the useful life period or decreasing the efficiency of operation. Much of this work has involved different geometrical configurations. Those tried out at the University of California are representative and are shown in the sketches of Fig. 20.4. One design that seemed most promising and was chosen to be used in the experimental work in the South Pacific is shown in more detail in Fig. 20.5. This still is seen to be a glass-covered design, with plastic basin liner, stainless steel distillate troughs, aluminum glass supports, and a frame of either wood or precast concrete members. The modular design gives flexibility of size, and since the addition of each module adds approximately 2.4 m^2 to the basin area, it increases the capacity by nearly 10 l/day with each additional module.

The cross section of some of the large solar stills are shown in Fig. 20.6. The Australian design, shown in Fig. 20.6a, has gone through several although essentially similar stages of development. In this model concrete side members contain grooves for the lower edges of the glass cover. The basin liner, which in the past has been polyethylene, but is presently butyl rubber sheet, is laid over the insulated bottom, formed into the distillate troughs and sealed by the glass edges. The distillate troughs are formed by the space between the concrete side members and an asbestos cement strip, the latter spaced away from the concrete by insulating material and supported on the basin side by the asbestos cement angles used as dams across the still basin. The ridge of the roof-shaped structure is formed by using silicone rubber sealant to form a hinge for the two glass panes opposite each other. In this design the flow of saline water through the still is continuous, requiring that the bottom of the still be sloped. To insure a small average depth of water of about 2 or 3 cm in a relatively long still, dams are placed across the still every 1 or $1\frac{1}{2}$ m so that there is a succession of shallow pools of saline water along the length of the sloping still. Various devices have been tried for lowering these dams during periods when the still is being flushed. This design has been quite successful and has been used with minor variations in several Australian installations. The basic expense in this design is the cost of glass. Horticultural glass was found to be lowest in price, but its rather stringent size limitations confined the width of bays to sizes of little more than one meter.

The second largest number of operating stills is found in the Greek islands, and these units have cross sections similar to Fig. 20.6b. These are glass covered with aluminum glass supports and distillate troughs. The side walls and curbs are concrete beams, with bays 1.5 wide arranged in pairs. The basin liner is butyl rubber sheet. It will be noted that these stills are unsymmetrical, with the long slope toward the equator. This arrangement was thought by the designers to be superior to the symmetrical type at the latitude of about 37°N. Although a similar design was proposed for West Pakistan, it was never approved and no further similar installations are known to the writers.

A single-sloped design is shown in Fig. 20.6c and is the one used by McGill University's Brace Research Institute for the installation at La Gonave, Haiti. This still is installed on a gentle slope, so that each bay is slightly lower than the one above it. The concrete curbs have cast-in distillate troughs, as well as second troughs for collecting rain water. The butyl rubber basin liner is glued to the concrete curbs, and this has caused some trouble because it does not extend to the edges of the glass cover, and hence tends to pull away from the curbs as the sand beneath the basin becomes compacted over a period of time.

The design used at Las Marinas, Spain, is shown in Fig. 20.6d and was derived from the *deep basin* concept tried out by the OSW at Daytona Beach, Florida. In this concept, a continuous basin, 28 m long and 31 m wide, is used, rather than separate bays, so that the basin liner can be continuous from one side of the still to the other. It was argued that this construction would avoid the need for accurate leveling of the ground, since the use of appreciable depths of water would prevent dry spots from appearing on the basin liner when evaporation occurred. However, after considerable experimentation, it was finally decided that the depth of water should be as small as possible, and so the Spanish installation used water depths of about 10 cm rather than the 30 cm used in the Florida experiments. This small water depth required fairly accurate leveling of the ground. The basin liner was an asphalt sheet laid in strips and hot-mopped to make it waterproof. Glass supports and distillate channels were formed in the precast elements. Asphaltic mastic caulking material was used to seal all glazing joints.

In addition to these flat horizontal basin-type distillers using solar heat directly, some research has involved the study of the tilted tray and the tilted wick solar stills. The tilted tray solar still is a basin-type unit in which the basin is broken up into narrow strips, with each strip on a different elevation, and arranged like a series of steps. This brings the water surface closer to the transparent cover than is possible with the flat basin-type and usually is accompanied with very shallow water depths. While this construction results in an increase of efficiency, the additional cost of construction has outweighed the increase in effectiveness. The tilted wick still is a glass-covered inclined plane which is faced with some sort of cloth wicking or other absorbent material. Saline water introduced along the upper edge of the inclined plane trickles down through the absorbent materials and maintains the latter in a wet condition. This device can work with a relatively

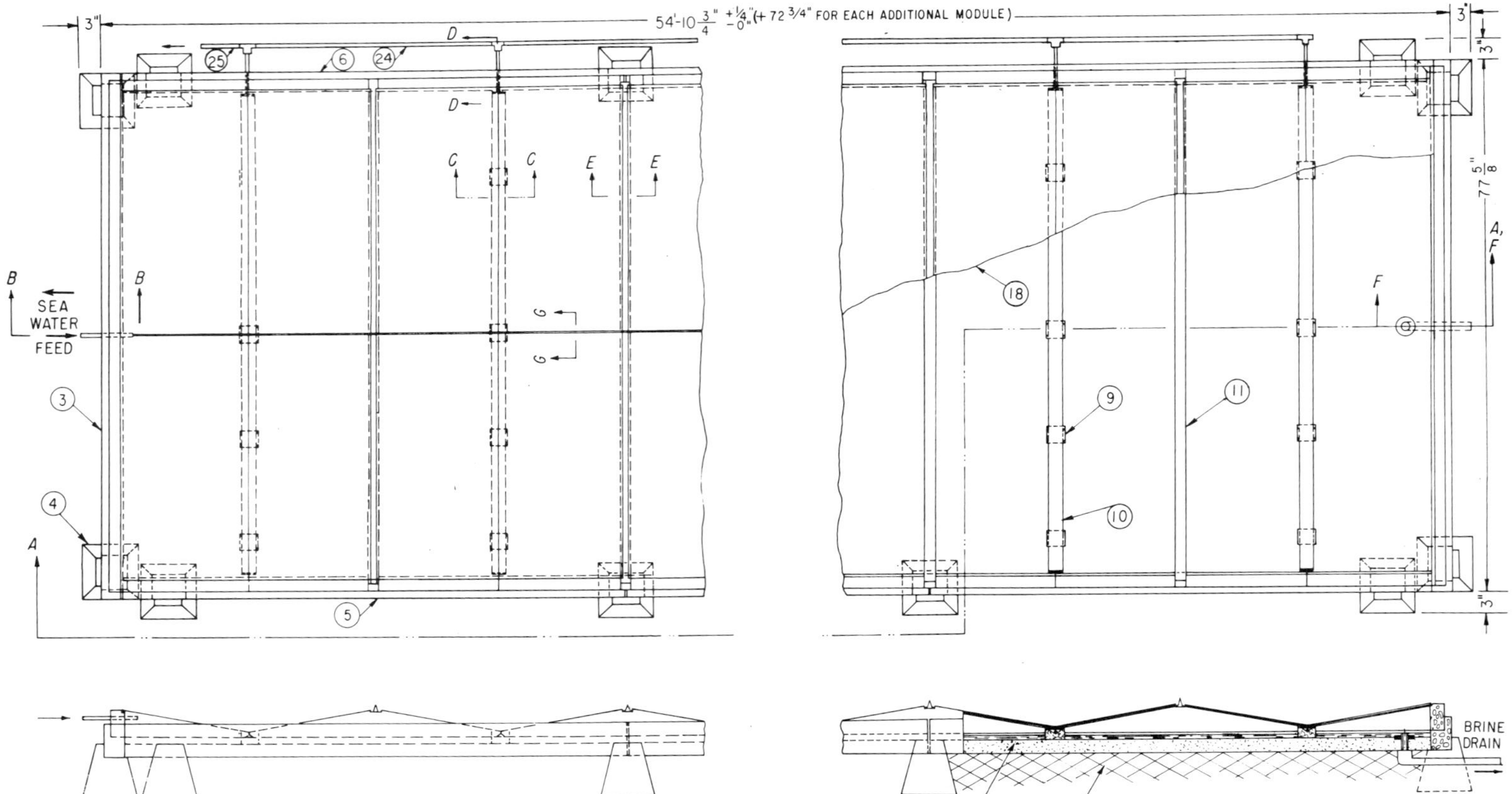

PLAN VIEW AND SECTION $A-A$

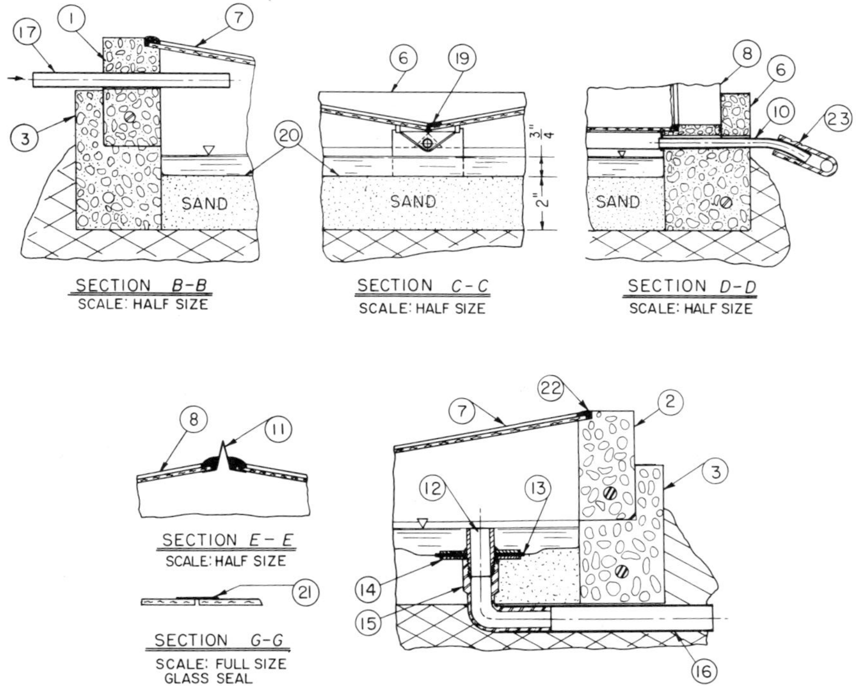

NOTE:

THIS DRAWING IS MADE FOR A 9 MODULES STILL. EACH MODULE IS $72\frac{3}{4}$"LG. TO ADD OR DELETE A MODULE DO THE FOLLOWING:

ITEM	ADD or DELETE	ITEM	ADD OR DELETE
4	2	11	2
5	1	18	8
6	1	19	4
8	4	20	$72\frac{3}{4}$" LENGTH
9	8	23	2
0	2	24	2

LIST OF MATERIAL		
ITEM	DESCRIPTION	REQ'D.
1	END GLASS SUPPORT	1
2	END GLASS SUPPORT	1
3	END FRAME	2
4	SUPPORT PIER	24
5	LOWER FRAME	9
6	LOWER FRAME	9
7	EXTREME SIDE GLASS SUPPORT	4
8	GLASS SUPPORT (SIDE)	34
9	PRODUCT TROUGH SUPPORT	72
10	PRODUCT TROUGH	18
11	METAL GLASS SUPPORT	17
12	DRAIN	1
13	RUBB'R GASKET 3"O.D.×$1\frac{1}{8}$"I.D. × $\frac{1}{16}$"T'K	2
14	COPPER WASHER $2\frac{3}{4}$ O.D.×$1\frac{1}{16}$ I.D.×$\frac{1}{16}$ T'K	1
15	$\frac{3}{4}$ F.P.T. $\frac{3}{4}$ TUBE PLASTIC ELBOW	1
16	$\frac{3}{4}$ DIA. PLASTIC PIPE 10' LONG	1
17	$\frac{1}{2}$ DIA. 12" LG. COPPER FEED TUBE	1
18	18" 36" DOUBLE STRENGTH GLASS	72
19	$\frac{1}{8}$ THK.×$\frac{3}{4}$ WIDE×3"LG. VINYL ADHESIVE TAPE	36
20	TYPE 65 GRIFFOLYN 8'W.×55'LG. (BLACK)	1
21	$\frac{3}{4}$"W. BLACK VINYL ELECTRICIAN TAPE	1 ROLL
22	"DAP" CAULKING COMPOUND	10 TUBES
23	$\frac{1}{2}$" PLASTIC TEE	18
24	$\frac{1}{2}$" PLASTIC PIPE × 35' LONG	17
25	$\frac{1}{2}$" PLASTIC PIPE × 10' LONG	1

FIG. 20.5 Assembly: sawtooth-type solar still.

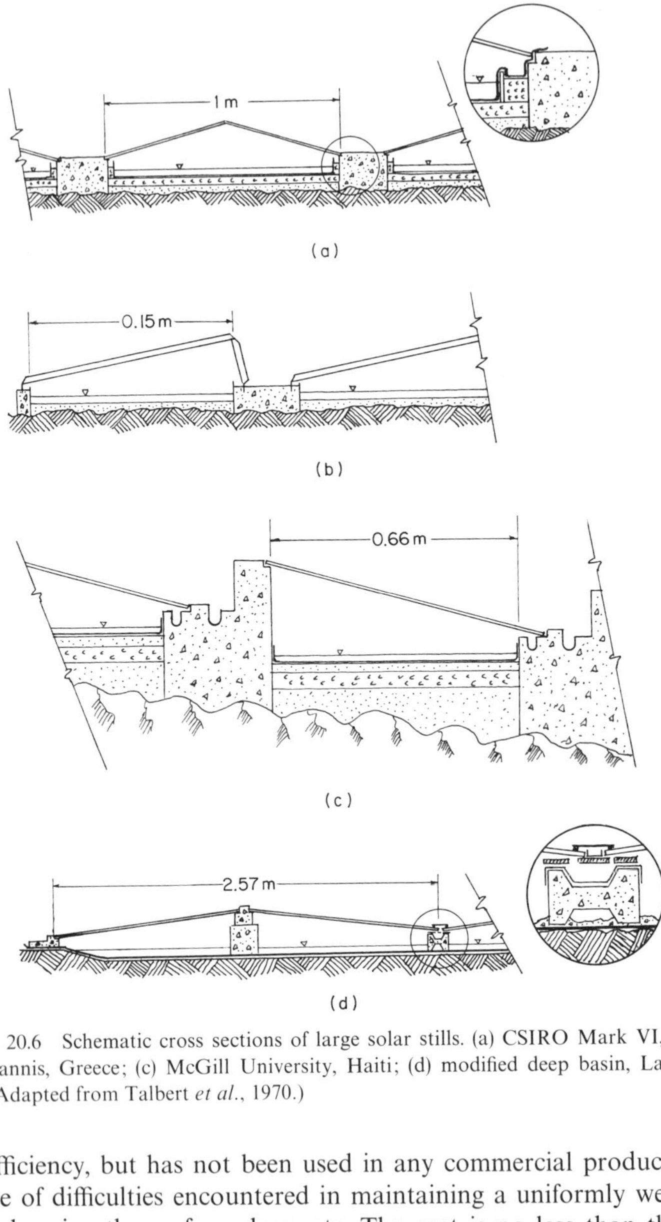

FIG. 20.6 Schematic cross sections of large solar stills. (a) CSIRO Mark VI, Australia; (b) Delyannis, Greece; (c) McGill University, Haiti; (d) modified deep basin, Las Marinas, Spain. (Adapted from Talbert *et al.*, 1970.)

high efficiency, but has not been used in any commercial production unit because of difficulties encountered in maintaining a uniformly wet surface and in keeping the surface clean, etc. The cost is no less than that of an equivalent basin-type unit, which is a further reason for the lack of prototype production units.

The foregoing descriptions will serve to show the variety of design concepts which have been explored. The primary consideration in all cases has been the cost of materials and labor for producing the stills as well as the costs of operation and maintenance, and the notable differences reflect the costs and availability of local materials and labor.

The following conclusions can be drawn relative to design features, based on the solar stills constructed to date. Conditions favoring high efficiency include small slope angles for the transparent cover, small water depths and low heat losses through the sides and bottom of the structure. When glass covers are used slope angles as low as 6° have been tried, with 10° angles being more widely favored. The problem with angles less than 10° is the danger that condensed water will drop back into the saline water basin rather than sliding down the slope into the distillate trough. Plastic covers generally require larger slope angles because these materials are relatively thin and tend to sag, making the minimum slope much less than the top-to-bottom average slope. Also, because plastics are less wettable than glass, the problems of small slope angles tend to be magnified due to dropwise condensation.

The use of small water depths in the basin requires accurate leveling of the material under the basin liner, since any humps in the liner could cause dry spots, thus decreasing the water surface area available for evaporation as well as deterioration of the basin liner. Practical minimum water depths seem to be of the order of 2 to 3 cm.

Heat losses through the sides and bottom can be decreased by applying insulation to these parts. Flexible closed cell plastic materials seem to be most favored for this use at the present time, with a 2 to 3 cm thickness being an acceptable amount.

There seems to be little data on the reevaporation of condensate from distillate troughs during the time elapsing between condensate formation and exit from the still enclosure. However, it is obvious that this time should be minimized and that the distillate trough should be shielded from direct solar radiation, if possible, if reevaporation is to be kept to a small value.

The research and development work done to date has also yielded some useful information on materials, particularly transparent cover materials, basin liners, and sealants. The choice of transparent cover materials lies between glass and plastic films. Glass possesses the virtues of high transmissivity for solar radiation, low transmissivity for low-temperature radiation, high wettability for water, and relatively high stability of properties over extended periods of time. Its chief drawback is its vulnerability to mechanical damage. Plastic films used for solar still covers have shown relatively high transmissivity for low-temperature radiation as well as for solar radiation, must be surface treated to be wettable with water, and are subject to ultraviolet degradation in various amounts. These materials seem to be less vulnerable to mechanical damage than glass, but have not yet been developed to be sufficiently long-lasting to make use of this desireable property.

Basin liner materials must absorb solar radiation readily, must be watertight, and, because the still may possibly run dry, should be capable of withstanding high temperatures without deleterious effects. Asphalt mats, similar to those used for lining stock tanks and roofing buildings, have been used to line large deep basin stills. While found satisfactory for these stills, they are too stiff for use in the shallow basins of stills with long narrow bays. For the latter types, basin liners of black butyl rubber sheet and black polyethylene sheet are favored, the latter being either plain or laminated and having a woven plastic mesh between two sheets. Butyl rubber has the capability of withstanding high temperatures, whereas there have been some burnouts of polyethylene liners when permitted to run dry.

The sealing of the transparent cover to prevent vapor leaks is most important, since leakage in this area can drastically curtail production rates. The most effective sealant used so far is silicone rubber sealant, which remains elastic for a considerable period of time. Various other sealants have been used, such as tars and tapes, but most of these deteriorate with time and become brittle so that cracks form and permit leakage of vapor.

20.5 PREDICTION OF PERFORMANCE OF BASIN-TYPE SOLAR STILLS

The performance of a solar still is generally expressed as the quantity of water produced by each unit of basin area in a day, i.e., cubic meters or liters of water per square meter of basin area per day. This quantity will vary with the design of the still, with the intensity of solar radiation and with the atmospheric conditions in the surroundings. Figure 20.7 shows the relation between production rate and the altitude of the sun at noon. This latter coordinate was chosen to emphasize the importance of weather variables. It will be noted that the points for clear days lie along a rather definite curve, and that the data points for nonclear days are scattered over the entire space below the clear-day curve. The solar flux for clear days is an almost linear function of the altitude of the sun at noon, so that the clear-day curve of production rate versus solar intensity is similar in shape to that of Fig. 20.7. While the scatter of points would be reduced somewhat in a plot of productivity versus solar radiation, as compared to Fig. 20.7, the use of measured values of solar intensity does not bring the cloudy day points onto any well-defined curve. Thus, in most published data the clear-day points are used to determine the performance curves. Since weather varies from one location to another, the use of regression curves by months, including cloudy day data, as recommended by Morse and Read (1966), is valid for only the location where the data were taken.

Figure 20.8 shows the clear-day production curves for a number of University of California solar stills of various designs. The fact that the

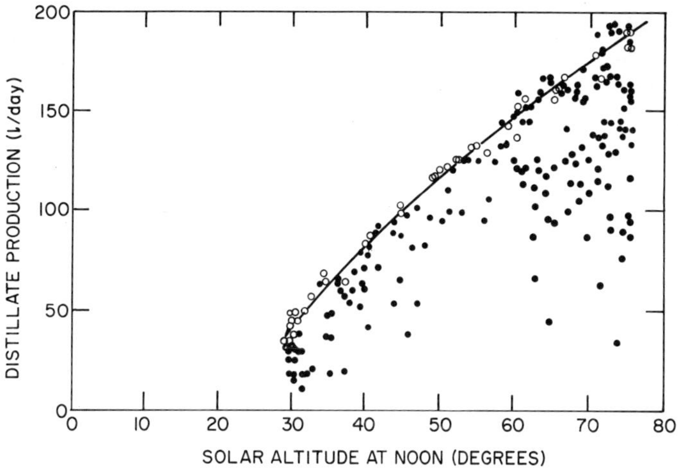

FIG 20.7 Effect of weather on solar still productivity. University of California Still No. 16, 38 m^2 area, 1958. ○, Clear day data; •, cloudy day data.

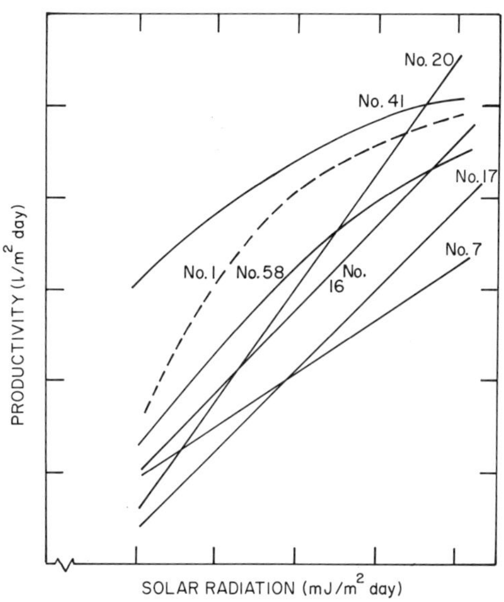

FIG. 20.8 Clear day productivity of several University of California solar stills. Still No. 1 used for comparison.

clear-day performance data define a curve and that the curves for different designs are distinct, suggests that it should be possible to predict the clear-day performance curves analytically. Analyses of solar still operation have been presented by many writers, the material which follows being adapted from that originated by Dunkle (1961), modified by Morse and Read (1968) and by Cooper and Read (1974).

The amount of distillate produced in a solar still depends primarily on the amount of solar radiation available but is affected also by several other factors. Figure 20.9a is a diagrammatic cross section of a solar still on which are drawn vectors to indicate the several heat and energy fluxes and their directions. In this figure, it is seen that an incoming ray of sunlight is absorbed in part by the transparent cover, the water and the basin liner, with a small fraction of it being reflected back to space by the glass and water surfaces. The portion of the energy absorbed in the basin which is available to produce vapor, and hence distillate, depends on the change in temperature of the basin water and on the heat losses from the water basin. Finally, the amount of vapor condensed, and thus the amount of distillate produced, is affected

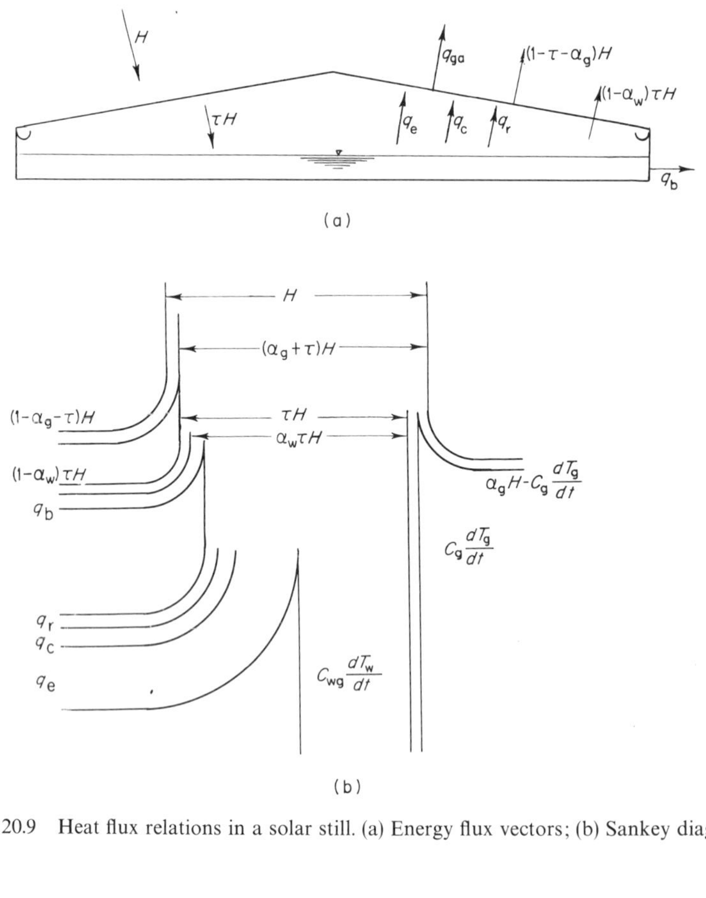

FIG. 20.9 Heat flux relations in a solar still. (a) Energy flux vectors; (b) Sankey diagram.

by heat and mass transfer to the transparent cover on the inside and the rate of heat dissipation from the cover on the outside.

Relations between the various heat fluxes are shown by the Sankey diagram of Fig. 20.9b. This diagram is qualitative only, although the widths of the several stems and branches are intended to be suggestive of relative magnitudes at some particular time. By referring to Fig. 20.9b, heat balance equations can be written for several conditions. (See List of Symbols, pages 431–432, for meaning of symbols.)

1. Heat balance on basin water

$$\alpha_w \tau H = q_b + C_{wb} \frac{dT_w}{dt} + q_e + q_r + q_c \tag{20.1}$$

2. Heat balance on glass cover

$$q_e + q_c + q_r + \alpha_g H = q_{ga} + C_{gs} \frac{dT_g}{dt} \tag{20.2}$$

3. Heat balance on the basin and cover assembly

$$\alpha_g H + \alpha_w \tau H = q_{ga} + q_b + C_{wb} \frac{dT_w}{dt} + C_{gs} \frac{dT_g}{dt} \tag{20.3}$$

It will be noted that these equations are formulated so that the reflected shortwave radiation, $(1 - \alpha_g - \tau)H$ and $(1 - \alpha_w)\tau H$, does not appear in them because this reflected radiation has no effect on vapor formation or condensation.

4. Heat flux from the water surface to the cover by radiation, q_r. This quantity can be estimated using the relation for heat transfer by radiation

$$q_r = F_s \sigma (T_w{}^4 - T_g{}^4) \tag{20.4}$$

In this expression, F_s is the configuration or shape factor, and depends on the geometry of the still and the nature of the radiation. The geometry can be approximated by two parallel planes. The radiation involved is diffuse radiation in long wavelengths, so that successive specular reflection between the transparent cover and water surface is negligible. As a result, the shape factor can be closely approximated (Boelter *et al.*, 1965, p. 684) by the emissivity of the water surface, usually taken as 0.9 for the conditions inside of the solar still. Thus, the heat flux by radiation between the cover and the water surface can be written as

$$q_r = 0.9 \sigma (T_w{}^4 - T_g{}^4) \tag{20.5}$$

5. *Heat flux from the water surface to the cover by free convection*, q_c This quantity can be estimated by using the heat transfer equation

$$q_c = h_c(T_w - T_g) \tag{20.6}$$

The value of h_c must be obtained from empirical data, which is usually correlated using dimensionless equations of the Nusselt–Grashof type, the one applying in this case being

$$h_c x/k = K[(x^3\rho^2 g\beta\,\Delta T/\mu^2)\cdot(\mu C_p/k)]^n \tag{20.7}$$

in which x is the glass-to-water spacing. The values of K and n depend on the magnitude of Grashof's number, $\mathrm{Gr} = x^3\rho^2 g\beta\,\Delta T/\mu^2$. The generally accepted values of K and n are as follows:

$\mathrm{Gr} < 10^3$;	$K = 1, \quad n = 0$
$10^4 < \mathrm{Gr} < 3.2 \times 10^5$;	$K = 0.21, \quad n = \frac{1}{4}$
$3.2 \times 10^5 < \mathrm{Gr} < 10^7$;	$K = 0.075, \quad n = \frac{1}{3}$

For the conditions of humidity, temperature, and geometry found in basin-type solar distillers with glass-to-water spacing of 75 mm or more, the value of Gr falls in the last of the above three ranges, so that

$$h_c x/k = 0.075[(x^3\rho^2 g\beta\,\Delta T/\mu^2)\cdot \mu C_p/k]^{1/3} \tag{20.8}$$

Since $(x^3)^{1/3} = x$, this quantity can be canceled out, indicating that h_c is independent of the size of the equipment for most stills.

The buoyancy factor $\beta\,\Delta T$ in Eq. (20.8) was originally evaluated for conditions of constant composition of the convecting fluid. However, in the solar still the humidity of the air, and therefore its composition, changes along the flow path. The vapor formed in the still is lighter than the dry air component and, therefore, increases the buoyant force and the rate of circulation of the air and associated vapor. This effect can be introduced into the equation by showing an increment to $\beta\,\Delta T$. The formulation used by Dunkle (1961) is based on a modification of $\beta\,\Delta T$ as given by Boelter *et al.* (1965), which is

$$\beta\,\Delta T' = (M_f T_w / M_w T_f) - 1 \tag{20.9}$$

where M is the molecular weight of the humid air mixture, T the temperature, subscript f indicates a distance far away from water surface, and w indicates a point at surface of water.

By using the perfect gas law for both the water vapor and dry air, and noting that the total pressure is the sum of the partial pressures of water vapor and dry air, algebraic manipulation results in the formulation

$$\beta\,\Delta T' = \left[\frac{T_w - T_g}{T_g} + \frac{p_w - p_{wg}}{[M_B/(M_B - M_A)]p_T - p_w}\cdot\frac{T_w}{T_g}\right] \tag{20.10}$$

where M_B is the molecular weight of dry air, 28.97; and M_A the molecular weight of water vapor, 18.02. The value of h_c can then be written as

$$h_c = \text{const}\left[T_w - T_g + \frac{p_w - p_{wg}}{[M_B/(M_B - M_A)]p_T - p_w} T_w\right]^{1/3} \qquad (20.11)$$

where the constant includes all of the fluid properties, evaluated at an average temperature, and depends for its magnitude on the system of units employed.

By inserting this value of h_c into Eq. (20.6), the heat flux by convection within the solar still becomes

$$q_c = 8.84 \times 10^{-4}\left[(T_w - T_g) + \frac{p_w - p_{wg}}{2.65p_T - p_w} T_w\right]^{1/3} (T_w - T_g) \qquad (20.12)$$

F. *Heat flux by evaporation and condensation*, q_e. The calculation noted in the previous paragraph relates to the amount of heat flux due to convection without condensation, so that a separate calculation must be made for the additional heat flux due to condensation. The amount of water transferred from the water surface in the basin to the condensate film on the cover can be estimated in terms of the analogy between heat transfer and mass transfer, as in Boelter *et al.* (1965). According to this analogy, the mass flowrate is proportional to the heat transfer coefficient and the driving potential, the latter being the difference in partial pressures of the material being transferred. The algebraic formulation of this quantity is

$$w_d = 18.41h_c(p_w - p_{wg}) \qquad (20.13)$$

The corresponding rate of heat flux due to condensation is then

$$q_e = w_d L \qquad (20.14)$$

$$= 6.8742 \times 10^{-9}[q_c/(T_w - T_g)](p_w - p_{wg})L \qquad (20.15)$$

G. *Heat loss through the base and perimeter of base*, q_b. This quantity is difficult to estimate because the temperature of the ground is generally unknown. However, there are enough experimental data to furnish a reasonable approximation, using the relation

$$q_b = h_b(T_w - T_a) \qquad (20.16)$$

H. *Heat flux from the transparent cover to the surroundings*, q_{ga}. This rate of heat dissipation depends both on radiation to the sky and on convection by air circulation. Radiation to the sky depends on the effective sky temperature, which is generally taken as 11°C less than the ambient temperature. The convective portion is taken as a function of the wind speed.

$$q_{ga} = \varepsilon_g \sigma [T_g{}^4 - (T_a - 11)^4] + h_c'(T_g - T_a) \tag{20.17}$$

$$h_c' x/k = 0.028\ \mathrm{Re}^{0.8} \tag{20.18}$$

in which h_c' is the convective heat transfer coefficient, transparent cover to air; x the width of solar still; k the thermal conductivity of air film on cover; and Re the Reynolds number $= vx\rho/\mu$ dimensionless.

It should be noted that the original derivation of these equations by Dunkle (1961) was based on assumed steady state conditions. The transient term in Eq. (20.1) for water heating was added by Morse and Read (1968), and the transient term for heating of the glass in Eq. (20.3) was added by the authors. The use of the equations for computing the hourly production of the solar still involves the assumption that the steady state equations, other than Eq. (20.1) and (20.3), can be used even though the temperature may be changing slightly during the hourly period. This is justifiable on the basis that the end result will have the accuracy tolerance of $\pm 20\%$ expected in most heat transfer calculations.

The forms of these simultaneous equations are such that explicit algebraic solution is quite involved. Many will choose to use digital computer techniques to find solutions for particular sets of conditions. Those who wish to solve them manually for particular conditions should consult Morse and Read (1968) or Howe (1974), both of which references contain charts for shortening the required trial and error techniques.

Useful information can be obtained from the equations without going through complete simultaneous solutions. From Eq. (20.1) certain conclusions can be drawn with respect to the depth of water and the use of bottom insulation. In this equation, the quantity C_{wb} represents the thermal capacity of the water basin, its contents and probably some of the materials under it, the chief component being the water contained in the basin. While it is true that the water must be heated up before evaporation, the less water there is to be heated, the more of the heat energy there will be to apply to evaporation. Thus, basin-type solar stills with very small depths of water will be more productive and have higher production rates per unit area than those with appreciable water depths. The case for bottom insulation is not quite so clear, since the equation representing this loss, Eq. (20.16), contains only a single coefficient. However, Morse and Read (1968) showed that this loss could be important under the conditions encountered in Australian installations, and recommended that plastic foam 2.5 cm thick should be used under the CSIRO models.

The importance of high temperatures of the water in the basin can be inferred from Eqs. (20.5), (20.12), and (20.15), which give the amounts of heat flux from the basin water surface to the cover surface. Each of these equations contains the quantities $(T_w - T_g)$ and T_w. One can calculate values for heat flux by each of the three modes using given magnitudes of $(T_w - T_g)$ for various magnitudes of T_w. When the results are summarized in terms of the fraction of heat energy transferred by each mode, the plots of such data appear as in Fig. 20.10. This type of plot was first presented by Cooper (1973), who pointed out that the results clearly indicated that the percentage of heat transferred by evaporation will increase with increasing water temperature T_w at a given value of $(T_w - T_g)$, so that the efficiency of the basin-type solar still will increase with an increase of water temperature.

Using this same set of equations, it is possible to determine the maximum efficiency that could be attained in an ideal basin-type solar still. For this determination, it is assumed that the sensible heat capacity of the water basin and contents is very small compared with the incoming solar radiation, and that the insulation beneath the basin is very effective. Under these conditions, all of the incoming solar radiation absorbed by the water must be transferred back from the water to the transparent cover and be dissipated therefrom. Computer solutions of the equations for these conditions have been made by Cooper (1973) for a wide range of incoming solar radiation and for various ambient temperatures. Except for the very low radiation intensities, he found a linear relation between the production rate and the incident solar radiation for any given ambient temperature. Since the solar still will operate at steady state conditions, this linear relation can be used in computing the total daily production. On this basis, Cooper showed that the maximum possible efficiency is about 60%, and the practical upper limit is about 50%.

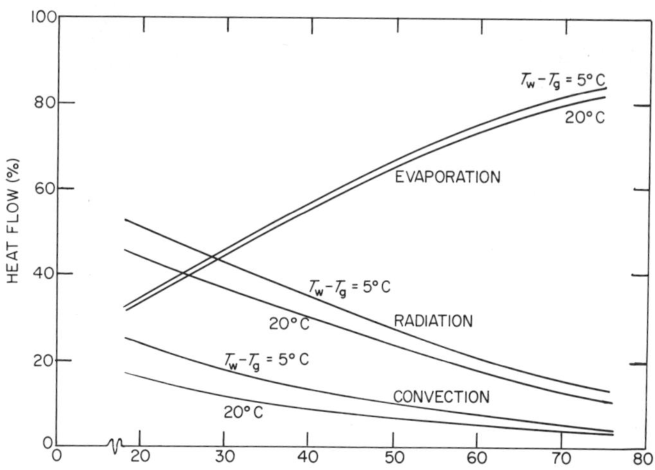

FIG. 20.10 Effect of water temperature on effectiveness of solar still.

20.6 SOLAR DISTILLATION WITH THE REUSE OF HEAT ENERGY

Because of the large area required to produce a given quantity of water by means of basin-type solar stills, investigators have attempted to devise new methods to use the solar energy more effectively. Basin-type solar stills use the collected solar heat only once since the heat of condensation of each kilogram of water produced is transmitted through the cover and discharged into the atmosphere by means of convection and radiation. This is similar to single-effect distillers using fuel-heat sources that are quite poor in economy and, therefore, not used for potable water production.

Conventional multiple-effect distillers, using heat from fuels, are more effective because they use the heat energy more than once. Figure 20.11 is a schematic flow diagram for a multiple-effect distillation plant, which consists of several similar evaporators, the necessary pumps and a final condenser. While many different types of evaporators could be used in such plants, the vertical-tube falling-film evaporator is shown for example. In this type of evaporator, the saline water is introduced at the top of each tube and is allowed to fall to the bottom inside the tube forming a thin film on the tube wall. For the sake of clarity, only one tube is shown in each evaporator, although many tubes would be used in the actual evaporator. In these evaporators, the steam is condensed on the outside of the tubes and vapor formed inside the tubes. The steam will condense on the outside of the tubes as long as its saturation temperature exceeds the temperature of the water film on the inside of the tube. The heat of condensation of the steam will be transferred through the tube wall and will cause evaporation of an equivalent amount of water from the saline water film. Because steam from a boiler is used in the first effect, this condensate, formed on the outside of the tube, is

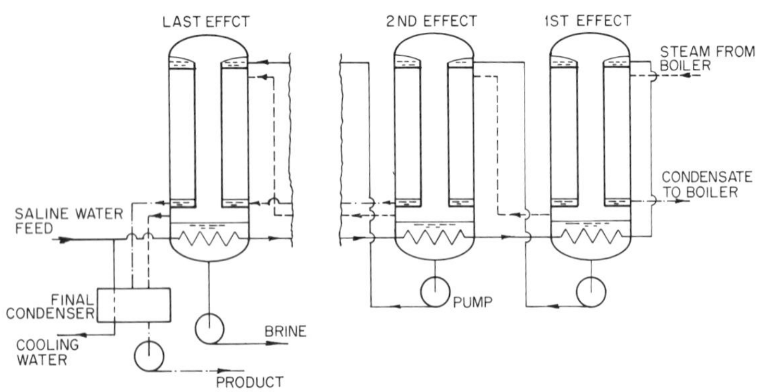

FIG. 20.11 Multiple-effect distillation "schematic."

returned to the boiler for reuse. The vapor formed inside the tubes of the first effect is used as the heating steam in the second effect. The unevaporated portion of the saline water in the bottom of the first effect is partially cooled as it preheats the saline water feed before it is introduced at the top of the second effect. The steam in the jacket of the second effect condenses on the outside of the tubes, transfers heat through the tube wall and causes evaporation of an equivalent amount of water from the falling film inside the tube. This process continues through all of the effects with decreasing temperatures until the last effect where the vapor formed is introduced into a condenser for final condensation. The condensate from each effect is passed into the steam space of the next effect where it contributes a little heat by flashing. The accumulated condensates, together with the vapor from the last effect, pass into the final condenser from which the total condensate is withdrawn as freshwater. Although Fig. 20.11 shows a water-cooled condenser, an air-cooled condenser could be used if needed.

The ratio of the mass of water produced in a multiple-effect still to the mass of steam used in the first effect is called the gained output ratio (GOR), and is a measure of the thermal performance of the distiller. The GOR is proportional to the number of effects and depends on the saturation temperature drop between effects, and the temperature of vapor entering the condenser. In general, it may be taken as 0.8 times the number of effects.

The application of the multiple-effect concept to the utilization of solar energy in atmospheric distillation of saline water has been investigated by Ginnings (1948), Telkes (1956b), Dunkle (1961) and Selçuk (1964a). Although the apparatus studied by each of the four investigators differed in constructional details, all used atmospheric pressure throughout the several effects.

Figure 20.12 is a schematic representation of a three-effect still adapted from Dunkle (1961). The still consists of a series of vertical parallel plates within an enclosure, with the first plate heated by solar-heated water flowing through a tube bonded to the plate. The last plate is cooled by cold saline water flowing through a tube similarly bonded to this plate. The saline water feed is preheated as it passes over the last plate and is then introduced at the top surface on the right-hand side of each plate, where part of it evaporates and the balance is collected at the bottom and discarded as concentrated brine. The vapor produced from the saline water sliding down the right-hand side of the first plate is condensed on the left-hand side of the second plate. The condensate produced is collected in a trough at the bottom of the plate and withdrawn from the enclosure as freshwater. The heat of condensation given up by the condensing vapor on the left-hand side of the second plate is conducted through this plate and furnishes heat to evaporate an equivalent amount of water from the saline water sliding down the right-hand side of this second plate. This process is repeated through all the plates except the last one, where the heat of condensation transmitted through this plate is used to preheat the saline water feed.

It will be realized that the solar heat delivered to the first plate is passed along to each of the succeeding plates and reused as many times as there are evaporator plates. Thus, the device operates on the multiple-effect principle described for a distiller using heat from a fuel source. The difference lies in the means used to produce the vapor pressure gradient through the plant which is required for maintaining the temperature gradient needed for heat transfer. In the fuel-operated plant, this was accomplished by a vacuum pump to exclude air from the entire system. In the atmospheric solar distiller, it is produced by using the partial pressure of air to supplement the vapor pressure of water in making up the total pressure of the atmosphere.

In the arrangement shown in Fig. 20.12, solar energy is used indirectly. Solar radiation is absorbed in the water passing through the solar collector and the heated water stored in an insulated storage tank. To operate the distiller, hot water is drawn from the top of this storage tank and is circulated through the tubes on the first plate, thus supplying heat to evaporate some of the saline water from the right-hand side of that plate. The circulating water is returned to the bottom of the storage tank after passing through the heating tube on the first plate. Had the left-hand side of the enclosure been made of transparent material which transmits solar radiation, the left-hand side of the first plate could have served as the solar collector, thus making

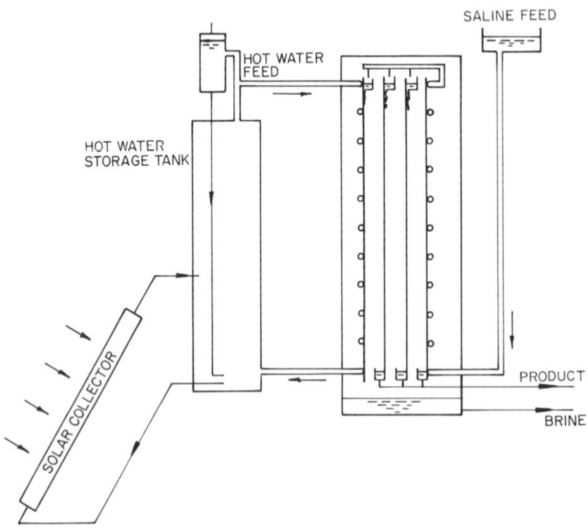

FIG. 20.12 Multiple-effect solar still (after Dunkle, 1961).

the operation of the distiller directly responsive to the solar energy input and taking the place of the separate solar collector and the storage tank.

An apparent advantage of an indirect solar collection system is that the multiple-effect still can operate as long as heat energy from the storage tank is available and, thus, the solar collector hot water storage system can be designed and sized for continuous operation of the still. In multiple-effect stills where the solar energy is used directly, the still will operate only during sunshine hours and cannot be used continuously. Again, it is obvious that the apparent advantage of the indirect system must be balanced against the additional cost of the collector water storage system.

Unfortunately, the additional production resulting from the multiple-effect atmospheric distiller as compared with that from the simple solar still has not compensated for the additional cost incurred in the construction of the more complicated multiple-effect still. The cost of product water from the simple solar still has been found to be lower than that from the multiple-effect still, and this accounts for the apparent disinterest in this type of multiple-effect solar distillation.

Another attempt to utilize solar energy more effectively in atmospheric distillation is the humidification–dehumidification process used in the 18.9 m^3/day (5000 US gal/day) Puerto-Penasco plant in Mexico (Hodges *et al.*, 1966). Figure 20.13 is a photograph of this plant, which consists of

FIG. 20.13 Humidification–dehumidification plant at Puerto-Penasco, Mexico (from Hodges, 1966, used with permission).

several bays of solar collector, a distiller in tower form, and two storage tanks. The photograph shows the five bays of solar collectors, each approximately 2 m wide by 100 m long, with a total collector area of 966 m^2. Saline water heated in these collectors was stored in an underground storage tank, from which it was pumped and delivered to the top of the packed tower seen in the photograph. This water was allowed to drip to the bottom of the tower thereby coming into direct contact with air moving up the tower through the packing countercurrently to the hot water. The resulting heated and humidified air was then passed to a second parallel tower through horizontal cross connections. Here it passed downward through the second tower and was cooled and partially dehumidified by coming in contact with finned tubes through which the incoming cold saline water was flowing. The condensate resulting from the partial dehumidification dripped off the finned tubes and was collected at the base of the tower as freshwater.

The power needed to drive the two-stage axial blower, pumps, and other auxiliary equipment was provided by a 50 kW diesel electric generating unit. Waste heat from the diesel engine of this unit was substituted for solar heat after the solar collectors had been irretrievably damaged in a storm. This source of waste heat has kept the plant operating without the solar collector for a number of years.

Although no exact figures have been given for the power required to circulate the air through the two towers of the distiller unit, it has been stated that a 30 kW electric motor was used to drive the blower, the latter being located in the bottom cross connection between the two towers. If the total capacity of this blower was used to pump the air, then the specific energy consumption for the blower alone would be approximately 38 kWh/m^3. This specific energy consumption exceeds that of compressors in small vapor-compression distillation plants, in which the specific energy requirements vary from 13 to 25 kWh/m^3, depending on the performance of the evaporator. For these units, the higher the performance of the evaporator, the lower will be the energy requirement of the vapor compressor. The high energy requirement of the humidification–dehumidification scheme described is due primarily to the large amount of air which must be circulated with the comparatively small amount of vapor represented by the humidity.

It is clear that in the humidification–dehumidification process, the transport of water vapor from the warm evaporation surface to the cooler condensation surface is coupled with cellular motions of air due to natural convections. Because air and water vapor are mixed, the cyclic heating and cooling of air in the enclosed space transfers a significant amount of energy that should be otherwise available for evaporation. Obviously, the elimination of air and other noncondensible gases from the enclosed space would improve the efficiency of the evaporator. However, this would require elaborate construction arrangements because the still would have to operate at subatmospheric pressures due to the low operating temperatures, thus necessitating provisions for preventing air leakage into the still and for maintaining the subatmospheric pressure should leakage occur.

These considerations have led several investigators (Eibling *et al.*, 1953; Baum, 1961; Brice, 1963; Weihe, 1972; Howe and Tleimat, 1974; Tleimat and Howe, 1975) to propose the use of solar energy to generate steam or provide hot water for driving conventional vacuum-type distillation plants.

In these schemes, Eibling and Weihe proposed the production of steam at pressures slightly above atmospheric for delivery to the distiller units by the use of concentrating collectors of the parabolic cylindrical type, with the heat receiver in the form of a pipe along the axis. Baum used a concentrating paraboloid-type collector to generate steam for driving a triple-effect distiller. In contrast, Brice, Howe, and Tleimat proposed the use of flat plate type collectors to heat water which could be expanded to generate steam at pressures below atmospheric and at temperatures of 60 to 70°C for delivery to the distiller units.

Of the six schemes cited above, four proposed the multiple-effect process while Brice and Weihe proposed the multiple-stage flash process. Since the performance of the multiple-effect or the multiple-stage flash distillation processes is a function of the temperature of the steam supplied to the distiller, the higher the steam temperature, the higher will be the performance GOR. Consequently, distillers using steam from concentrating collectors at temperatures between 105 and 110°C will have higher values of the GOR than identical distillers using steam from flat plate collectors at lower temperatures, with the same condenser temperature. However, this increase in distiller performance due to the use of concentrating collectors must be justified in terms of their capital costs and the effect of these costs on the total plant economy.

The GOR of a multiple-effect distiller increases with the number of effects, as well as with the temperature of the steam supply. Because increasing the number of effects decreases the temperature difference for heat transfer in each effect, the total heat transfer surface in all effects must be increased, thereby increasing the capital cost of the distiller. However, since an increase in the GOR will result in less steam being required, the solar collector section of the plant can be smaller and less costly. The decrease in cost of the solar collector section would need to be balanced against the increased cost of the additional effects in the distiller section. The above comments are based on heat transfer coefficients in the distiller section being the same in all cases. However, recently developed heat transfer surfaces have shown much improved heat transfer coefficients as compared with those previously available. Their use possibly could result in cheaper distilled water through the employment of more effects in the distiller and smaller solar collector areas in the steam production section.

For purposes of illustration, Fig. 20.14 adapted from Tleimat and Howe (1975), shows a schematic flow diagram for a 40 m^3/day plant using flat plate type collectors in the steam production portion and a multiple-effect distiller unit for seawater desalination. In the steam-producing section, water is taken from one of the storage tanks and introduced into the degasser. The pressure in the degasser is controlled to correspond to a saturation temperature slightly below the water temperature so that all dissolved noncondensible gases would be extracted by the vacuum pump. The water then enters the flash chamber where part of the water is flashed into vapor and the balance is cooled to the saturation temperature corresponding to the pressure in the flash chamber. The vapor formed in the flash chamber is then removed and passed into the first effect to be used as the driving steam for the multiple-effect distiller. The condensate from the first effect is returned to the water outflow from the flash chamber and pumped by pump 1 through the solar collector and through the jacket of the heat engine driving the electric generator to be reheated and then discharged into the second storage tank.

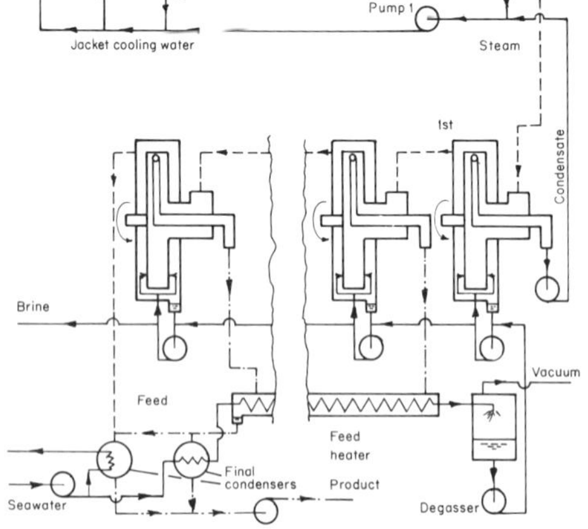

FIG. 20.14 Schematic diagram of a solar-assisted multiple-effect rotating-disk wiped-film distiller (after Tleimat and Howe, 1975).

During sunny hours when the thermal output of solar collectors exceeds the demand on the system, an additional quantity of water is extracted from the second tank by pump 2 and is added to the water returned from the flash tank. During hours when solar energy is not available, valve 7 would be closed and pump 2 shut down so that the water from the flash tank is pumped through the jacket of the engine driven generator and is discharged into the second storage tank. When the level of water in the first storage tank falls below a predetermined level, the water would then be pumped from the second storage tank by opening and closing the appropriate valves in the system. It must be noted here that the total arrangement in the top section of Fig. 20.14 is basically a boiler where the sun provides the heat energy through the solar collector and the heat engine driven generator provides the mechanical energy to drive the various pumps and mechanical equipment in the plant. It should be noted that the heat extracted from the jacket of the engine was found to be just sufficient to compensate for the heat loss from the storage tanks (Tleimat and Howe, 1975).

The unique part of this scheme is the proposed use in the multiple-effect system of the Tleimat wiped-film-rotating-disk evaporator, a very high performance evaporator developed at the Sea Water Conversion Laboratory, University of California, Berkeley. Overall heat transfer coefficients of the order of 20 to 50 kW/m^2 °C were obtained experimentally on a laboratory model. Because of the high heat transfer coefficients, small temperature drops between effects can be used in the multiple-effect distiller. This permits a large number of effects to be used and, consequently, could result in a high performance distiller. For the conditions of 60°C steam from the solar boiler and 20°C cooling seawater, Tleimat and Howe (1975) were able to propose 15 effects with a resulting GOR of 12.8. Also, because of the relatively low operating temperature of the solar collectors, higher collection efficiency could be obtained. As compared to the low productivity of simple stills, these two factors could be responsible for reducing the total area of the solar stills required for a given capacity by a factor of approximately 15.

20.7 THE USE OF WASTE HEAT IN SOLAR STILLS

Freshwater production from simple solar stills depends on the conversion of solar energy into heat to drive the humidification–dehumidification process and, therefore, produces only during a portion of a day. Since the most significant component of the cost of freshwater from solar stills is due to the capital cost of the still, it is important to develop methods that lead to the increase of the still's productivity, thereby reducing the capital investment per unit of product water. One possible source of increased production is the use of waste heat. In many places where waste heat from gasoline or diesel engines is available, the saline feedwater for the solar still can be heated by that source so as to keep it operating during sunless periods and thus increase its productivity.

Three different approaches for the utilization of waste heat to increase the productivity of solar stills have been investigated. The first (Tleimat and Howe, 1966), used the cooling water containing the waste heat as the saline water feed for the still during sunless periods. The second approach (Proctor, 1973), used the waste heat contained in cooling water to heat the saline water in the still basin during sunless hours by passing the waste hot water through metal tubes immersed in the basin water. The third approach, Malik and Tran (1973), was a *batch* process operation in which saline water containing waste heat was used as the saline water feed to be introduced in the still basin at the start of the sunless period. Thus, the first and third approaches used the waste cooling water directly in the still, while the second one used a heat exchanger so that the cooling water was conserved and reused for its major purpose of cooling an engine or process.

The experimental work done by these investigators has given information that can be used to estimate the possible increase in production to be achieved by the use of waste heat.

Tleimat and Howe (1966) obtained their experimental data from a small still composed of a glass tube 12.5 cm in diameter and 117 cm long with a black painted shallow tray 10 cm wide, 1.2 cm deep, and 115 cm long. The glass tube was sealed at both ends by gasketed wood blocks held together by tie rods. During the operation when waste heat was used, heated water at controlled flowrate and inlet temperature was run through the still with the condensate being collected at the bottom of the tube and withdrawn from one end. Experimental data on the productivity of this still were correlated experimentally with an accuracy of $\pm 10\%$ by the equation

$$\dot{m} = 0.165 W_{\mathrm{w}}^{0.3}(T_{\mathrm{in}} - T_{\mathrm{a}} - 14.4) \qquad (20.19)$$

where $\dot{m}$ is the productivity of still (kg/day m^2 tray area), W_{w} the water flowrate (kg/hm^2 tray area), T_{in} the water inlet temperature (°K), T_{a} = minimum ambient air temperature during the night (°K), for the following conditions: water flowrate, 32–77 l/hm^2 tray area; inlet water temperature, 35–65°C and ambient air temperature, 11–16°C.

From Eq. (20.19), it appears that $(T_{\mathrm{in}} - T_{\mathrm{a}})$ must be greater than 14.4 in order to have any supplemental production of freshwater. Furthermore, since the flowrate of saline water W_{w} appears in the equation to the 0.3 power, the flowrates must be large in proportion to the output. For example, if $(T_{\mathrm{in}} - T_{\mathrm{a}})$ is 30°, the flowrate of warm cooling water would be 40 l/hm^2 to yield 5 l/m^2 day of additional water.

Proctor (1973) obtained his experimental results on an 18.5 m^2 basin-type solar still with a 17.30 m long hard drawn copper tube, 22.23 mm outside diameter, and 19.74 mm inside diameter, submerged in the saline water and extending for the full length of the still. The water from the waste heat source was circulated inside the copper tube while the saline water in the still was circulated at constant flowrate concurrently to the water inside the tube. Those interested in the detailed information regarding the operation, design, and results of such a system should consult the given reference.

Malik and Tran (1973) obtained their experimental results on a 3.66 m × 1.22 m solar still covered with an inflated polyvinyl chloride plastic film. The frame of the still was constructed of wood while the bottom was made of hardboard and was covered with a basin liner made from a 0.76 mm thick black butyl rubber sheet. At the start of the experiment, a measured quantity of heated water at a given temperature was introduced into the still basin. Experiments were started at 8 P.M. and continued just past midnight. Theoretical analysis on the still indicated that the product m was a function of the initial quantity of hot water added to the still M, the initial water temperature T_0, and the water temperature at the end of the experiment T, and was given by

$$m = (T_0 - T)M/(\lambda - T)$$

where λ was a constant whose value depended on the initial water temperature T_0. The values of λ as a function of T_0 shown in Table 20.2 were given by Malik and Tran.

TABLE 20.2

T_0	F	120	130	140	150
	C	48.9	54.4	60.0	65.6
λ	F	2455.5	2268.2	2246.8	2107.5
	C	1364.2	1260.1	1248.2	1170.8

Test results were compared with those predicted from Eq. (20.20) and were found to correlate with an accuracy of $\pm 15\%$.

20.8 CONCLUSION

Solar distillation was the subject of extensive research and development efforts in various parts of the world during the 1950s and 1960s. These efforts dwindled and have almost disappeared because the investigators were unable to design solar stills that would produce large quantities of water as cheaply as other types of distillation. Solar stills operating presently include about 20 basin-type installations with capacities varying from about 400 l/day to 25,000 l/day and an unknown number of smaller units.

The basin-type solar still has produced distilled water at a cost per unit of product lower than other types of solar equipment and is the only type in operation. Numerous configurations of basin-type units have been tried in efforts to reduce capital costs and increase efficiencies. Operating efficiencies of 35 to 50% for basin-type stills have been achieved in practical units, as compared with a theoretical maximum of slightly more than 60%. These practical units are designed to collect solar radiation, use it once, and discard it as heat into the atmosphere so that solar energy amounting to several times the latent heat of the water produced is required.

The production rate of a basin-type solar still can be predicted for a given solar radiation input using heat balances and other analytical equations. This analysis shows that the effectiveness of operation increases with the intensity of solar radiation flux and with the temperature of water in the basin. Thus, designers strive to maximize basin water temperatures.

During the period of active research and development in solar distillation, the cost of fossil fuels was low, and consequently, distillers producing large amounts of water and using energy from this source could produce distilled water more cheaply than could solar stills. However, the recent energy crisis has resulted in fuel costs sufficiently great to raise the possibility that solar distilled water could become competitive in more cases than in the past. These comments, which apply to basin-type and other simple solar stills, lead to the suggestion that devices in which heat is reused should receive more attention. A review of some of these devices is presented in Section 20.6.

21

The Economics of Solar Energy

A. A. M. SAYIGH

MECHANICAL ENGINEERING DEPARTMENT
COLLEGE OF ENGINEERING
RIYADH UNIVERSITY
RIYADH, SAUDI ARABIA

21.1 INTRODUCTION

Solar energy research has been confined to the enthusiastic and pure researcher since 1870. Since then several solar energy utilizing devices have been made and tested, such as solar stoves, solar heaters, solar stills, solar dryers, solar refrigerators, solar pumps, and various solar cells for generating electricity. It is only in the last ten years that more intensive work has been done in the solar energy field. But for the ordinary man, the use of solar energy in place of conventional fuels is still strange. This is only natural because of the low cost, availability, and reliability of other energy appliances, such as electric or gas stoves, heaters, etc. Since the late nineteenth century and the early twentieth century, men have experimented with various types of energy and concluded that oil was the best and cheapest as well as the easiest to obtain. We can get an idea of the advantage that men saw in oil as a fuel by looking at Table 21.1 which represents the various fuels and their calorific values. Due to various factors, some of which have been mentioned in Chapter 1, oil prices have risen fourfold since the early 1960s, so that oil is no longer the most obvious answer to our energy needs. When people were faced with the problem of not having enough fuel to operate their heaters in winter, not to feel the comfort of cool air in hot summer, or not being able to drive their cars, they started to search for an alternative—and one of the alternatives considered was solar energy. During the last ten years, over five hundred houses have been built

TABLE 21.1

Calorific Value of Various Fuels

Fuels	Btu/lb	kJ/kg
Peat	1000–3000	2300–6900
Oil shales	1700–5000	3900–11500
Lignites	5000–9000	11500–20700
Bituminous coals	10200–14500	23500–33400
Anthracite	14900–15200	34400–35000
Bunker C oil	18300 gross	42000 gross
Diesel Oil	19200 gross	44000 gross
Petrol (gasoline)	20500 gross	47100 gross

The net values for the liquid fuels are about 6% less.

Fuel gases	Btu/ft^3	kJ/m^3
Blast furnace gas	90	3350
Producer gas (coke)	130	4840
Producer gas (coal)	160	5950
Blue Water gas	290	10800
Coal gas	450	16700
Coke oven gas	525	19500
Natural gas	950–1050	35,300–39,200
Acetylene gas	1550	58000
Refinery tail gases	2750	65300
Butane	3200	119500

in the U.S.A. that are run partially or totally by solar energy. In some countries, 80% of the domestic heating is done by solar energy and at least ten large solar stills are in operation all over the world, the minimum water yield being 1000 m^3/day. But the average man will wait to see more people using solar appliances before he will use them—socially and psychologically, he needs to be convinced before he will use such appliances with any ease.

21.2 ECONOMIC CRITERIA

The fact that solar energy is free does not mean that it will automatically have a low utilizing cost. The interest lies in generating, say, electricity, and the equipment that converts solar energy to electricity through some devices (see Chapter 14 and Fig. 21.1) has to be of low or reasonable cost before it can replace hydroelectricity or a steam-generating plant.

Another negative characteristic of solar energy is that it cannot be used at will and until this problem is solved, whether by solar hydrogen production

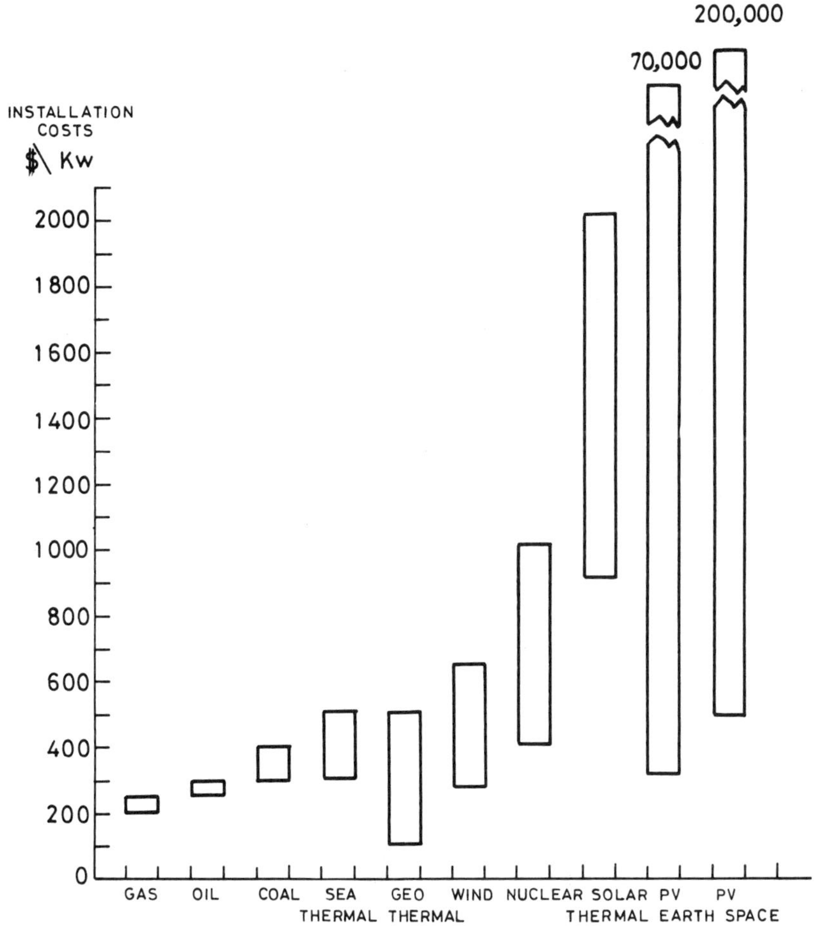

FIG. 21.1 Estimated installation costs for electric generation plants versus primary energy sources (Anderson, 1975).

or other means (see Chapter 18), the cost of solar energy in general will stay high. Finally, the amount of solar energy available is about 1.2 kW/m^2 on a sunny day in the United States, and clearly some method has to be evolved to concentrate this energy. Therefore, when considering the cost analysis, the prime consideration should be given to the capital investment. Prior to 1955, solar energy devices were more expensive than any fuel-burning device, but now, due to the rising cost of conventional fuels, more countries are thinking of and competing in using solar energy as an alternative fuel. Since the element of competition has arisen, this will certainly lead to lower costs for producing less expensive solar appliances, and the solar devices will eventually gain ground in the market place. Two variables must be borne in mind when considering the cost analysis of two competing commodities (e.g., an electric cooker and a solar cooker), in any given society or location. First is the cost of the fuel consumed by such a device, and second is the initial cost of production.

It is worth stating that the efficiency parameter of a solar device is not of prime importance while it is of great importance in a conventional power plant design. Since the cost of solar energy is nil, the cost of the power output is relatively independent of the conversion efficiency and is almost totally dominated by capital cost. This fact is demonstrated in the cost of a solar cell which was produced by Bell Telephone Laboratories. This has an efficiency of 15% and its cost in 1956 was \$100/kWh, while in the same year, the solar electric generator developed in Italy by Somor Corporation had an efficiency of 0.5%, but produced electricity at a cost of \$1/kWh (Kastens, 1963).

Solar energy is best used in small power station generation, say 10 to 100 kW sizes and no larger than 1000 kW. This is due to the almost constant cost of electricity generation by solar energy, irrespective of the size of the station. This is not the case, for example, in a hydroelectric generating station where the cost of electricity drops as low as 30% or less when the station size is ten times larger.

21.3 COST OF SOME SOLAR APPLIANCES

Several solar devices are now available in the market and these devices vary in price from place to place and country to country. Their efficiency also varies. Therefore it is difficult to name a fixed price for a solar article, however, a broad line can be drawn regarding the market for each kind of solar equipment available now.

A. Solar Pumps

Up to 1960, there was only one company producing solar pumps. This was the Somor Corporation in Italy. Several sizes were produced by this Company, namely, from 0.1 to 5 hp. Their prices were about \$1000/hp. A French Company is now in operation, CFP/TOTAL of Paris (Clemot and Fiatte, 1975), which produces various sizes of solar pumps (1–25 kW). Twenty of them are now in operation in remote areas of Africa and South America. In another publication, Fiatte and Clemot (1975) compared the advantages and disadvantages of thermal energy plants with solar energy plants as shown in Table 21.2.

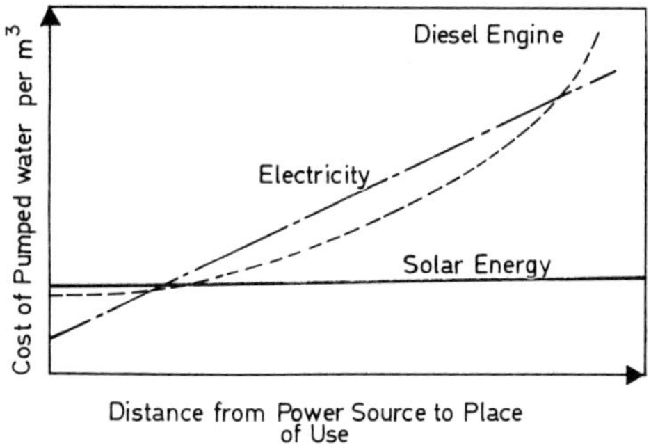

FIG. 21.2 Cost prediction of one cubic meter of pumped water versus station distance from power source (Fiatte *et al.* 1975).

TABLE 21.2

Comparison between Thermal Power Plant (Diesel Engine) and Solar Energy Plant

Source of power	Advantages	Disadvantages and difficulties
Thermal energy (diesel engine) [Cost of 1 m^3 of water is 2–2.2 French francs from 1 kW station.]	1 Low investment 2 Great flexibility of use and of installation	1 High operating cost 2 Imports of raw material 3 Fuel transportation 4 Fuel storage 5 Spare parts 6 Consumption of natural resources 7 Localized pollution (exhaust gases) 8 Need for skilled manpower
Solar energy [Cost of 1 m^3 of water is 2.5–3 French francs from 1 kW station.]	1 Low operating cost 2 Autonomy 3 Long equipment lifetime (10–20 years) 4 No degradation of natural resources 5 Pollution-free 6 The use of refrigeration which is underneath the collectors 7 No skilled manpower is required 8 Maintenance reduced to caretaking	1 High investment 2 Discontinuous performance requiring storage tank 3 More initial civil engineering work

Figure 21.2 shows a comparison between various methods of pumping water versus the distance from power source to place of use (Fiatte and Clemot, 1975).

B. Solar Stoves

These were well received all over the world. In India, the cost of a solar stove is reported to be \$8 and similarly, in Lebanon. In the U.S., Daniels used a plastic stove which costs about \$2. In Burma, a do-it-yourself solar stove kit costs about \$0.50, but involves about 8 h of manual work to assemble. In 1975, at an International Solar Energy Conference in Los Angeles, Davis Instruments exhibited a picnic beach and camp stove called solar cooket. This stove costs about \$14. The cooket is small and needs to be focused every 20 min and can only cook for one person.

C. Solar Heating and Cooling

The cost of conventional heating is shown in Fig. 21.3, solar heating costs are much less than that if the initial cost of the collector is excluded. At the present moment, many solar heating/cooling systems are available in the market in the U.S. Alcone (1975) presented the cost of such a system in Table 21.3

Solar water heating has been used for some time in most of the countries of the world. An average family heater costs \$2.4 to \$4.3/million kJ in the U.S. (Sousson and Siefert, 1975). In Lebanon and Cyprus, a heater costs about \$200. In Turkey, several companies produce solar water heaters on a commercial scale, one of these companies is Gunsa Gunesu su Isiticisi in Ankara. The cost of such a heater is about \$200. In Japan, over half a million heaters are in use. In Algeria, Jordan, Australia, and South Africa

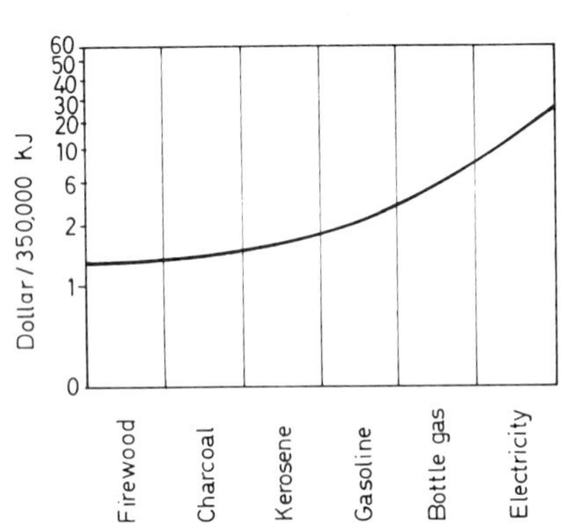

FIG. 21.3 Retail energy prices in most countries (based on study done by Stanford Research Institute in 1962 and uprated by factor of 3 to be at the present cost).

TABLE 21.3

Cost of Heating/Cooling System[a]

Component	Cost of material ($)	Labor ($)	Total ($)
Air-to-air heat pump and controls	1150	100	1250
Variable speed blower with control	150	25	175
Duct work and controls	250	—	250
Storage, $20/m^3	1260	—	1260
Heat exchanger and controls for water heating	75	25	100
Total cost of solar/heat pump system	2885	150	3035
Conventional unitary packaged system	600	100	700

[a] From Alcone (1975).

hundreds of domestic heaters are in use. The author and others built a cheap solar heater in Riyadh, Saudi Arabia, in 1972, which cost only $150, with an absorber area of one meter square (Sabbagh *et al.*, 1973).

D. Solar Power Plants

The majority of solar energy work has been recently devoted to the generation of electricity and its cost analysis. Figure 21.4 shows the cost of power generation ($/kW) for different types of fuel (Anderson, 1975), while Fig. 21.5 shows the electricity cost per kilowatt hour versus the load factor for various means of electricity generation. The load factor (L) was calculated from the equation (Kakac and Veziroglu, 1976)

$$G_c = \frac{10^3 R(PC)}{8760L} \quad \text{mills/kWh}$$

where G_c is the capital cost component of energy, R the total fixed charge rate per year, (PC) the plant capital investment cost ($/kW), and L the load factor (%).

In summer of 1975 in Los Angeles, a paper was presented by Selçuk *et al.* (1975) concerning "preliminary technical and economic data on solar thermal power conversion systems." Five groups of terrestrial solar plants were discussed, viz, flat plate (90–125°C), central receiver (500°C), and photovoltaics. A formula was proposed for the energy cost $(0.0224/L)(1/P)$

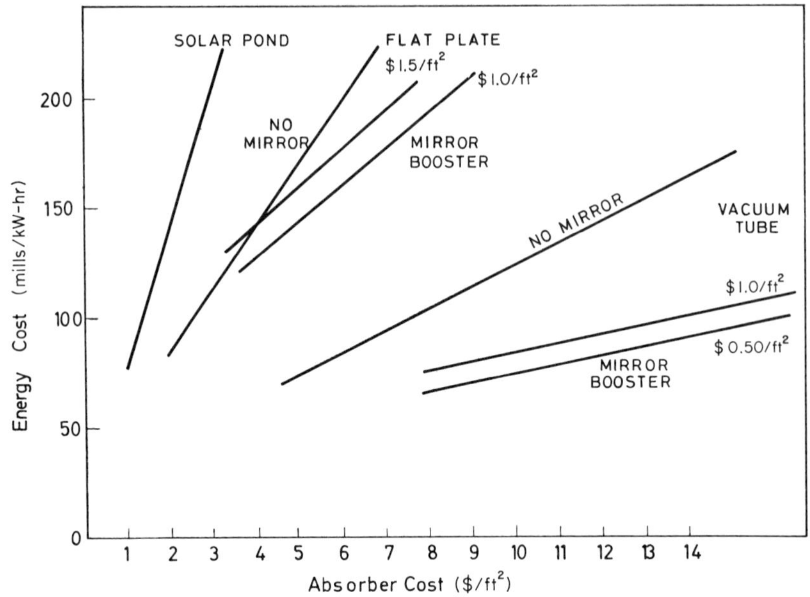

FIG. 21.4 Cost comparison of flat plate collector concepts for various thermal solar power generation (Selçuk *et al.*, 1975).

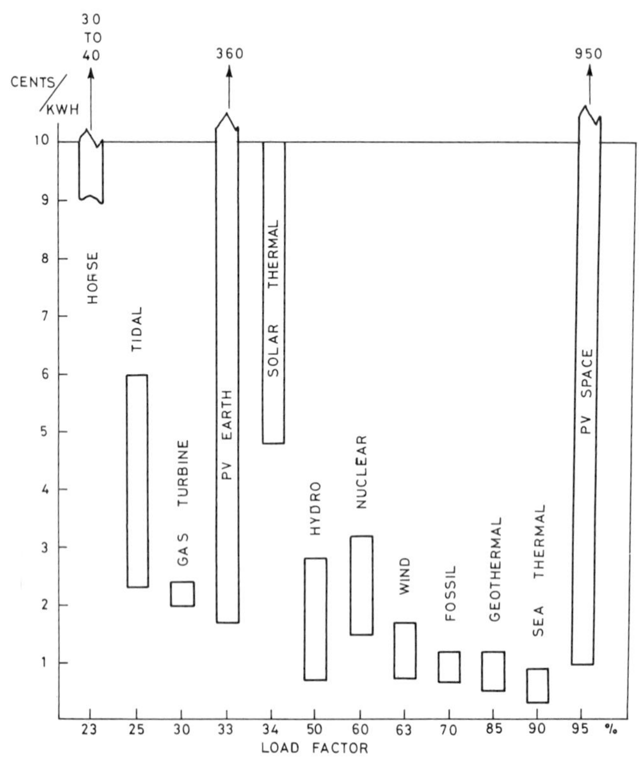

FIG. 21.5 Estimated electric power costs versus primary energy sources (Anderson, 1975).

TABLE 21.4

Solar Plants Cost Summary, 100–150 MW

	Direct capital cost ($/kW)						
			Dish				
Items	Flate plate	2-D	Brayton	Steam	Chemical	Central	Photovoltaic
Collectors	950	940	760	640	640	470	1075
Receiver	—	250	95	30	85	180	—
Energy transport	155	260	60	160	110	20	75
Power conversion	210	180	400	160	180	150	—
Other	105	100	125	100	100	150	105
Total	1420	1730	1440	1090	1120	970	1250
Energy cost (mills/kWh)	95	86	72	54	56	48	85
Estimated cost comparison	6.6	6.9	5.7	6.6	6.5	7.0	3.9
Overall efficiency	9.5	18.1	22.5	26.3	26.5	24.3	8

which excludes operation and maintenance costs, the cost is at 1975 dollars value, where $1/P$ is direct capital costs in \$/kW and L is the average annual load factor based on 6% long-term inflation over 25 years plant life, 10.5% interest on capital, and 5% of capital for insurance, taxes, depreciation, profits, and so on. Figure 21.4 shows cost comparison among flat plate collectors by using the previous concept of solar thermal power generation. Table 21.4 compares the cost summary of 100–150 MW of various solar plants. Hildebrandt and Vant-Hull (1974) proposed a tower-top focus solar energy collector. The station consists of 450-m tower with 2.6 km^2 mirror field, receiver, and heat exchanger and costs $7,900,000. The station will produce over 400 MW of heat and the cost of one thermal kilowatt hour is $0.077. This will compete with oil and gas heat in cost and efficiency.

E. Solar Stills

One of the most attractive uses of solar energy is obtaining potable water from brakish water. On a small scale, several countries were producing solar stills (Sayigh and El-Salam, 1976, Bartoli *et al.*, 1975, Boldrin and Scalabrin, 1975). The yield of these small stills is about 3–5 l/m^2/day and their cost is $50–150, which is high in comparison with conventional stills.

The possibility of using a solar-assisted distillation plant (Tleimat and Howe, 1975) that can produce 13,2000 m^3/year of fresh water has an operation cost which includes interest on money invested, fuel cost, spare parts, labor, etc., of \$2.65/m^3 of fresh water. Table 21.5 shows some of the large solar distillation plants (Taylor, 1975), while Table 21.6 shows a cost comparison between various methods of desalination.

TABLE 21.5

Some of the Large Solar Distillation Plants in the World

Country	Name of station	Date installed	Area (m^2)	Cover material
Chile	—	1870	5574	Glass
Australia	Caiguna	1966	419	Glass
	Coober Pedy	1966	3532	Glass
	Hamelin Pool			
	Pool	1966	627	Glass
	Muresk	1963	419	Glass
Greece	Kimolos	1968	2557	Glass
	Nisyros	1968	2100	Glass
	Patmos	1967	8857	Glass
	Regna	1965	1487	Plastic
	Salamis	1966	381	Plastic
	Siemy	1964	2770	Plastic
Spain	Las Marinas	1966	1069	Plastic
Tunisia	Chekmou	1967	432	Plastic
	Mahdia, Petit St.	1968	1273	Plastic
	Vincent	1968	1766	Plastic
Haiti	Source Philippe	1969	~1000	Glass

TABLE 21.6

A Cost Comparison between Various Methods of the Desalination Process

Method	Cost (\$/m^3)
Pressure distillation (fuel)	2.5
Three-stage distillation (fuel)	5.5
Single-stage distillation (fuel)	12.5
Electrolysis (fuel)	5.0
Freezing process	6.5
Solar freezing process	8.0
Solar distillation	3–6

21.4 CONCLUSION

One of the major advantages of solar energy is that it is available near the place of demand. For example, solar pumps that are used in irrigation schemes and solar distillation can be used at the point of consumption. This is a great economic advantage. Another economic advantage associated with the use of solar appliances is that minimum care after construction is needed and only a small number, if any at all, of spare parts is required.

The use of solar energy and its economics can only be judged relatively, in other words, they depend on the initial assumptions. For example, Tleimat and Howe (1975), in discussing the cost of fresh water production from sea water by using a solar assisted distillation plant, propose three estimates. First, the collector cost at $40/m^2, system life of 30 years, labor cost at $4000/year, and the cost of water at $1.14/m^3. A second estimate which is more realistic uses cost of collector as $80/m^2, system life of 20 years, and labor cost of $20,000/year, which results in water cost of $2.65/m^3. A third uses the collector cost as $120/m^2, system life of 10 years, and labor cost of $20,000/year, thus the cost of water is $3.34/m^3.

Collector cost depends on method of construction, the type of material used, and the exact requirements for a particular collector (Davis, 1975). In Table 21.7 Davis lists the various collectors' costs. Another example of cost

TABLE 21.7

Collectors' Costs as Based on 1974 Price Minus 30%

Name and temperature range	Construction	Cost ($/ft^2)	Availability
Collector-1	Metal box (150 psi)		
(0–176.7°C)	AL absorber	6.20	1974
	Cu absorber	8.50	1974
	Steel absorber	6.70	1974
	Al–Cu absorber	8.00	1974
		less 30%	1980
	Integral absorber & structural foam steel with selective coating (10 psi)	3.00	1978
	Nonmetallic (10 psi)	2.00	1980
Collector-2 (0–400°C)	Evacuated glass cylinder	10.00	1978
Collector-3 (0.65.7°C)	Plastic (10 psi)	1.60	1976
Collector-4 (0–260°C)	Type 1 with selective coating	7.20	1976

analysis is that which is given by Löf and Tybout (1974) concerned with heating and cooling of dwellings in eight different locations in the US. The analysis was based on collector cost of \$21.50/m^2, 20-year life, and an interest rate of 8%. The paper shows that for cities with high cooling loads, such as Albuquerque, Charleston, Miami, Omaha, and Phoenix, the cost of solar energy for the combined operation is less than any of the separate heating or cooling processes, while in cities like Boston and Seattle, the cost of solar energy for heating is less than the cost of the combined processes. For generating electricity by the photovoltaic process, the cost is still ten to twenty times that of conventional methods.

Recently more scientists and engineers are working to reduce this cost. Therefore one can predict that by 1980 solar energy will be a viable source of energy, while by 1990 it will be much cheaper than any presently existing form of energy.

References

Abbot, C. G. (1963). Solar variation and weather, *Smithson. Misc. Collect.* **146** (3), 1–71.

Abbot, C. G. (1966). Solar variation a weather element, *Proc. Nat. Ac. Sci.* **56** (6), 1627–1634.

Abbot, C. G., and Aldrich, L. B. (1932). An improved water-flow pyrheliometer and the standard radiation, *Smithson. Misc. Collect.* **87** (5).

Abbot, C. G., Fowle, F. E., and Aldrich, L. B. (1922). On the distribution of radiation over the sun's disk, *Ann. Astrophys. Obs. Smithson. Inst.* **4**, 221–257.

Abdel-Aal, and Nazmy, M. Y. (1976). On the storage of hydrogen generated by solar energy, *in* "Heliotechnique and Development" (M. A. Kettani and J. E. Soussou, eds.), DAA Press, Cambridge, Massachusetts.

Ackerman (1915). The utilization of solar energy, Ann. Rep. of the Smithsonian Inst.; 141–146.

Adler, S., Levite, G., and Tabor, H. (1961). The Altenkirch solar-cooled house, *UN Conf. New Sources Energy Rome* **6**, 60–65.

Akyurt, M. *et al.* (1972). A solar dehydrator for orchards, *Comples Conf., Instanbul, Turkey, September.*

Akyurt, M., and Selçuk, M. K. (1973). A solar drier supplemented with auxiliary heating systems for continuous operation, *Sol. Energy* **14**, No. 3, 313–320.

Alcone, J. M. (1975). Low cost solar augmented heat pump system for residential heating and cooling, Solar Energy Div., ASME, Houston, Texas, November.

Aldrich, L. B. (1949). The Abbot silver-disk pyrheliometer, *Smithson. Misc. Collect.* **3**, No. 14.

Aldrich, L. B., and Hoover, W. H. (1964). *Ann. Astrophys. Obs., Smithson. Inst.* **7**.

Allen. C. W. (1958). Solar radiation, *Quart. J. R. Meteorol. Soc.* **84**, No. 362, 307–318.

Allison, J. F., Arndt, R., and Meulenberg, A. (1975). A comparison of the Comsat violet and nonreflective solar cells, 10th *Intersoc. Energy Conversion Eng. Conf. 10th, Newark, Delaware, August.*

Ambrose, C. W., and Bandopadhayay, P. C. (1970). Asymmetrical heating in non-circular ducts, *Proc. Int. Sol. Energy Soc. Conf., Melbourne* Paper no. 7/17.

Anderson, J. H. (1975). Sea thermal power as a hydrogen and methanol generator, *in* "Hydrogen Energy" (T. N. Veziroglu, ed.), Part A. Plenum Press, New York.

Angrist, S. W. (1971). "Direct Energy Conversion," 2nd ed. Allyn and Bacon, Boston, Massachusetts.

Ångström, A. K. (1898). The absolute determination of the radiation of heat with electrical compensation pyrheliometer with examples of the application of the instrument, *Astrophys. J.* **9**, 334.

Ångström, A. K. (1919). Note on the comparisons between pyrheliometers and the difference between the Ångström standard and the Smithsonian standard, Monthly Weather Review, U. S. Weather Bureau, November. 1919.

Ångström, A. K. (1924). Solar and atmospheric radiation Q, *J. R. Meteorol. Soc.* **50**, 121.

Anonymous (1952). Utilization of solar energy, (Report of NPL committee), *Research* **5**, 522–529.

Anonymous (1958). IGY Instruction Manual, Part VI, Radiation Instruments and Measurement, Int. Assoc. Meteorol., Radiat. Comm., Comm. for Radiat. Instructions, *Ann. Int. Geophys. Year*. **5**, No. 6. Pergamon, Oxford.

Anonymous (1960). "Handbook of Geophysics," revised ed., US Air Force. Macmillan, New York.

Anonymous (1961). Calculation of the direct insolation on plane surfaces over specified time intervals, Brace Res. Inst. Tech. Rep., T35. McGill Univ., Montreal.

Anonymous (1971a). IVHE Guide, IVHE London.

Anonymous (1971b). Guide to Meteorological Instrumentation and Observing Practices, Secretariat of the World Meteorological Organization, Geneva, Switzerland, 4th ed.

Anonymous, (1971c). Solar electromagnetic radiation, NASA SP8005, Washington, D.C., May.

Anonymous, (1972a). Heat and mass transfer between a water surface and the atmosphere, Water Resources Res. Lab. Rep. No. 14. Tennessee Valley Authority, Norris, Tennessee, April.

Anonymous, (1972b). "Tables of Temperature, Relative Humidity, Precipitation and Sunshine for the World Met. Office." HMSO, London.

Anonymous, (1974a). Standard specification for solar constant and air mass zero solar spectral irradiance, ASTM Standard, E490–73a. Annual Book of ASTM Standards, Part 41, ASTM, Philadelphia, Pennsylvania.

Anonymous, (1974b). Solar energy in developing countries: perspectives and prospects, *Ekistics* **225**, 88–92.

Anonymous, (1975a). Solar-oriented architecture, Coll. of Arch., Arizona State Univ. and AIA Res. Foundation, ASU, Tempe, Arizona.

Anonymous, (1975b). Eppley Black and White Pyranometer Model 8–48 or 8–48A, Brochure distributed by the Eppley Lab., Inc., Newport, Rhode Island.

Anonymous, (1975c). "Energy Primer." Portola Press, Menlo Park, California.

Apfeldorfer, C. *et al.* (1973). Study of spectral correlations at Jerusalem and at Washington, Paper E9, Congr. The Sun in the Service of Mankind, UNESCO.

Arashi, H., and Sakurai, T. (1976). Infrared pyrometry for measuring target temperature, *High Temp.-High Press.* **8**, 349–356.

Asher, N. G., and Reti, A. R. (1961). Engineering and economic study of the use of solar energy especially for space cooling in India and Pakistan, *UN Conf. New Sources Energy, Rome* **6**, 66–74.

ASHRAE (1972). "Handbook of Engineering Fundamentals." ASHRAE, New York.

Babcock, H. W. (1949). Unpublished experiments at the Mount Wilson Observatory, Pasadena, California.

Backus, C. E. (1972). Laser activation of solar cells, *IEEE Photovoltaic Spec. Conf. Rec. 9th, Silver Spring, Maryland, May*.

Backus, C. E. (1975). Solar energy conversion at high solar intensities, *J. Vac. Sci. Tech.* **12**, No. 5.

Backus, C. E. (1976). "Solar Cells." IEEE Press, New York.

Backus, C. E., Evans, D. L., and Ralph, E. L. (1976). Photovoltaic systems using sunlight concentration, *in* "Heliotechnique and Development" (M. A. Kettani and J. E. Soussou, eds.), DAA Press, Cambridge, Massachusetts.

Badylkes, I. S. (1966). Dimensionless numbers for the thermodynamic evaluation of working mixtures for absorption refrigeration machines, *Proc. Int. Inst. Refrig., Trondheim*, **1**, 107–113. Pergamon, Oxford.

Bahadori, M. N. (1973). A feasibility study of solar heating in Iran, *Sol. Energy* **15**, 3–26.

Baker, L. H. (1967). Film heat-transfer coefficients in solar collector tubes at low Reynolds numbers, *Sol. Energy* **8**, 78–85.

Baldwin, C. M. *et al.* (1975). Paper 32/2, *ISES Conf., Los Angeles, July.*

Ball, J. (1933). The Qattara depression of the Libyan desert and the possibility of its utilization for power production, *Geograph. J.* **82**, No. 4.

Bartoli, B. *et al.* (1976). A nonconventional solar still, *in* "Heliotechnique and Development" (M. A. Kettani and J. E. Soussou, eds.), DAA Press, Cambridge, Massachusetts.

Batchelor, G. K. (1954). Heat transfer by free convection across a closed cavity between vertical boundaries at different temperatures, *Quart. J. Appl. Math.* **12**, 209.

Baum, V. A. (1961). Solar distillers, *UN Conf. New Sources Energy, Rome* **6**.

Baum, V. A., and Kakahaev, A. (1972). Utilizing solar energy and evaporative cooling methods for air conditioning, *Appl. Sol. Energy* (*Geliotekhnika*) **8**, 57–60.

Baum, V. A., Aparasi, R. R., and Garf, B. A. (1957). High power solar installations, *Sol. Energy* **1**, 6–12.

Baum, V. A., Kakabaev, A., and Khandurdyev, A. (1972). Efficiency of a solar cooler with an open flat solution desorber, *Appl. Sol. Energy* (*Geliotekhnika*) **8**, 57–60.

Becker, C. F., and Boyd, J. S. (1957). Solar radiation availability on surfaces in the US as affected by season, orientation, latitude, altitude, and cloudiness, *Sol. Energy*, **1**, No. 1, 13.

Beckman, W. A. (1968). Radiation and convection heat transfer in a porous bed, *J. Eng. Power*, **Jan.**, 51–54, ASME.

Beckman, W. A., Mitchell, J., and Porter, W. (1973). *J. Heat Transfer Trans ASME*, **95**, 257.

Bénard, H. (1901). Les Tourbillons Cellulaires dans une Nappe Liquide Transportant de la Chaleur Par Convection en Regime permanent, *Ann. Chem. Phys.* 7th Ser. 62–144.

Bennett, H., Bennett, J., and Nagel, M. (1960). *J. Opt. Soc. Am.* **50**.

Bennett, I. (1965). Monthly maps of mean daily insolation for the United States, *Sol. Energy* **9**, No. 3.

Bennett, I. (1967). Frequency of daily insolation in Anglo North America during June and December, *Sol. Energy* **11**, No. 1.

Bennett, I. (1969). Correlation of daily insolation with daily total sky cover, opaque cover and percentage of possible sunshine, *Sol. Energy* **12**, No. 3.

Bifano, W. J. (1968). NASA Tech. Note D4727.

Bilgen, E. (1975). On the feasibility of direct dissociation of water using solar energy, Presented at the *Hydrogen Energy Fundamentals Symp. Miami Beach, March.*

Bilham, E. G. (1929). An instrument for the optical examination of sunshine recorder lenses, *J. Sci. Instrum.* **6**.

Billington, N. S. (1967–1968) The working environment, Proc. I, Mech. E., 182, Part 3E—Heating and Ventilation for a Human Environment.

Bliss, R. W. (1955) Solar house heating—A panel, *Proc. World Symp. Appl. Sol. Energy, Phoenix.* 151–158.

Bliss, R. W. (1959). *Sol. Energy* **3**.

Bliss, R. W. (1961). Atmospheric radiation near the surface of the ground, *Sol. Energy* **5**, 103–120.

Boer, K. W. *et al.* (1973). A combined solar thermal and electrical house system, *UNESCO Conf. Sun Service Mankind*, Paper 108.

Bockris, J. O'M (1975). On methods for the large scale production of hydrogen from water, *in* "Hydrogen Energy" (T. N. Veziroglu, ed.), Part A. Plenum Press, New York.

Boelter, L. M. K., Cherry, V. H., Johnson, H. A., and Martinelli, R. C. (1965). "Heat Transfer Notes." McGraw–Hill, New York.

Boldrin, B., and Scalabrin, G. (1976). An improved continuous solar still with distinct heating and condensing partitions, *in* "Heliotechnique and Development" (M. A. Kettani and J. E. Soussou, eds.), DAA Press, Cambridge, Massachusetts.

Bonhomme, R., and Ometto, J. C. (1975). Estimation of the net-radiation flux in tropical area, *ISES Conf., Los Angeles*.

Bossalasco, M. *et al.* (1965). Solar constants and sunspots, *Pure Appl. Geophys.* (*Pageophys.*) **62**, 207–214.

Boussinesq (1903). Theorie Analitique de la Chaleur, Paris, t. ii, 172.

Bramson, M. A. (1958). "Infrared Radiation." Plenum Press, New York.

Brandhorst, H. W. *et al.* (1975). Interim solar cell testing procedures for terrestrial applications, NASA TM X-71771, NASA, Lewis Res. Center, Cleveland, Ohio.

Brice, D. B. (1963). Saline water conversion by flash evaporation utilizing solar energy, *Adv. Chem. Ser.* **38**, 99–116.

Brinkworth, B. J. (1972). "Solar Energy for Man." Compton Press, London and Halstead Press, New York.

Brinkworth, B. J. (1975). Selection of design parameters for closed-circuit forced-circulation heating systems *Sol. Energy* **17**. 1975.

Broda, E. (1975). Solar energy utilization—the photochemical approach, *in* "Heliotechnique and Development" (M. A. Kettani and J. E. Soussou, eds.), DAA Press, Cambridge, Massachusetts.

Brooks, F. A., and Miller, W. (1963). "Availability of Solar Energy," Univ. of Calif. Eng. and Sci. Extension Ser., 1st ed. McGraw-Hill, New York.

Brosens, J. P. (1965). Solar thermionic generators for space power, *ASME Ser. A, J. Eng. Power*, **86**, No. 6, 281.

Brunt, D. (1932). *Quart. J. R. Met. Soc.* **58**, 389.

Buchberg, H. *et al.* (1968). Design consideration for cellular solar collectors, ASME, Paper No. 68-WA/Sol-3.

Buchberg, H. *et al.* (1971). Performance characteristics of rectangular honeycomb solar thermal converter, *Sol. Energy* **13**, 193–221.

Buchberg, H. *et al.* (1974). Natural convection in enclosed spaces: A review of application to solar energy collection, ASME Paper No. 74-WA/HT-12.

Budyko, M. I. (1956). The heat balance of the earth's surface, translated by the office of Tech. Services, US Dept. of Commerce, Washington, D.C.

Buelow, F. H., and Sobel, A. T. (1963). Galvanized steel roof construction for solar heating, *Agr. Eng.* **44**, No. 6, 312–313, 316–317.

Butz, L. W., Beckman, W. A., and Duffie, J. A. (1974). Simulation of a solar heating and cooling system, *Sol. Energy* **16**, 129–136.

Bylander, E. G. (1971). "Materials for Semiconductor Functions." Hayden Book Co., New York.

Cabelli, A., and De Vahl Davis, G. (1970). Rep. No. 1970/FMT/2, Univ. of New South Wales, Australia.

Cadiergues (1952). *Rev. Chaleur Ind.* No. 329, 395.

Cadiergues (1953a). *Rev. Chaleur Ind.* No. 330, 17.

Cadiergues (1953b). *Rev. Chaleur Ind.* No. 331, 43.

Campen, C. E. (ed.) (1960). "Handbook of Geophysics," pp. 15–16. Macmillan, New York.

Cane, R. L. D. *et al.* (1975). *ISES Conf., Los Angeles* Paper 32/5.

Carslaw, H. S., and Jaeger, J. C. (1959). "Conduction of Heat in Solid." Oxford Univ. Press, New York.

Catton, I. (1966). *Phys. Fluids* **9**, 2521–2522.

Catton, I. (1970). *Trans. ASME J. Heat Transfer* No. 92, 186.

Catton. I., and Edwards, D. K. (1967). Effect of side walls on natural convection between horizontal plates heated from below, *Trans. ASME J. Heat Transfer Ser. C* No. 89, 295–299.

Chandrasekhar, S. (1961). "Hydrodynamics and Hydromagnetic Stability." Oxford Univ. Press (Clarendon), London and New York.

Chang, K., and Hong, S. (1972). *Appl. Sci. Res.* **27**, 19.

Chao, R. E., and Cox, K. E. (1975). An analysis of hydrogen production via closed-cycle schemes, *in* "Hydrogen Energy" (T. N. Veziroglu, ed.), Part A. Plenum Press, New York.

Charters, W. W. S. (1970). A study of heat transfer and pressure drop for turbulent flow in rectangular ducts, *Proc. Thermodyn. Conf.* Inst. of Eng., Adelaide, Australia.

Charters, W. W. S. (1971). Some aspects of flow duct design for solar air heater applications, *Sol. Energy* **13**, No. 2, 283–288.

Charters, W. W. S., and MacDonald, R. W. G. (1973). Heat transfer effects in solar air heaters, *UNESCO Conf. Sun Serv. Mankind Paris* Paper E37.

Charters, W. W. S., and Peterson, L. F. (1972). Free convection suppression using honeycomb cellular materials, *Sol. Energy*, **13**, No. 4, 353–361.

Cheng, K. C., and Hong, S. W. (1972). Effect of tube inclination on laminar convection in uniformly heated tubes for flat-plate solar collectors, *Sol. Energy* **13**, 363–371.

Chepurniy, N. (1975). Conduction heat transfer from a heated disk at the surface of a semi-infinite medium, Fluid Mech. Lab. Tech. Rep. No. 75–1, Dept. Civil Eng. and Appl. Mech., McGill Univ., Montreal, June.

Chepurniy, N., and Savage, S. B. (1974). An analytical and experimental investigation of a laboratory solar pond model, Paper 74-WA/Sol-3. ASME Annual Meeting, New York, November.

Chepurniy, N., and Savage, S. B. (1975). Effect of diffusion on concentration profiles in solar ponds, *Sol. Energy* **17**, No. 3, 203–205.

Cherne, J. M. (1974). Flat plate solar air collectors for agricultural use, *Proc. Workshop Sol. Collectors Heating Cooling of Buildings* NSF/RANN 75-019, 119–130.

Chinnappa, J. C. V. (1962). Performance of an intermittent refrigerator operated by a flat plate collector, *Sol. Energy* **6**, 143–150.

Chinnappa, J. C. V. (1967). Experimental study of the intermittent vapour absorption refrigeration cycle employing the refrigerant-absorbent systems of ammonia-water and ammonia-lithium nitrate, *Sol. Energy* **1**, 1–18.

Chinnappa, J. C. V. (1974). Solar operation of ammonia-water multistage air conditioning cycles in the tropics, *Sol. Energy* **16**.

Chiou, J. P. *et al.* (1965). A slit-and-expanded aluminum foil matrix solar collector, *Sol. Energy* **9**, No. 2, 73–80.

Choudhury, NK. D. (1963). Solar radiation at New Delhi, *Sol. Energy* **2**, No. 7, 44–51.

Christiansen, C. (1883). Wiedermanns, Ann 19.

Chung, R., and Duffie, J. A. (1961), Cooling with solar energy, *UN Conf. New Sources Energy, Rome* **6**, 20–28.

Clark, A. F. *et al.* (1975). Experimental performance of shallow solar ponds, *ISES Conf, Los Angeles* (see also Stengel, R. F., Solar ponds promise 10 MW range electric power, *Design News* **30**, No. 3, 32–33.

Clemont, M., and Fiatte, F. (1976). Solar energy applied to water pumping, *in* "Heliotechnique and Development" (M. A. Kettani and J. E. Soussou, eds.), DAA Press, Cambridge, Massachusetts.

Clerk, R. C. (1963). The Utilization of Fly Wheel Energy, SAE Paper 711A.

Close, D. J. (1963). Solar air heaters for low and moderate temperature applications, *Sol. Energy* **7**, No. 3, 117–124.

Close, D. J. (1967). A design approach for solar processes, *Sol. Energy* **7**, 112–122.

Coffari, E. (1976). A mathematical model for the digital computation of the hours of sunshine on an inclined plane, *in* "Heliotechnique and Development" (M. A. Kettani and J. E. Soussou, eds.), DAA Press, Cambridge, Massachusetts.

Cohen, R. K., and Hiester, N. K. (1957). A survey of solar furnace installations in the United States, *J. Sol. Energy Sci. Eng.* **1**, No. 2, 3, 105–118.

Colebrook, C. (1939). *J. Inst. Civil Eng.* **11**, 133.

Collins, R. E., and Davis, K. E. (1975). Geothermal storage of solar energy, *in* "Heliotechnique and Development" (M. A. Kettani and J. E. Soussou, eds.), DAA Press, Cambridge, Massachusetts.

Conn, W. M. (1951). A Solar furnace of 120 inches diameter for studying highly refractory substances, *Rev. Sci. Instrum.* **22**, 945–951.

Conn, W. M. (1954). Gesichtpunkte für den Entwurf von Sonnen-Spiegelöfen für Hochtemperaturforschung und -Entwicklung, *Z. ang. Phys.* **6**, 284–291.

Conn, W. M., and Braught, G. (1954). Separation of incident and emitted radiations in a solar furnace by means of rotating sectors, *J. Opt. Soc. Am.* **44**, 45–47.

Cooper, P. I. (1973). The maximum efficiency of single-effect solar stills, *Sol. Energy* **15**, 205–217.

Cooper, P. I., and Read, W. R. W. (1974). Design philosophy and operating experience for Australian solar stills, *Sol. Energy* **16**, 1–8.

Coulson, K. L. (1975). "Solar and Terrestrial Radiation, Methods and Measurements." Academic Press, New York.

Cuomo, V. (1974). Solar radiation, its correlation with the atmospheric factors in Italy, and the influence of orientation, *Int. College on Applied Phys. 1st Course Sol. Energy Conversion, Procida, Italy.*

Currin, C. G. *et al.* (1972). Feasibility of low cost silicon solar cells, *IEEE Photovoltaic Specialists Conf. Rec., 9th, Silver Springs, Maryland, May.*

Curtin, D. J., and Statler, R. L. (1975). Review of radiation damage to silicon solar cells, *IEEE Trans. Aerospace Electron. Syst.* **AES-11**, No. 4.

Dake, J. M. K. (1973). The solar pond: Analytical and laboratory studies, *UNESCO Congr. Sun Serv. Mankind, Paris.*

Dalibot, B. (1974). Generateurs solaires pour applications terrestres, *Photovoltaic Power Generat. Conf. Proc., Hamburg, Germany.*

Dallaire, G. (1975). Tougher pollution laws spur use of impermeable liners, *ASCE Civil Eng.* **May**, 63–67.

Damagnez, J. A. (1968). Modifications du bilau d'energie d'une surface evporante par les abris-serres plastique, *Acta Horticultural* **May**, No. 9.

Daniels, F. (1974). "Direct Use of the Sun's Energy." Ballantine Press, New York.

Dannies, J. H. (1959). Solar air conditioning and solar refrigeration, *Sol. Energy* **3**, 34–39.

Davies, J. M., and Cotton, E. S. (1957). Design of the Quartermaster solar furnace, *J. Sol. Energy Sci. Eng.* **1**, No. 2–3, 16–22.

Davis, E. S. (1975). Solar-assisted gas-energy water-heating feasibility for apartment, *Sol. Energy* **17**, No. 4.

Davis, S. H. (1967). *J. Fluid. Mech.* **30**, 465–478.

De Beni, G. (1975). Considerations if iron–chlorine–oxygen reactions in relation to thermochemical water splitting, *in* "Hydrogen Energy" (T. N. Veziroglu, ed.), Part A. Plenum Press, New York.

Defant, A. (1961). "Physical Oceanography," Pergamon, Oxford.

De Graaf, J. G. A., and Van Der Held, E. F. M. (1953). The relation between the heat transfer and the convection phenomena in enclosed plane air layers, *Appl. Sci. Res.* **A3**, 393–409.

Dietz, A. G. H., and Czapek, E. L. (1950). Solar heating of houses by vertical south wall storage panels, ASHVE, J. Sect., Heating, Piping, and Air Conditioning, ASHVE 56, Annu. Meeting, Dallas, Texas, pp. 1–8.

Dirmhirn, I. (1958). Untersuchungan an sternpyranometern, *Arch. Met. Geophys. Bicklim., Vienna* **B.9**(2), 124–148.

Drew, T., Nagle, W., and Smith, W. (1935). *Trans. AIchE* **31**, 605.

Dropkin, D., and Somerscales, E. (1965). Heat transfer by natural convection in liquids confined by two parallel plates which are inclined at various angles with respect to the horizontal, *Trans. ASME J. Heat Transfer* **87**, 77–84.

Drummond, A. J. (1956). On the measurement of sky radiation, *Arch. Met. Vienna B* **7**, 414–436.

Drummond, A. J. (1967). The Calibration of Thermal Radiation Detectors. The Eppley Lab., Reprint Ser. No. 35.

Drummond, A. J., and Greer, H. W. (1966). An integrating hemisphere (artificial sky) for the calibration of meteorological pyranometers, *Sol. Energy* **10**, No. 8.

Drummond, A. J., and Hickey, J. R. (1968). The Eppley-JPL solar constant measurement program, *Sol. Energy* **12**, No. 2, 217–232.

Drummond, A. J., and Thekaekara, M. P. (1973). The Extraterrestrial Solar Spectrum, Inst. of Environm. Sci., Mt. Prospect, Illinois.

Drummond, A. J. *et al.* (1967). Multichannel radiometer measurement of solar irradiance, *J. Space. Rocket* **4**, No. 9, 1200–1206.

Drummond, A. J. *et al.* (1968). New value of the solar constant of radiation, *Nature (London)* **218**, No. 5138, 259–261.

Duffie, J. A., and Beckman, W. A. (1974). "Solar Energy Thermal Processes." Wiley, New York.

Duffie, J. A., and Sheridan, B. E. (1965). Lithium bromide refrigerators for solar operation, *Mech. Chem. Trans. Instrum. Eng., Aust.*, 79–88.

Dunkle, R. V. (1961). Solar water distillation: The roof-type still and a multiple-effect diffusion still, *Int. Dev. Heat Transfer, ASME*, 895–902.

Dunkle, R. V. (1965) A method of solar air conditioning, *Mech. Chem Trans. Instrum. Eng.* **MC 1**, No. 1, 73–78.

Dunkle, R. V., and Davey, E. T. (1970). Flow distribution on absorber banks, Paper presented at the *Melbourne ISES Conf.*

Dunkleman, L., and Scolnik, R. (1959). Solar spectral irradiance and vertical attenuation in the visible and ultraviolet, *Opt. Soc. J. Am.* **49**, No. 4, 356–367.

Duwez, P. (1963). Concentration of solar energy, *in* "Introduction to the Utilization of Solar Energy" (A. M. Zarem and D. D. Erway, eds.). McGraw-Hill, New York.

Eckert, E. (1942). VDI-Forschungsh, No. 416, pp. 1.

Eckert, E. R. G., and Carlson, W. O. (1961). Natural convection in an air layer enclosed between two vertical plates with different temperatures, *Int. J. Heat Mass Transfer* **2**, 106–120.

Eckert, E., and Drewitz, O. (1943). NACA Tech. Mem. 1045.

Eckert, E. Diaguila, A. J., and Livingood, J. (1955). NACA Tech. 3584.

Edinger, J., and Geyer, J. (1965). Heat Exchange in the Environment. John Hopkins Univ. Res. Proj. RP-49, Baltimore, Maryland.

Edwards, D. K. (1969). Suppression of cellular convection by lateral walls, *Trans. ASME Heat Transfer* 145–150.

Edwards, D. K., and Catton, I. (1969). *Int. J. Heat Mass Transfer* **12**, 23–30.

Edwards, D. K., and Sun, W. M. (1970). Prediction of the onset of natural convection in rectangular structures, *ISES Conf., Melbourne, Australia* Paper 7/62.

Edwards, D. K. *et al.* (1975). *ISES Conf., Los Angeles* Paper 32/4.

Eibling, J. A., Thomas, R. E., and Landry, B. A. (1953). An investigation of multiple-effect evaporation of saline waters by steam from solar radiation, Office of Saline Water, US Department of the Interior, Res. and Dev. Rep. No. 2.

Elata, C., and Levin, O. (1965). Hydraulics of the solar pond, *Congr. Int. Assoc. Hydraulic Res., 11th, Leningrad.*

Elbaz, C. (1966). CNES, Note Tech., NT 3 Paris.

Elder, J. W. (1965). *J. Fluid Mech.* **23**, 77–88.

Elder, J. W. (1966). Numerical experiments with free convection in a vertical slot, *J. Fluid Mech.* **24**, 823–843.

Eliseev, V. N., Usmanov, Yu, U., and Teslenko, L. N. (1971). Theoretical investigation of the thermal regime of a solar pond, *Geliotekhnika* **7**, No. 4, 17–23.

Eliseev, V. N., Usmanov, Yu, U., and Umarov, G. Ya. (1973). On determining the efficiency of a salt solar pond, *Geliotkhnika* **9**, No. 1, 44–46.

Elsasser, W. M. (1942). Harvard Meteorological Studies, No. 6.

Elterman, L. UV. (1968). Visible and IR Attenuation for Altitudes to 50 km, AFCRL-68-0153. Office of Aerospace Res., U.S. Air Force.

Escher, W. J. D., and Hanson, J. A. (1975). Ocean based solar-to-hydrogen energy conversion macro system, "Hydrogen Energy" (T. N. Veziroglu, ed.). Part A. Plenum Press, New York.

Ewe, H. H. Justi, E. W. and Schmitt, A. F. (1972). Further progress in the electrochemical storage of solar energy by three electrodes storing cells, *Comples Internat. Bull.*, June.

Farber, E. A. (1970). Design and performance of a compact solar refrigeration system, *ISES Conf., Melbourne* Paper 6/58.

Farber, E. A. (1973). Univ. of Florida Solar Energy Lab. Proc., *UNESCO Congr. Sun Service Mankind,* Paper EH106.

Farber, E. A., and Prescott, F. L. (1973). Solar Powered V2 Vapor Engine, *UNESCO Congr. Sun Service Mankind,* Paper e7.

Faugere, J. F., and Delpont, J. P. (1966). CNES, Note Technique, NT 4 Paris.

Fiatte, F., and Clemot, M. (1975). Water pumping—a practical application of solar energy, CFP/Total Publ., October.

Fischer, H. (1974). Physics and technology of photovoltaic solar energy conversion, "Advances in Solid State Physics," Vol. 14. Pergamon, Oxford.

Fishenden, M., and Saunders, O. A. (1950). "Heat Transfer." Oxford Univ. Press, London and New York.

Florschuetz, L. W. (1975). On heat rejection from terrestrial solar cell arrays with sunlight concentration, *Photovoltaic Specialists Conf. Record, 11th, Scottsdale, Arizona, May.*

Flowers, E. C. (1974). The "so-called" Parson's black problem with old style Eppley pyranometers, Report and Recommendations of Solar Energy Data Workshop, November 1973, Nat. Sci. Foundation (RANN Grant No. AG-495).

Foex, M. (1960). *Proc. Int. Symp. High Temp. Technol., California* p. 313. McGraw-Hill, New York.

Francia, G. (1961). *Conf. New Energy Sources, Rome* **35**, 5, 71.

Fritz, S. (1951). Solar radiation energy and its modification by the earth and its atmosphere, *Comp. Meteorol., Am. Met. Soc.*

Fröhlich, C. *et al.* (1973). The third international comparisons of pyrheliometers and a comparison of radiometric scales, *Sol. Energy* **14**.

Fry, J. F. (1967). Evaluation of a SET VIII thermionic generator, *Intersoc. Energy Conversion Eng. Conf. Proc., 2nd, Miami, Florida.*

Funk, J. E., and Reinstrom, R. M. (1966). Energy requirements in the production of hydrogen from water, *I and EC Process Design Develop.* **5**, No. 3.

Funk, J. E., Conger, W. L., and Carty, R. H. (1975). Evaluation of multi-step thermochemical processes for the production of hydrogen from water, *in* "Hydrogen Energy" (T. N. Veziroglu, ed.), Part A. Plenum Press, New York.

Fuschillo, N. R. *et al.* (1966). Flat plate solar thermoelectric generator for near earth orbits, *Adv. Energy Convers.* **6**, 103.

Ginnings, D. C. (1948). Multiple-effect solar still, US Patent No. 2,445,350.

Gac, A. M. *et al.* (1968). Etude Experimental Concernment le Renouvellement de l'atmosphere et le Bilen Energetique de Serre Maraicheres Revetues de Plastique et de Verre, Centre National d'Etudes Techniques et de Recherches Technologiques pour l'Agriculture, les Forets de l'Equipment Rura, Republic Francaise, Ministere de l'Agriculture, No. 95.

Gallet, P. (1967). *C. R. Acad. Sci. Paris* **264**, Ser. B, 292.

Gallet, P. (1968). *Ann. Faculte d Sci. Marseille* **41**, 23.

Gallet, P. (1973). Etudes des Propriétés Radiatives des Structures Cellulaires. These Docteur Es Sciences Physiques, Univ. de Provence, France.

Garbell, M. A. (1963). The sea that spills into a desert, *Sci. Am.* **209**, No. 2.

Gates, D. M. (1960). Near infrared atmospheric transmission to solar radiation, *J. Opt. Soc. Am.* **50**, 1299.

Gates, D. (1965). Meteorol. Monogr. 28, Pl. Am. Meteorol. Soc., Boston, Massachusetts.

Gates D. M., and Harrop, W. J. (1963). Infrared transmission of the atmospheric to solar radiation, *Appl. Opt.* **2**, 887.

Gebhart, B. (1971). "Heat Transfer." McGraw-Hill, New York.

Geiger, R. (1965). "The Climate Near the Ground." Harvard Univ. Press, Cambridge, Massachusetts.

Geist, J. (1971). NBS Tech. Note No. 594, US Dept. of Commerce, Nat. Bur. of Std.

Ginnings, D. C. (1948). Multiple-effect solar still, US Patent No. 2,445,350.

Girardier, J. P. *et al.* (1973). Les Moteurs Solaires et l'Habitat Pour Les Zones Arides, *UNESCO Congr. Sun Service Mankind, Paris.*

Glaser, P. E. (1968). The future of power from the sun, *Intersoc. Energy Conversion Eng. Conf. Proc., 3rd, Boulder, Colorado.*

Globe, S., and Dropkin, D. (1959). Natural convection heat transfer in liquids by two horizontal plates and heated from below, *Trans ASME J. Heat Transfer* **81**, 24–30.

Golueke, G., and Oswald, W. J. (1963a). Power from solar energy via algae-produced methane, *Sol. Energy* **7**, 86.

Golueke, C. G., and Oswald, W. J. (1963b). Power from solar energy conversion and storage, Oklahoma State Univ.

Goody, R. M. (1958). "The Physics of the Atmosphere." Cambridge Univ. Press, London and New York.

Gosh, B. N. (1973). A new glass roof dryer for cocoa beans and other crops, UNESCO Congr. Sun in the Service of Mankind, Paris, Paper no. V-30.

Gould, R. F. (ed.) (1967). Regenerative EMF cells, *Adv. Chem. Ser.* No. 64.

Green, L. (1974). An open gas-turbine cycle for solar energy conversion, *UNESCO Congr. Sun Service Mankind, Paris.*

Griffiths, P., and Lee, M. (1967). *Int. J. Heat Mass Transfer* **10**, 697.

Guide to Meteorological Instrumentation and Observing Practices (1971). Secretariat of the World Meteorological Organization, Geneva, Switzerland, 4th ed.

Gupta, C. L., and Garg, H. P. (1967). Performance studies on solar air heaters, *Sol. Energy* **II**, No. 1, 25–31.

Halmos, G. B. (1975). Solar protection—costs and benefits, *Heating Air Cond. J.* **44** (520), 24–25.

Hamit, Y. H., and Beckman, W. A. (1969). Transpiration cooling and radiatively heated porous bed, ASME Paper No. 69-WA/Sol-6.

Hamit, Y. H., and Beckman, W. A. (1971). Performance of air cooled radiatively heated screen matrices, *J. Eng. for Power*, April, 221–224.

Hanson, K. J. (1974). Comments on the quality of the NWS pyranometer network data from 1954 to present. Report and Recommendations of Solar Energy Data Workshop, November, NSF (RANN Grant no. AG-495).

Harding, Josiah (1883). Apparatus for solar distillation, *Proc. Inst. Civil Eng.* **73**, 284–288.

Haurwitz, B. (1948). Insolation in relation to cloud type, *J. Meteorol.* **5**(3), 110.

Hausen, H. (1943). "Darstellung des Warmeubergangges in Rohren Durch Verallgemeinerte Potenzbeziehungen," *VDI Z.* No. 4, 91.

Hay, H., and Yellott, J. I. (1969), Natural air conditioning with roof ponds and moveable insulation, *ASHRAE Trans.* **75**, Part 1, 165.

Haynos, J. *et al.* (1974). The COMSAT nonreflective silicon solar cell: A second generation improved cell, *Photovoltaic Power Generation Conf. Proc., Hamburg, Germany.*

Hensley, J. C., Jr., and Young, P. G. (1972). Molded plastic solar still, US Patent No. 3,655,517.

Heywood, H. (1953). Solar energy: Past, present and future applications, *Engineering* **176**, 377–380, 409–411.

Heywood, H. (1970). A general equation for calculating total radiation on inclined surfaces, *ISES Conf.* Paper 2/31.

Hickey, J. R. (1974). Future instrumentation for solar energy measurement, Report and Recommendations of Solar Energy Data Workshop, November, NSF (RANN Grant No. AG-495).

Hickman, R. G., Krikorian, O. H., and Ramsey, W. J. (1975). Thermochemical hydrogen production research at Lawrence Livermore Laboratory, *in* "Hydrogen Energy" (T. N. Veziroglu, ed.), Part A. Plenum Press, New York.

Hiester, N. K. *et al.* (1957). Theoretical considerations on performance characteristics of solar furnaces, *Jet Propul.* **27**, 507–513.

Hildebrandt, A. F. (1975). Solar tower thermochemical energy cycles, *in Hydrogen Energy Fundamentals Symp. Proc.* (T. N. Veziroglu, ed.). Univ. of Miami, Florida.

Hildebrandt, A. F., and Vant-Hull, L. L. (1974). A tower top focus solar energy collector, ASME Paper No. 9, Vol. 96.

Hill, A. N. (1966). Calibration of solar radiation equipment at the U. S. Weather Bureau, *Sol. Energy* **10**, No. 4.

Hilpert, R. (1933). *Forsch. Gebiete Ingenieurw.* **4**. 215.

Hirschmann, J. R. (1970). Salt flats as solar-heat collectors for industrial purposes (review article), *Sol. Energy* **13**, No. 1, 83–97.

Hirschmann, J. R. (1974). The cosine function as a mathematical expression for the processes of solar energy, *Sol. Energy* **16**, 117–124.

Hirschmann, J. R., and Gaete, W. E. (1962). Colector solar de poza abiert con soluciones de cloruro de magnesio, *Rev. Sci.* No. 120, 43–95.

Hodges, C. N., Thompson, T. L., Groh, J. E., and Frieling, D. H. (1966). Solar distillation utilizing multiple-effect humidification, Office of Saline Water, US Department of the Interior, Res. and Dev. Rep. No. 194.

Hoffman, M. E. (1973). Evaluation of Solar Radiation, Climatic Conditions and Building Materials Influence in Building Temperatures Design by the Total Thermal Time Constant Method, *UNESCO Congr. Sun Service Mankind, Paris*, Paper EH.75.

Hogg, F. C. (1971). A switched-bed regenerative cooling system, *Proc. Int. Conf. Refrig., 13th* **4**, 41–46.

Hollands, K. G. T. (1963a). Directional selectivity; the emittance and absorptance properties of vee corrugated specular surfaces, *Sol. Energy* **7**, No. 3, 108–116.

Hollands, K. G. T. (1963b). The regeneration of lithium chloride brine in a solar still for use in solar air conditioning, *Sol. Energy* **7**, 39–43.

Hollands, K. G. T. (1965a). Honeycomb devices in flat plate solar collectors, *Sol. Energy* **9**, No. 3, 159–164.

Hollands, K. G. T. (1965b). *Phys. Fluids* **8**, 389–390.

Hollands, K. G. T. (1966). Natural convection in a horizontal layer of air with internal constraints, Ph.D. Thesis, McGill Univ., Canada.

Hollands, K. G. T. (1972). *AIChe-ASME Heat Transfer Conf.* Paper 72-HT-60.

Hollands, K. G. T. (1973). Natural convection in horizontal thin walled honeycomb panels, *J. Heat Transfer* **95**, 439–444.

Hollands, K. G. T. (1975). Studies on methods of reducing heat losses from flat plate solar collectors, Semiannual Progr. Rep., Univ. of Waterloo.

Hottel, H. C., and Unger, T. A. (1958). Colloque Int., CNSR, Montlouis, France.

Hottel, H. C., and Whillier, A. (1955). Evaluation of flate plate solar collector performance, *Trans. Conf. Use Sol. Energy, Univ. of Arizona* **II**, 74–104.

Hottel, H. C., and Woertz, B. B. (1942). The performance of flat plate solar heat collectors, *Trans. ASME* **64**, 91–104.

Howe, E. D. (1974). "Fundamentals of Water Desalination," Science and Technology Series, Vol. 1, 360 pp. Dekker, New York.

Howe, E. D., and Tleimat, B. W. (1974). Twenty years of work on solar distillation at the University of California, *Sol. Energy* **16**, 97–105.

Hoyt, D. V. (1974). A review of presently available solar radiation instruments, Report and Recommendations of Solar Energy Data Workshop, NSF (RANN Grant No. AG-495).

Hudson, R. D. (1975). "Infrared Detectors." Halstead Press, New York.

Hukuo, N., and Mii, H. (1957). Design problems of a solar furnace, *J. Sol. Energy Sci. Eng.* **1**, No. 2, 3, 108–114.

Idso, S. B., and Jackson, R. D. (1969). *J. Geophys. Res.* **74**, 5397.

IGY Instruction Manual, Part VI, Radiation Instruments and Measurement, Int. Assoc. Meteorol. Radiat. Comm., Comm. for Radiat. Instructions, *Ann. Int. Geophys. Year* **5**, No. 6, Pergamon, Oxford.

Inville, W. L. (ed.) (1963). *Proc. Conf. Sol. Energy Conversion Storage, Oklahoma State University.*

Iqbal, M. (1966). Free convection effects inside tubes of flat plate solar collectors, *Sol. Energy* **10**, 207–211.

Iqbal, M. (1976). Solar and wind induced internal and external coefficients from greenhouses, *in* "Heliotechnique and Development" (M. A. Kettani and J. E. Soussou, eds.), DAA Press, Cambridge, Massachusetts.

Iqbal, M., and Stachiewicz, J. (1966). *J. Heat Transfer Trans. ASME.* **88**, 109.

Jackson, T. W., Harrison, W. B., and Boteler, W. C. (1958). *Trans. ASME.* **80**, 739.

Jain, G. C. (1973). Heating of solar pond, *UNESCO Congr. Sun Service Mankind, Paris.*

Jakob, M. (1946). Free convection through enclosed plane gas layers. *Trans. ASME* **68**, 189–194.

Jakob, M. (1949a). *Trans. ASME.* **6**.

Jakob, M. (1949b). "Heat Transfer," Vol. 1. Wiley, New York.

James, L. W., and Moon, R. L. (1975). GaAs concentrator solar cells, *IEEE Photovoltaic Specialists Conf. Rec, Scottsdale, Arizona.*

Jannot, M., and Mordchelles-Regnier, G. (1967). *Euratom Conf, Brussels.*

Jarvinen, P. O. (1975). Solar-heated-air turbine generating systems, IECEC Record, 583–592.

Jeffreys, H. (1926). *Phil. Mag.* **2**, 833–844.

Jensen, A. S., and Limensky, I. (1960). Photoemissive energy conversion, *Power Sources Conf. Proc., 14th*, PSC Publ. Committee, Red Bank, New Jersey.

Jessup, E. (1974). A brief history of the solar radiation program, Report and Recommendations of Solar Energy Data Workshop, November 1973, NSF (RANN Grant No. AG-495).

Johnson, F. S. (1954). The solar constant, *J. Meteorol.* **11**, No. 6, 431–439.

Joly, F. (1975). Economic criteria of selection for closed cycle thermo-chemical water splitting processes, *in* "Hydrogen Energy" (T. N. Veziroglu, ed.), Part A. Plenum Press, New York.

Jones, W. P. (1973). "Air Conditioning Engineering," 2nd ed. Arnold, New York.

Jose, P. D. (1957). The design of the heliostat mirror for a solar furnace, *J. Sol. Energy Sci. Eng.* **1**, No. 2-3, 23–27.

Juston-Coumet, F., and Duburg. Le four Solaire de la Bouzareah, *Bulletin des engrais*, No. 402.

Kakabaev, A., and Golaev, M. (1971). Properties of refrigeration installations with a sprayer chamber and solar heaters, *Appl. Sol. Energy* (*Gelioteknika*) **7**, 36–41.

Kakabaev, A., Klyshchaeva, O., and Khandurdiev, A. (1972). Refrigeration capacity of an absorption solar refrigeration plant with flat glazed solution regenerator, *Appl. Sol. Energy* (*Gelioteknika*) **8**, 90–95.

Kakac, S., and Veziroglu, T. N. (1976). Solar production of synthetic fuels as a means of storing solar energy, *Int. Conf Mech. Eng.* (*Energy*) Univ. of Eng. and Tech., Lahore, Pakistan.

Kamada, O. (1964). Method of measuring target temperature in a solar furnace, *Appl. Opt.* **3**, 1397–1400.

Kamada, O. (1965). Theoretical concentration and attainable temperature in solar furnaces, *Sol. Energy* **9**, 39–47.

Kamada, O., Takizawa, T., and Sakurai, T. (1971). A high temperature x-ray diffractometer using a solar furnace, *J. Appl. Phys.* **10**, 485–490.

Kastens, M. L. (1963). The economics of solar energy, introduction to "The Utilization of Solar Energy," McGraw-Hill, New York.

Kapur, J. C. (1960). A report on the utilization of solar energy for refrigeration and air conditioning applications, *Sol. Energy* **4**, 39–47.

Kays, W. M. (1955). *Trans. ASME* **77**, 1265.

Kays, W. M. (1966). "Convection Heat and Mass Transfer." McGraw-Hill, New York.

Keene, H. B. (1913). *Proc. Roy. Soc.* **88A**, 60.

Keith, D. F. (1930). Cooling with portable units, *Refrig. Eng.* **20**, 361.

Kemeny, G. A., and Somers, E. V. (1962). Combined free and forced-convective flow in vertical circular tubes—experiments with water and oil, *Trans. ASME. J. Heat Transfer* 339–346.

Kendell, J. M., and Berdahl, C. M. (1970). *Appl. Opt.* No. 9, 1082.

Kent, A. C., and Bowyer, J. M. (1972). Natural convection by air in small cells heated from below, *Ind. Eng. Chem. Fundamentals* **11**, No. 3.

Kettani, M. A. (1970). "Direct Energy Conversion." Addison-Wesley, Reading, Massachusetts.

Kettani, M. A. (1973). Attempts at mapping the solar intensity distribution for the Arabian peninsula, *UNESCO Congr., Sun Serv. Mankind, Paris* Paper E10.

Kettani, M. A. (1976). Storage of solar energy in the form of potential hydraulic energy, *in* "Heliotechnique and Development" (M. A. Kettani and J. E. Soussou, eds.), DAA Press, Cambridge, Massachusetts.

Kettani, M. A., and Gonsalves, L. M. (1972). HHE power generation, *Sol. Energy* **14**, No. 1.

Kettani, M. A., and Scott, R. E. (1974). A Red Sea HHE plant, *Comples Int. J.*, 19–25.

Kettani, M. A. and Soussou, J. E., eds. (1976). "Heliotechnique and Development," Proceedings of the International Conference of COMPLES, Nov. 2–6, 1976, Dhahran, Saudi Arabia, 2 vols., DAA Press, Cambridge, Massachusetts.

Khanna, M. L., and Singh, N. M. (1967). Industrial solar drying, *Sol. Energy* **II**, No. 2.

Kikoin, I. K., and Noskov, MM. (1934). A new photoelectric effect in cuprous oxide, *Phys. Zh. Sovjet.* **5**, 586.

Kimball, H. H., and Hobbs, H. E. (1923). A new form of thermoelectric recording pyrheliometer, *J. Opt. Soc. Am.* **7**, No. 9.

Kirpichev, M. V., and Baum, V. A. (1954). Exploitation of the sun's rays, *Privoda* (*Nature*) **43**, 45–53.

Klarsfeld, S. M. (1970). *Rev. General Therm.* No. 108, 1403–1423.

Klein, S. A. (1973). The effects of thermal capacitance upon the performance of flat plate solar collectors, M.Sc. Thesis, Univ. of Wisconsin.

Klein, S. A. (1975). Calculation of flat plate collector loss coefficients, *Sol. Energy* **17**, No. 1.

Kondratyev, K. (1969). "Radiation in the Atmosphere," Academic Press, New York.

Kondratyev, K., and Nikolsky, G. A. (1970). Solar radiation and solar activity, *Q. J. Roy. Meteorol Soc* **96**, No. 3, 509–522.

Koutsoheras, W. (1976). "Natural Convection Phenomena in Inclined Cells with Finite Side Walls—A Numerical Solution." M. Eng. Sc. Thesis, University of Melbourne.

Koutsoheras, W., and Charters, W. W. S. (1975). Natural convection phenomena in inclined cells with finite side walls—A numerical solution, *ISES Conf. Los Angeles* Paper 32/3.

Kraus, A. D. (1964). "Extended Surfaces." Spartan Books, Baltimore, Maryland and Cleaverhum Press, London.

Kreith, F. (1968). "Principles of Heat Transfer." International Textbook, Scranton, Pennsylvania.

Kuo, H. L., and Platzman, G. W. (1961). A normal mode nonlinear solution of the Rayleigh problem, *J. Fluid Mech.* **10**, 611–634.

Kuo, H. L., and Platzman, G. W. (1967). *Bietr. Phys. Atmos.* **33**, 137–168.

Kupper, A., Hauptmann, E. G., and Iqbal, M. (1969). Combined free and forced convection in a horizontal tube under uniform heat flux, *Sol. Energy* **12**, 439–446.

Kurzweg, U. H. (1970). Stability of natural convective motion in an inclined channel, *Trans. ASME. J. Heat Transfer* 190–191.

Kuzay, T. M., Malik, M. A. S., and Boer, K. W. (1974). Solar collectors of solar one. *Proc. Workshop Sol. Collectors Heating Cooling Buildings NSF* (*RANN*), 99–108.

Labs, D., and Neckel, H. (1968). The radiation of the solar photosphere from 2000 A to 100μ, *Z. Astophys.* **69**, 1–73.

Labs, D., and Neckel, H. (1970). Transformation of the absolute solar radiation data into the international pyrheliometric temperature scale of 1968, *Sol. Phys.* **15**, 79–87.

Lalude, O. A., and Buchberg, H. (1971). Design and application of honeycomb porous-bed solar air heaters, *Sol. Energy* **13**, 223–242.

Laszlo, T. S., de Dufour, W. F., and Erdell, J. (1958). A guiding system for solar furnaces, *J. Sol. Energy Sci. Eng.* **2**, No. 1, 18–20.

Latimer, J. R. (1966). Calibration program of the Canadian Meteorological Service, *Sol. Energy* **10**, No. 4.

Latimer, J. R. (1971). Radiation Measurements, IFYGL, Tech. Manual Ser. No. 2, NRC, Canada.

Lawand, T. A. (1963). Solar driers for farm produce, Tech. Rep. T.4, Brace Res. Inst., Brace Experiment Station, St. James Barbados.

Lawand, T. A. (1966). How to make a solar cabinet dryer for agricultural produce, Brace Experiment Station, St. James Barbados, 1–9.

Lawand, T. A. (1967). The operation of a large scale solar agricultural drier, Tech, Rep. T.33, Brace Experiment Station.

Lawand, T. A. *et al.* (1973). An investigation of the contribution of solar energy in heating greenhouses in Quebec, *J. Eng. Power ASME* 114–118.

Liang, S. F., and Acrivos, A. (1970). Stability of buoyancy driven convection in a tilted slot, *Int. J. Heat Mass Transfer* **13**, 449–458.

Lin, B., and Jordan, R. (1960). *Sol. Energy* **4**.

Lindmayer, J., and Allison, J. F. (1972). The violet cell: An improved silicon solar cell, *IEEE Photovoltaic Specialists Conf. Record, 9th*, Silver Spring, Maryland.

List, R. J. (1971). "Smithsonian Meteorological Tables." Smithsonian Inst. Press, Washington, D.C.

Liu, B. Y. H., and Jordan, R. C. (1960). The interelationship and characteristic distribution of direct, diffuse and total solar radiation, *Sol. Energy* **4**, No. 3.

Liu, B. Y. H., and Jordan, R. C. (1963). The long term average performance of flat plate solar energy collectors, *Sol. Energy* **7**, No. 2, 53.

Lloyd, W. W. (1975). Fabrication of an improved vertical multijunction solar cell, *IEEE Photovoltaic Specialist Conf. Record, 11th, Scottsdale, Arizona.*

Löf, G. O. G. (1950). Performance of solar energy collectors of overlapped glass-plate type, *Proc. Course Symp. Space Heating Sol. Energy, MIT* pp. 72–83.

Löf, G. O. G. (1956). Solar house heating panel II, *Proc. World Symp. Appl. Sol. Energy*. 131.

Löf, G. O. G. (1963). The heating and cooling of buildings with solar energy, *in* "Introduction to the Utilization of Solar Energy" (A. M. Zarem, and D. D. Erway, eds.). McGraw-Hill, New York.

Löf, G. O. G., and Neves, T. D. (1953). Heating of air by solar energy, *Ohio J. Sci.* **53**, 272–280.

Löf, G. O. G., and Tybout, R. A. (1974). The design and cost of optimized systems for residential heating and cooling by solar energy, *Sol. Energy* **16**, 9–18.

Löf, G. O. G. *et al.* (1946). Solar energy utilization for house heating, Rep. PB 25375, Office of Production R & D, War Production Board.

Löf, G. O. G. *et al.* (1961). Design and performance of domestic heating system employing solar heated air . . . The Colorado house, *Proc. UN Conf. New Sources Energy, Rome* Paper no. E/Conf 35/S/114.

Löf, G. O. G. *et al.* (1963). Residential heating with solar heated air—the Colorado house, *ASHRAE* **77**.

Löf, G. O. G. *et al.* (1966). World distribution of solar radiation, Rep. 21, Solar Energy Lab., Univ. Wisconsin.

Loferski, J. J. (1963). Recent research on photovoltaic solar energy converters, *Proc. IEEE.*

Loh, E., Hiester, N. K., and Tietz, T. E. (1957). Heat flux measurements at the sun image of the California Institute, of Technology lens-type solar furnace, *J. Sol. Energy Sci. Eng.* **1**, No. 4, 23–26.

Lundberg, R. E., Reynolds, W. C., and Kays, W. M. (1963). NASA Tech. D-1972.

MacDonald, T. H. (1951). Some characteristics of the eppley pyrheliometer, *Mon. Weather Rev.* **79**, No. 8.

Majumdar, N. C., Mather, B. L., and Kaushik, S. B. (1972). Prediction of direct solar radiation for low atmospheric turbidity, *Sol. Energy* **13**, No. 4.

Malik, M. A. S., and Buelow, F. H. (1973). Heat transfer characteristics of a solar drier, *UNESCO Congr. Sun Service Mankind, Paris* Paper V25.

Malik, M. A. S., and Tran, V. V. (1973). Simplified mathematical model for predicting the nocturnal output of a solar still, *Sol. Energy* **14**, 371–385.

Malkus, W. V. R., and Veronis, G. (1958). Finite amplitude cellular convection, *J. Fluid Mech.* **4**, 225–260.

Mani, A., and Chacko, O. (1963). Diffuse solar radiation at Delhi and Poona, *Ind. J. Met. Geophys.* **14** (4), 416–432.

Marshall, K. W. *et al.* (1975). *ISES Conf. Los Angeles* Paper 32/1.

Martinelli, R. C., and Boelter, L. M. K. (1942). The analytical prediction of superposed free and forced viscous convection in a vertical pipe, *Univ. Calif. Publ. Eng.* **5** (2), 23–58.

Masson, H. (1966). Quantitative estimation of solar radiation, *Sol. Energy* **10**, No. 3.

Matthael, A. (1975). Capital and running costs of air conditioning and solar protection, *Heating Air. Cond. J.* **44** (S 22), 12–17.

McAdams, W. H. (1954). "Heat Transmission." McGraw-Hill, New York.

Meinel, A. B., and M. P. (1976). "Applied Solar Energy: An Introduction," Chapters 4–7. Addison-Wesley, Reading, Massachusetts.

Meunier, R. M. (1973). Caracteristiques et objectifs des fours solaires du Laboratoire Central de l'Armement, *Rev. Int. Hautes Temp. Refract.* **10**. 297–302.

Miller, K. W. (1943). Solar heat trap, Memorandum to Office of Production Res. and Develop., W. P. B., July 12.

Mitsui, A. (1975). Photoproduction of hydrogen via microbial and biochemical processes, *in* "Hydrogen Energy Fundamentals Symposium Proceedings" (T. N. Veziroglu, ed.). Univ. of Miami Press, Miami, Florida.

Mockovciak, J. (1972). A systems engineering overview of the satellite power Station, *Intersoc. Energy Conversion Eng. Conf. Proc. 7th San Diego, California.*

Moody, L. F. (1944). Friction factors for pipe flow, *Trans. ASME J. Heat Transfer* **66**, 671.

Moon, P. (1940). Proposed standard solar radiation curves for eng. use, *J. Franklin Inst.* **230**(5), 583–617.

Morgan, C., and Craig, G. T. (1975). A comparison of predicted solar insolation with measured values at San Diego, California, *ISES Conf., Los Angeles.*

Morse, R. N., and Read, W. R. W. (1968). A rational basis for the engineering development of a solar still, *Sol. Energy* **12**, 5–18.

Morton, B. (1959). *Quart. J. Mech. Appl. Math.* **12**, 410.

Mull, W., and Reiher, H. (1930). Der Wameshutz von Luftschiechten, Gesundheits-Ingenieur, Reihel, Heft 28, Munich and Berlin.

Mullick, S. C., and Gupta, M. C. (1974). Solar desorption of absorbent solutions, *Sol. Energy* **16**, 19–23.

Muradov, D., and Shadiev, O. (1971). Testing of a solar refrigerator, *Appl. Sol. Energy* (Geliotekhnika) **7**, No. 3, 33.

Nakagawa, Y. (1960). Heat transport by convection, *Phys. Fluids* **3**, 82–86.

Neuwirth, O. S. (1958). The photolysis of nitrosyl chloride and the storage of solar energy, *Trans. Conf. Use Sol. Energy, Sci. Basis* **4**, 115. Univ. Arizona Press, Tucson, Arizona.

Nicolet, M. (1948). La mesure du rayonnment solaire, *Inst. Ray. Met. Belg. Miscofasc.* **21**.

Numan, M. Y. (1975). Standardization and Prediction of Irradiance and Illuminance of Surfaces of Different Inclinations and Orientations in Tropical Regions, Comples, 1st Semestre, pp. 9–18.

Ohta, T. *et al.* (1975). Water splitting system hybridised by photochemical thermoelectric utilizations of solar energy, *Proc. U.S.—Japan Seminar Key Technol. Hydrogen Energy Syst., Univ. Yokohama, Japan.*

Oliver, D. R. (1962). The effect of natural convection on viscous flow heat transfer in horizontal tubes, *Chem. Eng. Sci.* **17**, 335–350.

Ostrach, S. (1953). NACA Rep. 1111.

Ostrach, S., and Pnueli, D. (1963). *Trans. ASME J. Heat Transfer* **85**, 1346–1354.

Oustromov, G. A. (1958). NACA TM 1407.

Owen, P., and Thomson, W. (1963). *J. Fluid Mech.* **15**, 321.

Oyediran, O. (1973). Solar radiation, net radiation and temperature in Argonne, Chicago, *Sol. Energy* **15**, 155–170.

Ozoe, H. *et al.* (1974). *Int. J. Heat Mass Transfer* **17**, 401–406.

Ozoe, H. *et al.* (1974). *Int. J. Heat Mass Transfer* **17**, 1209–1216.

Oztoker, U., and Selçuk, M. K. (1971). Theoretical analysis of a system combining a solar still with a controlled environmental greenhouse, ASME Paper no. 71/WA/SOL-9.

Page, J. K. (1964). The estimation of monthly mean values of daily total short-wave radiation on vertical and inclined surfaces from sunshine records for latitude 40°N–40°S, *Proc. U.N. Conf. New Sources Energy* **4**, 378.

Paleocrasses, S. (1975). Photolysis of water as a solar energy conversion process, *in* "Hydrogen Energy" (T. N. Veziroglu, ed.), Part A. Plenum Press, New York.

Pangborn, J. R., and Sharer, J. C. (1975). Analysis of thermo-chemical water splitting cycles, *in* "Hydrogen Energy" (T. N. Veziroglu, ed.), Part A. Plenum Press, New York.

Pellew, A., and Southwell, R. V. (1940). On maintained convective motion in a fluid heated from below, *Proc. R. Soc. London* **A176**, 312–343.

Perrot, M. *et al.* (1962). Remarques théoriques sur la structure cellulaire de G. francia, *Comples Bull.* No. 3, 20–33

Perrot, M. *et al.* (1963). *C. R. Acad. Sci. Paris* **t256**, 5383.

Perrot, M. *et al.* (1965). *C. R. Acad. Sci. Paris* **t256**, 2383.

Perrot, M. *et al.* (1958). Colloque International, CNRS, Montlouis.

Perrot, M. *et al.* (1963). *Bull. Comples* **4**.

Perrot, M. *et al.* (1966). *Bull. Comples* **10**, 91.

Perrot, M. *et al.* (1967). Les structures cellulaires antiraonnantes et leurs applications industrielle, *Sol. Energy* **11**, 34–40.

Peterson, L. F. (1971). An examination of the stability criterion for fluid layers subject to adverse temperature gradients. M. Eng. Sci. Thesis, Melbourne Univ., Australia.

Pivic, N. W. (1958). The use of higher plant for storing of energy, *Trans. Conf. Use Sol. Energy, The Scientific Basis.* **4**, 182. Univ. of Arizona Press, Tucson, Arizona.

Pneuli, D. (1964). *Trans. ASME* **E31**, 376.

Polhausen, E., and Angew, Z. (1921). *Math. Mech.* **1**, 115.

Pozin, G. N. (1971). Solution of the system heat balance equations for hot houses in the unsteady regime, *Geliotekhnika* **7**. No. 2, 47–55.

Priestley, C. (1959). "Turbulent Transfer in the Lower Atmosphere." Univ. Chicago Press, Chicago, Illinois.

Prigmore, D., and Barber, R. (1975). Design considerations and test data for a rankine cycle prototype, *Sol. Energy* **17**, 185–192.

Proctor, D. (1973). The use of waste heat in a solar still, *Sol. Energy* **14**, 433–449.

Rabl, A., and Nielsen, C. E. (1975). Solar ponds for space heating, *Sol. Energy* **17**, No. 1, 1–12.

Rahilly, W. P. (1972). Vertical multijunction solar cells, *IEEE Photovoltaic Specialist Conf. Record, 9th, Silver Spring, Maryland.*

Rappaport, P. (1959). The photovoltaic effect and its utilization, *RCA Rev.* **20**, 373.

Rayleigh, Lord (1916). On convection currents in a horizontal layer of fluid when the higher temperature is on the underside, *Phil. Mag.* **S6**, 32, 529–546.

Read, W. R. (1974). A solar timber kiln, *Sol. Energy* **15**, 309–316.

Reddy, S. J. (1971). An empirical method for the estimation of total solar radiation, *Sol. Energy* **13**, 289–292.

Reynolds, O. (1901). "Scientific Papers of Osborne Reynolds," Vol II. Cambridge Univ. Press, London and New York.

Rich, B. R. (1953). An investigation of heat transfer from an inclined flat plate in free convection, *Trans. ASME* **75**, 489–497.

Riches, M. (1975). Insolation in the United States: An Overview, AIAA, Paper no. 75–114, *AIAA/AAS Sol. Energy Earth Conf.*

Robeson, K. A. (1970). Evaluation of evaporative rock-bed regenerative coolers, *Aust. Refrig. Air. Cond. Heating* **24**, Pt.1(2), Pt.2(3).

Robinson, G. (1947). *Quart. J. R. Met. Soc.* **73**, 315

Robinson, N. (ed.) (1966). "Solar Radiation." Elsevier, Amsterdam.

Rose, A. (1956). *Photocond. Conf.* Wiley, New York.

Rowley, F. B., and Eckley, W. A. (1932). Surface coefficients as affected by direction of wind, *ASHVE Trans.* **38**, 33–45.

Rowley, F. B., Algren, A. B., and Blackshaw, J. L. (1930). Surface conductances as affected by air velocity, temperature and character of surface, *ASHVE Trans.* **36**, 501–508.

Russell, J. R. (1974). Central Station Solar Power: Making a Viable System Economic, Power Eng. No. 1974.

Russell, J. R. *et al.* (1974). Principles of the fixed mirror solar concentrator, Rep. no. GA-A12902. General At. Co., San Diego, California.

Sabbagh, J. A., Sayigh, A. A. M., and El-Salam, E. M. A. (1973a). Correlation of solar radiation and sunshine duration in Riyadh, Saudi Arabia, *Pakistan J. Sci. Ind. Res.* **16**, No. 6.

Sabbagh, J. A., Sayigh, A. A. M., and El-Salam, E. M. A. (1973b). Solar water heaters in Saudi Arabia, *Arab Sci Congr., 7th, Cairo* **II**, 213–230.

Sabbagh, J. A., Sayigh, A. A. M., and El-Salam, E. M. A. (1975). Estimation of the total solar radiation from meteorological data, *ISES Conf., Los Angeles.*

Sabbagh, J. A. *et al.* (1976). To be published in *J. Heat Transfer, ASME.*

Sakurai, T., and Takizawa, T. Thermal expansion of SnO_2 at high temperatures, (1971). *High Temp.—High Press.* **3**, 325–331.

Sakurai, T., Kamada, O., Shishido, K., and Inagaki, K. (1964). Construction of a large solar furnace, *Sol. Energy* **8**, 117–126.

Sakurai, T., Mochizuki, S., and Ishigame, M. (1975). Measurement of electrical conductivity of oxides at high temperatures by microwave ellipsometry, *High Temp.-High Press.* **7**, 411–417.

Samuels, M. R., and Churchill, S. W. (1967). *AIChE J.* **13**, 77–85.

Sargent, S. L., and Beckman, W. A. (1968). Theoretical performance of an ammonia-sodium thiocyanate intermittent absorption refrigerator, *Solar Energy* **12**, 137–146.

Saucer, K. M. (1958). Photogalvanic cells, *Trans. Conf. Use Sol. Energy: The Scientific Basis* **5**, 43. Univ. Of Arizona Press, Tucson, Arizona.

Saulnier, B. *et al.* (1975). Field testing of a solar pond, *ISES Conf., Los Angeles.*

Sayigh, A. A. M., and El-Salam, E. M. A. (1976). Optimum design of a single slope solar still in Riyadh, Saudi Arabia, *Arab Phys. Soc. Conf., Benghazi.*

Schlichting, H., and Kestin, J. (1968). "Boundary Layer Theory," 6th ed. McGraw-Hill, New York.

Schluter, A., Lortz, D., and Busse, F. (1965). *J. Fluid Mech.* **23**, 129.

Schmidt, E., and Silveston, P. L. (1959). Natural convection in horizontal liquid layers, *Chem. Eng. Prog., Symp. Ser.* **29**, 163–169.

Schroder, J. (1974). Thermal energy storage and control, ASMF Winter Annual Meeting, New York, November.

Segel, L. A., and Stuart, J. T. (1962). *J. Fluid Mech.* **13**, 289–306.

Seginer, I. and Levav, N. (1971). Models as tools in greenhouse climate design, Technion-Israel Inst. of Tech., Agric. Eng. Faculty, Publ. no. 115, Haifa, Israel.

Seifert, W. W., Bakr, M. A., and Kettani, M. A. (1973). "Energy and Development: A Case Study." MIT Press, Cambridge, Massachusetts.

Selçuk, M. K. (1964a). Design and performance evaluation of a multiple-effect, tilted solar distillation unit, *Sol. Energy* **8**, 23–30.

Selçuk, M. K. (1964b). The flat plate collector performance at high temperature, *Sol. Energy* **8**, 57–62.

Selçuk, M. K. (1966). Solar air preheater and fruit drier, *Comples Bull. No. 10*, 79–82.

Selçuk, M. K. (1967). Final report of the solar fruit drier, *Comples Bull. No. 12*, 91–101.

Selçuk, M. K. (1970). The Effect of Solar Radiation on the Energy Balance of Controlled Environment Greenhouses, ASME Paper No. 70–WA/SOL-3.

Selçuk, M. K. (1971). Thermal and economic analysis of the overlapped glass plate solar air heaters, *Sol. Energy* **13**, No. 2. 165–191.

Selçuk, M. K., and Yellott, J. I. (1962). Measurement of solar radiation with silicon photovoltaic cells, *Sol. Energy* **6**, No. 4.

Selçuk, M. K., Caputo, R. S., and Truscello, V. C. (1975a). Preliminary technical and economic data on solar thermal power conversion systems, *Int. Solar Energy Conf., Los Angeles.*

Selçuk, M. K., Ersay, O., and Akyurt, M. (1975b). Development, theoretical analysis and performance evaluation of shelf-type solar driers, *Sol. Energy* **16**, No. 2.

Sellers, W. D. (1965). "Physical Climatology." University of Chicago Press, Chicago, Illinois.

Sellars, J. R., Tribus, M., and Klein, J. S. (1956). *ASME Trans.* 441.

Sharma, M. R., and Pal, R. S. (1965). Interrelationship between total, direct and diffuse solar radiation in the tropics, *Sol. Energy*, **9**, No. 4, 183–192.

Shelton, J. (1975). Underground storage of heat in solar heating systems, *Sol. Energy* **17**, No. 2, 137–143.

Sheridan, N. R. (1961). Solar air conditioning, *J. Inst. Eng. Aust.* **33**, 47–52.

Shirland, F. A. (1966). The history, desgin, fabrication and performance of Cds thin film solar cells, *Adv. Energy Conversion* **6**, 201.

Shoemaker, M. J. (1962). Notes on a solar collector with unique air permeable media, *Sol. Energy* **5**, No. 4, 138–141.

Somerscales, E. F. C., and Gadza, I. W. (1969). *Int. J. Heat Mass Transfer* **12**, 1491.

Soussou, J. E., and Seifert, W. W. (1976). Formulation of energy policies in developing countries, *in* "Heliotechnique and Development" (M. A. Kettani and J. E. Soussou, eds.), DAA Press, Cambridge, Massachusetts.

Sparrow, E., and Gregg, J. (1959). *Trans. ASME. J. Heat Transfer* **81**, 13, 291.

Stair, R., and Ellis, H. T. (1968). The solar constant based on new spectral irradiance data from 3100 to 5300 angstroms, *J. Appl. Meteorol.* **7**, No. 8, 635.

Stanhill, G. (1966). Diffuse sky and cloud radiation in Israel, *Sol. Energy* **10**, No. 2, 96–101.

Steward, W. G., and Krieth, F. (1975). Optical design characteristics of a stationary concentrating reflector cum tracking absorber of a solar energy collector, *Appl. Opt.* **14**, No. 7, 1509–1512.

Stirn, R. J., and Yeh, Y. M. (1973). Solar and laser energy conversion with schottky barrier solar cells, *IEEE Photovoltaic Specialists Conf. Record, 10th, Palo Alto, California.*

Stoecker, W. F. (1958). "Refrigeration and Air Conditioning." McGraw-Hill, New York.

Stolzenbach, K. D. *et al.* (1968). Prediction of temperatures in solar ponds, *Annu. Meeting Sol. Energy Soc.*, Palo Alto, California.

Straubel, H. (1949). Der Sonnen-Schmelzspiegel Z. *Angew. Phys.* **1**, 542–545.

Sverdrup, H. U., Johnson, M. W., and Fleming, R. H. (1942). "The Oceans." Prentice-Hall, Englewood Cliffs, New Jersey.

Swartman, R. K., and Ogunlade, O. (1966). Solar radiation estimates from common parameters, *Sol. Energy Conf., Tempe, Arizona.*

Swartman, R. K., and Ogunlade, O. (1967). *Sol. Energy* **11**, 170.

Swartman, R. K., Ha, V., and Swaminathan, C. (1973a). Comparison of ammonia-sodium thiocynanate in a solar refrigeration system, UNESCO Congress Sun in the Service of Mankind, Paper E5.

Swartman, R. K., Ha, V., and Newton, A. J. (1973b). Survey of solar-powered refrigeration, ASME Winter Annu. Meeting, Detroit, Paper 73-WA/SOL-6.

Swet, C. J. (1974). Thermochemical water cracking using solar heat, *Theme Conf. Proc.* (T. N. Veziroglu, ed.). Univ. of Miami.

Swet, C. J. (1975). Thermochemical water cracking using solar heat, *in* "Hydrogen Energy" (T. N. Veziroglu, ed.), Part A. Plenum Press, New York.

Swinbank, W. (1963). *Quart. J. Roy. Met. Soc.* **89**, 339.

Swinbank, W. (1964). *Quart. J. Roy. Met. Soc.* **90**, 119.

Szego, G. C. *et al.* (1972). The energy plantation, Paper No. 729168, IECEC.

Szulmayer, W. (1973). Thermodynamics of sun drying, UNESCO Congr. Sun in the Service of Mankind, Paris, Paper No. V24.

Tabor, H. (1955). *Bull. Res. Council Israel*, **5A/5C**, 1.

Tabor, H. (1958a). Radiation, convection and conduction coefficients in solar collectors, *Bull. Res. Council Israel* **6C**, 155–176.

Tabor, H. (1958b). Stationary mirror systems for solar collectors, *Sol. Energy* **2**, No. 3–4, 27–33.

Tabor, H. (1959). Solar collector developments, *Sol. Energy* **3**, No. 3, 8–9.

Tabor, H. (1961). Large area solar collectors (solar ponds) for power production, *UN Conf. New Sources Energy* Publication S/47, New York.

Tabor, H. (1962). Use of solar energy for cooling purposes, *Sol. Energy* **6**, 136–142.

Tabor, H. (1963). Solar ponds, *Sol. Energy* **7**, No. 4, 189–194.

Tabor, H. (1964). Solar ponds, *Electron. Power* 296–299.

Tabor, H. (1969). *Sol. Energy* **12**, 549–552.

Tabor, H., and Matz, R. (1965). Solar pond project, *Sol. Energy* **9**, No. 4, 177–182.

Tabor, H., and Zeimer, H. (1962). Low cost focussing collector for solar power units, *Sol. Energy* **4**, 55–59.

Takakura, T., Jordan, R. A., and Boyd, L. L. (1969). Dynamic simulation of plant growth and environment in the greenhouse, Winter Meeting ASAE, Chicago, Paper No. 69–942.

Talbert, S. G., Eibling, J. A., and Löf, G. O. G. (1970). Manual on solar distillation of saline water, Office of Saline Water, US Department of the Interior, Res. and Dev. Rep. No. 546.

Tanner, D., Potter, C., and Pope, D. (1965). *Int. J. Heat Mass Transfer* **8**, 419, 427.

Taylor, C. S. (1976). Problems of a solar concentration and high temperature storage system, *in* "Heliotechnique and Development" (M. A. Kettani and J. E. Soussou, eds.), DAA Press, Cambridge, Massachusetts.

Taylor, F. B. (1975). Desalting water by solar distillation, Modern Government/National Development, June/July 1975.

Teagan, W. P., and Sargent, S. L. (1973). A solar powered combined heating/cooling system with the air conditioning unit driven by an organic rankine cycle engine, *UNESCO Congr. Sun in the Service of Mankind, Paris.*

Telkes, M. (1955). "Solar Energy Research," Univ. of Wisconsin Press, Madison, Wisconsin.

Telkes, M., and Raymond, E. (1946). Chemical storage of heat, Heating and Ventilating Magazine, November, p. 80.

Telkes, Maria (1956a). Solar stills, *Proc. World Symp. Appl. Sol. Energy, Phoenix, Arizona,* 73–79.

Telkes, Maria (1956b). Research on methods for solar distillation, Office of Saline Water, US Department of the Interior, Res. and Dev. Rep. No. 13.

Telkes, M., and Raymond, E. (1949). Storing solar heat in chemicals—a report on the Dover house, Heating and Ventilating Reference Section, pp. 80–86.

Thekaekara, M. P. (1970). Proposed standard values of the solar constant and solar spectrum, *J. Environm. Sci.* **13**, No. 4, 6–9.

Thekaekara, M. P. (1973). Solar energy outside the earth's atmosphere, *Sol. Energy* **14**, No. 2, 109–127.

Thekaekara, M. P. (1974a). Data on incident solar energy, the energy crisis and energy from the sun, Inst. Environ. Sci., Mount Prospect, Illinois, pp. 21–49.

Thekaekara, M. P. (1974b). Extraterrestrial solar spectrum, 3000–6100 A, at 1 A intervals, *Appl. Opt.* **13**, No. 3, 518–522.

Thekaekara, M. P., Callingbourne, R. H., and Drummond, A. J. (1972). A comparison of working standard pyranometers. The Eppley Lab. Reprint Ser. No. 57, *Bull. Amer. Met. Soc.* **53**, No. 1.

Thompson, H. A., and Sogin, H. H. (1966). *J. Fluid Mech.* **24**, 451–479.

Thomson, J. T. (1881). *Proc. R. Soc. Glasgow* **13**, 464–470.

Tleimat, B. W. and Howe, E. D. (1966). Nocturnal production of solar distillers, *Sol. Energy* **10**, 61–66.

Tleimat, B. W., and Howe, E. D. (1976). Solar-assisted distillation of sea water, *in* "Heliotechnique and Development" (M. A. Kettani and J. E. Soussou, eds.), DAA Press, Cambridge, Press, Cambridge, Massachusetts.

Trombe, F. (1961). Les installations de Mont-Louis et le four solaire de 1000 kW d'Odeillo Font-Romeu, Applications Thermiques de l'Energie Solaire dans le Domaine de la Recherche et de l'Industrie, C.N.R.S., No. LXXXV.

Trombe, F., and Ducarrior, M. (1972). Perspectives on cooling and refrigeration through the use of solar energy and other natural means, Paper presented at the UNESCO Seminar on Solar Energy, Niamey, Niger, October.

Trombe, F., and Le Phat Vinh, A. (1973). Thousand kW solar furnace, built by the National Center of Scientific Research in Odeillo (France), *Sol. Energy* **15**, No. 1, 57–61.

Trombe, F. (1952). Sur la realization d'un four solaire semi-industriel au Laboratoire de Mont-Louis (Pyrenees-Orientales), *Comptes Rend.* **235**, 704–705; Sur quelques details de montage du four solaire semi-industriel de Montlouis, *Comptes Rend*, **235**, 1211–1213.

Trombe, F., and Foex, M. (1957). The production of cold by means of solar radiation, *Sol. Energy* **1**, 51–52.

Trombe, F., and Le Phat Vinh, A. (1971). Methodes de concentration de l'energie solaire, *Cahiers AFEDES* No. 3.

Trombe, F., Foex, M., and La Blanchetais, C. H. (1947). Concentration de l'énergie solaire pour atteindre très haute température, *Ann. Chim.* (12) **2**, 385–419.

Trombe, F. *et al.* (1973). First results obtained with the 1000 kW solar furnace, *Sol. Energy* **15**, No. 1, 63–66.

Turner, J. S. (1973). "Buoyancy Effects in Fluids." Cambridge Univ. Press, London and New York.

Usmanov, Yu. U., and Eliseev, V. (1973). Effect of evaporation on the thermal state of a solar pond, *Geliotekhnika* **9**, No. 5, 45–47.

Usmanov, Yu. U, Umarov, G. Ya., and Zakhidov, R. A. (1969). Salt water ponds as solar energy accumulators, *Geliotekhnika* **5**, No. 2, 49–55.

Usmanov, Yu. U., Eliseev, V., and Umarov, G. (1971a). Optical characteristics of a solar reservoir, *Geliotekhnika* **7**, No. 1, 28–32.

Usmanov, Yu. U., Eliseev, V., and Umarov, G. (1971b). On the optical characteristics of a solar pond, *Geliotekhnika* **7**, No. 3, 78–81.

Usmanov, Yu. U. Eliseev, V., and Umarov, G. (1971c). Experimental investigation of the thermal regime of a model of a solar pond, *Geliotekhnika* **7**, No. 4, 24–28.

Usmanov, Yu. U., Eliseev, V., and Umarov, G. (1973). Experimental study of the removal of heat from a solar salt water pond, *Geliotekhnika* **9**, No. 6, 23–26.

Vartanyan, A. V., Shermazanyan, Ya, T., and Arutyunyan, V. V. (1974a). Kinematics of direct-tracking installation, *Geliotekhnika* **10**, No. 3, 7–12.

Vartanyan, A. V., Shermazanyan, Ya. T., and Arutyunyan, V. V. (1974b). Analysis of the kinematics of heliostats, *Geliotekhnike* **10**, No. 3, 47–53.

Vasilevskis, S. (1966). Large telescope design criteria, *in* "Telescope Design" (D. L. Crawford, ed.), pp. 115–118. Academic Press, New York.

Veinberg, V. B. (1959). "Optics in equipment for the utilization of solar energy," Transl. AEC-tr-4471. State Publ. House, Moscow.

Vest, C. M., and Arpaci, V. S. (1969). *J. Fluid Mech* **36**, 1.

Veziroglu, T. N. *et al.* (1975). Fossil/hydrogen energy mix and population control, *Proc. US/Jpn. Seminar Key Technol. Hydrogen Energy Syst., Univ. Yokohama, Japan.*

Veziroglu, T. N., and Kakac, S. (1976). Solar production of hydrogen as a means of storing solar energy, *in* "Heliotechnique and Development" (M. A. Kettani and J. E. Soussou, eds.), DAA Press, Cambridge, Massachusetts.

Vliet, G. C. (1969). *Trans. ASME J. Heat Transfer* 511–516.

von Kaleczinsky, A. (1902). Ueber die ungarischen wärmen und heissen koschsalzseen als naturliche wärmaccumulatoren sowie über die herstellung von wärmen salszeen wärmaccumulatoren, *Ann. Phys. IV Folge* No. 7, 408–416.

von Karman, T. (1921). NACA Tech. Mem. 1092.

Ward, G. T. (1961). Determination of the sun's position, Tech. Rep. T34. Brace Res. Inst., McGill Univ., Montreal.

Ward, D. S., and Löf, G. O. G. (1975). Design and construction of a residiential solar heating and cooling system, *Sol. Energy* **17** 13–20.

Watmuff, J. H. (1975). An experimental and theoretical investigation of a flat plate solar collector, M. Eng. Sci. Thesis, Univ. Melbourne.

Weihe, H. (1972). Fresh water from sea water, distilling by solar energy, *Sol. Energy* **13**, 439–444.

Weinberger, H. (1964). The physics of the solar pond, *Sol. Energy* **8**, No. 2, 45–56.

Westwater, J. (1968). *Proc. Annu. Southeastern Seminar Therm. Sci., 4th, Univ. of Tennessee.*

Whillier, A. (1953). Solar energy collection and its utilization for house heating, Sc.D. Thesis, Mech. Eng. Dept., MIT Cambridge Massachusetts.

Whillier, A. F. (1956). The determination of hourly values of total solar radiation from daily summations, *Arch. Meteorol. Geophys. Bioklimatol. Vienna Ser. B* **7**, 197–244.

Whillier, A. F. (1963). Lecture Notes on Solar Energy, Tech. Rep. T. 63. Brace Res. Inst. of McGill Univ., Montreal.

Whillier, A. F. (1964). Performance of black-painted solar air heaters of conventional design, *Sol. Energy* **8**, No. 1, 31–37.

Whillier, A. F. (1965). Solar radiation graphs, *Sol. Energy* **9**, No. 3, 164.

Whillier, A. F. (1967). Design factors influencing solar collector performance, Low Temperature Engineering Applications of Solar Energy, ASHRAE, pp. 27–40, New York.

Whillier, A. F., and Hottel, H. C. (1955). The performance of flat plate solar energy collectors, *Trans. Conf. Uses Sol. Energy* Vol. III, Thermal Processes, Part 2, pp. 70–104. Univ. of Arizona Press, Tucson, Arizona.

Whillier, A. F., and Saluja, G. (1965). Effect of materials and construction details on the thermal performance of solar water heaters, *Sol. Energy* **9**, No. 1, 21–26.

Wiederhold, P. R. (1963). Energy storage for high power discharges, *Astronaut. Aerosp. Eng.* **1**, No. 4.

Wilbur, P. J., and Mitchell, C. E. (1975). Solar absorption air conditioning alternatives, *Sol. Energy* **17**, 193–199.

Wilkes, G. B., and Peterson, C. M. F. (1937). Radiation and convection across air spaces in frame construction, *Heat. Piping Air Condit.* **9**, 505–510.

Wilkes, J. O., and Churchill, S. W. (1966). *AIChe J.* **12**, 161–166.

Williams, C. S., and Becklund, O. A. (1972). "Optics: A Short Course for Engineers and Scientists." Wiley (Interscience), New York.

Williams, D. A. *et al.* (1957). Intermittent absorption cooling systems with solar regeneration, ASME Annual Meeting, Paper 57–A–260.

Williams, D. *et al.* (1963). *J. Eng. Power Trans. ASME* **85**, 213.

Williams, L. G. (1964). Applications of solar collectors to hay and corn drying, Agr. Eng. Dept., Univ. Idaho Rep. October 25.

Willson, R. C. (1971). The active cavity radiometric scale, the international pyrheliometric scale and the solar constant, J. *Geophys. Res.* **76**.

Willson, R. C. (1974). Advance in pyrheliometric instrumentation, Report and Recommendations of Solar Energy Data Workshop, NSF–RA–N–74–062 [see also, New radiometric techniques and solar constant measurements, *Sol. Energy* **14**, (1973)].

Winston, R. (1974). Principles of solar concentrators of a novel design, *Sol. Energy* **16**, No. 2.

Wolf, M. (1971). A new look at silicon solar cell performance, *Energy Conversion* **11**, June, 63–73.

Wolf, M. (1974). Historic development of photovoltaic power generation, *Photovoltaic Power Generation Conf. Proc., Hamburg, Germany.*

Wolf, M. (1975). Outlook for Si photovoltaic devices for terrestrial solar energy utilization, *J. Vacuum Sci. Technol.* **12**, No. 5.

Wolfe, W. L. (ed.) (1965). "Handbook of Military Infrared Technology," "Office of Naval Res., Dept. of the Navy, Superintendant of Documents, U.S. Govt. Printing Office, Washington, D.C.

Woodall, J. M., and Hovel, H. J. (1972). *Appl. Phys. Lett.* **21**, 379.

Woodall, J. M., and Hovel, H. J. (1975). Outlooks for GaAs terrestrial photovoltaics, *J. Vacuum Sci. Technol.* **12**, No. 5.

Yamamoto, G. (1952). *Sci. Rep. Tohoku Univ. 5th Ser.* **4**, No. 1.

Yellott, J. I. (1960). Japan applies solar energy, *Mech. Eng.* **March**, 46.

Yellott, J. I. (1961). Transmission and absorption of solar radiation in glass, *Proc. Int. Seminar Solar Aeolian Energy, Greece* 263–281.

Yellott, J. I. (1964). Solar Radiation Device, U.S. Patent No. 3, 145,568.

Yellott, J. I. (1967). The measurement of solar radiation, Low Temperature Engineering Application of Solar Energy, ASHRAE.

Yellott, J. I. (1974). "Solar Energy Utilization for Heating and Cooling," NSF Document no. NSF 74–41. U.S. Govt. Printing Office, Washington, D.C., Stock No. 3800–00188; see also ASHRAE Handbook of Applications, Chapter 59.

Yellott, J. I., and Hay, H. (1969). Thermal analysis of a building with natural air conditioning, *ASHRAE Trans.* **75**, Part 1, 185.

Yellott, J. I., Kokoropoulos, P., and Fogle, T. (1964). Use of silicon cells as pyrheliometers, *Proc. 1961 Sounion Conf.* Plenum Press, New York.

Yellott, J. I. *et al.* (1975). "Solar Oriented Architecture," Architecture Foundation, Arizona State Univ., Tempe.

Yih, C. S. (1959). *Quart. J. Appl. Math.* **17**, 25–42.

Zarem, A., and Erway, D. (1963). "Introduction to the Utilization of Solar Energy." McGraw-Hill, New York.

Index

D

E

F

G

H

T

W

B
C 8
D 9
E 0
F 1
G 2
H 3
I 4
J 5